ADVANCES IN CHEMICAL PHYSICS

VOLUME XCIV

Advances in
CHEMICAL PHYSICS
Polymeric Systems

Edited by

I. PRIGOGINE

University of Brussels
Brussels, Belgium
and
University of Texas
Austin, Texas

and

STUART A. RICE

Department of Chemistry
and
The James Franck Institute
The University of Chicago
Chicago, Illinois

VOLUME XCIV

AN INTERSCIENCE® PUBLICATION
JOHN WILEY & SONS, INC.
NEW YORK • CHICHESTER • BRISBANE • TORONTO • SINGAPORE

CONTRIBUTORS TO VOLUME XCIV

JEAN-LOUIS BARRAT, Département de Physique des Matériaux, Université Claude Bernard-Lyon I, Villeurbanne, France

A. BAUMGÄRTNER, Institut für Festkörperforschung, Jülich, Germany

M. A. CARIGNANO, Department of Chemistry, Purdue University, West Lafayette, Indiana

LEWIS J. FETTERS, Corporate Research Science, Laboratories, Exxon Research and Engineering Company, Annandale, New Jersey

SANDRA C. GREER, Department of Chemical Engineering, University of Maryland at College Park, College Park, Maryland

GARY S. GREST, Corporate Research Science Laboratories, Exxon Research and Engineering Company, Annandale, New Jersey

JOHN S. HUANG, Corporate Research Science Laboratories, Exxon Research and Engineering Company, Annandale, New Jersey

JEAN-FRANÇOIS JOANNY, Institut Charles Sadron, Strasbourg, France

MICHAEL, E. G. LYONS, Electroactive Polymer Research Group, Physical Chemistry Laboratory, University of Dublin, Dublin, Ireland

M. MUTHUKUMAR, Department of Polymer Science, University of Massachusetts, Amherst, Massachusetts

DIETER RICHTER, Institut für Festkörperforschung, Jülich, Germany

I. SZLEIFER, Department of Chemistry, Purdue University, West Lafayette, Indiana

v

INTRODUCTION

Few of us can any longer keep up with the flood of scientific literature, even in specialized subfields. Any attempt to do more and be broadly educated with respect to a large domain of science has the appearance of tilting at windmills. Yet the synthesis of ideas drawn from different subjects into new, powerful, general concepts is as valuable as ever, and the desire to remain educated persists in all scientists. This series, *Advances in Chemical Physics*, is devoted to helping the reader obtain general information about a wide variety of topics in chemical physics, a field that we interpret very broadly. Our intent is to have experts present comprehensive analyses of subjects of interest and to encourage the expression of individual points of view. We hope that this approach to the presentation of an overview of a subject will both stimulate new research and serve as a personalized learning text for beginners in a field.

I. PRIGOGINE
STUART A. RICE

CONTENTS

THEORY OF POLYELECTROLYTE SOLUTIONS

JEAN-LOUIS BARRAT

Département de Physique des Matériaux (URA CNRS 172)
Université Claude Bernard Lyon I Villeurbanne, France

JEAN-FRANÇOIS JOANNY

Institut Charles Sadron (UPR CNRS 022) Strasbourg, France

CONTENTS

Advances in Chemical Physics, Volume XCIV, Edited by I. Prigogine and Stuart A. Rice.
ISBN 0-471-14324-3 © 1996 John Wiley & Sons, Inc.

I.   INTRODUCTION

Polyelectrolytes are polymers bearing ionizable groups, which, in polar solvents, can dissociate into charged polymer chains (macroions) and small *counterions*. Well-known examples of such systems are proteins, nucleic acids and synthetic systems such as sulfonated polystyrene or polyacrylic acid. Very often the polar solvent is water. Because of their fundamental importance in biology and biochemistry, and because of their hydrosolubility, ionizable polymers have been the object of continued interest since the early days of polymer science (see, e.g., [1, 2] and references therein). It is, however, a widely acknowledged fact [3] that they remain among the least understood systems in macromolecular science. This situation is in sharp contrast with the case of neutral polymers. In fact, even the properties that are now routinely used to "characterize" neutral polymers (e.g., light scattering, osmometry, viscosity) are still very poorly understood in polyelectrolyte solutions.

Briefly speaking, the origin of this difference with neutral polymers can be traced back to the difficulty in applying renormalization group theories and scaling ideas to systems in which long-range (Coulomb) forces are present. The success of the modern theories (and, although this was not always realized as they were developed, of many older approaches) of neutral polymer solutions is based on the fact that the range of the interactions between molecules is much smaller than the scale determining the physical properties of the solution, which is the size of the polymer chain or the correlation length. If, as it is in general the case in polymer physics, the main issue is the variation of solution properties with molecular weight, the interactions only affect prefactors that can be adjusted to experimental results. The theories [4–6] developed in this spirit have proven extremely powerful for the interpretation of experimental data on neutral polymer solutions. Polyelectrolyte solutions are more complex—both short-range (excluded volume) and long-range (Coulombic) interactions are simultaneously present. The screening of the electrostatic interactions introduces an intermediate length scale in the problem, which can be comparable to the chain size or to the correlation length. Moreover, the details of the local chain structure are important and control the phenomenon of counterion condensation [7] (see Section III). This condensation in turn modifies the long-range part of the interaction. The long-range interactions actually also have a nontrivial influence on the *local structure* [8–10] (stiffness) of the polymer chains (Section IV). The implication of this complicated coupling between small and large length scales is that the theoretical results for polyelectrolytes can be expected to be more "model dependent" than for neutral polymers. In particular, the functional dependence of several physical properties on molecular weight or concentration depends in some cases on the local description that is used for the polymer chain. Such situations never occur in neutral polymers.

Intrinsically related to this theoretical problem is the more practical difficulty in comparing experiments and theory for polyelectrolyte solutions. In neutral polymers, the comparison is made possible by adjusting interaction-dependent prefactors and fitting experimental results to the theoretically predicted dependence on, for example, molecular weight. In polyelectrolytes, "microscopic" parameters can have a bigger influence and a much more precise modeling might be necessary in order to interpret experimental results. This often involves the introduction of more adjustable parameters in the interpretation, which sometimes becomes inconclusive.

In spite of these considerations, the modern approach to neutral polymer physics has inspired a number of contributions to the polyelectrolyte problem in the last two decades. By modern, we mean here those theoretical approaches that attempt to use as much as possible "coarse-grained" models, with a minimal number of microscopic parameters. The goal of this chapter is to summarize these approaches, with the hope that they could provide a useful conceptual framework for further development. The chapter is organized as follows. Sections II–IV are concerned with single-chain properties. Section II deals with the academic problem of a charged chain in the absence of counterions. The phenomenon of screening, and the influence of the small ions on the interaction between monomers, is discussed in Section III. The influence of long-range interaction on the local stiffness (electrostatic rigidity) is described in Section IV. In Sections V and VI, the properties of interacting polyelectrolyte chains are considered. The simpler case of gels and brushes—in which the structure is controlled by the preparation of the system—is described first. Semidilute and concentrated solutions are the subject of Section VI. Finally, some results on dynamical properties (diffusion, electrophoretic mobility, and viscosity) are presented in Section VII.

Some important subjects have been left out either because they have been reviewed recently elsewhere or because our understanding is too fragmentary. Among those are the dielectric properties of polyelectrolyte solutions [11], their rheological properties, computer simulation results [12], and all the problems related to polyelectrolyte adsorption [13]. Finally, we systematically refer the reader to the original studies for detailed comparison with the experimental results. The reason, already mentioned, is that such comparisons are, in general, not simple and involve the determination of several parameters.

II. CHARGED CHAINS AT INFINITE DILUTION: ASYMPTOTIC PROPERTIES

A. Definition of Model and Flory-like Calculation

In the theoretical limit of infinite dilution and zero salt concentration, a polyelectrolyte chain can be modeled as a connected sequence of charged and uncharged monomers in a *dielectric vacuum* that represents the solvent. This model is not very realistic to describe actual polyelectrolyte solutions, as will become clear in our discussion of ion condensation and screening, but it serves here to introduce some important concepts relative to charged polymer chains.

The energy for such an isolated polymer chain, made up of N monomers, is written as:

$$H = H_0 + \tfrac{1}{2} k_B T \sum_{i=1,N} \sum_{j \neq i} \frac{l_B z_i z_j}{|\mathbf{r}_i - \mathbf{r}_j|} \tag{2.1}$$

where H_0 is the energy of the neutral polymer chain that would be obtained by "switching off" the electrostatic interaction, and the double summation on the left hand side includes all monomer pairs. Here z_i is the charge (in units of the electronic charge e) of monomer i, T is the absolute temperature, and $l_B = e^2/(4\pi\varepsilon_0\varepsilon_r k_B T)$ is the Bjerrum length, which characterizes the strength of electrostatic interactions in the solvent. For water at $T = 300$ K, $l_B \simeq 0.7$ nm.

The *neutral polymer energy* H_0 contains terms describing the chemical bonding of the monomers as well as short-range excluded volume terms. In this section, we focus on the very simple case where H_0 is the free energy of a Gaussian, or bead–spring, chain, that is,

$$H_0 = \frac{3k_B T}{2b^2} \sum_{i=1,N-1} (\mathbf{r}_i - \mathbf{r}_{i+1})^2 \tag{2.2}$$

The *monomer size b* is such that the mean-squared averaged end-to-end distance for the neutral chain is $R_0^2 = Nb^2$. The physical situation described by this model is that of a very weakly charged, flexible chain. In that case, each *spring* actually represents a flexible subchain of several neutral monomers. Moreover, the absence of excluded volume interactions implies theta conditions for the neutral backbone. Although these conditions are rather unlikely to be achieved in experiments, the model contains the two key ingredients that monitor the asymptotic behavior of charged polymers in the absence of screening, namely long-range interactions and monomer connectivity. Thee are the only relevant contributions as can be seen from the form of the Flory free energy for a chain of N monomers, carrying a total charge fNe [i.e., f is the fraction of charged monomers, $(1 - f)N$ monomers are neutral]. The Flory energy [14] for a chain* of size R is the sum of an elastic energy $k_B T R^2 / Nb^2$ and of a Coulombic energy $k_B T (Nf)^2 l_B / R$,

$$E_{\text{Flory}} = k_B T \left(\frac{R^2}{Nb^2} + \frac{(Nf)^2 l_B}{R} \right) \qquad (2.3)$$

which upon minimization with respect to R gives an equilibrium size

$$R \sim Nf^{2/3}(l_B b^2)^{1/3} \qquad (2.4)$$

This establishes the well-known result that the chain size is proportional to the number of monomers and to a one-third power of the charge fraction f. This result was first obtained by Katchalsky and co-workers [15] using a distribution function approach, shortly before Flory published an equivalent calculation for chains with short-range excluded volume.

The electrostatic interactions tend to swell the chain, therefore the equilibrium radius of a weakly charged polyelectrolyte chain must be larger than the Gaussian radius R_0. This condition defines a minimum charge fraction $f_G \simeq N^{-3/4}(b/lb)^{1/2}$ above which the electrostatic interactions are relevant and the radius is given by (2.4). If the fraction of charges f is smaller than f_G, the polyelectrolyte chain essentially behaves as a neutral chain and has a Gaussian radius R_0. In the following we always implicitly assume that $f > f_G$.

In contrast to Flory's calculation, whose success is known [5] to arise from an uncontrolled cancellation of errors, Katchalsky's approach is now

---

* The Flory free energy ignores numerical prefactors of order unity; the prefactor of the electrostatic energy depends on the actual distribution of the charges in the spheres of radius R. It is equal to $\frac{3}{5}$ if the charges are uniformly distributed.

believed to yield an essentially exact scaling result, as will become clear in the next sections. A more refined calculation of the electrostatic energy, accounting for the fact that the shape of the chain is rodlike rather than spherical, only yields a logarithmic correction to (2.4)

$$R \sim N f^{2/3} [l_B b^2 \ln(N)]^{1/3} \qquad (2.5)$$

Because of this rodlike character, the polyelectrolyte chain is often described as being a "fully extended" object. This statement, however, somewhat overinterprets the actual meaning of relation (2.4). Short-range molecular flexibility does not have to be frozen in a locally stretched configuration to obtain such relations.

To understand this point, and to get a deeper insight into the meaning of Katchalsky's calculation, it is useful to introduce the concept of electrostatic blob [14]. An electrostatic blob is a chain subunit within which the electrostatic interactions can be considered as a weak perturbation. The size ξ_e and the number of monomers g in such a subunit are thus related by

$$\frac{(fg)^2 l_B}{\xi_e} \simeq 1 \qquad (2.6)$$

Another relation between ξ_e and g is obtained from the Gaussian statistics of the chains on the scale ξ_e, $\xi_e^2 \simeq gb^2$. This gives

$$\xi_e \sim b \left(\frac{f^2 l_B}{b} \right)^{-1/3} \qquad g \sim \left(\frac{f^2 l_B}{b} \right)^{-2/3} \qquad (2.7)$$

The size of the chain [Eq. (2.4)] is therefore $(N/g)\xi_e$, so that the chain can be thought of as a linear (in the sense that its lateral dimensions are vanishingly small compared to its length) string of *electrostatic blobs* (Fig. 1).

The blob picture is particularly useful to discuss the structure of a charged chain in a good or poor solvent. In a good solvent, the only modification compared to the theta solvent case is that good solvent

Figure 1. Schematic representation of the blob structure of a charged polymer chain. The chain is Gaussian up to the scale ξ_e, stretched on larger length scales.

statistics characterized by a swelling exponent $\nu \neq \frac{1}{2}$ must be used in the relation that relates the blob size to the number of monomers in a blob, $\xi_e \sim g^\nu$. The result is still a linear string of blobs, $R \sim Nf^y$, but the exponent y that describes the variation with the charge fraction is now $(2 - 2\nu)/(2 - \nu)$. In a poor solvent, the problem is more subtle and has been analyzed in detail by Khokhlov [16]. In that case, the chain would be completely collapsed into a globular state in the absence of electrostatic interactions. The effect of these interactions is to break up the globular structure into a string of *globular blobs*. The criterion that determines the blob size is now that the electrostatic energy within the blob is sufficient to overcome the interfacial energy that must be paid when monomers are exposed to the solvent, that is,

$$\frac{(fg)^2 l_B}{\xi_e} \sim \frac{\Gamma \xi_e^2}{k_B T} \tag{2.8}$$

where Γ is the interfacial tension between the collapsed phase and the pure solvent. A second relation is obtained by expressing that the blob has the same structure as a small section of collapsed polymer,

$$g \sim n\xi_e^3 \tag{2.9}$$

Here n is the density of the collapsed chain. In the simple case where the second and third virial coefficients between monomers are, respectively, $b^3(T - \theta)/T = -b^3\tau$, b^6, $n \sim \tau b^{-3}$ and $\Gamma = k_B T \tau^2 b^{-2}$. Equations (2.7) and (2.4) are replaced by

$$\xi_e \sim b \left(\frac{f^2 l_B}{b} \right)^{-1/3}$$

$$g \sim \tau \left(\frac{f^2 l_B}{b} \right)^{-1} \tag{2.10}$$

$$R \sim \frac{N}{g} \xi_e \sim N\tau^{-1} \left(\frac{f^2 l_B}{b} \right)^{2/3}$$

B. Variational Approaches

The Flory-type calculation discussed above is a typical mean-field approach. Similar approaches that have been successfully applied to the charged bead–spring model of Section II.A include the *chain under tension* model [14, 17] and the *Gaussian variational method* [5, 17–19]. Both methods are mean-field approaches, formulated in a variational fashion. The chain is approximated by a simpler system described by a trial Hamiltonian H_t, and the "best" trial Hamiltonian is obtained by minimizing the variational free energy

$$F_{\text{var}} = \langle H \rangle_t - TS_t \qquad (2.11)$$

where $\langle H \rangle_t$ is the average value of the energy for the true system, taken with a statistical weight $\exp(-H_t/k_B T)$, and S_t is the entropy of the trial system.

In the chain under tension model, the trial Hamiltonian is that of a noninteracting Gaussian chain subject to an external tension $\mathbf{F}$,

$$H_t(\mathbf{F}) = \frac{3k_B T}{2b^2} \sum_{i=1}^{N-1} (\mathbf{r}_{i+1} - \mathbf{r}_i)^2 + \mathbf{F}(\mathbf{r}_N - \mathbf{r}_1) \qquad (2.12)$$

This model is simple enough to allow explicit calculations of many properties of the chain. In particular, it can be shown that the minimization of the corresponding variational free energy yields a force

$$F \sim \left(\frac{f^2 l_B}{b} \right)^{1/3} \ln(FN^{1/2})^{1/3} \qquad (2.13)$$

and a chain size

$$R \sim N \left(\frac{f^2 l_B}{b} \right)^{1/3} \left[\ln \left(N \left(\frac{f^2 l_B}{b} \right)^{2/3} \right) \right]^{1/3} \qquad (2.14)$$

The results for the chain size and chain structure factor were shown to agree *quantitatively* with Monte Carlo calculations [17]. This model can also be used to give a quantitative meaning to the electrostatic blob size introduced above. This is done by noting that in the chain under tension, the mean-squared distance between two monomers separated by n monomers along the chain is $R(n)^2 = nb^2 + n^2(b^2 F/3k_B T)^2$. The blob size can be obtained by identifying the second term with $[(n/g)\xi_e]^2$, $\xi_e = 3k_B T/F$, in agreement with (2.7). The blob size is therefore directly related to the prefactor of n^2 in the expression of $R(n)^2$.

In the Gaussian variational approach, the trial Hamiltonian is a general quadratic form of all monomer coordinates. This method, which is tractable for cyclic chains only, was first introduced by des Cloizeaux and Jannink [5] as an alternative to Flory's theory for chains with short-range excluded volume. In that case, however, the method gives a swelling exponent $\nu = \frac{2}{3}$ in three dimensions and is therefore inferior to Flory's approach. For long-range interactions it was shown [18] that the variational Gaussian approach predicts the same asymptotic behavior as

Flory's theory and as the chain under tension method. As for the chain under tension model, the quantitative agreement with numerical simulation results is good [17].

C. Renormalization Group Calculations

The coincidence of the predictions from the Flory approach and the variational Gaussian method suggests that the predicted asymptotic behavior is indeed exact, except for logarithmic corrections. Exact renormalization group calculations were performed for space dimensions $d > 4$ [20, 21], showing that for $4 < d < 6$ the swelling exponent of the chain is $\nu = 2/(d-2)$ (as does the variational calculation of Ref. 18). No such calculation is available for $d = 3$, but the result for $d = 4$ suggests that the maximum value $\nu = 1$ is obtained for $d < 4$, while Gaussian behavior is recovered for $d > 6$. Moreover, a "naive" real-space renormalization argument [4] also predicts $\nu = 1$. Note that a naive extension of the Flory theory to d dimensions predicts an upper critical dimension where the chains are Gaussian equal to 6, but a lower critical dimension where $\nu = 1$ equal to 3 in disagreement with the renormalization group results.

The success of variational and mean-field methods for chains with long-range interactions is not unexpected. As pointed out elsewhere [18], these methods usually fail because a poor approximation of the monomer–monomer correlation function is used in the calculation of the potential energy. A Gaussian correlation function will be maximum when two monomers are in contact, whereas the true correlation function vanishes in that case. For long-range forces, most of the potential energy comes from configurations in which the monomers are far apart, so that the Gaussian approximation becomes much better.

D. Screening of Electrostatic Interactions

When a finite concentration of salt is present in the solution—which is always the case in experiments, at least because of water dissociation—the solution becomes a conducting rather than dielectric medium. The fundamental implication is that the Coulomb interaction between charged monomers is screened by the salt solution, that is, the electrostatic potential created by a monomer, or a group of monomers, falls off exponentially rather than algebraically with distance. Thus distant parts of the chain do not interact, and the chain can be expected to behave, in the asymptotic limit $N \to \infty$, as a random walk with short-ranged repulsive interactions.

It is a general result from the theory of charged solutions that charge fluctuations become uncorrelated (or, equivalently, that an external charge distribution is screened) over a typical distance κ_s^{-1}, where κ_s is related to the thermodynamic properties of the solution [22]. The screened interaction between monomers, mediated by the salt solution, can formally be obtained by integrating out the degrees of freedom associated to the salt ions in the partition function as shown in Appendix A. The resulting interaction is in general a complex function of the monomer coordinates, involving many body interactions. In the case of dilute salt solutions and sufficiently weak perturbations, however, the linear response (or Debye–Hückel) theory can be used, and the salt can be treated as an ideal gas. The screening length κ^{-1} is then given by

$$\kappa^2 = 4\pi l_B I \tag{2.15}$$

where I is the ionic strength of the solution defined as $I = \Sigma_{\text{ionic species}} Z_i^2 c_i$, Z_i being the valence and c_i the concentration of species i.

The effective interaction can in this limit be written as a sum of pairwise additive interactions between the monomers. The effective pair potential is given by the Debye–Hückel formula [23]

$$v_{\text{DH}}(r) = k_B T \frac{l_B}{r} \exp(-\kappa r) \tag{2.16}$$

This short-range interaction potential is the starting point of many theoretical studies of polyelectrolyte solutions. The definition of an effective pair potential requires, however, two conditions. The first condition, that the ionic solution is dilute, is usually satisfied in polyelectrolyte solutions. The second condition of a weak external perturbation introduced by the polymer on the small ion solution must be considered more carefully and is addressed in the next section.

Although the potential (2.16) is short ranged, its range κ^{-1} can be much larger than the monomer size, which is the typical interaction range in neutral polymers. Depending on the ionic strength, κ^{-1} can vary typically from less than 1 nm to more than 100 nm. At distances smaller than κ^{-1}, the screened potential (2.16) is close to the unscreened Coulomb potential discussed in Section II.A. The behavior and properties of polyelectrolyte chains can therefore be markedly different from those of neutral chains over rather large length scales. In fact, Section IV will show that in some cases the effect of electrostatic interactions extends over length scales much larger than κ^{-1}.

Finally, it should be noted that the concept of effective interaction between monomers, obtained in integrating out the degrees of freedom of the small ions, is useful only for the calculation of *static* properties. Dynamical properties such as diffusion constants or viscosities do not have to be identical to those of a hypothetical polymer solution in which the monomer–monomer interaction would be given by (2.16). The ionic degrees of freedom can play a nontrivial role in the dynamics, as they do in suspensions of charged colloïds [24].

E. Annealed and Quenched Polyelectrolytes

An important degree of freedom of polyelectrolyte chains is the distribution of the charges along the chemical sequence. This distribution can be either quenched or annealed [25]. When a weakly charged polyelectrolyte is obtained by copolymerization of a neutral and of a charged monomer, the number of charges on each chain and the positions of the charges along the chain are fixed, the distribution is quenched. If the charges are sufficiently regularly distributed along the chain and do not have a tendency to form long blocks, the precise distribution of the charges does not seem to be a relevant variable for the chain statistics. At the level of scaling laws that we use, one can expect that only the numerical prefactors are different when polyelectrolytes with a random or a periodic distribution of charges are compared.

Polyacids or polybases are polymers where the monomers can dissociate depending on the pH of the solution and acquire a charge. The dissociation of an H^+ ion from a polyacid gives rise to the apparition of a COO^- group and thus of a negative charge. This is an annealed process, the total number of charges on a given chain is not fixed but the chemical potential of the H^+ ions and thus the chemical potential of the charges is imposed (by the pH of the solution). The positions of the charges along the chain are also not fixed; the charges can move by recombination and redissociation of an H^+. The chemical potential of the charges μ depends on the fraction f of dissociated acid groups and is related to the pH of the solution by [2]

$$\text{pH} = pK_0 + \mu(f) \qquad (2.17)$$

The charge chemical potential has two contributions, an entropic contribution related to the mixing of charged and non charged groups along the chains $k_B T \ln[f/(1-f)]$ and an electrostatic contribution $\mu_{el}(f) = N^{-1} \partial F_{el}/\partial f$, which is the derivative of the electrostatic free energy of the chain. In a good or θ solvent, the electrostatic contribution μ_{el} is an increasing monotonic function of f, and for a given pH the fraction of charges f is well defined; the properties of the chains are very similar to those of a quenched chain with a charge fraction f. In a poor solvent, the blob model introduced by Khokhlov and discussed earlier leads to a nonmonotonic variation of μ_{el} and thus of μ with f. This nonmonotonic variation indicates a conformational transition of the chain between a collapsed weakly charged conformation and a stretched strongly charged conformation. The collapse transition of a polyelectrolyte with varying f is thus predicted to be a first-order transition [25]. The discontinuous collapse of the chains could explain titration curves of polyacids or polybases in a poor solvent (e.g., polymethacrylic acid) which show a plateau in the pH–f curve [26].

III. LOCAL ASPECTS OF SCREENING

The Flory-like model presented in the previous section considers only the interactions between charged monomers along the polyelectrolyte chain and ignores the role of the small counterions that neutralize the polyelectrolyte charge. When the polyelectrolyte is strongly charged, the electrostatic potential on the chain is large and some of the counterions remain bound to the chain. This phenomenon is known as counterion condensation, or Manning condensation [7]. For many purposes one must then consider that the chain has an effective charge due to the charged monomers and to the sheath of bound, or condensed, counterions. The effective charge is lower than the nominal charge of the monomers, Counterion condensation can be effective even at infinite dilution and reduces the electrostatic interactions between monomers. We first give here the more qualitative description of the condensation due to Manning [7] and then present an earlier discussion based on the Poisson–Boltzmann equation due to Fuoss et al. [27]. The condensed counterion sheath around a polyelectrolyte chain is electrically polarizable; this can induce attractive interactions between polyelectrolyte chains. We discuss in Section III.A possible mechanisms for attractive electrostatic interactions in polyelectrolyte solutions.

A. Counterion Condensation

For simplicity, we consider a negatively charged polyelectrolyte chain, in a salt-free solution, that has locally a rodlike conformation and can thus be considered as a rod over a distance L much larger than the monomer size b. The distance between charges along the chain is A and the fraction of charged monomers is $f = b/A$. If we ignore the contribution of the counterions, the electrostatic potential at a distance $r \ll L$ from the chain is obtained from Gauss theorem as $\psi(r) = 2l_B/A \ln(r)$ (we use here as unit of the electrostatic potential $k_B T/e$, where e is the elementary charge). The distribution of the counterions around the chain is given by Boltzmann statistics for monovalent counterions, $n(r) = n_0 \exp - \psi(r) \simeq r^{-(2l_B/A)}$, where n_0 is a constant. The total number of counterions per unit length within a distance r from the chain is then

$$p(r) = \int_0^r 2\pi r' \, dr' \, n(r') = \int_0^r 2\pi n_0 r'^{(1-2l_B/A)} \, dr' \qquad (3.1)$$

When the charge parameter $u = l_B/A$ is smaller than one, the integral giving $p(r)$ is dominated by its upper bound and $p(r)$ decreases to zero as the distance r gets small. In this case where the charge fraction is small $f < b/l_B$, there is no counterion condensation on the polyelectrolyte chain. In the opposite limit where the charge is large $u = l_B/A > 1$, the integral giving $p(r)$ diverges at its lower bound, indicating a strong condensation of the counterions on the polyelectrolyte chains. As the condensation proceeds, the effective value of the charge parameter decreases. The counterion condensation stops when the effective charge parameter is equal to one. The polyelectrolyte chain and the condensed counterion sheath are then equivalent to a polyelectrolyte with a distance between charges $A = l_B$ or to a fraction of charged monomers $f = b/l_B$. The remaining monomeric charges are neutralized by the condensed counterions.

One of the directly measurable quantities that strongly depends on the condensation of the counterions is the osmotic pressure of the polyelectrolyte solution. As shown below, in many cases the osmotic pressure of a polyelectrolyte solution is dominated by the counterions, the contribution of the polyelectrolyte chains being only a very small correction. Below the Manning condensation threshold, the counterions are not condensed and are essentially free. At leading order, the solution can be considered as an ideal gas of counterions of concentration fc where c is the total monomer concentration and the osmotic pressure is [28]

$$\pi = k_B T f c \qquad (3.2)$$

The first correction to the ideal-gas behavior is due to the polarization of the counterion gas by the polyelectrolyte chains. The polarization energy is estimated using the Debye–Hückel approximation for the gas of counterions [29]. The screening is due to the counterions and $\kappa^2 = 4\pi l_B f c$. The polarization free energy per unit volume reads

$$F_{\text{pol}} = -k_B T f c u \ln(\kappa) = -\frac{k_B T}{2} c \left(\frac{b l_B}{A^2}\right) \ln\left(\frac{4\pi l_B c b}{A}\right) \qquad (3.3)$$

The osmotic pressure of the solution is directly calculated from this relation

$$\pi = k_B Tfc(1 - \tfrac{1}{2}u) \tag{3.4}$$

In many cases, for weakly charged polyelectrolytes, $l_B \simeq b$, so that $u \simeq f$, and the polarization correction is small. In most of the following, we use the ideal-gas expression for the osmotic pressure of the counterions in a weakly charged polyelectrolyte solution.

Above Manning condensation threshold $u > 1$, the bound counterions do not contribute to the osmotic pressure. The remaining counterions, at a concentration $cb/l_B = fc/u$, behave as a Debye–Hückel gas, polarized by a chain that has an effective charge parameter equal to unity. The osmotic pressure is given by Eq. (3.4), where fc is replaced by cb/l_B and u by 1,

$$\pi = \frac{k_B Tcb}{2l_B} = \frac{k_B Tfc}{2u} \tag{3.5}$$

Several other physical quantities such as the electrical conductivity of the solution or the electrophoretic mobility of the chains [30] strongly depend on the condensation of the counterions and are dominated by the free counterions. Their determination is, in general, in good agreement with Manning condensation theory.

B. Poisson–Boltzmann Approach

Fuoss et al. [27] have studied the interaction between the polyelectrolyte chains and their counterion clouds using the so-called Poisson–Boltzmann equation. They consider a solution of infinite rodlike molecules, which are all parallel. The number of molecules per unit area is Γ. Around each of these molecules there exists an equipotential surface where the electric field vanishes. On average, this surface can be approximated by a cylinder parallel to the polyelectrolyte chains with a radius R defined such that $\Gamma \pi R^2 = 1$. The average density of counterions in the cylinder is $c_i = fc = 1/(\pi R^2 A)$, where A is the distance between charges along the chain. From an electrostatic point of view, each chain with its counterions in the cylinder of radius R is independent of all the other chains. The electrostatic problem that must be solved is therefore that of a single infinite chain with its counterions embedded in a cylinder of radius R with the boundary condition that the electric field vanishes on the surface of the cylinder.

The electrostatic potential in this cell satisfies the Poisson equation $\Delta\psi = -4\pi l_B c_i(r)$, where $c_i(r)$ is the local counterion concentration at a distance r from the chain. The local counterion concentration is obtained from Boltzmann statistics $c_i(r) = c_i \exp[-\psi(r)]$. This leads to the Poisson–Boltzmann equation

$$\Delta\psi(r) = -\kappa^2 \exp[-\psi(r)] \tag{3.6}$$

The Debye screening length is defined here as $\kappa^2 = 4\pi l_B c_i$; it is such that $\kappa^2 R^2 = 4u$ where u is the charge parameter defined above. The Poisson–Boltzmann equation has been solved exactly in this geometry in Ref. 27. The expression of the potential critically depends on the value of the charge parameter. Below the Manning condensation threshold where $u < 1$, the electrostatic potential is given by

$$\psi = \ln\left[\frac{\kappa^2 r^2}{2\beta^2} \sinh^2\left(-\operatorname{arctanh}\beta + \beta \ln\left(\frac{r}{R}\right)\right)\right] \tag{3.7}$$

where β is an integration constant related to u by $u(1-\beta^2)(1-\beta/\tanh[\beta \ln(b/R)]^{-1}$.

Above the Manning condensation threshold the electrostatic potential reads

$$\psi = \ln\left[\frac{\kappa^2 r^2}{2\beta^2} \sin^2\left(-\arctan\beta + \beta \ln\left(\frac{r}{R}\right)\right)\right] \tag{3.8}$$

and β is given by $u = (1-\beta^2)(1-\beta/\tan[\beta \ln(b/R)]^{-1}$. The distribution of counterions around each polyion can then be determined using Boltzmann's statistics.

At each point of the solution, the pressure has two contributions: an ideal-gas contribution and an electrostatic contribution proportional to the square of the local electric field. At the edge of the cell $(r = R)$, the electric field vanishes and there is only an ideal-gas contribution. The pressure being uniform throughout the solution can be calculated at this point, $\pi = k_B T c_i(r = R) = k_B T c_i \exp - [\psi(r = R)]$. Both above the Manning condensation threshold (3.4) and below the threshold (3.5) the calculated pressure is in agreement with that calculated from Manning theory.

Manning theory describes the condensation of the counterions as a transition between two states, a bound state and a free state. In the Poisson–Boltzmann approach, the distribution of the counterions is continuous and there is no bound state. However, close to the polyelectrolyte chain, the interaction energy between one counterion and the polyelectrolyte chain becomes much larger than $k_B T$ (in the region where $\psi \gg 1$). One could then divide space into two regions: (1) a region close to the chain where the interaction energy is larger than $k_B T$ (with a size of the order of a few times the monomer size b) and where the counterions can be considered as bound and (2) a region further away from the chain where the interaction between a counterion and the chain is smaller than $k_B T$ and where the counterions can be considered as unbound.

C. Attractive Electrostatic Interactions

The counterions condensed on a polyelectrolyte chain move essentially freely along the chain; they more or less form a one-dimensional gas of average density g/b where $g = f - b/l_B$ is the fraction of condensed counterions. The charge density along the chain is therefore not a frozen variable and shows thermal fluctuations due to the mobility of the counterions. When two chains are sufficiently close, the charge density fluctuations on the two chains are coupled by the electrostatic interactions; this leads to attractions between the polymers that are very similar in nature to the van der Waals interactions between polarizable molecules. A fluctuation-induced attraction can be expected for all polyelectrolytes where the charges are mobile along the chain. This is, for example, the case above Manning condensation threshold but also for annealed polyelectrolytes, which are polyacids or polybases, where the charge can be monitored by tuning the pH of the solution [25]. The ionized groups are not fixed on these chains, only the chemical potential of the charges is imposed and the charge density also has thermal fluctuations.

We first calculate the attractive interaction for two parallel rodlike molecules of length L at a distance x in a solution where the salt concentration is n, following the ideas introduced by Oosawa [31]. When a charge fluctuation $\delta c_1(z)$ occurs along the first chain, it polarizes the second chain where a charge fluctuation of opposite sign $\delta c_2(z)$ appears. The attraction between the two chains is due to the Coulomb interaction between these charge fluctuations. Within a linear response approximation, the free energy of the charge fluctuations can be written in Fourier space as

$$H = \frac{k_B T}{2} \sum_q \sum_{i,j} S_{ij}^{-1} \delta c_i(q) \delta c_j(-q) \tag{3.9}$$

Here, $\delta c_i(q)$ is the Fourier transform at a wave vector q of the charge fluctuation $\delta c_i(z)$ ($i = 1, 2$). The charge structure factor on one of the chains is $S_{11}(q) = g/b$ and corresponds to a one-dimensional ideal gas of charges. The crossed structure factor is due to the electrostatic interactions between the two chains obtained by summing the Debye–Hückel interactions

$$S_{12}^{-1}(q) = 4\pi l_B \int \frac{d^2k}{4\pi^2} \frac{e^{iqx\cos\theta}}{k^2 + q^2 + \kappa^2} = 2l_B K_0[x(q^2 + \kappa^2)^{1/2}] \quad (3.10)$$

where $K_0[x]$ is a modified Bessel function of the second kind [32]. The fluctuation contribution to the interaction free energy between the two rods is obtained by summing the partition function $\exp(-H/k_B T)$ over all fluctuations. We obtain

$$F_{att}(x) = L \frac{k_B T}{2} \int \frac{dq}{2\pi} \ln\left(1 - 4\frac{g^2 l_B^2}{b^2} K_0[x(q^2 + \kappa^2)^{1/2}]\right) \quad (3.11)$$

This expression can be simplified by assuming that the fraction g of recondensed counterions is smaller than one and by looking for the asymptotic limits where the distance x between the rods is smaller or larger than the screening length κ^{-1}. For short distances between the rods, the interaction energy is

$$F_{att}(x) = -\alpha k_B T \frac{g^2 l_B^2}{b^2} \frac{L}{x} \quad (3.12)$$

where α is a numerical constant. For large distances, the attractive interaction decays as

$$F_{att}(x) = -\beta k_B T \frac{g^2 l_B^2}{b^2} \frac{L}{x^{3/2} \kappa^{1/2}} \exp(-2\kappa x) \quad (3.13)$$

β being another numerical constant. As expected, the interaction is attractive. It is proportional to the square of the Bjerrum length, indicating that this is a second-order electrostatic effect: The fluctuation on the first rod creates an electric field that polarizes the second rod; the charge fluctuation on the second rod then creates an electric field that interacts with the charge fluctuation of the first rod. For parallel rods, the interaction is also proportional to the length of the rod, each segment interacts mainly with the segment of the other rod, which is directly facing it. For the same reason, the attractive interaction is screened over a length $\frac{1}{2}\kappa^{-1}$, that is, over half the screening length of the repulsive Coulombic interaction.

When the two rods make an angle θ $(0 < \theta < \frac{1}{2}\pi)$, the attractive electrostatic interaction cannot be calculated from the fluctuation free energy (3.10). Instead, we directly discuss it in terms of the polarizability of the two polyelectrolytes.* If a charge fluctuation $\delta c_1(q)$ is created on the first rod, it creates an electrostatic potential on the second rod $\delta\psi(q') = K(q, q')\delta c_1(q)$, where the kernel is defined as

$$K(q, q') = \frac{2\pi l_B}{\sin\theta} \frac{\exp(-px)}{p}$$

with a wave vector p related to q and q' by

$$\kappa^2 = \kappa^2 + \frac{q^2 + q'^2 - 2qq'\cos\theta}{\sin^2\theta}$$

The response function of the one-dimensional gas of recondensed counterions on the second rod is g/b and the induced charge fluctuation on the second rod is $\delta c_2(q') = -(g/b)K(q, q')\delta c_1(q)$. This charge fluctuation creates a potential on the first rod $\delta\psi'(q) = K(q, q')\delta c_2(q')$. The attractive interaction energy between the two rods forming an angle θ at a distance x is then estimated as

$$F_{att}(x) = -k_B T \frac{g}{b} \int \frac{dq\,dq'}{4\pi^2} K^2(q, q')\langle\delta c_1^2(q)\rangle \tag{3.14}$$

The average value of the charge fluctuation is given by the ideal-gas statistics $\langle\delta c_1^2(q)\rangle = g/b$. This leads to an attractive interaction between nonparallel rods

$$F_{att}(x) = -k_B T \frac{g^2 l_B^2}{b^2 \sin\theta} (\pi - \theta)\text{Ei}(2\kappa x) \tag{3.15}$$

where $\text{Ei}(u)$ is the exponential integral function [32]. As for parallel rods, the attractive interaction is a second-order electrostatic effect, it is proportional to l_B^2, and it decays at large distances as $\exp(-2\kappa x)$. When the angle between the two rods is not very small, the attractive force is independent of the length of the rods. In this geometry only the chain segments in the vicinity of the crossing point interact. If the distance between the rods is smaller than the screening length, the crossover to the parallel rod behavior occurs when the angle θ is smaller than x/L. At short distances $x < \kappa^{-1}$, $\text{Ei}(u) = \ln u$ and the attractive interaction energy varies logarithmically. As the attraction is independent of the chain length, the attractive interaction between two polymers that are rodlike

---

* This is equivalent to performing a perturbation expansion to second order in the crossed term that appears in Eq. (3.10).

over a distance larger than the screening length is also given in Eq. (3.14).

The fluctuation attractive interaction must be compared to the repulsive Coulombic interaction between the chains. For two rods at a distance x that make a finite angle, the repulsive Coulomb interaction is of order $Fel \simeq k_B T 2\pi l_B f^2/(\kappa b^2) \exp(-\kappa x)$ where f is the effective fraction of charged monomers (equal to b/l_B above the condensation threshold). At large distances, as already mentioned, the screening of the attractive interaction is stronger and the repulsive interaction is dominant. At small distances the attractive interaction is dominant. The distance at which the attractive force is larger than the repulsive force is of order $x \simeq l_B(g/f)^2$. In general this corresponds to very short distances of the order of the Bjerrum length. The attractive fluctuation interaction is thus expected to be important only in strong coupling situations where the Bjerrum length is large (e.g., in the presence of multivalent ions) or if some external constraint imposes very small distances between chains.

The attractive interaction can also be considered within the framework of the Poisson–Boltzmann approach. The Poisson–Boltzmann equation is derived from a mean-field theory that considers only the average concentration profile of the counterions around each polyelectrolyte chain. A more refined theory should take into account the thermal fluctuations around this concentration profile. Such a theory has been built to study the interactions between charged colloidal particles and also leads to attractive interactions [33, 34]. Qualitatively, the conclusions are the same as the one presented here. The attractive interactions are dominant at short distances and become relevant only in the limit of strong electrostatic coupling.

Experimentally, the addition of multivalent ions to polyelectrolyte solutions often provokes a precipitation. This can be explained by electrostatic interactions [35] or by complexation of the polymer by the multivalent ions [36].

The interaction between charge fluctuations is not the only mechanism that leads to attractive interactions between polyelectrolyte chains. Recently, Ray and Manning [37] have presented a model for attractive interactions based on the overlap between the condensed counterions sheaths around the two interacting polymers. In the simple version of Manning condensation theory presented here, the polyelectrolyte chains are infinitely thin lines, and the pointlike counterions condense directly on the line. A more refined version allows for a finite condensation volume that can be determined by minimization of the free energy of the condensed counterions. When two parallel rods come close to one another, the condensation volume around each rod expands in the space between the two rods and the two condensation volumes overlap. This leads to the formation of a polyelectrolyte dimer with a single condensation sheath. The expansion of the condensation volume leads to an increase in translational entropy of the counterions and thus to an attractive interaction. The model shares thus some analogy with the capillary condensation between thin films wetting parallel cylinders [38]. An attractive interaction is predicted at intermediate distances much larger than the polyelectrolyte diameter but much smaller than the electrostatic screening length. This interaction has been calculated only for parallel polyelectrolyte rods; it has not been calculated for rods intersecting at a finite angle.

IV. ELECTROSTATIC RIGIDITY

In this section, we assume that the interaction between charged monomers can be described by the Debye–Hückel potential, Eq. (2.16). Within this approximation, we discuss the structure of the polymer chain at length scales intermediate between the short length scales discussed in Section III and the chain size. It has already been mentioned in Section II.D that at very large scales, a charged chain has the same structure as a neutral polymer chain in a good solvent, since the potential (2.16) is short ranged. The radius increases with molecular weight as $R \simeq N^{3/5}$ (in the Flory approximation). At short scales ($L < \kappa^{-1}$), on the other hand, screening is inoperant, so that the chain is expected to take the rodlike structure described in Section I. The simplest assumption concerning the chain structure at intermediate length scales, which was made in all early work in the field [39], is that the crossover from rodlike to Flory-like behavior takes place above a length scale κ^{-1}, equal to the range of the interactions. This simple assumption was challenged by the works of Odijk [8, 10] and Skolnick and Fixman [9], who showed how the Debye–Hückel interaction can induce a rodlike conformation at length scales much larger than the interaction range. Their theory, and its limitations, are presented here.

A. Odijk–Skolnick–Fixman Theory

In their calculations, Odijk, Skolnick, and Fixman (OSF) consider a semiflexible chain with total contour length L characterized by its "bare" persistence length l_0, and that carries charges separated by a distance A along its contour. The interaction between the charged monomers is given by (2.16), and only electrostatic interactions are taken into account (the polymer backbone is in a theta solvent). For strongly charged chains with $A < l_B$ Manning condensation can be in a first approximation be accounted for by replacing A with l_B. In the limit $l_0 \ll A$ of a weakly charged chain, this model is equivalent to the Gaussian chain model discussed in Section II.A, with $N = L/A$, $b^2 = Al_0$, and $f = 1$. We consider here the general case where l_0/A is not small [40].

The total energy of the chain is the sum of the intrinsic curvature energy and of the (screened) electrostatic energy. If the polymer is described as a planar curve of curvilinear length L, a chain configuration is specified by the function $\theta(s)$ defined for $-\tfrac{1}{2}L < s < \tfrac{1}{2}L$ by $\cos[\theta(s)] = \mathbf{t}(s) \cdot \mathbf{t}(0)$, where $\mathbf{t}(s)$ is the unit vector tangent to the chain at the point of curvilinear abscissa s. Assuming that $\theta(s)$ remains small, the electrostatic potential between two charges located at s_1 and s_2 on the chain may be expanded around the value obtained for a rodlike configuration, $v_{DH}(|s_2 - s_1|)$. The following expression is then obtained for the total energy $H[\theta]$ of a given configuration $\{\theta(s)\}$:

$$H[\theta] = H_0 + \tfrac{1}{2} k_B T \int_{-L/2}^{L/2} ds \int_{-L/2}^{L/2} ds' \, \frac{d\theta(s)}{ds} \left[l_0 \delta(s - s') + K(s, s') \right] \frac{d\theta(s')}{ds}$$

$$(4.1)$$

Here H_0 is the electrostatic energy of a rod, the term proportional to l_0 is the bare curvature energy of a noninteracting semiflexible chain, and the contribution of the electrostatic interactions is described by the kernel $K(s, s')$, which reads (for $s > s'$)

$$K(s, s') = \frac{1}{A^2} \int_{-L/2}^{s'} ds_1 \int_s^{L/2} ds_2 \, v'_{DH}(s_2 - s_1) \frac{(s_2 - s)(s' - s_1)}{s_2 - s_1} \quad (4.2)$$

with $v'_{DH}(s) = dv_{DH}/ds$. The only assumption made in obtaining this result is that, within the range κ^{-1} of the interaction potential, the chain remains in an almost rodlike configuration, that is, $|s_2 - s_1| - |\mathbf{R}(s_1) - \mathbf{R}| \ll |s_2 - s_1|$ for $|s_2 - s_1| < \kappa^{-1}$. For long chains, the integration in (4.2) can be extended to infinity, and $K(s, s')$ becomes a function of $s - s'$,

$$K(s) = \frac{1}{6A^2} \int_0^\infty dx \, \frac{x^3}{x + s} v'_{DH}(x + s) \qquad (4.3)$$

The statistical properties of the chain are obtained by integrating the Boltzmann factor $\exp(-H[\theta]/k_B T)$ over all possible configurations, that is, over all functions $\theta(s)$. The integration can be carried out analytically, since the energy is a quadratic function of θ. The calculation is simplified by introducing the Fourier transform of the kernel (4.3),

$$\tilde{K}(q) = \int_0^\infty ds \, \exp(iqs)K(s)$$

$$= l_{OSF} \frac{2\kappa^2}{q^2} \left[\frac{\kappa^2 + q^2}{q^2} \ln\left(\frac{\kappa^2 + q^2}{\kappa^2}\right) - 1 \right] \qquad (4.4)$$

where

$$l_{OSF} = \frac{l_B}{4A^2\kappa^2} \qquad (4.5)$$

has the dimension of a length and is known as the Odijk–Skolnick–Fixman length. The local flexibility of the chain can be characterized by the mean-squared angle $\langle \theta(s)^2 \rangle$ between the chain direction at the origin and after a contour length s. For a neutral semiflexible chain, $\langle \theta(s)^2 \rangle = s/l_0$ varies linearly with s. The chain configuration can be described as resulting from an *angular diffusion* process, with a diffusion constant l_0^{-1}. For the charged chain, with the approximate energy (4.1), one gets

$$\langle \theta(s)^2 \rangle = \frac{4}{\pi} \int_0^\infty dq \, \frac{\sin^2(qs/2)}{q^2} \frac{1}{l_0 + \tilde{K}(q)} \qquad (4.6)$$

This expression simplifies in the limits of small s and large s. For large s,

$$\langle \theta(s)^2 \rangle = \frac{s}{l_0 + l_{OSF}} \qquad (4.7)$$

which expresses the fact that at large scales, the chain conformation can be described by an effective persistence length $l_0 + l_{OSF}$, which is the sum of a bare and of an electrostatic contribution. This is the well-known result first obtained elsewhere [8, 9]. It indicates that the influence of the screened electrostatic interactions can extend much beyond their range κ^{-1}, since l_{OSF} is, for weakly screened solutions, much larger than κ^{-1}. The persistence length also decreases with the salt concentration as $l_{OSF} \simeq n^{-1}$ whereas the Debye screening length has a slower decay $\kappa^{-1} \simeq n^{-1/2}$.

For small values of s, (4.6) reduces to

$$\langle \theta(s)^2 \rangle = \frac{s}{l_0} \qquad (4.8)$$

The chain statistics at short scales are not modified by electrostatic interactions. The crossover between the *intrinsic* regime described by (4.8) and the *electrostatic* regime described by (4.7) takes place when the electrostatic interactions become strong enough to perturb the statistics of the neutral flexible chain. The crossover length s_c can be obtained qualitatively from the following argument.* If a small-chain section, of length $s < \kappa^{-1}$, is bent to form an angle θ, the cost in bare curvature energy is $k_B T l_0 \theta^2 / s$, while the cost in electrostatic energy is $k_B T l_B (s/A)^2 (\theta^2 / s)$. The two energies are comparable for $s \simeq s_c$, which gives $s_c \sim A(l_0/l_B)^{1/2}$.

A more detailed treatment of the crossover regime is possible by ignoring logarithmic factors and replacing the exact kernel (4.4) by the approximate expression, $\tilde{K}(q) = l_{OSF} \kappa^2 / (\kappa^2 + 2q^2)$. The integral (4.6) can then be computed analytically and yields

$$\langle \theta(s)^2 \rangle = \frac{s}{l_{OSF} + l_0} + \frac{l_{OSF}}{\kappa l_0^{1/2}(l_0 + l_{OSF})^{3/2}} \left[1 - \exp\left(-\frac{s}{s_c} \right) \right] \quad (4.9)$$

with $s_c = \kappa^{-1}(l_{OSF} + l_0)^{-1/2} l_0^{1/2}$. This formula crosses over from the intrinsic regime toward the electrostatic regime for $s \sim s_c$. In the weak screening limit, $l_0 \ll l_{OSF}$ and the simple result $s_c \sim A(l_0/l_B)^{1/2}$ is recovered.

The picture that emerges from this calculation is that the chain flexibility depends on the length scale. At short scales, $s < s_c$, the chain structure is determined by its bare rigidity l_0, while the electrostatic rigidity (4.5) dominates at large scales. Large-scale properties, such as the giration radius or the structure factor at small wave vector, can be determined by applying the standard formula for semiflexible chains of persistence length l_{OSF}. Excluded volume effects between Kuhn segments of length $2l_{OSF}$ are accounted for by assigning a diameter κ^{-1} to each segment. The excluded volume between Kuhn segments is $l_{OSF}^2 \kappa^{-1}$ [10].

The only approximation required to obtain the Odijk–Skolnick–Fixman length is the expansion that yields Eq. (4.1). The calculation is therefore consistent if the angle $\langle \theta(\kappa^{-1})^2 \rangle$ is small compared to unity. It is easily checked that for $s \sim \kappa^{-1}$ the second term dominates in the right-hand side of (4.9). The requirement $\langle \theta(\kappa^{-1})^2 \rangle \ll 1$ is then equivalent to $s_c l_{OSF} / [l_0(l_0 + l_{OSF})] \simeq s_c/l_0 \ll 1$, that is, the angular deflection of the chain must be small when the crossover region is reached. In other words, the angular fluctuations that take place before the electrostatic interactions can come into play and rigidify the chain should not be too strong. If these fluctuations are too large, that is, if the chain is too

---

* From its definition, s_c can be described as the contour length of an *orientational blob*.

flexible, the perturbation expansion that underlies the OSF calculation breaks down. The criterion for the validity of the calculation can be simply written, in the limit of weak screening, as $l_0 > A^2/l_B$. This implies that the OSF calculation should be directly applicable to stiff chains, such as DNA ($l_0 \sim 50$ nm), but has to be reconsidered for flexible chains, such as polystyrene sulfonate ($l_0 \sim 1$ nm). Indeed, numerical simulations clearly confirm the behavior described by Eqs. (4.4) and (4.6) for values of s_c/l_0 smaller than 0.2 [40]. Deviations from this behavior appear for larger values of s_c/l_0.

Before closing this section, it is worth mentioning that the electrostatic rigidity of a charged chain was computed numerically by Le Bret [41] and Fixman [42] using the Poisson–Boltzmann equation rather than the Debye–Hückel approximation. At high ionic strength, the results deviate significantly from the OSF prediction, and the electrostatic rigidity tends to behave as κ^{-1} rather than κ^{-2}.

B. Alternative Calculations for Flexible Chains

The calculation presented in the previous section shows that the Odijk–Skolnick–Fixman theory in its original form is not applicable to flexible or weakly charged chains as such. A generalization of this theory was nevertheless used by Khokhlov and Katchaturian [43]. These authors propose that expression (4.5) can be used for flexible chains, with the only modification that the distance between charges A had to be replaced by $\xi_e/(fg)$ [see Eq. (2.7)]. The bare persistence length l_0 is also replaced by the blob size ξ_e. (A recent calculation of Li and Witten, including the thermal fluctuations of the chain, gives a more quantitative basis to this approach [44].) As a result, the total persistence length of the chain reads

$$l_{KK} = \xi_e + \frac{1}{4\kappa^2 \xi_e} \qquad (4.10)$$

The arguments of Section IV.B show, however, that the extension of the OSF formula is rather speculative, since the condition that was derived for the validity of the OSF theory is not met for a chain characterized by the parameters $l_0 = \xi_e$ and $A = \xi_e/fg$. When the persistence length is dominated by the electrostatic interactions, the radius of the chain is given by the Flory statistics

$$R_{KK} = N^{3/5} \xi_e^{-4/5} \kappa^{-3/5} b^{6/5} \qquad (4.11)$$

This generalization of the Odijk theory is consistent if the Debye screening length is larger than the electrostatic blob size ξ_e. At higher ionic strength, in the weak coupling limit, there is no electrostatic rigidity and the polyelectrolyte chain behaves as a flexible chain with short-range interactions given by the Debye–Hückel formula. Its radius is given by Eq. (4.12).

A number of approximate theories, based on different variational ansatz, have been proposed to describe the structure of flexible chains [40, 45–47]. These calculations differ at the technical level. The basic idea is to describe a flexible charged chain (with screened interactions) by some model of noninteracting semiflexible chain and to variationally optimize the persistence length of the noninteracting system. All these variational calculations produce for flexible chains a persistence length that scales as κ^{-1}, that is, the "naive" result of Katchalsky and Pfeuty. An example of such a variational calculation can be given for the particularly simple case of a freely jointed charged chain, with $l_0 = A$. The variational ansatz used to represent the system is that of a non-interacting chain in which the angle between neighboring segments is bounded by a maximum value θ_{max}, so that the persistence length is $l_p = 2A/\theta_{max}^{-2}$. The variational entropy per segment is then $-\ln(\theta_{max})$. The variational energy can be expressed as a function of the structure factor of a chain with a persistence length l_p, $S_{ni}(\mathbf{q})$ as $\int d^3\mathbf{q} S_{ni}(\mathbf{q}) v_{DH}(q)$. Using the expression of the structure factor obtained by des Cloizeaux [48], the energy is the sum of the electrostatic energy of a rod and of a correction due to the bending that can be approximated by $k_B T(l_B/A)(1/\kappa l_p)$. Minimizing the sum of these two terms with respect to θ_{max}, one finds $l_p \sim \kappa^{-1}$. This approach can be straightforwardly generalized to the case where each rod in the freely jointed chain represents an electrostatic blob [40], in which case it applies to the charged Gaussian chain considered in Section II. The variational model is then that of a *semiflexible chain of electrostatic blobs*, analogous to the chain under tension model (Section II), except for the fact that the direction of the tension is now fluctuating and becomes uncorrelated over a persistence length κ^{-1}. In that case, it must be realized that the *persistence length* κ^{-1} is defined in reference to the distance in space, rather than the contour length along the chain. The radius of the chain scales in this case as

$$R_P = N^{3/5} \xi_e^{-3/5} \kappa^{-2/5} b^{6/5} \tag{4.12}$$

Variational calculations have the obvious drawback that the result crucially depends on the variational ansatz. An ansatz too far from the actual structure can yield incorrect results. Also, their range of validity is difficult to assess. The variational calculation for Gaussian chains briefly described above is consistent only as long as the electrostatic blob concept can be used. This gives an upper limit for the charge density of the chain for which its description by a semiflexible chain of electrostatic blobs is possible, $A^{-1} < (l_0 l_B)^{-1/2}$ [this corresponds to $g > 1$ in Eq. (2.7)]. This upper limit corresponds to the lower charge density for which the OSF calculation is expected to be consistent. The coincidence, however, might be only fortunate. A complete calculation that would crossover from the κ^{-2} behavior of the electrostatic stiffness, which is well understood for rigid chains, to the κ^{-1} behavior that is suspected for flexible chains on the basis of variational calculations is still missing.

C.   Case of Poor Solvents

The preceeding discussion focused on the case where the neutral chain backbone is in a good or theta solvent. A rather different result is obtained for the case of a flexible, weakly charged chain in a poor solvent. This case was considered in the absence of screening in Section II.A, where the chain was described as a linear string of *poor solvent blobs*, each of size $\xi_e = b(f^2 l_B/b)^{-1/3}$ and containing $\tau f(f^2 l_B/b)^{-1}$ charged monomers [cf. Eq. (2.10)]. Following the procedure of Khokhlov and Khachaturian [43], one can attempt to apply the OSF theory to the chain of blobs. The theory is valid if the linear charge density along the chain of blobs, fg/ξ_e, is larger than the critical value $(\xi_e l_B)^{-1/2}$. This condition can be rewritten as $\tau > (f^2 l_B/b)^{1/3}$. For weakly charged chains $(f \ll 1)$, it is thus satisfied even for moderately poor solvents. The electrostatic persistence length that results from applying the OSF theory is

$$
\begin{aligned}
l_p &= \xi_e + \frac{f^2 g^2 l_B}{\kappa^2 \xi_e^2} \\[2mm]
&= \xi_e + \left(\frac{\tau^2}{b\kappa^2}\right)\left(\frac{f^2 l_B}{b}\right)^{-1/3}
\end{aligned}
\tag{4.13}
$$

The surprising implication of this result is that the persistence length (and therefore the radius of gyration of the chain) increases as the solvent quality decreases. This increase is the consequence of an increase in the linear charge density, which is itself induced by the collapse of the monomers into dense blobs. This effect, however, exists only as long as the charge density exceeds the value l_B^{-1} at which Manning condensation takes place. This gives an upper limit for τ, $\tau < f(f^2 l_B/b)^{-1/3}$. The conclusion is that an increase in the persistence length as τ increases should occur in the temperature range $f^{1/3}(b/l_B)^{1/3} > \tau > f^{2/3}(l_B/b)^{2/3}$.

V. CHARGED GELS AND BRUSHES

Many properties of polyelectrolyte solutions are dominated not by the chain conformation but by the counterions. We discuss two examples of this situation, the polyelectrolyte gel and the polyelectrolyte brush. In both problems, the polyelectrolyte chains occupy a small region of space (close to the grafting surface for the brush) surrounded by pure solvent. The small ions in the solution are free to diffuse in and out the polyelectrolyte region, and the external solvent region acts as a reservoir that imposes the small ion chemical potential. A Donnan equilibrium [49, 50] is reached that relates the concentrations of the small ions inside the polyelectrolyte region to their imposed concentration in the reservoir. If the monomer concentration in the polyelectrolyte region is c, the charge density due to the polymer is fc. The reservoir contains salt at a density n, and the Debye–Hückel screening length κ^{-1} is given by $\kappa^2 = 8\pi n l_B$. For simplicity we assume in the following that the counterion of the polymer is identical to the positive ion of the salt. We call n_+ and n_- the concentration of positive and negative ions, respectively, in the polyelectrolyte region.

The polyelectrolyte charge creates an electrostatic potential difference between the polyelectrolyte region and the reservoir U. The chemical potentials of the small ions in the polyelectrolyte region are $\mu_+ = k_B T \ln n_+ + U$ and $\mu_- = k_B T \ln n_- - U$. At equilibrium these chemical potentials are equal to the salt chemical potential in the reservoir $\mu = k_B T \ln n$. This leads to a chemical equilibrium law for the small ions

$$n_+ n_- = n^2 \tag{5.1}$$

A second relation between the concentrations n_+ and n_- is provided by the electroneutrality condition $n_+ = n_- + fc$. The salt concentration inside the polyelectrolyte region (given by n_-) is smaller than that of the reservoir. The difference in osmotic pressure between the two regions is $\pi = k_B T(n_+ + n_- - 2n) = [(fc)^2 + 4n^2]^{1/2} - 2n$. When the salt concentration n is smaller than the counterion concentration fc, the difference in salt concentration is small and the osmotic pressure is that of the ideal gas of counterions discussed in Section III.A, $\pi = k_B T fc$. In the limit where the salt concentration is larger than the counterion concentration, the osmotic pressure is given by

$$\pi = k_B T \frac{(fc)^2}{4n} = k_B T \frac{c^2}{2} \frac{4\pi l_B f^2}{\kappa^2} \tag{5.2}$$

This osmotic pressure defines an effective virial coefficient between monomers $v_{el} = f^2/2n = 4\pi l_B f^2 \kappa^{-2}$. This is identical to the excluded volume that can be calculated from the Debye–Hückel interaction (2.16) between monomers.

The major assumption made here is that the ion densities are uniform. This is the case when regions of size κ^{-1} around each polyelectrolyte, where the counterion concentration is increased, overlap, that is, when the Debye–Hückel screening length is larger than the distance between chains. This is always true in the absence of salt. The osmotic pressure calculated from the Donnan equilibrium is of purely entropic origin. If the ion densities are uniform the electrostatic contribution to the osmotic pressure is small compared to the translational entropy of the small ions as checked below for a gel. The osmotic pressure is thus independent of the strength of the electrostatic interaction characterized by the Bjerrum length l_B. The only role of the electrostatic interaction is to enforce electrical neutrality.

A. Grafted Polyelectrolyte Layers

Grafted polymer layers (polymer brushes) [51] are obtained by attaching one of the polymer end points on a planar solid surface. If the distance between chains is smaller than their natural size, a thick layer with a size proportional to the chain molecular weight is formed. The chains are stretched by the repulsive interactions between monomers.

Polymer brushes are often very efficient to enhance colloidal stabilization [52]. Grafted polyelectrolyte layers have been described theoretically by several authors [53–56]. We follow here the lines of the original work of Pinus that assumes that the polymer concentration is uniform throughout the thickness h of the brush. A similar approximation for neutral polymer brushes was introduced by Alexander [57] and de Gennes [58]. We also assume in this section that the local electrostatic screening length is larger than the distance between chains imposed by the distance D between grafting points so that the counterion concentration is roughly constant in the layer.

The electrostatic field and the distribution of counterions in the vicinity of a charged solid surface with a surface charge density ρ can be determined from the Poisson–Boltzmann equation. In the absence of added salt, the counterion density decays as a power law of the distance from the surface, and the counterions are confined in the vicinity of the surface over the Gouy–Chapman length $\lambda = (2\pi\rho l_B)^{-1}$. If the grafting surface is neutral, the charge density of a grafted polyelectrolyte with $\sigma = D^{-2}$ grafted chains per unit area layer is due to the monomer charges

$\rho = \sigma N f$, and the Gouy–Chapman length [23] of the polymer brush is

$$\lambda \simeq \frac{1}{\sigma f N l_B} \qquad (5.3)$$

Two cases must then be considered. When the Gouy–Chapman length is larger than the brush size h, most of the counterions are outside the brush, and the brush is charged. When λ is smaller than the brush height, the counterions are confined in the grafted layer and the brush is neutral.

In the neutral limit, the brush is swollen by the osmotic pressure of the counterions $\pi = k_B T f c$. This pressure is balanced by the elasticity of the polymer chains. For a Gaussian chain, the elastic pressure is $\pi_{el} = \sigma h / (N b^2)$. The monomer concentration in the brush being $c = \sigma N / h$ the thickness of the brush is given by

$$h = N b f^{1/2} \qquad (5.4)$$

The thickness of the brush is in this regime independent of the strength of the electrostatic intractions since the stretching of the chains is due only to entropic effects. The thickness given by Eq. (5.4) is larger than the size of an individual chain in a dilute solution given by (2.5). This result has been obtained by assuming that the monomer concentration is constant inside the adsorbed polymer layer. Similar scaling results are obtained if this constraint is released. Using a self-consistent field method, Zhulina et al. [54] have shown that the concentration profile has then a Gaussian decay.

In the charged limit, one can in a first approximation assume that the counterion concentration is constant up to the Gouy–Chapman length, it is equal to $c_i = (fch)/\lambda$. The counterion pressure that stretches the chain is therefore $\pi = k_B T f c h / \lambda$. The thickness of the brush is then

$$h = N^3 l_B b^2 \sigma f^2 \qquad (5.5)$$

In this regime the electrostatic interactions have a contribution of the same order as the entropic contribution and the thickness depends on their strength. The thickness of the layer grows faster than linearly with molecular weight; however, one should keep in mind that the layer thickness remains smaller than the Gouy–Chapman length and thus that this thickness remains smaller than the thickness in the neutral regime (5.4). A precise determination of the counterion profile in this regime has been done by Pincus [53]. The same scaling laws are obtained when this profile is taken into account.

When salt is added to the layer, the polyelectrolyte brush behaves as a neutral polymer brush with an effective excluded volume $v_{el} = f^2/(2n)$; the thickness of the brush is then

$$h = Nb(\sigma b^2)^{1/3} f^{2/3} (nb^3)^{-1/3} \qquad (5.6)$$

The monomer concentration profile has also been calculated in this limit by Zhulina et al. [54]. The important result is that the crossover to the neutral brush regime (5.4) occurs when the salt concentration is larger than the counterion concentration in the brush. This leads in general to a high ionic strength, and the thickness of a grafted polymer layer is rather insensitive to ionic strength over a broad range of ionic strength. This makes grafted polyelectrolyte layers particularly interesting to promote colloidal stabilization.

If the polyelectrolyte is annealed, the pH inside the grafted layer can be significantly different from outside. This could induce a nonmonotonic variation of the thickness with grafting density [59].

B. Polyelectrolyte Gels

A model based on the Donnan equilibrium, very similar to that of polymer brushes, can be made for charged polymeric gels [60]. The polymer gel is at equilibrium with a reservoir of solvent if the osmotic pressure in the gel is equal to the osmotic pressure of the salt in the reservoir, that is, if the swelling osmotic pressure due to the entropy of the counterions is balanced by the pressure due to the elasticity of the polymer chains. If the chains between the crosslinks of the gel are Gaussian chains and have N monomers and an end-to-end distance (mesh size) R, the elastic pressure is of order $\pi \simeq k_B T(c/N)(R^2/Nb^2)$. If we now make the c^* assumption proposed by de Gennes [4] that, in the swollen gel at equilibrium, the chains are just at the overlap concentration, the mesh size of the gel is such that $c = N/R^3$. In the absence of salt, the mesh size is then given by Eq. (5.4):

$$R = Nbf^{1/2} \qquad (5.7)$$

In the presence of salt, the mesh size is given by

$$R = N^{3/5} b \left(\frac{f^2}{2n} \right)^{1/5} \qquad (5.8)$$

The important point to notice is that, in contrast to neutral gels, the mesh size of the gel is different from the radius of an isolated polymer chain in a dilute solution. As explained above, these results are valid when the salt concentration is small enough that the Debye–Hückel screening length is larger than the gel mesh size, that is, when the mesh size is larger than the radius of an isolated chain in solution given by (2.4). The chains in this case should be viewed as chains under strong tension. The small ion osmotic pressure is exerted at the surface of the gel, and the force is transmitted to the internal chains via the crosslinks.

The mesh size and thus the equilibrium swelling of the gel do not depend on the strength of the electrostatic interaction l_B. When the monomer concentration in the gel is uniform, the electrostatic energy of the gel vanishes. The actual concentration in the gel is not homogeneous, and the inhomonogeneity can be characterized by the structure factor $S(q)$. The electrostatic energy per unit volume of the gel can then be expressed as

$$E_{el} = c \frac{f^2}{2} \int d\mathbf{q}\, S(q) \frac{4\pi l_B}{q^2 + \kappa^2} \tag{5.9}$$

The structure factor of the charged gel is not known exactly, but one can expect it to have a sharp peak at a position $q^* \simeq R^{-1}$, the height of the peak being of order N. The integral giving the electrostatic energy of the gel is dominated by the peak of the structure factor and can be estimated up to numerical prefactors:

$$E_{el} = \frac{c k_B T}{N} \frac{f^2 N^2 l_B}{R} \frac{1}{1 + \kappa^2 R^2} \tag{5.10}$$

It can then be directly checked that, in the weak screening limit where $\kappa R \ll 1$, the electrostatic energy is smaller than the translational entropy of the counterions.

In the strong screening limit $\kappa R \gg 1$, the electrostatic interactions are dominant. The charged monomers in the gel interact via a Debye–Hückel potential, and the electrostatic interactions contribute to the rigidity of the chains between crosslinks as discussed in the previous section. At very high ionic strength the electrostatic persistence length is small, and the chains behave as neutral chains with an electrostatic excluded volume $v_{el} = f^2/2n$. The same scaling result as Eq. (5.8) is then found for the mesh size.

The elastic shear modulus of the gel can be calculated by imposing a uniform deformation γ to the crosslinks. The increase of the square of the mesh size imposed by the deformation is of order $(\gamma R)^2$ and if the chains are Gaussian, the elastic energy stored in the gel is

$$F_{el} = k_B T \frac{c}{N} \frac{\gamma^2 R^2}{Nb^2} \tag{5.11a}$$

This leads to a shear modulus

$$G = k_B T \frac{c}{N} \frac{R^2}{Nb^2}$$

$$\simeq k_B T f c \quad \text{(salt-free case)} \tag{5.11b}$$

$$\simeq k_B T \frac{f^2 c^2}{n} \quad \text{(with added salt)}$$

In all cases, the shear modulus of the gel is thus of the order of the osmotic pressure of the small ions (counterions and salt).

The last property of the gel that we discuss is the cooperative diffusion constant, which can be obtained from a two-fluid model [61]. The displacement field of the gel $u(\mathbf{r}, t)$ results from a balance between the elastic restoring force and a viscous force due to the solvent drag. The elastic force per unit volume of the gel is $G\nabla^2 u$. The viscous force exerted by a solvent flowing through a porous medium of pore size R is given by the Brinkman equation [62] and is of order $\eta R^{-2}(\partial u / \partial t)$ where η is the solvent viscosity. The force balance on the gel is then written as

$$\eta R^{-2} \frac{\partial u}{\partial t} = G\nabla^2 u \tag{5.12}$$

The relaxation of the gel deformation is diffusive with a cooperative diffusion constant

$$D = \frac{GR^2}{\eta} \tag{5.13}$$

The cooperative diffusion constant is independent of the molecular weight in the absence of salt $(D \simeq f^{1/2})$. It decreases with N in the presence of salt $[D \simeq N^{-2/5}(f^2/n)^{1/5}]$.

Experimentally, the properties of charged gels, as well as the properties of neutral gels, strongly depend on the preparation condition of the gels. The simple results presented here (based on the so-called c^* theorem) implicitly assume that the gel was prepared by crosslinking chains in the vicinity of their overlap concentration. A more refined theory [63] based on an affine deformation hypothesis first proposed by Flory [64] and then developed by Panyukov [65] has recently been constructed. It provides an explanation for the experiments of Skouri and co-workers [66] that show that the elastic shear modulus of polyelectrolyte gels with a fixed concentration smaller than the equilibrium swelling concentration decreases with the fraction of charged monomers.

VI. SEMIDILUTE SOLUTIONS

The overlap concentration of polyelectrolyte solutions in the absence of salt is very low; the chain radius given by Eq. (2.4) increases linearly with molecular weight, and the overlap concentration decreases as $c^* \simeq b^4/(l_B N^2 f^2)$. Most of the experiments with long chains are thus performed in the semidilute range where the electrostatic interactions between different chains are strong. Our understanding of the conformation of polyelectrolyte chains in semidilute solutions is, however, rather poor and no general view is available. We discuss in this section a few aspects of the static properties of interacting polyelectrolyte chains.

A. Ordering Transitions in Polyelectrolyte Solutions

The models presented in Section II to describe the conformation of polyelectrolytes in salt-free solutions consider only a single chain and ignore the screening due to the counterions. If the polymer concentration is finite, the counterion concentration is finite, and the screening length due to the counterions κ^{-1} is given by $\kappa^2 = 4\pi l_B fc$. In a dilute solution the screening length is always larger than the average distance between chains $d \simeq (N/c)^{1/3}$. Different chains thus interact via a long-range pure Coulomb potential. The polyelectrolyte chains are strongly charged objects, and this long-range interaction can lead to the formation of Wigner crystals, very similar to the colloidal crystals observed in solutions of charged spherical particles [67] or charged elongated particles (viruses) with a mesoscopic size. The centers of mass of the particles are then regularly distributed on a periodic lattice. The solution has a tendency to crystallize when the interaction between neighboring chains $k_B T N^2 f^2 l_B/d$ is larger than the thermal excitation $k_B T$. This naive estimate predicts crystal formation at a very low concentration, in the dilute range $c \simeq 1/(N^5 f^6 l_B^3)$. The crystal is expected to melt when screening is important, in the vicinity of the overlap concentration. There is no clear experimental evidence for this crystallization in flexible polyelectrolyte solutions. This may be due to the weak elastic resistance of this crystal that could be destroyed by any kind of perturbation (mechanical perturbation). In any case we expect strong interactions in dilute polyelectrolyte solutions that lead to a structuration of the solution and thus to the appearance of a peak in the structure factor $S(q)$ of the solution corresponding either to liquidlike or to solidlike order. The wavevector q^* at the peak is of the order of the inverse distance between chains $q^* \simeq 2\pi/d \simeq (c/N)^{1/3}$ [68]. The peak is observed experimentally and the variation of its position with concentrations is in good agreement with this prediction. It can be suppressed by adding salt to the solution, so that the screening length κ^{-1} becomes smaller than the distance between chains.

As in dilute solutions, polyelectrolytes have, in a semidilute solution, a locally rodlike conformation that can be characterized by a persistence length l_p. Quite similarly to rodlike molecules, semiflexible macromolecules undergo an Onsager transition [69, 70] between an isotropic liquid phase and a nematic ordered phase where the molecules are parallel. The Onsager concentration where this transition takes place has been calculated by Semenov and Khokhlov [71] and Grosberg and Khokhlov [72]. For neutral molecules of diameter d it scales as $c_o \simeq 1/(l_p db)$. For charged molecules, the diameter of the molecule must be replaced by an effective diameter equal to the range of the electrostatic interaction κ^{-1} [70], and the Onsager concentration is of order $c_o \simeq 1/(l_p \kappa^{-1} b)$. All of the models for polyelectrolyte conformation based on the Odijk–Skolnick–Fixman approach predict an isotropic-nematic transition for polyelectrolyte solutions in the presence of salt. Experimentally, the transition is only observed for very rigid polymers (such as DNA where a cholesteric phase is observed) where the intrinsic persistence length is very large and dominates the electrostatic persistence length. There is no definite evidence for an Onsager transition in flexible polyelectrolyte solutions. This is an unresolved issue. Possible explanations are based on the idea that the scaling theories ignore some prefactors that may be large and shift the transition to an unobservable value. The isotropic-nematic transition is also expected only for anisotropic enough objects, the ratio between the persistence length and effective diameter κ^{-1} must be larger than a finite number $\simeq 5$ [73]. This criterion is not always met experimentally. A thorough theoretical study of possible isotropic-nematic transition in polyelectrolyte solutions has been made by Nyrkova [74]. In the following we will ignore this possibility and consider that a semidilute polyelectrolyte solution remains isotopic at any concentration.

B. Correlation Length and Osmotic Pressure of Semidilute Polyelectrolyte Solutions

In a dilute solution, polyelectrolyte chains have a rodlike conformation at a local scale smaller than the persistence length l_p, and the persistence length is always larger (or of the same order of magnitude) than the screening length of the electrostatic interactions. In a semidilute solution, the polymer chains overlap and, if we assume that they do not form ordered phases, they form a temporary network with a mesh size ξ. In the absence of salt, the chains also have a rodlike conformation at the scale ξ, and their persistence length is, as shown below, larger than the mesh size. The mesh size can then be calculated from a scaling argument by imposing that it is equal to the isolated chain radius (2.4) at the overlap concentration c^*. (For simplicity, we suppose here that locally the polymer shows local Gaussian statistics; the other cases can be treated in a similar way.) Equivalently, a pure geometrical argument can be used, by imposing a close packing condition for chain subunits of size ξ, $c = (\xi/\xi_e)g/\xi^3$. In any case, ξ is given by [14, 39]

$$\xi \simeq b^{-1}c^{-1/2}\xi_e^{1/2} \tag{6.1}$$

The number of monomers within a volume of size ξ^3, or correlation blob, is then $G_\xi = c^{-1/2}\xi_e^{3/2}b^{-3}$.

For weakly charged polyelectrolytes, the counterions are free and behave roughly as an ideal gas. The screening length due to the counterions is then $\kappa_i^{-1} = (4\pi l_B fc)^{-1/2}$. This screening length is of the order of the mesh size for strongly charged polyelectrolytes ($f \simeq 1$), but it is larger than ξ for weakly charged polyelectrolytes. The polymer itself, however, contributes to the screening of the electrostatic interactions and the actual screening length is smaller than the counterion screening length. From a purely electrostatic point of view each cell of size ξ containing G_ξ monomers is neutral and can be considered as independent of all the others. It is thus reasonable to assume that the effective screening length is of the order of the mesh size ξ [43, 75].*

As salt is added to the solution, the screening length decreases and becomes dominated by salt ($\kappa^2 = 8\pi n l_B$) when the Debye–Hückel screening length of the salt κ^{-1} is smaller than the mesh size ξ. When the screening is dominated by the salt, the electrostatic persistence length of the chains decreases with the salt concentration. As long as the persistence length is larger than the mesh size ξ, the mesh size is still given by Eq. (6.1). If the persistence length is smaller than the mesh size, the polyelectrolyte solution behaves as a neutral semiflexible polymer solution and the mesh size or correlation length can be calculated from the radius in a dilute solution using scaling arguments. If the Odijk–Skolnick–Fixman statistics is assumed [43, 76],

$$\xi \simeq b\xi_e c^{-3/4}\kappa^{3/4}b^{-3/2} \tag{6.2}$$

---

*This is somewhat similar to the assumption that, in neutral polymers, the *hydrodynamic* screening length is equal to the static correlation length.

while if the persistence length is assumed equal to the screening length [14, 39]

$$\xi \simeq \xi_e^{3/4} c^{-3/4} \kappa^{1/2} b^{-3/2} \tag{6.3}$$

Experimentally, the important feature of semidilute polyelectrolyte solutions is the fact that, if the salt concentration is low enough, the structure factor $S(q)$ has a peak at a finite wave vector q^*. At small wave vectors, the structure factor, because of the electroneutrality of the solution, is dominated by the small ions, which behave as an ideal gas. In the absence of salt, $S(q=0) = 1/f$. If the salt concentration n is larger than the counterion concentration, $S(q=0) = 2n/(f^2 c)$. At large wave vectors, corresponding to distances smaller than the mesh size ξ, the polyelectrolyte has a rodlike behavior and the structure factor is given by $S(q) = \xi_e/(qb^2)$. When the wavelength q^{-1} is equal to the mesh size, the value of the structure factor is $S^* = G_\xi = c^{-1/2} \xi_e^{3/2} b^{-3}$. It can be checked explicitly that this value is larger than the thermodynamic value $S(q=0)$, and the structure factor must thus have a peak at a finite wave vector q^*. The existence of the peak is therefore related to the very small compressibility of the small ion gas that gives $S(q=0)$ due to the electroneutrality constraint. This discussion is valid as long as the counterions and the salt ions are uniformly distributed throughout the solution, that is, as long as the screening length is larger than the mesh size of the solution. In the salt-dominated regime, where the salt screening length is smaller than the mesh size of the solution, the counterions are confined in a sheath of size κ^{-1} around each polymer, and the solution essentially behaves as a neutral polymer solution where the structure factor decays monotonically. The peak thus disappears in the salt-dominated regime. At the crossover between the two regimes, the $q=0$ value of the structure factor increases very sharply. When the effect of salt is not dominant, the peak position q^* defines the correlation length of the solution and one expects that it is of the order of the inverse of the mesh size $q^* \simeq 1/\xi$. The position of the peak of the structure factor increases thus as $c^{1/2}$ in a semidilute solution (and as $c^{1/3}$ in a dilute solution as explained above). It is also important to note that the correlation length ξ and thus the peak position are roughly independent of the salt concentration as long as the Debye–Hückel screening length is smaller than ξj. These results are in rather good agreement with experiments.

In the osmotic regime where the Debye–Hückel screening length is larger than the mesh size, the small ions are uniformly distributed and the osmotic pressure of the solution is the same as that of a gel as discussed in Section V; it is dominated by the counterions. It is equal to $\pi = k_B Tfc$ if the counterion concentration is smaller than the salt concentration, $(fc > n)$ and to $\pi = k_B T f^2 c^2 / 4n$ if $fc < n$. In the salt-dominated regime $\kappa^{-1} < \xi$, the polyelectrolyte chains behave as neutral semiflexible chains with a persistence length l_p and an effective diameter equal to the screening length κ^{-1} (up to logarithmic corrections). The excluded volume between Kuhn segments of size l_p is of order $l_p^2 \kappa^{-1}$. If the persistence length is larger than the mesh size (or of the same order), the osmotic pressure varies as

$$\frac{\pi}{k_B T} = c^2 \kappa^{-1} \xi_e^{-2} b^4 \tag{6.4}$$

At the crossover between the osmotic and the salt-dominated regime, the osmotic pressure varies very rapidly and the crossover is not smooth. When the persistence length becomes small enough compared to the mesh size, the excluded volume correlations are relevant and the osmotic pressure increases as $\pi \simeq c^{9/4}$. At very high ionic strength, the electrostatic interaction is very weak and cannot be approximated by an excluded volume between Kuhn segments. In this weak coupling regime there no longer is an electrostatic persistence length (it is shorter than the screening length), and the polyelectrolyte chains should be considered as flexible chains where the monomers interact via the short-range Debye–Hückel potential with an effective excluded volume $v_{el} = f^2/2n$. The smooth crossover between the salt-dominated and the weak coupling regimes occur when $\kappa \xi_e \simeq 1$, that is, when the screening length becomes equal to the electrostatic blob size.

C. Electrostatic Rigidity in Semidilute Solutions

In this section we discuss the conformation of polyelectrolyte chains in a semidilute solution and the effect of the interactions between different chains on the persistence length of a polyelectrolyte. For simplicity we consider only a strongly charged polyelectrolyte for which the bare persistence length l_0 and the distance between charges A are such that $A^2 > l_0 l_B$ so that the Odijk–Skolnick–Fixman theory can be applied in a dilute solution. We also assume that the bare persistence length is much smaller than the mesh size of the semidilute solution. The results can easily be extended to weakly charged polyelectrolytes if necessary by renormalizing the bare persistence length to the blob size.

Two types of theories have been proposed to describe chain conformation in a semidilute solution. The early work of Odijk [76] and Hayter et al. [77] makes the assumption that the interchain interactions have a negligible influence on the persistence length so that the persistence length is given by the usual Odijk–Skolnick–Fixman theory with the relevant screening length. A scaling theory for the chain conformation has been built on this assumption. It was, however, later pointed out by Witten and Pincus [78] that if a polyelectrolyte solution is isotropic, the electrostatic interaction energy between two chains can be reduced by a bending of the chains to avoid each other and thus that the interaction energy between different chains reduces the persistence length. A persistence length equal to the mesh size of the solution is obtained in the osmotic regime where the Debye–Hückel screening length of the salt is larger than the solution mesh size. In a revised version of this theory, in the salt-dominated regime, the corrections to the persistence length due to the interactions between chains are found to be small.

The simplest approach [79] to the persistence length of a polymer chain in a semidilute solution is to use linear response theory. Within the framework of linear response, an effective pair interaction between monomers can be introduced. The Fourier transform of the effective interaction is related to the structure factor $S(q)$ of the solution by $\hat{v}_{\text{eff}}(q) = \hat{v}_{\text{DH}}(q)[1 - cS(q)\hat{v}_{\text{DH}}(q)]$. Following the lines of the Odijk argument, the effective persistence length can then be calculated as

$$l_p = l_0 + \frac{\hat{v}_{\text{DH}}(q=0)}{16\pi k_B T A^2}\left[1 - cS(q=0)\hat{v}_{\text{DH}}(q=0)\right] \qquad (6.5)$$

This relation gives the persistence length as a function of the $q = 0$ value of the structure factor, which is a thermodynamic quantity. If the explicit form of the Debye–Hückel potential is substituted, the persistence length is $l_p = l_0 + l_{\text{OSF}}[1 - 4\pi l_B cS(q=0)/\kappa^2]$. A reduction of the persistence length from the single-chain value induced by interchain interactions is therefore predicted. However, this result can be used only in the limit of linear response, that is, if the interaction between chains is smaller than $k_B T$. The interaction between two rods crossing at perpendicular angle is

$$\beta = \frac{2\pi l_B}{\kappa A^2} \qquad (6.6)$$

The linear response result is thus valid if $\beta \ll 1$, which corresponds to the weak coupling regime for rigid polyelectrolytes.

No direct calculation of the effect of interchain interactions on the persistence length seems available in the strong coupling regime $\beta \gg 1$. Some insight into the problem can, however, be gained by looking at the following simpler two-dimensional problem: one semiflexible test chain lies in a plane and interacts with all the other chains replaced by infinite rods perpendicular to this plane with a concentration per unit area Γ. It is a good approximation to replace the other chains by infinite rods as long as their persistence length is larger than the screening length. In the limit where the Debye screening length is smaller than the distance between the chains, the solution can be considered as a quasi ideal gas of rods. When the test chain bends, the distribution of the rods around the test chain is changed, and this contributes to the bending energy. The persistence length can be calculated following the lines of the work of Odijk in a dilute solution; the result is

$$l_p = l_{OSF} - \Gamma\kappa^{-3}f(\beta) \tag{6.7}$$

where the function $f(\beta)$ is defined as

$$f(\beta) = \frac{\beta}{2}\int_0^{+\infty} dz \, \exp[-\beta\exp(-z)]$$
$$\times [-\beta z^2 \exp(-2z) + \exp(-z)(-z^2 + z + 1)] \tag{6.8}$$

The correction to the Odijk value is always negative, which corresponds also to a reduction of the persistence length from the isolated chain value. In the weak coupling limit, $f(\beta) = 3\beta^2/8$ in agreement with the linear response theory. In the strong coupling limit, the interaction is very similar to a hard sphere interaction and $f(\beta) = (\ln \beta)^2$ varies only logarithmically with the interaction strength. The correction to the single-chain persistence length due to interchain interactions is small.

These results obtained in two dimensions find a simple interpretation in terms of individual deflections of the test chain by the perpendicular rod. When the test chain interacts with one rod at a distance r, it bends and its orientation changes over a distance κ^{-1} by a finite angle that can be shown to be equal to

$$\theta \simeq \kappa r \exp(-\kappa r) \tag{6.9}$$

The persistence length can then be obtained by summing the individual deflections weighed by the Boltzmann factor related to the direct electrostatic interactions between the chains. The persistence length given by Eq. (6.7) is then found up to a numerical prefactor. The independent deflection model is easily generalizable to real three-dimensional chains. The chains do not always cross at a finite angle, but the conformations where two chains are almost parallel are energetically costly, since the electrostatic interaction is very high in this case (it increases linearly with chain length). This leads to a persistence length

$$l_p^{-1} = l_{OSF}^{-1} \left[1 + \ln^2(\beta) \frac{\alpha c A^3}{\kappa l_B} \right] \tag{6.10}$$

where α is a numerical constant of order unity. The important result is that the correction is small, and thus that in the salt-dominated regime, the reduction of the persistence length due to the interactions between different chains can be neglected.

The situation is very different in the osmotic regime, the chains interact very strongly, and the distance between interacting chains is always of the order of the mesh size $\xi \simeq (cb)^{-1/2}$; the screening length κ^{-1} is also of order ξ. There is no precise theory of the persistence length in the osmotic regime, the two following qualitative arguments suggest, however, that the persistence length is of the order of the mesh size ξ. As the distance r between interacting chains is of order κ^{-1}, the deflection angle given by Eq. (6.9) is always of order 1. Whenever two chains cross, they bend by an angle of order unity and loose the memory of their orientation. The persistence length is thus of the order of the distance between crossings, that is, of the order of the mesh size ξ. The second argument is based on the strong structure of the solution in the osmotic regime. The structure factor has a strong peak, and with a very rough approximation the chains can be thought of as lying on a lattice with a lattice constant of order ξ. The total electrostatic energy of this lattice does not depend on the bending of the chains over distances larger than ξ. When the chains are bent on a scale ξ, entropy is gained but at the expense of bending energy associated with the bare persistence length l_0. The entropy is clearly dominant when $l_0 < \xi$, which leads to a persistence length of order ξ. An important consequence is that in salt-free semi-dilute solutions, both flexible and rigid polyelectrolytes have a persistence length of the order of the correlation length. It is also important to notice that the persistence length of rigid polyelectrolytes is predicted to vary nonmonotonically with ionic strength. In the osmotic regime, it is dominated by interchain interactions and increases with ionic strength. In the salt-dominated regime, it is dominated by interchain interactions and decreases with ionic strength. There is no clear-cut experimental evidence of this nonmonotonic variation of the persistence length with ionic strength [80].

D. Concentrated Solutions of Flexible Polyelectrolytes

The semidilute regime discussed in the two previous subsections corre-sponds to chains that are rodlike at intermediate length scales and that can be described locally by the electrostatic bloc model of Section II. At a higher concentration, however, the mesh size of the solution is smaller than the electrostatic blob size. Electrostatic interactions only play a minor role in this regime and the polyelectrolyte chains remain Gaussian at all length scales. The electrostatic blob size ξ_e is equal to the correlation length ξ when the electrostatic blobs are close packed. If the local structure of the blobs is Gaussian, this occurs at a concentration

$$c^{**} = b^{-3}f^{2/3}\left(\frac{l_B}{b}\right)^{1/3} \tag{6.11}$$

In the concentrated regime $c > c^{**}$, the electrostatic interactions are weak, and the correlations in the solution can be studied within the framework of the so-called random-phase approximation (RPA), as first suggested by Borue and Erukhimovich [81] and Joanny and Leibler [82]. This leads to a structure factor (monomer–monomer concentration correlation function)

$$\frac{1}{S(q)} = \frac{1}{S_0(q)} + v + w^2c + \frac{4\pi l_B f^2}{q^2 + \kappa^2} \tag{6.12}$$

Here $S_0(q)$ is the structure factor of Gaussian chains, given by the Deybe function, and in the following it is written as $S_0(q)^{-1} = (Nc)^{-1}[1 + g(Nq^2b^2/6)]$. The second and third virial coefficients that describe the interactions between *neutral* monomers (i.e., that would describe the interactions for $f = 0$) are, respectively, v and w. The second virial coefficient v is the excluded volume v and vanishes at the θ temperature. The third viral coefficient w^2 is assumed here to be positive. The last term comes from the screened electrostatic interactions. The screening is due to all the small ions and the Debye–Hückel screening length is defined here by $\kappa^2 = 4\pi l_B(fc + 2n)$. The statistics of the electrostatic blobs remain Gaussian if the excluded volume is small enough, namely if the electrostatic blob size ξ_e is smaller than the so-called thermal blob size $\xi_t \approx b^4/v$.

Within the RPA approximation (6.12), the inverse osmotic compressibility of the neutral solution is $1/Nc + v + w^2c$. That of the charged solution is $1/Nc + v + w^2c + 4\pi l_B f^2/\kappa^2$. Hence, in the small wave vector limit, the electrostatic interactions give rise to an additional excluded volume $v_{el} = 4\pi l_B f^2/\kappa^2$.

At high wave vectors, the chains are Gaussian and the structure factor decays as $S(q) \approx 12/q^2b^2$. At small salt concentrations, the structure factor given by (6.12) has a peak at a wave vector q^* given by

$$(q^{*2} + \kappa^2)^2 = \frac{24\pi l_B f^2 c}{g' b^2} \tag{6.13}$$

where g' is the derivative of g with respect to $\frac{1}{6}Nq^2b^2$, equal to $\frac{1}{2}$ when $\frac{1}{6}Nq^{*2}b^2$ is large and to $\frac{1}{3}$ when it is small. The peak position is independent of the solvent quality. In the absence of salt, it scales as $q^* \simeq f^{1/2}c^{1/4}$ and crosses over smoothly to the peak position in the semidilute range $q^* \simeq 1/\xi$ given by Eq. (6.1) at the concentration c^{**}. As the salt concentration is increased, the peak shifts toward zero wave vector and disappears when $\kappa^4 > 24\pi l_B f^2 c/g' b^2$. At higher ionic strength the polymer solution behaves as a neutral polymer solution and the structure factor decreases monotonically with the wave vector. The structure factor given by Eq. (6.12) is in good agreement with neutron scattering experiments on weakly charged polyacrylic acid or poly-methacrylic acid solutions in water [83].

The osmotic pressure of the solution can also be calculated from the RPA (one-loop) approximation. In the absence of salt it is equal to

$$\frac{\pi}{k_B T} = fc + \frac{1}{2}vc^2 + \frac{1}{3}w^2c^3 - Af^{3/2}c^{3/4}l_B^{3/4}b^{-3/2} \tag{6.14}$$

where A is a numerical constant of order unity [81]. The first term is the osmotic pressure of the counterions discussed extensively above; the two following terms are due to the nonelectrostatic interactions between monomers. The last term is due to the charge fluctuations and is very similar to the so-called polarization pressure of simple electrolytes.

Finally, the cooperative diffusion constant of the solution can be calculated from the structure factor using linear response theory and describing the hydrodynamic interactions by the so-called Oseen tensor [84]. When the electrostatic interactions dominate over the excluded volume interactions, the cooperative diffusion constant is given by

$$D = \frac{k_B T}{6\pi\eta} \frac{\alpha^3}{\kappa^2} \left(1 + \frac{\kappa^2}{\alpha^2}\right)\left(2 + \frac{\kappa^2}{\alpha^2}\right)^{-1/2} \tag{6.15}$$

where η is the solvent viscosity and where the wave vector α is defined as $\alpha^4 = (q^{*2} + \kappa^2)^2$. In a first approximation, the cooperative diffusion constant varies as $D \simeq f^{3/2}c^{3/4}(fc + 2n)^{-1}$. It is thus a nonmonotonic function of the monomer concentration c. It increases at low concentration, reaches a maximum when the counterion concentration is of the order of the salt concentration, and decreases at higher concentration. The unusual decrease at high concentration is due to the coupling to the counterion gas, which dominates the compressibility of the solution. This might explain the unusual variation of the diffusion constant observed on polyacrylic acid solutions and gels.

A summary of the various regimes expected for polyelectrolyte solutions in the semidilute and concentrated ranges is given in Fig. 2.

E. Mesophase Formation in Poor Solvents

In a poor solvent below the θ temperature, the polyelectrolyte chains are subject to two antagonist forces, the attractive (van der Waals–like) interaction and the repulsive Coulombic interaction. As explained for isolated chains in Section II, the attractive interaction is dominant at short length scales and induces a local collapse of the solution. If the salt concentration is not too high, phase separation is not expected at a macroscopic scale but at a mesoscopic scale. The polymer solution is expected to form mesophases where the polyelectrolyte concentration is not homogeneous and where the polymer dense and polymer dilute

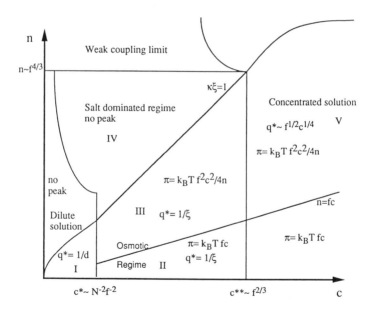

Figure 2. Schematic state diagram for a polyelectrolyte solution, in the plane c (polymer concentration), n (concentration of monovalent added salt). The boundaries between the different regions of the plane indicate crossovers, not sharp transitions. Region I (dilute) is discussed in Sections II, III, and VI.A. Regions II, III, and IV (semidilute) are discussed in Sections VI.B and VI.C. Region V (concentrated) is discussed in Section VI.E.

regions are periodically arranged. These phases are very similar to the mesophases observed in surfactant or block copolymer systems and various symmetries are possible (lamellar, cubic, and hexagonal phases) [85, 86]. Qualitatively, the mesophases are stabilized by the translational entropy of the counterions. If a macroscopic phase separation occurs, the separated phases must be electrically neutral and the translational entropy of the counterions is low. When a mesophase is formed, the electroneutrality can be locally violated. This is associated to a cost in electrostatic energy, but the translational entropy of the counterions is tremendously increased compared to a neutral dense state.

In the vicinity of the θ temperature, mesophase formation can be investigated using the same RPA as in the previous section. The structure factor is given by Eq. (6.12) where the excluded volume $v = -b^3\tau$ is now negative, τ being the temperature shift from the θ point. The structure factor has a strong peak at a wave vector q^*, and this peak diverges as the temperature is lowered (v becoming more negative), indicating the formation of a periodic structure with a period $2\pi/q^*$. In the absence of salt, the spinodal for mesophase formation obtained from the divergence of the peak of the structure factor occurs when

$$\frac{1}{N} + vc + w^2c^2 + \frac{1}{N}g\left(\frac{Nq^{*2}b^2}{6}\right) + \frac{g'b^2}{6}\left(\frac{24\pi l_B f^2 c}{g'b^2}\right)^{1/2} = 0 \qquad (6.16)$$

The qualitative shape of this spinodal line in a temperature concentration plane is shown in Fig. 3. The temperature and the concentration at the maximum of the spinodal line are given by $\tau_m \simeq f^{2/3}(l_B/b)^{1/3}$ and $c_m \simeq b^{-3}f^{2/3}(l_B/b)^{1/3}$. Upon addition of salt, the period of the mesophase increases and when $q^* = 0$ ($\kappa^4 = 24\pi l_B f^2 c/g'b^2$), the mesophase transition becomes a macroscopic phase separation very similar to the demixing transition of a polymer in a poor solvent.

In this weak segregation limit the symmetry of the phases and the topology of the phase diagram have been calculated by the Moscow group [87] by studying the nonlinear fluctuations of the solution.

At a temperature much smaller than the θ temperature, in the strong segregation regime, the interface between the dilute and dense polymer regions is sharp and can be described in terms of a surface tension. In a very dilute solution micelles are expected to form and their characteristics have been studied elsewhere [82, 88]. At higher concentrations periodic phases are predicted. A partial phase diagram considering only the

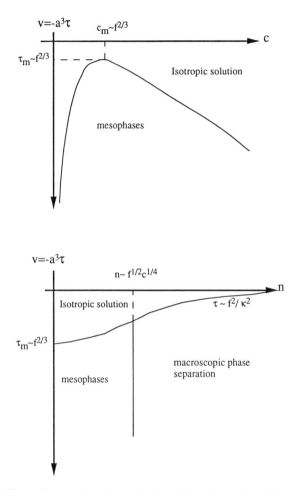

Figure 3. Phase diagram of a dense polyelectrolyte solution in a bad solvent. Upper part: phase diagram in the concentration and temperature plane for a fixed salt concentration. Lower part: phase diagram as a function of salt concentration and temperature for a given polymer concentration.

lamellar ordering has been determined by Nyrkova et al. [89] by a direct minimization of the mean-field free energy.

Most experimental results seem to be obtained outside the region of the phase diagram where mesophases form. To our knowledge, mesophase formation has been reported only in one case [90], and no detailed study of the mesophase structure seems to exist.

VII. DYNAMICAL PROPERTIES

In this section, we briefly discuss some dynamic properties of polyelectrolyte solutions. Only two quantities are considered: the mobility and the electrophoretic mobility of a chain in a very dilute solution and the viscosity of the solution. In an electrophoresis experiment, the hydrodynamic interactions are screened and the electrophoretic mobility is independent of the molecular weight. The separation of molecules of different size (such as DNA) is thus not possible by simple electrophoresis. The concentration variation of the viscosity of a polyelectrolyte solution is very different from that of an organic neutral polymer solution. This effect has been known experimentally for a long time (it is sometimes referred to as the polyelectrolyte effect), but it remains poorly understood.

A. Mobility and Electrophoretic Mobility of a Single Charged Chain

The theory of transport properties in dilute electrolyte solution has been worked out long ago by Debye, Onsager, and Falkenhagen (see [91], [92], and historical references therein). The existence of long-range Coulomb interactions between the species in solution results in a deviation of the transport coefficients from the pure solvent values (ionic mobility, conductivity, electrophoretic mobility, and viscosity) that increases as the square root of the concentration, rather than the linear increase that is the rule for neutral species. The origin of this anomalous behavior can be traced back to two effects known as the relaxation effect and the electrophoretic effects. Although these effects have different physical origins, they both produce a variation of the transport coefficients as the square root of the ionic concentration.

The relaxation effect can be described as follows. A test charge placed in an ionic solution is surrounded by a spherically symmetric polarization cloud where charges of opposite sign are predominant, which can be described by the standard Debye–Hückel theory. If the test charge moves at a constant velocity $\mathbf{U}$ with respect to the solvent, the polarization cloud is slightly distorted, and the charge experiences an additional electric field that tends to slow down it motion. The resulting force therefore appears as an electrostatic friction on the test charge. More generally, the moving charge Q (position $\mathbf{R}$, velocity $\mathbf{U}$) in an electrolyte solution can be shown (see Appendix B) to produce an electric field inside the solution of the form

$$\mathbf{E}(\mathbf{r}) = -\mathbf{T}(\mathbf{r} - \mathbf{R})\mathbf{U} \qquad (7.1)$$

where the α, β component of the matrix $\mathbf{T}$ is

$$T_{\alpha\beta}(\mathbf{R}) = \frac{1}{(2\pi)^3} \int d^3\mathbf{k} \exp(i\mathbf{k} \cdot \mathbf{R}) \frac{k_\alpha k_\beta}{k^2} \frac{Q\zeta_0 \kappa^2}{k_B T(k^2 + \kappa^2)^2} \qquad (7.2)$$

The ionic mobility ζ_0^{-1} is for simplicity assumed to be identical for small ions.

The electrophoretic effect is important only for the description of ionic motion driven by an external electric field (electrophoresis). In that case, the force that drives a charge motion also acts on the surrounding polarization cloud. The solvent velocity field that results from this body force acting on the polarization cloud modifies the relative velocity of the charge with respect to the solvent creating an increased dissipation. In other words, the volume of solvent within a distance κ^{-1} of the moving charge is dragged along as the charge moves, with a friction coefficient of order $\eta\kappa^{-1}$, where η is the solvent viscosity. In the language used to study polymer dynamics, the electrophoretic effect can be described as a screening of the hydrodynamic interactions *when the motion is created by an electric field*. The hydrodynamic interactions are usually described in terms of the Oseen tensor $\mathbf{H}(\mathbf{r})$ [6], which gives the velocity field that results from the application of a local force $\mathbf{F}_0\delta(\mathbf{r})$ localized at the origin. When the force results from an electric field $\mathbf{E}_0$ applied to a charge Q located at the origin, it is easily shown (see Appendix B) that, if the polarization cloud is described within the Debye–Hückel approximation, the velocity field is given by

$$\exp(-\kappa r)\mathbf{H}(\mathbf{r})Q\mathbf{E}_0 \qquad (7.3)$$

This velocity field is screened in a similar way as the electrostatic interaction and decays as $\exp(-\kappa r)/r$. The hydrodynamic interaction with neighboring particles is also screened over a distance κ^{-1}.

The extension of these concepts to macroions has been the subject of numerous contributions. Much of this work, however, is concerned with spherical ([24] and references therein) or rodlike [93] particles. The case of flexible polymers has been considered by Hermans [94], Manning [92], and Muthukumar [95]. In Manning [92], the relaxation and electrophoretic effects are treated on a different footing. In Muthukumar [95], the relaxation effect is ignored. In the following, we present a simplified description that allows for a simultaneous treatment of the two effects at

a microscopic level. Our conclusions are essentially similar* to Manning's
[92]. They differ from those of Muthukumar [95], which we believe to be
in error.

We consider here two types of experiments. In a sedimentation
experiment, the external force is the same on all monomers and has no
effects on the small ions. The chain mobility μ relates the velocity of the
center of mass of the molecule $\mathbf{v}_G$ to the total force acting on the
molecule $\mathbf{F}$: $\mathbf{v}_G = \mu\mathbf{F}$. In the electrophoresis experiment, the external
force is due to the applied electric field $\mathbf{E}$ acting both on the charged
monomers and on the free small ions. The electrophoretic mobility is
defined by $\mathbf{v}_G = \mu_e\mathbf{E}$.

The model that we use to describe the dynamics of the charged
polymer is a simple extension of the standard Zimm model of neutral
polymers [6]. The velocity and position of monomer i are denoted,
respectively, by $\mathbf{R}_i$ and $\mathbf{v}_i$; the velocity field of the solvent is $\mathbf{u}(r)$. The
equation of motion of monomer i, bearing a charge fe, results from a
balance between a friction force and all the other forces acting on the
monomer:

$$\zeta[\mathbf{v}_i - \mathbf{u}(\mathbf{R}_i)] = \mathbf{F}_i + fe\mathbf{E}_r(\mathbf{R}_i) \tag{7.4}$$

The monomer solvent friction coefficient is ζ. The force $\mathbf{F}_i$ is the sum of
an external force F_{ext} (the sedimentation force or the applied electric
field), a Langevin random force θ_i, and of an intramolecular force $\mathbf{F}_{pol,i}$.
The last term $\mathbf{E}_r$ is the electric relaxation field.

The velocity field of the solvent is given by an Oseen-type relation:

$$\mathbf{u}(\mathbf{R}_i) = \sum_j \mathbf{H}(R_i - R_j)[s(R_i - R_j)\mathbf{F}_{ext} + \mathbf{F}_{pol,j} + \theta_j]$$
$$+ \mathbf{H}_s(R_i - R_j)fe\mathbf{E}_r(\mathbf{R}_j)] \tag{7.5}$$

the screening factor $s(r)$ is equal to $\exp(-\kappa r)$ in the case of electrophoresis and to unity otherwise. If we assume that the solvent velocity
field varies slowly over a scale κ^{-1}, the relaxation field $\mathbf{E}_r(\mathbf{R}_i)$ on
monomer i is obtained from (7.1) as

---

* The result that is arrived in the literature [92] can be obtained within our analysis by
ignoring the second term on the right-hand side of Eq. (7.6). The result obtained for the
electrophoretic mobility is then $\mu_e = fe(H_{s0})/[1 + fe(H_{s0})T_0]$, which is only slightly different
from (7.9), since (H_{s0}) is indepenent of molecular weight. A problem would arise, however,
if the same approximation were made in the calculation of the mobility.

$$E_r(\mathbf{R}_i) = -\sum_j \mathbf{T}(\mathbf{R}_i - \mathbf{R}_j)[\mathbf{v}_j - \mathbf{u}(\mathbf{R}_j)] \tag{7.6}$$

These equations can be treated in the spirit of the Kirkwood–Riseman preaveraging approximation, by replacing the tensors $\mathbf{H}(\mathbf{R}_i - \mathbf{R}_j)$, $s(\mathbf{R}_i - \mathbf{R}_j)\mathbf{H}(\mathbf{R}_i - \mathbf{R}_j)$ and $\mathbf{T}(\mathbf{R}_i - \mathbf{R}_j)$ by their equilibrium average values. Introducing

$$H_0 = N^{-1}\left\langle \sum_{ij} \mathbf{H}(\mathbf{R}_i - \mathbf{R}_j) \right\rangle$$

$$H_{s0} = N^{-1}\left\langle \sum_{ij} \exp(-\kappa|\mathbf{R}_i - \mathbf{R}_j|)\mathbf{H}(\mathbf{R}_i - \mathbf{R}_j) \right\rangle \tag{7.7}$$

$$T_0 = N^{-1}\left\langle \sum_{ij} \mathbf{T}(\mathbf{R}_i - \mathbf{R}_j) \right\rangle$$

the mobility μ and electrophoretic mobility μ_e are expressed as

$$\mu = \frac{1}{N}\left[H_0 - H_{s0}fe\frac{T_0}{\zeta}\left(1 + fe\frac{T_0}{\zeta}\right)^{-1} \right] \tag{7.8}$$

$$\mu_e = feH_{s0}\left(1 - fe\frac{T_0}{\zeta}\right)^{-1} \tag{7.9}$$

The mobility is given by the usual Kirkwood–Riseman result [6], with a correction that arises from the relaxation effect. This relaxation correction reduces the mobility. Using Eq. (7.2), T_0 can be computed as

$$
\begin{aligned}
fe\frac{T_0}{\zeta} &= f^2 l_B \frac{\zeta_0}{3\zeta} \int \frac{d^3\mathbf{q}}{(2\pi)^3} S(q) \frac{\kappa^2}{(q^2 + \kappa^2)^2} \\
&= f^2 \kappa l_B \frac{\zeta_0}{3\zeta} N^{-1}\left\langle \sum_{ij} \exp(-\kappa|\mathbf{R}_i - \mathbf{R}_j|) \right\rangle
\end{aligned}
\tag{7.10}
$$

where $S(q)$ is the structure factor of the chain [normalized so that $S(q = 0) = N$]. From the second line of (7.10) it is seen that the effect of the relaxation field is independent of molecular weight whatever the chain conformation and is small when the charge fraction f is small. For example, if the polymer is rodlike over a distance κ^{-1}, $feT_0/\zeta = f^2 l_B/(3b)(\zeta_0/\zeta)$. The averaged screened Oseen tensor, H_{s0}, is also independent of molecular weight and proportional to $-\ln(\kappa b)$. The mobility is thus dominated by the Kirkwood–Riseman contribution and is essentially the same as for a neutral chain with the same conformation. The relaxation correction becomes important for strongly charged rodlike chains in the vicinity of Manning condensation threshold. In this case, the mobility is reduced by a factor of order unity from the Kirkwood–Riseman value. The approximations that we have made are only valid for long chains, $R \gg \kappa^{-1}$. In view of the complex form of Eq. (7.8), it is not surprising that experimental results for polyelectrolyte diffusion constants are often intermediate between the *nondraining* and *free-draining* limits [96].

The electrophoretic mobility μ_e, on the other hand, is in any case independent of molecular weight because of the screening of hydrodynamic interactions. For weakly charged polyelectrolytes, the relaxation correction is small and the electrophoretic mobility can be rewritten in terms of the structure factor of the chains as

$$\mu_e = \frac{fe}{3\eta\pi^2} \int dk \, S(k) \frac{k^2}{k^2 + \kappa^2} \qquad (7.11)$$

For a rodlike molecule, the electrophoretic mobility decreases logarithmically as the ionic strength is increased $\mu_e = -(fe/3\eta\pi^2 b)\ln \kappa b$. For a flexible chain, using the stretched chain model of Section II to describe the local conformation of the chains, we obtain $\mu_e = -(fb/l_B)^{1/3}(e/\eta b)\ln \kappa \xi_e$; the increase with the charge fraction is weaker and the electrophoretic mobility also decreases logarithmically with ionic strength. The general features of these theoretical predictions are in reasonable agreement with experimental observations. A detailed comparison between theoretical and experimental results is given elsewhere [92].

B. Viscosity of Polyelectrolyte Solutions

The viscosity of polyelectrolyte solutions has been known very early to exhibit a behavior qualitatively different from that of neutral polymer solutions (see [1] and references therein). This behavior was first described by Fuoss [97] using an empirical law, which gives the reduced viscosity $\eta_r[\eta(c) - \eta_s]/\eta_s c$ as a function of the polymer concentration c

$$\eta_r = \frac{A}{1 + Bc^{1/2}} \qquad (7.12)$$

Detailed studies [98] show that the constant A, obtained by extrapolating the results to vanishing concentration, is proportional to $N^2 f^2$, while the ratio A/B is proportional to Nf.

It was later realized that this law is satisfactory only for high enough concentrations and in salt-free solutions, in which case it reduces to $\eta_r \sim c^{-1/2}$. At low concentration, the behavior of η_r is actually non-monotonic [99]. The reduced viscosity η_r presents a peak at a finite concentration, of the order of the salt concentration. This peak disappears when enough salt is added to the solution. A phenomenological law, recently proposed by Rabin et al. [100], seems to account well for the experimental observations. According to these authors, the intrinsic viscosity is given by

$$\eta_r = \frac{A_1 c}{\kappa^3} \tag{7.13}$$

where κ is the Debye screening parameter, and A_1 is proportional to the molecular weight.

Finally, it is important to note that this anomalous behavior of polyelectrolyte solutions can be rationalized at low concentrations by using the method of isoionic dilution [1, 101]. In this method, the viscosity is studied as a function of polymer concentration for a given value of the ionic strength (i.e., as the solution is diluted, salt is added in order to keep the ionic strength constant). The increase of the reduced viscosity at low c is then linear, as for neutral polymers. In neutral polymer solutions, this linear behavior is usually represented in the form $\eta_r = [\eta] + k_H [\eta]^2 c$, where $[\eta]$ is the intrinsic viscosity and k_H is the Huggins constant. In polyelectrolyte solutions, $[\eta]$ is large and varies strongly with the ionic strength I. The data of Pals and Hermans [101], for example, can be represented in the form $[\eta] = [\eta](I = \infty) + CI^{-1/2}$ [102]. The slope of the linear term, $k_H [\eta]^2$, is also much larger than the corresponding quantity in neutral polymers. The Huggins constant k_H varies as $I^{-1/2}$ [102].* This indicates that the importance of interactions between chains, and also shows that, as c is increased a crossover to the "interaction-dominated" behavior described by (7.13) rapidly takes place.

In short, three different concentration regimes can be distinguished in the behavior of the reduced viscosity as a function of c (Fig. 4).† In the low concentration region (left side of the peak), η_r is an increasing function of c. This region is most usefully characterized by isoionic

---

* In fact, the Huggins constant is not a particularly useful concept in a system where interactions that are not of hydrodynamic origin are expected to be important.

† This distinction is a simplification of the description proposed by Wolff [102] and seems valid at all but the highest concentrations.

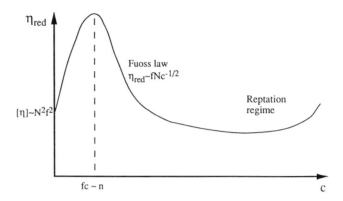

Figure 4. Schematic representation of the reduced viscosity as a function of polymer concentration c at a fixed salt concentration n.

dilution curves. The peak region corresponds to comparable salt and counterions concentrations ($fc \simeq n$). Finally, in the more concentrated region, η_r is a decreasing function of c, $\eta_r \sim c^{-y}$ with $y < 1$. Obviously, this description is valid only in solutions where the salt concentration is small enough, since if n is large the behavior of usual neutral polymers must be recovered.

At present, a comprehensive theory of polyelectrolyte viscosity is lacking. A number of interpretations, however, have been put forward to explain the unusual behavior of these solutions. We will distinguish here two types of models based either on the conformation of the polymers or on the interactions between different chains.

In conformation-based models [14, 103–105] (which have been formulated for flexible polyelectrolytes), it is assumed that the viscosity of the polyelectrolyte solution is only a consequence of the structural properties of the solution. The calculation proceeds by analogy with the well-understood case of neutral polymers. For semidilute solutions, the viscosity can be computed by applying the Rouse, or reptation, theories to a chain of correlation blobs of size ξ, containing G_ξ monomers. The correlation length of polyelectrolyte solutions has a variation with concentration very different from that of neutral polymer solutions, so that the behavior of the viscosity is also qualitatively different. This idea was first applied by de Gennes and co-workers [14]. These authors interpreted the $[\eta] \sim Nc^{-1/2}$ behavior in salt-free semidilute solutions given by the Fuoss law as corresponding to Rouse dynamics for a chain of correlation blobs of size $\xi \sim \kappa^{-1}$ [Eq. (6.1)] with $G_\xi = g(\xi/\xi_e)$ monomers.

The viscosity is then given by

$$\eta = \eta_s \frac{c}{N} \left(\frac{N}{G_\xi}\right)^2 \xi^3$$

$$= N\eta_s c^{1/2} \xi_e^{-3/2} b^3$$

(7.14)

This analysis was extended by Dobrynin et al. to the case where salt is added to the solution [75]. In that case, ξ is given by (6.3), $G_\xi = c\xi^3$ and the viscosity is

$$\eta = N\eta_s c^{5/4} \kappa^{-3/2} \xi_e^{-9/4} b^{9/2}$$

(7.15)

The crossover between (7.14) and (7.15) takes place when the mesh size is equal to κ^{-1}, like the crossover between (6.1) and (6.3). Note that Eq. (7.15) predicts a maximum in the reduced viscosity $\eta_r(c)$ for $fc = n$.

Dobrynin and co-workers [75] also give an empirical justification for the observation that semidilute polyelectrolyte solutions obey Rouse dynamics over a broader range of concentrations above the overlap concentration than neutral polymer solutions. In neutral polymers, it is observed that entanglements become important when each chain interacts with about 10 other chains [106]. If this criterion is applied to polyelectrolytes, the concentration c_{rept} for the onset of reptation behavior is given by

$$\frac{c_{\text{rept}}}{N} \left(\frac{N}{G_\xi} \xi(c_{\text{rept}})^2\right)^{3/2} \simeq 10$$

(7.16)

This can be rewritten as $c_{\text{rept}} \xi(c_{\text{rept}})^3 \simeq N/100$. The overlap concentration c^*, on the other hand, is given by $c^* \xi(c^*)^3 \simeq N$. In polyelectrolyte solutions, $c\xi^3(c)$ decreases much more slowly with increasing concentration c than in neutral polymer solutions, so that the onset of reptation is expected at higher concentrations. If Eq. (6.1) is used, the ratio c_{rept}/c^* is of order 10^4, while for neutral polymers it is of order 10^2.

Another useful way of computing the viscosity of neutral polymer solutions, which allows a description of the dilute/semidilute crossover, is to write it in a scaling form:

$$\eta = \eta_s F\left(\frac{c}{c^*}\right)$$

(7.17)

where η_s is the solvent viscosity, F is a scaling function, and c^* is the overlap concentration. The function F can be expanded as a power series $[F(x) = 1 + a_1 x + a_2 x^2 \cdots]$ for small values of its argument, that is, in dilute solutions. In the semidilute regime, the functional form of F depends on the mechanism (Rouse or reptation) that governs stress relaxation. In this formulation, the specificity of polyelectrolyte solutions is related to the fact that the interactions between monomers depend on the polymer concentration. Therefore c^* in (7.17) must now be interpreted as the overlap concentration corresponding to a polymer having the same interactions as the charged polymer at concentration c. For example, when the mesh size is larger than κ^{-1}, we can write

$$c^*(c) = \frac{N}{R(c)^3} \qquad (7.18)$$

where $R(c) = \kappa^{-1}[(N/g)\xi_e\kappa]^{3/5}$, given by Eq. (4.12), is the radius of an isolated chain with this interaction. The viscosity is obtained by writing $\eta = \eta_s F[c/c^*(c)]$, and assuming Rouse behavior $\eta \sim N$. This gives

$$\eta = \eta_s\left[\frac{c}{c^*(c)}\right]^{5/4} = N\eta_s c^{5/4}\kappa^{-3/2}(\xi_e)^{9/4}b^{-9/2} \qquad (7.19)$$

which is identical to (7.15).

Another "conformational" interpretation of polyelectrolyte viscosity was given by Reed [104] using similar arguments. His calculation assumes that c is in the vicinity of $c^*(c)$, so that the x^2 contribution in the function $F(x)$ dominates. The formula used for c^* is appropriate for short chains or large κ^{-1}. It corresponds to an ideal behavior for a chain with a persistence length κ^{-1}:

$$R(c)^2 = 2\left(\frac{N}{g}\right)\xi_e\kappa^{-1} \qquad (7.20)$$

This leads to a viscosity

$$\eta \sim \eta_s\left(\frac{c}{c^*}\right)^2 \sim \eta_s N\frac{c^2}{\kappa^3}\left(\frac{\xi_e}{g}\right)^3 \qquad (7.21)$$

[Note that the same result would be obtained by requiring Rouse behavior for the viscosity, i.e., $F(c/c^*) \sim N$.] Again, the behavior $\eta_r \sim fNc^{-1/2}$ is obtained in the absence of salt. With added salt, the experimentally observed behavior (7.13) is reproduced. The main difference between the interpretations of Reed and of Dobrynin et al. is the expression used for $R(c)$. The crossover from (7.21) to (7.15) can be expected to take place as c increases and $R(c)$ crosses over from the ideal expression (7.20) to the Flory expression (2.5).

Finally, we note that in all conformation-based models the $c^{-1/2}$ behavior of η_r, corresponds to situations where the role of added salt is negligible. It is interesting that the value of the constant A in the Fuoss law (7.12), which corresponds to a (generally incorrect) extrapolation of this behavior to $c = 0$, yields a value $A \sim N/R^3 \sim N^2 f^2$ that corresponds to the intrinsic viscosity for a salt-free, dilute solution of polyelectrolytes $(R \sim Nf^{2/3})$. This coincidence, however, is not explained theoretically. At high concentrations, entanglements become important, and the reduced viscosity increases with concentration [75, 102].

The conformation-based models appear to provide a qualitatively correct description of flexible polyelectrolyte viscosity (see refs. 74 and 104 for a comparison with experimental data). They only apply, however, to sufficiently concentrated solutions, near or above overlap concentration. The "polyelectrolyte behavior" described by equations (7.12) or (7.13) is also observed in solutions of short chains or even for charged spherical particles [107]. Elsewhere [108] it was suggested that this behavior could be described on the basis of the liquid-state "mode-coupling" theories developed for the study of spherical charged colloids [109]. Unfortunately, these theories are analytically tractable only in the limit of weak electrostatic coupling between the particles, which is certainly not realized in dilute polyelectrolyte solutions (see Section VI.A). In this weak coupling regime, it was shown that the viscosity of a suspension of spherical charged particles (charge Z, concentration c_0, hydrodynamic radius R) is given by

$$\eta \sim \eta_s R c_0^2 Z^4 l_B^2 \kappa^{-3} \qquad (7.22)$$

Cohen et al. [108] noticed that this behavior is similar to the experimentally observed one (7.13), if the polymer chain is assumed to bear an effective charge $Z^* = (fN)^{1/2}$. The particle concentration is replaced by c/N, and the hydrodynamic radius is proportional to N. The physical origin of the effective charge Z^*, however, is unclear.

This discussion indicates that a full theoretical understanding of the viscosity of polyelectrolyte solutions is far from being achieved. The main results can be summarized as follows. In all models, a maximum in η_r is predicted when the salt and the counterion concentration are of the same order of magnitude, $fc \simeq n$. The maximum can occur, depending on chain length and salt concentrations, either in the dilute or in the semidilute region. For short chains or low salt concentrations, it thus takes place in dilute solution. The appropriate theory to describe the peak should be similar to that found in [108, 109]. At concentration higher than the peak concentration, the reduced viscosity behaves as $fNc^{-1/2}$. Although the behavior is similar above and below the overlap concentration, the interpretations that have been proposed are very different. For longer chains, or higher salt concentrations, $fc \simeq n$ corresponds to a semidilute solution. In that case, the viscosity can be described using Eq. (7.15) (for c larger than c^*) or (7.21) (for c near c^*). Below c^*, the appropriate description is again that found in the literature [108, 109].

VIII.   CONCLUSIONS

In this review we have presented the recent theories that describe the conformational and dynamical properties of charged polymer chains. Much of the focus has been put on the long-range character of the electrostatic interactions and on the role of the small counterions that ensure the electrical neutrality. Little attention has been paid to interactions of nonelectrostatic origin [110], which have only been introduced at some places in terms of virial coefficients. Most experiments on polyelectrolyte solutions are made in water, which is not a good solvent for polyelectrolytes, which are organic polymers. The nonelectrostatic interactions between molecules dissolved in water are not simple van der Waals interactions and have components due to hydrogen bonding or to the hydrophobic effect [23]. It is not obvious that these very specific interactions can be modeled simply in terms of virial coefficients. Experimentally these strong attractive interactions lead in some cases to the formation of aggregates, which can be responsible for the slow modes [96, 111, 112] sometimes observed when measuring the relaxation of concentration fluctuations via quasi-elastic light scattering. The simple theoretical models that we have described do not take into account these complicated effects that are in many cases dominant. This makes a direct comparison between the theory and the experimental results rather difficult and we have not attempted here any quantitative comparison.

Despite this difficulty, some experimental features seem to be rather general in polyelectrolyte solutions. The structure factor of a polyelectrolyte solution always shows a peak at low ionic strength. The position of the peak has been discussed in Section VI both in dilute and semidilute solutions. Most experiments are also consistent with a rodlike behavior of the polyelectrolyte chains at small length scales. The rigidity of the chain is then characterized by an electrostatic persistence length. The variation of the persistence length with ionic strength or with the polymer concentration is, however, still a matter of controversy as seen in Section IV, and more complete theories are certainly needed. The effect of heterogeneities of the charge distribution or of nonelectrostatic interactions on the persistence length do not seem to have been discussed. A good agreement between theory and experiment seems to exist only for rigid polymers where the theory of Odijk or Skolnick and Fixman [113, 114] gives a quantitative description when counterion condensation is properly taken into account. Fuoss law and the variation of the viscosity of polyelectrolyte solutions with the concentration have also been extensively confirmed experimentally. The theoretical results presented in Section VII remain at the level of scaling laws and are not based on very systematic arguments. It should be also noted that the derivation of the Fuoss law in semidilute solution based on the Rouse model implicitly assumes that there is only one length scale and therefore that the chain persistence length is of the order of the correlation length ξ as suggested by some of the theoretical models.

An alternative way to study the properties of polyelectrolyte solutions is numerical simulations [12]. All the parameters are well controlled in the simulations and they should allow sorting out the perspective roles of the electrostatic and nonelectrostatic interactions and thus to provide quantitative tests of the theoretical models. Many simulations have been performed on isolated polyelectrolyte chains in a salt-free solution. There is in general quantitative agreement between the results and the electrostatic blob model. The simulations in the presence of salt often performed by using the Debye–Hückel potential are less conclusive. For rigid polyelectrolytes, the agreement with the Odijk–Skolnick and Fixman theory is good; for flexible weakly charged polyelectrolytes, there is no good separation of length scales between the screening length and the persistence length and a quantitative test of the theories is difficult. Recently, Stevens and Kremer [115] have performed simulations on semidilute polyelectrolyte solutions, taking into account explicitly the discrete counterions. The polymers are strongly charged flexible chains in the vicinity of the Manning condensation threshold. This is a very heavy numerical work, but it should allow a very detailed description of the influence of electrostatic interactions in polyelectrolyte solutions. The extension of these simulations to weakly charged or more rigid polyelectrolytes for which the analytical theories have been constructed would provide strong tests of these theories.

The discussion in this chapter has also been limited to simple polyelectrolyte chains comprising only monomers with the same charge and neutral monomers. There are several other polymeric systems where electrostatic interactions play an important role. Polyampholytes [116, 117] are polymers that carry charges of both signs. Due to the attractions between opposite charges, they are often insoluble in salt-free water and can only be dissolved at a finite ionic strength. They also show an antipolyelectrolyte effect, the reduced viscosity increasing with ionic strength. Polysoaps are comblike polymers, often polyelectrolytes, where hydrophobic side chains are grafted on the backbone [1, 118]. They have properties combining those of polyelectrolyte chains and of small ionic surfactants and form, for example, intramolecular micelles. Another system of interest is mixtures between neutral and charged polymers [119, 120], which have been studied both experimentally and theoretically. The addition of a charged polymer can increase the solubility of a neutral polymer in water. The formation of mesophases is also expected in these systems.

Throughout this chapter we also have implicitly assumed that the solvent (essentially water) is a continuous dielectric medium with a uniform dielectric constant. At high polymer concentration the dielectric constant crosses over from that of the polar solvent to that of an organic medium (essentially polymer) [121]. If the dielectric constant is low, charges of opposite signs are not dissociated and the polymer has an ionomer behavior dominated by attactive interactions between dipoles. Even in a more dilute solution, it is not clear that locally the dielectric constant has the solvent value; this could turn out to be important in Manning condensation theory, for example. There seems to be little theoretical work on these issues. Also, the small ions have been considered as pointlike; all specific interactions due to hydration or to the finite size of the ions have been neglected. Small differences between ions due to these effects are usually observed.

One of the essential conclusions of this review is that polyelectrolytes remain a poorly understood state of matter and that much work is still needed both from the experimental and from the theoretical point of view to reach a degree of understanding equivalent to that of neutral polymers. From a theoretical viewpoint, polyelectrolytes are in most cases strongly coupled systems, involving many different length scales. It might well be that in such situations, an appropriate description of the system can only be obtained by the use of sophisticated liquid state theories, such as those recently developed for dense neutral polymers [122]. Experimentally, it is surprising that relatively few systematic studies on charged polymers, using modern investigation tools, in particular neutron scattering, are available. Such studies were initiated in the late seventies (see, e.g., ref. 123) but have not been as conclusive as the equivalent studies on neutral polymers. Finally, it is obvious that the area of polyelectrolyte solutions offers a number of problems that can be approached by simulation. Simulations that explicitly include counterions can be used to assess the accuracy of the Debye–Hückel or Poisson–Boltzmann approximations [124]. Density functional-based simulation methods [125] for classical charged systems can be used to go beyond such approximations, even in the presence of added salt. Finally, the dynamical properties of charged polymers have not, up to now, been investigated in numerical studies.

APPENDIX A: EFFECTIVE INTERACTION BETWEEN CHARGED MONOMERS

In this appendix, it is shown that the effective interaction between charged monomers can, within the linear response approximation, rigorously be written as a sum of pairwise additive interactions [Eq. (2.16)]. As a by-product of the calculation, the various structure factors (ion–ion, polymer–polymer, and polymer–ion) are linked within this approximation by simple relations.

The system is a polymer solution with a concentration c of (positively) charged monomers and a concentration n of monovalent added salt. For simplicity we assume that the counterions are identical to the salt cations. The densities of charged monomers, charged positive and negative ions are denoted by $\rho_P(\mathbf{r})$, $n_+(\mathbf{r})$ and $n_-(\mathbf{r})$, respectively. By integrating out the coordinates of the small ions, the partition function of the system can formally be written as

$$Z = \text{Tr}_P \exp\left(-\frac{H_P + F_i[\psi]}{k_B T}\right) \tag{A.1}$$

where Tr_P denotes an integration over monomer coordinates, and H_P is the energy of the polymer (2.1). Here, $F_i[\psi]$ is the free energy of the small ions in a potential that is the sum of the external potential $\psi(\mathbf{r})$ created by the polymer:

$$\psi(\mathbf{r}) = k_B T l_B \int d\mathbf{r}' \frac{\rho_P(\mathbf{r}')}{|\mathbf{r} - \mathbf{r}'|} \tag{A.2}$$

The free energy F_i is obtained by minimizing with respect to the ionic densities $n_+(\mathbf{r})$ and $n_-(\mathbf{r})$ a free-energy functional, which in the mean-field, or Poisson–Boltzmann, approximation reads

$$k_B T \int d\mathbf{r}\{n_+(\mathbf{r}) \ln[n_+(\mathbf{r})] + n_-(\mathbf{r}) \ln[n_-(\mathbf{r})]\}$$

$$+ \tfrac{1}{2} l_B K_B T \int d\mathbf{r}\, d\mathbf{r}' \frac{[n_+(\mathbf{r}) - n_-(\mathbf{r})][n_+(\mathbf{r}') - n_-(\mathbf{r}')]}{|\mathbf{r} - \mathbf{r}'|}$$

$$+ \int d\mathbf{r}[n_+(\mathbf{r}) - n_-(\mathbf{r})]\psi(\mathbf{r}) \tag{A.3}$$

The linear response, or Debye–Hückel, approximation involves a further simplification of Eq. (A.3), by expanding the first two terms to second order in the deviations of n_+ and n_- from their average values. The minimization is easily carried out, and the resulting expression for F_i is

$$F_i[\psi] = - \int d^3\mathbf{k} \, \frac{1}{2} \frac{4\pi l_B}{k^2} \frac{\kappa^2}{k^2 + \kappa^2} \, \rho_P(\mathbf{k})\rho_P(-\mathbf{k}) \qquad (A.4)$$

with $\kappa^2 = 4\pi l_B(2c + n)$. The interaction that results from adding this contribution from the ionic *free energy* to the electrostatic repulsion between the monomers,

$$\int d^3\mathbf{k} \, \frac{1}{2} \frac{4\pi l_B}{k^2} \, \rho_P(\mathbf{k})\rho_P(-\mathbf{k})$$

gives the screened pair potential (2.16).

The Debye–Hückel calculation can easily be extended to compute the response of various densities to external fields, or equivalently the partial structure factors. The results are simple in the case of a salt concentration much larger than the counterion concentration $(n \gg c)$. Defining the ionic charge density as $\rho_Z = n_+ - n_-$, the following relationships are obtained between the Fourier transforms of the polymer–polymer, polymer–charge, and charge–charge correlation functions (denoted by S_{PP}, S_{PZ}, and S_{ZZ}, respectively)

$$S_{PZ}(k) = - \frac{\kappa^2}{k^2 + \kappa^2} S_{PP}(k)$$

$$S_{ZZ}(k) = 2n \frac{k^2}{k^2 + \kappa^2} + \left(\frac{\kappa^2}{k^2 + \kappa^2} \right)^2 S_{PP}(k) \qquad (A.5)$$

The total charge structure factor $S_{PP} + S_{ZZ} - 2S_{PZ}$ vanishes as $2nk^2/\kappa^2$ in the small-k limit, as implied by the Stillinger and Lovett sum rules [22].

APPENDIX B: RELAXATION AND ELECTROPHORETIC EFFECTS

In this appendix, we derive Eqs. (7.1) and (7.3), describing, respectively, the relaxation and electrophoretic effects. As usual, the two effects are considered independently [91], since the coupling between them is relevant only at a level that goes beyond the limits of the Debye–Hückel linearized theory.

B.1. Relaxation Field

The relaxation field is the electric field created by a charged particle (charge Qe) moving in a monovalent salt solution (concentration n) with a constant velocity $\mathbf{V}$. As explained in Section VII.A, this relaxation field arises from the deformation of the Debye–Hückel polarization cloud when the charge is set into motion. At equilibrium, the densities $n_+(r)$ of positive and $n(r)$ of negative ions that surround the charged particle (placed at the origin) are

$$n_{\pm}^{(\text{eq})}(r) = n\left(1 \mp Q\,\frac{l_B}{r}\exp(-\kappa r)\right) \tag{B.1}$$

In a time-dependent situation, however, the Boltzmann statistic that is used to obtain these densities must be replaced with a diffusion equation:

$$\zeta_0\,\frac{\partial n_{\pm}(\mathbf{r}, t)}{\partial t} = \nabla\cdot[k_B T\,\nabla n_{\pm}(\mathbf{r}, t) \mp n_{\pm}(\mathbf{r}, t)\mathbf{E}(\mathbf{r}, t)] \tag{B.2}$$

where ζ_0 is the ionic mobility. The electric field $\mathbf{E}$ is the sum of the field $\mathbf{E}_P(\mathbf{r} - \mathbf{V}t)$ created by the moving particle and of the electric field created by the ionic charge density, $\mathbf{E}_i$, which verifies Poisson equation:

$$\nabla\cdot\mathbf{E}_i(\mathbf{r}, t) = 4\pi l_B[n_+(\mathbf{r}, t) - n_-(\mathbf{r}, t)] \tag{B.3}$$

The solution of (B.2) in the stationary state is of the form $n_{\pm}(\mathbf{r}, t) = n_{\pm}^{\text{eq}}(\mathbf{r} - \mathbf{V}t) + \delta n_{\pm}(\mathbf{r} - \mathbf{V}t)$. Inserting this form into (B.2) and (B.3), and linearizing with respect to $\mathbf{V}$ and Q, the following equation for $\delta n_+ - \delta n_-$ is obtained:

$$(\nabla^2 + \kappa^2)[\delta n_+(\mathbf{r}) - \delta n_-(\mathbf{r})] = \frac{\zeta_0}{k_B T}\,\mathbf{V}\cdot\nabla\left[\left(2nQ\,\frac{l_B}{r}\exp(-\kappa r)\right)\right] \tag{B.4}$$

The resulting ionic charge density can be expressed as

$$\delta n_+(\mathbf{r}) - \delta n_-(\mathbf{r}) = \frac{1}{(2\pi)^3}\int d^3\mathbf{k}\,\frac{\zeta_0 Qe\kappa^2}{k_B T}\,\frac{i\mathbf{k}\cdot\mathbf{V}}{(k^2 - \kappa^2)^2}\exp(-i\mathbf{k}\cdot\mathbf{r}) \tag{B.5}$$

and the resulting electric field is given by Eq. (7.1).

B.2. Electrophoretic Effect

We now consider the calculation of the hydrodynamic velocity field around a charge Q at the origin, when the charge experiences a force $Q\mathbf{E}_{\text{ext}}$ created by an external electric field $\mathbf{E}_{\text{ext}}$. The solvent experiences a force $Q\mathbf{E}_{ext}\delta(\mathbf{r})$ created by the charge at the origin, and a body force $e[n_{+}(\mathbf{r}) - n_{-}(\mathbf{r})]\mathbf{E}_{ext}$. The equation that determines the hydrodynamic velocity field $\mathbf{U}(\mathbf{r})$ in the solvent is

$$-\eta_s\nabla^2\mathbf{U} + \nabla P = Q\mathbf{E}_{ext}\delta(\mathbf{r}) + e[n_{+}(\mathbf{r}) - n_{-}(\mathbf{r})]\mathbf{E}_{ext} \qquad (B.6)$$

where η_s is the viscosity and P the pressure. Here, $n_{+}(\mathbf{r}) - n_{-}(\mathbf{r})$ is the charge density within the *equilibrium* Debye–Hückel polarization cloud. In Fourier space, the velocity field is given by a generalization of Oseen's formula,

$$U_i(\mathbf{k}) = \frac{1}{\eta_s(k^2 + \kappa^2)}\left(\delta_{ij} - \frac{k_ik_j}{k^2}\right)(QE_j) \qquad (B.7)$$

which gives, in real space,

$$U(\mathbf{r}) = \mathbf{H}_s(\mathbf{r})(Q\mathbf{E}_{ext}) \qquad (B.8)$$

with the screened Oseen tensor H_s being

$$H_{s,ij} = \frac{\exp(-\kappa r)}{8\pi\eta_s r}\left(\delta_{ij} + \frac{r_ir_j}{r^2}\right) \qquad (B.9)$$

REFERENCES

1. C. Tanford, *Physical Chemistry of Macromolecules*, Wiley, New York, 1961.

2. See, e.g., the contribution by M. Mandel in *Encyclopedia of Polymer Science and Engineering*, Herman F. Mark, et al. ed., Wiley, New York, 1990.

3. K. S. Schmitz in *Macro-ion Characterization*, K. S. Schmitz, ed., American Chemical Society, Washington, D.C., 1994.

4. P. G. de Gennes, *Scaling Concepts in Polymer Physics*, Cornel University Press, Ithaca, NY, 1979.

5. J. des Cloiseaux and G. Jannink, *Les polymères en solution, leur modélisation et leur structure*, Editions de Physique, Paris, 1985.

6. M. Doi and S. F. Edwards, *The Theory of Polymer dynamics*, Oxford University Press, Oxford, 1986.

7. G. S. Manning, *J. Chem. Phys.* **51**, 954 (1969).

8. T. Odijk, *J. Polym. Sci.* **15**, 477 (1977).

9. J. Skolnick and M. Fixman, *Macromolecules* **10**, 944 (1977).

10. T. Odijk and A.C. Houwaart, *J. Polym. Sci.* **16**, 627 (1978).

11. M. Mandel and T. Odijk, *Ann. Rev. Phys. Chem.* **35**, 75 (1984).

12. B. Dünweg, M. J. Stevens and K. Kremer in *Computer Simulation in Polymer Physics*, K. Binder, eds., Oxford University Press, Oxford, 1995.

13. R. Varoqui, *J. Physique II (France)* **3**, 1097 (1993).

14. P. G. de Gennes, P. Pincus, R. M. Velasco, F. Brochard, *J. Phys.* **37**, 1461 (1976).

15. W. Kuhn, O. Künzle, and A. Katchalsky, *Helv. Chim. Acta* **31**, 1994 (1948).

16. A. R. Khokhlov, *J. Phys. A* **13**, 979 (1980).

17. J.-L. Barrat and D. Boyer, *J. Phys. II* **3**, 343 (1993).

18. J.-P. Bouchaud, M. Mézard, G. Parisi, and J. S. Yedidia, *J. Phys. A* **24**, L1025 (1992).

19. B. Jonsson, T. Peterson, and B. Soderberg, *J. Phys. Chem.* **99**, 1251 (1994).

20. P. Pfeuty, R. Velasco, and P. G. de Gennes, *J. Phys. Lett.* **38**, L5 (1977).

21. P. G. de Gennes, *Nuovo Cimento* **7**, 363 (1977).

22. J.-P. Hansen and I. R. McDonald, *Theory of Simple Liquids*, Academic Press, New York, 1986).

23. J. N. Israelachvili, *Intermolecular and Surface Forces*, Academic Press, London, 1985.

24. W. B. Russel, D. A. Saville, and W. R. Schowalter, *Colloidal Dispersions*, Cambridge University Press, Cambridge, 1991.

25. E. Raphael and J. F. Joanny, *Europhys. Lett.* **13**, 623 (1990).

26. X. Auvray, R. Anthore, C. Petipas, J. Huguet, and M. Vert, *J. Phys. (France)* **47**, 893 (1986).

27. R. M. Fuoss, A. Katchalsky, S. Lifson, *Proc. Natl. Acad. Sci. USA* **37**, 579 (1951).

28. J.-F. Joanny and P. Pincus, *Polymer* **21**, 274 (1980).

29. L. Landau and E. M. Lifchitz, *Physique Statistique*, Mir, Moscow, 1967.

30. F. Wall and P. Grieger, *J. Chem. Phys.* **20**, 1200 (1952).

31. F. Oosawa, *Polyelectrolytes*, Dekker, New York, 1971.

32. M. Abramovitz and I. A. Stegun, *Handbook of Mathematical Functions*, Dover, New York, 1965.

33. See, e.g., the contribution by S. Marcelja in *Liquids at Interfaces*, J. Charvolin, J.-F. Joanny, and J. Zinn-Justin, eds., North-Holland, Amsterdam, 1990.

34. M. J. Stevens and M. O. Robbins, *Europhys. Lett.* **12**, 81 (1990).

35. M. Olvera de la Cruz, L. Belloni, M. Delsanti, J. P. Dalbiez, O. Spalla, and M. Drifford, *J. Chem. Phys.* (1995).

36. J. Wittmer, A. Johner, and J.-F. Joanny, *J. Phys. II* **5**, 635 (1995).

37. J. Ray and G. S. Manning, *Langmuir* **10**, 2450 (1994).

38. H. T. Dobbs and J. M. Yeomans, *Mol. Phys.* **80**, 877 (1993).

39. P. Pfeuty, *J. Phys.* **39**, C2-149 (1978).

40. J.-L. Barrat and J.-F. Joanny, *Europhys. Lett.* **24**, 333 (1993).

41. M. Le Bret, *J. Chem. Phys.* **76**, 6243 (1982).

42. M. Fixman, *J. Chem. Phys.* **76**, 6346 (1982).

43. A. R. Khokhlov and K. A. Khachaturian, *Polymer* **23** (1982) 1793.

44. H. Li and T. A. Witten, *Fluctuations and Persistence Length of Charged Flexible Polymers* (preprint 1995).

45. M. Schmidt, *Macromolecules* **24**, 5361 (1991).

46. D. Bratko and K. Dawson, *Macromol. Theory Simul.* **3**, 79 (1994); *J. Chem. Phys.* **99**, 5732 (1993).

47. B. Y. Ha and D. Thirumalai, *Macromolecules* **28**, 577 (1995).

48. J. des Cloizeaux, *Macromolecules* **6**, 403 (1973).

49. T. L. Hill, *Introduction to Statistical Thermodynamics*, Dover, New York, 1986.

50. F. Donnan, *Z. Physik. Chem. A* **168**, 369 (1934); F. Donnan, E. Guggenheim, *Z. Physik. Chem. A* **122**, 346 (1932).

51. S. T. Milner, *Science* **251**, 905 (1991).

52. D. H. Napper, *Polymeric Stabilizations of Colloidal Dispersions*, Academic Press, New York, 1983.

53. P. Pincus, *Macromolecules* **24**, 2912 (1991).

54. E. B. Zhulina, O. V. Borisov, and T. M. Birshtein, *J. Phys. II* **2**, 63 (1992).

55. S. Miklavic and S. Marcelja, *J. Phys. Chem.* **92**, 6718 (1988).

56. S. Misra, S. Varanasi, and P. Varanasi, *Macromolecules* **22**, 5173 (1989).

57. S. Alexander, *J. Phys.* **38**, 983 (1977).

58. P. G. de Gennes, *Macromolecules* **13**, 1069 (1980).

59. E. B. Zhulina, O. V. Borisov, and T. M. Birshtein, *Macromolecules* **28**, 1491 (1995).

60. J.-L. Barrat, J.-F. Joanny, and P. Pincus, *J. Phys. II* **2**, 1531 (1992).

61. T. Tanaka, L. O. Hocker, and G. B. Benedek, *J. Chem. Phys.* **59**, 5151 (1973).

62. H. C. Brinkman, *Appl. Sci. Res.* **A1**, 27 (1947).

63. R. Colby, M. Rubinstein, A. Dobrynin, and J.-F. Joanny, *Elastic Modulus and Equilibrium Swelling of Polyelectrolyte Gels* (preprint 1995).

64. P. J. Flory, *Principles of Polymer Chemistry*, Cornell University Press, Ithaca, NY, 1954.

65. S. V. Panyukov, *Sov. Phys. JETP* **71**, 372 (1990).

66. R. Skouri, F. Schosseler, J. P. Munch, and S. J. Candau, *Macromolecules* **28**, 197 (1995).

67. See, e.g., the contribution by P. Chaikin in *Physics of Complex and Supermolecular Fluids*, S. A. Safran and N. A. Clark, eds., Wiley, New York, 1987.

68. K. Kaji, K. Urakawa, T. Kanata, and R. Kitamaru, *J. Physique (France)* **49**, 993 (1988).

69. P. G. de Gennes and J. Prost, *The Physics of Liquid Crystals*, Oxford University Press, Oxford, 1993.

70. L. Onsager, *Ann. N.-Y. Acad. Sci.* **51**, 627 (1949).

71. A. N. Semenov and A. R. Kokhlov, *Soviet Physics Uspekhi* **31**, 988 (1988).

72. A. Y. Grosberg and A. R. Khokhlov, *Statistical Physics of Macromolecules*, AIP Press, New York, 1994.

73. D. Frenkel, H. Lekkerkerker, and A. Stroobants, *Nature* **332**, 82 (1988).

74. I. Nyrkova, N. Shusharina and A. Khokhlov, *Polym. Prepr.* **34**, 939 (1993).

75. A. V. Dobrynin, R. H. Colby and M. Rubinstein, *Macromolecules* **28**, 1859 (1995).

76. T. Odijk, *Macromolecules* **12**, 688 (1979).

77. J. Hayter, G. Jannink, F. Brochard-Wyart, and P. G. de Gennes, *J. Phys. Lett.* **41**, L-451 (1980).

78. T. A. Witten and P. Pincus, *Europhys. Lett.* **3**, 315 (1987).

79. J.-L. Barrat and J.-F. Joanny, *J. Phys. II* **4**, 1089 (1994).

80. M. Oostwal and T. Odijk, *Macromolecules* **26**, 6489 (1993).

81. V. Borue and I. Erukhimovich, *Macromolecules* **21**, 3240 (1988).

82. J.-F. Joanny and L. Leibler, *J. Phys.* **51**, 545 (1990).

83. A. Moussaid, F. Schosseler, J. P. Munch, and S. J. Candau, *J. Phys. II (France)* **3**, 573 (1993).
84. A. Ajdari, L. Leibler and J.-F. Joanny, *J. Chem. Phys.* **95**, 4580 (1991).
85. L. Leibler, *Macromolecules* **13**, 1602 (1980).
86. F. S. Bates and G. H. Fredrickson, *Ann. Rev. Phys. Chem.* **41**, 525 (1990).
87. E. E. Dormidontova, I. Y. Erukhimovich, and A. R. Khokhlov, *Colloid Polym. Sci.* **272**, 1486 (1994).
88. J. F. Marko and Y. Rabin, *Macromolecules* **25**, 1503 (1992).
89. I. A. Nyrkova, A. R. Khokhlov, and M. Doi, *Macromolecules* **27**, 422 (1994).
90. M. Shibayama, T. Tanaka, and C. C. Han, *J. Phys. IV* **3**, 25 (1993).
91. P. M. Résibois, *Electrolyte Theory*, Harper & Row, New York, 1968.
92. G. S. Manning, *J. Phys. Chem.* **85**, 1506 (1981).
93. J. D. Sherwood, *J. Chem. Soc. Faraday Trans. 2* **79**, 1091 (1982).
94. J. J. Hermans, *J. Polym. Sci.* **18**, 527 (1955).
95. M Muthukumar, *Macromol. Theory Simul.* **3**, 71 (1994).
96. See, for example, the contribution by W. F. Reed in *Maro-ion Characterization*, K. S. Schmitz, ed., American Chemical Society, Washington, D.C., 1994.
97. R. M. Fuoss, *Discuss. Farad. Soc.* **11**, 125 (1951).
98. M. W. Kim and D. G. Peiffer, *Europhys. Lett.* **5**, 321 (1988).
99. H. Eisenberg and J. Pouyet, *J. Pol. Sci.* **13**, 85 (1954).
100. Y. Rabin, J. Cohen, and Z. Priel, *J. Pol. Sci. C* **26**, 397 (1988).
101. D. T. Pals and J. J. Hermans, *Rev. Trav. Chim.* **71**, 433 (1952).
102. C. Wolff, *J. Phys. Colloques* **39**, C2-169 (1978).
103. R. M. Davis and W. B. Russel, *J. Polym. Sci. B* **24**, 511 (1986).
104. W. F. Reed, *J. Chem. Phys.* **101**, 2515 (1994).
105. M. Rubinstein, R. H. Colby, and A. V. Dobrynin, *Phys. Rev. Lett.* **73**, 2776 (1994).
106. Y. H. Lin, *Macromolecules* **20**, 3080 (1987).
107. J. Yamanaka, H. Matsuoka, H. Kitano, and N. Ise, *J. Coll. Int. Sci.* **134**, 92 (1990).
108. J. Cohen, Z. Priel, and Y. Rabin, *J. Chem. Phys.* **88**, 7111 (1988).
109. W. Hess and R. Klein, *Adv. Phys.* **32**, 173 (1983).
110. T. Odijk, *Macromolecules* **27**, 4998 (1994).
111. M. Sedlak, *Macromolecules* **28**, 793 (1995).
112. S. Foester, M. Schmidt, and M. Antonietti, *Polymer* **31**, 781 (1990).
113. G. Maret and G. Weill, *Biopolymers* **22**, 2727 (1983).
114. P. Hagerman, *Biopolymers* **20**, 251 (1981).
115. M. J. Stevens and K. Kremer, *Phys. Rev. Lett.* **71**, 2229 (1993); *Macromolecules* **26**, 4717 (1993); *J. Chem. Phys.* **103**, 1669 (1995).
116. P. G. Higgs and J.-F. Joanny, *J. Chem. Phys.* **94**, 1543 (1991).
117. Y. Kantor, H. Li, and M. Kardar, *Phys. Rev. Lett.* **69**, 61 (1992).
118. M. S. Turner and J.-F. Joanny, *J. Phys. Chem.* **97**, 4825 (1993).
119. A. R. Khokhlov and I. A. Nyrkova, *Macromolecules* **25**, 1493 (1992).
120. M. Perreau, I. Iliopoulos, and R. Audebert, *Polymer* **30**, 2112 (1989).
121. A. Khokhlov and E. Kramarenko, *Macromol. Theory Simul.* **3**, 45 (1994).
122. J. Melenkevitz, K. S. Schweizer, and J. G. Curro, *Macromolecules* **26**, 6190 (1993).

123. See, e.g., the contribution by G. Jannink in *Physics and Chemistry of Aqueous Ionic Solutions*, M. Bellisent-Funel and G. W. Neilson, eds., Dordrecht, Reidel, 1987.
124. C. E. Woodward and B. Jönsson, *Chem. Phys.* **155**, 207 (1991).
125. H. Löwen, J-P. Hansen, and P. A. Madden, *J. Chem. Phys.* **98**, 3275 (1993).

STAR POLYMERS: EXPERIMENT, THEORY, AND SIMULATION

GARY S. GREST, LEWIS J. FETTERS, AND JOHN S. HUANG

Corporate Research Science Laboratories, Exxon Research and Engineering Company, Annandale, New Jersey

DIETER RICHTER

Institut für Festkörperforschung, Jülich, Germany

CONTENTS

Advances in Chemical Physics, Volume XCIV, Edited by I. Prigogine and Stuart A. Rice.
ISBN 0-471-14324-3 © 1996 John Wiley & Sons, Inc.

I. INTRODUCTION

The occurrence of polymers branched in a random fashion is common. Chain transfer reaction can cause short- and long-chain branching in polymerizations such as the high-pressure polymerization of ethylene. Branching can also be introduced intentionally by the use of a polyfunctional monomer in end-linking polymerizations. Similar branching can be produced in addition to polymerizations by the use of a small amount of difunctional monomer, for example, divinylbenzene. There also has been much interest in graft polymerization by which long-chain branches can be introduced onto a backbone, which is often a different polymer from the branches.

The properties of branched polymers can be quite different from those of linear polymers of the same molecular weight. For example, bulk viscosities as well as concentrated and dilute solution viscosities can be lower for branched polymers than for a linear material of equivalent molecular weight. As an example, the melt processing behavior of polymers can be manipulated by alterations in the average molecular weight, molecular weight distribution, and the frequency and length of long branches in the molecules. Thus, there is an obvious need to correlate and characterize the type and degree of branching in a polymer with its effect on the physical properties in solution or melt.

In all of these examples of branching, there is a mixture of branched and unbranched material. The unbranched and branched polymers can have a wide molecular weight distribution, as can the branches themselves. Also, the frequency of branches and the segment lengths between branch points can vary. Hence, the physical properties of such materials represent some average of the properties of all the different species present.

Such great diversity poses a problem with regard to the quantitative interpretation of the effects of branching. Theoretical models can be set up for simple branched architectures. However, calculations involving even these structures can be quite complex, and thus the attempt to describe a system of a mixture of complicated branched polymers is a burden that is best avoided. The true test of any theory dealing with branched polymer behavior is, of course, dependent on its success in predicting and explaining the physical parameter in question. However, for a fair evaluation the exact nature of the experimental polymer must be known.

A theoretical approach might seem quite sound, yet an actual molecule might have properties that vary from the predictions, owing to effects not accounted for by the simplified model. Thus, it is desirable to test a theory with the simplest possible system. For branched polymers, this is a macromolecule with a single branching unit with all of the branches being of equal molecular weight, that is, a star-branched polymer. There are three variables in such a system: the number of branches, the molecular weight of the branches, and the polymer type. The variation of these parameters can be decreased by comparing the star-branched polymer results to those for linear polymers of the same type and molecular weight. In this way, the effect of branching on the behavior of a polymer can be examined by itself.

Anionic polymerization can produce branches with narrow molecular weight distributions. Postpolymerization linking involving the still active chain ends can then be done to produce branched polymers with a predetermined number of arms. These species are then excellent model compounds for testing the theories.

The foregoing is from a 1978 review of star polymers [1]. Since that time considerable strides have been made in synthesis, characterization, and theory regarding these model branched polymers. With the advent of modern supercomputers and fast workstations, computer simulations of many arm stars are now feasible. Using carbosilane dendrimers, well-characterized stars with as many as 128 arms attached to a central core have been synthesized [2]. Other examples of starlike structures are formed by polymers with one ionic end group that naturally associate into starlike aggregates in a low dielectric media [3, 4]. Asymmetric diblock copolymers in a selective good solvent for one of the blocks can also form starlike micelles in dilute solution [5–7]. The scaling approach developed by Daoud and Cotton [8] and Birshtein and Zhulina [9] has been particularly important in furthering our understanding of both the static and dynamic properties of star polymers. This picture can be used to understand how end-grafted chains stabilize colloidal particles [10]. While in most cases studied the arms have equal length, it is also possible to make asymmetric length arms [11] or stars where the arms are of two or more different polymer types [12]. In the dilute limit, light and neutron scattering and viscometry have been very useful in obtaining new insight into how the size of a star depends on the functionality, arm molecular weight, and solvent quality. Recent advances in neutron spin echo scattering [13] techniques have led to a much better understanding of the dynamics of the entire star as well as its individual arms. We believe that there has been a lot of progress since 1978, and in this chapter, in the main, we cover many of the developments in this area that have occurred since then.

An outline of the chapter is as follows. In the next section, we describe some new synthesis pathways that allow one to produce high functionality stars in a controlled fashion. In Section III, we discuss several measurements used to determine the size of a star polymer. We also compare these results to the predictions of the random-walk (RW) model. In Section IV, we discuss the scaling approach [8, 9], which has been very important in understanding the effect of excluded volume interactions on size of a star. In Section V, we review the progress in computer simulations for star polymers. These results are compared to both the scaling theory and to experiments on stars under a variety of solvent conditions. In Section VI, we discuss static scattering methods, which have been very important in determining the shape of a star polymer and giving some additional insight into their internal structure. Up to this point, most of the discussion has been for stars in dilute solution. In Section VII, we describe both theory and experiment on more concentrated solutions, including the ordering that can occur, in some cases, at higher concentrations for large functionality stars. The internal dynamics of stars is considered in Section VIII, including a discussion of the dynamic structure factor $S(Q, t)$. The diffusion and viscoelastic properties of a star in a melt are considered in Section IX. Finally, in Section X, we briefly review some new developments and discuss some applications of star polymers.

II. SYNTHESIS

A. Chlorosilanes

Anionic polymerizations involving the lithium counterion have proven to be the most versatile system for the preparation of star-branched polymers. The chain head groups readily undergo metathesis with the electronegatively substituted silanes,

$$SiCl_4 + 4PLi \rightarrow SiP_4 + 4LiCl$$

where P denotes a polymer chain. If properly performed, the above procedure will lead to the preparation of materials of uniform branching in combination with the near-monodisperse molecular weight distributions and predictable molecular weights. This chemistry was first exploited in the early 1960s for the three- and four-arm polystyrene (PS) [14] and polybutadiene (PBd) stars [15].

Roovers and Bywater [16] were the first to highlight the influence of steric restraints in obtaining high yields when the styryllithium head group was used. This difficulty is avoided if the head group is converted to butadienyllithium by the simple process of adding butadiene in a stoichiometric amount relative to the aryllithium head group or by limiting the number of chlorines per silicon to two and separating the silicone by ethylene spaces. This basic chemistry has allowed the preparation model star-shaped structures ranging in functionalities from 3 to ca. 270 arms.

The advent of high functionality dendrimer carbosilanes commenced with the synthesis of octa-, dodeca, and octa-doca materials [17, 18]. Some 10 years later, the insights gained in the synthesis of dendritic molecules [19, 20] were used to prepare carbosilane dendrimers with functionalities of 32, 64, and 128 [2, 21]. A 270 functionalized material was prepared [22] via the hydrosilation of a 10^4 molecular weight 1,2-polybutadiene. The basic chemical process is outlined in Fig. 1 where the alternate use of a chlorosilane and vinyl magnesium bromide led to the preparation of fourth-generation (4G) dendrimers [23]. The reaction of CH_3HSiCl_2 with the 4G material results in the formation of the 128 functionalized linking agent. Zhou and Roovers [23] have shown that this process is essentially side reaction free; although α-addition of hydrosilation, dehydrogenation, and methy silylation can occur.

The reluctance of the styryllithium head group to react with more than two chlorines per silicon atom was partly overcome by adding a polar cosolvent to the mixture [24]. It is known that the styryl- and dienyllithium active centers aggregate as dimers and a fraction of these dimers in turn self-assemble to yield prolate ellipsoids [24]. Polar solvents disrupt this aggregation via head group solvation. Pennisi and Fetters [25] found that the presence of a polar solvent allowed styryllithium to react in a quantitative fashion with all three silicon–chlorine bonds in CH_3SiCl_3 where the arm molecular weight $M_a \leq 3.5 \times 10^4 \, \mathrm{g \, mol^{-1}}$. Higher molecular weight PS arms required the addition of 1,3-butadiene in order to effect complete reaction of the Si–Cl bonds.

The reluctance of the styryllithium head group to react in a quantitative fashion with CH_3SiCl_3 and $SiCl_4$ was exploited by Mays [26] who prepared polyisoprene-g-polystyrene branched structures. This "simple" graft copolymer was prepared as follows:

$$PSLi + CH_3SiCl_3 \rightarrow PS(CH_3)SiCl_2 + CH_3SiCl_3\uparrow + LiCl$$

where $[CH_3SiCl_2]/[PSLi] = 12$. Following removal of the excess chloro-

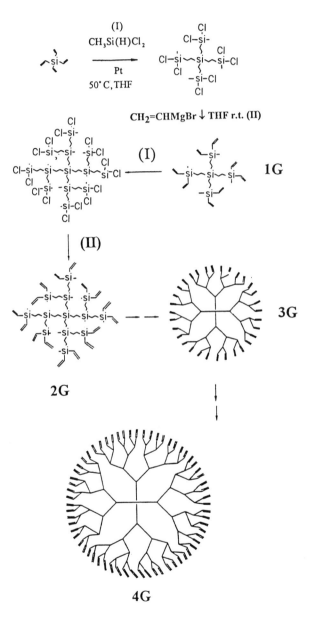

Figure 1. Synthetic scheme for chlorosilane dendrimer preparation.

silane, the polyisoprenyllithium was added to the chlorosilane capped polystyrene chain to form the polyisoprene-g-polystyrene simple graft. Star copolymers made of arms with more than one polymer type have since become known as mikto (for mixed) arm stars [12]. Most of these contain stars with types of arms, though recently Iatrou and Hadji-christidis [12] have used the chlorosilane approach to prepare three-miktoarm star terpolymers with three different arms and four-miktoarm star quaterpolymers with four different arms. In the later case, a star of type ABCD was prepared where A is PS, B is PBd, C is polyisoprene (PI), and D is poly(4-methylstyrene).

Heteroarm block copolymers were prepared via multi-step anionic reactions by Jerome et al. [27]. The respective arms were lipohilic-polystyrene or a polydiene (A)- and lipophobic-polyoxirane (B). Three armed materials were prepared of the $PA(PB)_2$ type. However, no information is available regarding, for example, their solution properties, bulk morphology or rheology.

B. Divinylbenzene

The earliest mention of the use the divinylbenzene (DVB) isomers as a linking route to star polymer formation was in 1964 by Milkovich [28]. Therein the attempt was made to take polystyrene–polyisoprenyllithium diblocks and link them together with a crosslinked DVB nodule, formed by the homopolymerization of DVB. However, the characterization data presented demonstrated that the primary product formed was linear triblock, not the claimed star (see Section X). Similar results were given in 1966 by Zilliox et al. [29] for homopolyisoprene; the resultant "star" PI sample assayed to have a M_W only twice that of the parent material. These combined findings demonstrate the nonobvious nature of PI star formation via the DVB linking chemistry. The first high-yield synthesis of polystyrene stars via the DVB route was that of Worsfold and co-workers in 1969 [30]. Using DVB–head group ratios of ~2–21 resulted in functionalities in the range of 6–15.5.

The elusive DVB-linked polydiene stars were first reported by Bi and Fetters [31, 32]. The potential difficulties in the preparation of the polydiene stars is highlighted in Fig. 2 where the propagation rate behavior of the four processes in the styrene–diene copolymerization [33] are given as a function of active center ([PLi]) concentration. It is sufficient to state that the propagation of styrene (analogous to DVB) is notably faster than the initial crossover step involving the iso-prenyllithium head group and styrene. Qualitatively, this implies that at low DVB–head group ratios relatively few of those head groups would react with DVB due to its low concentration and the fact that relatively

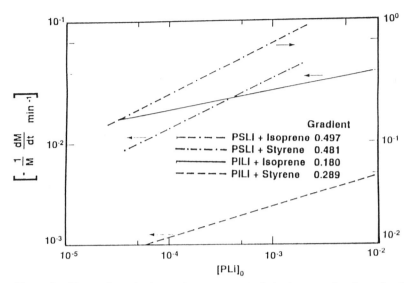

Figure 2. Homopolymerization and crossover rate behavior as a function of active center [PLi] concentration in cyclohexane at 40°C.

rapid homopolymerization of DVB would deplete the remaining mono-mer prior to completion of the crossover event. However, under the proper reaction conditions [31–34] the DVB isomers and their commer-cial mixtures are one route to the facile high-yield (>90%) production of polydiene star. Nonetheless, the chlorosilane route remains the "academ-ic" linking chemistry of choice since it allows the preparation of star families where functionality remains constant (within experimental error) while arm molecular weight is allowed to vary.

III. LENGTH SCALES OF A STAR

Static light scattering [35] is a convenient tool for determining a number of macroscopic properties of a star polymer, including the weight-aver-aged molecular weight M_W, the mean-squared radius of gyration $\langle R_G^2 \rangle$, and the second virial coefficient A_2. Quasi-elastic light scattering is commonly used for determining the diffusion coefficient at infinite dilution D_0. From viscometric measurements, one can determine the intrinsic viscosity $[\eta]$ and Huggins coefficient k_H,

$$\frac{\eta - \eta_s}{\eta_s c} = [\eta](1 + k_H[\eta]c + \cdots) \tag{1}$$

where η_s is the solvent viscosity and c is the concentration. From these relatively simple measurements, one can already obtain significant insight into the effect changing the functionality f has on the properties of a star polymer. In this section, we concentrate on a variety of ways to characterize the size of a star and investigate how they depend on solvent quality.

In addition to the radius of gyration, $\langle R_G^2 \rangle^{1/2}$, which is determined directly by static scattering, several other measures of the size of a polymer can be defined. These include the hydrodynamic radius,

$$\langle R_H \rangle = \frac{k_B T}{6 \pi \eta_s D_0} \qquad (2)$$

and the viscosity radius R_V,

$$\langle R_V \rangle = \left(\frac{3}{10 \pi N_A} [\eta] M_W \right)^{1/2} = 5.41 \times 10^{-9} ([\eta] M_W)^{1/3} \qquad (3)$$

where k_B is the Boltzmann constant, T the absolute temperature, and N_A is Avogadro's number. Here, $\langle R_V \rangle$ is radius of an equivalent sphere and is based on the Stokes–Einstein relation for the viscosity of a suspension of spheres. For $A_2 > 0$, one can also define an effective radius from the second virial coefficient A_2, according to

$$\langle R_T \rangle = \left(\frac{3 A_2 M_W^2}{16 \pi N_A} \right)^{1/3} = 4.63 \times 10^{-9} (A_2 M_W^2)^{1/3} \qquad (4)$$

In contrast to $\langle R_V \rangle$ and $\langle R_H \rangle$, which are derived from zero concentration intercepts, $\langle R_T \rangle$ is derived from the concentration dependence of light scattered.

In order to study the dependence of the star dimensions on the functionality f of the star and to compare different experimental systems with each other as well as with theory, it is convenient to normalize these various radii by either the corresponding result for a single arm $\langle R \rangle_a$ or by $\langle R \rangle_{lin}$, the value for a linear chain of the same total molecular weight as that of the star. Both normalizations have been used extensively in the literature. For the later case, one often defines the quantities g and h:

$$g = \frac{\langle R_G^2 \rangle}{\langle R_G^2 \rangle_{lin}} \qquad (5)$$

$$h = \frac{\langle R_H \rangle}{\langle R_H \rangle_{\text{lin}}} \tag{6}$$

One can also define a corresponding ratio for the intrinsic viscosity,

$$g' = \frac{[\eta]}{[\eta]_{\text{lin}}} \tag{7}$$

Of course, the linear and single arm "benchmark" dilute solution behavior of the intrinsic viscosity $[\eta]$, viscosity radius $\langle R_V \rangle$, hydrodynamic radius $\langle R_H \rangle$, mean-squared radius of gyration $\langle R_G^2 \rangle^{1/2}$, all under a variety of solvent conditions, as well as the good solvent limit (GSL) for the thermodynamic radius $\langle R_T \rangle$ must be known. The respective power laws for these parameters, and the data sets upon which they are based, are available in a recent, extensive compilation [36].

These characteristic ratios can be easily calculated for a star polymer assuming that each chain segment adopts a random walk. A brief description of the derivation $\langle R_G^2 \rangle$ is presented in Appendix A. These calculations [37] involve only the functionality f, the number of monomers N per arm, and their effective bond length. The later two parameters disappear when one compares the results for a star to that of its linear analog at constant molecular weight. For the radius of gyration, Zimm and Stockmayer [37] found, Eq. (A7),

$$g_{\text{rw}} = \frac{\langle R_G^2 \rangle}{f \langle R_G^2 \rangle_a} = \frac{3f - 2}{f^2} \tag{8}$$

where rw stands for random walk. The hydrodynamic parameters including the hydrodynamic radius $\langle R_H \rangle$, and the intrinsic viscosity $[\eta]$ have also been determined for a RW star [38, 39],

$$h_{\text{rw}} = f^{1/2}[2 - f + 2^{1/2}(f - 1)]^{-1} \tag{9}$$

and

$$g'_{\text{rw}} = \frac{(2/f)^{3/2}[0.396(f - 1) + 0.196]}{0.586} \tag{10}$$

The classical theories predicting the dependencies of intrinsic viscosities, radii of gyration, and hydrodynamic radii have been in existence for many years. The last 15 years or so have seen developments in synthesis procedures that have allowed the preparation of the model materials needed for the evaluations of these theories.

The experimental data upon which the following developments are based can be found in Refs. 2, 11, 16, 21, 22, and 40–44 where functionalities ranged from 3 to 270 and are summarized in Tables I and II. As this is a compilation of different data sets, it is difficult to assign error bars. However, from the spread in the available data, we believe that a reasonable estimate of the error bars is ±5%. The GSL results for g, h, and g' are very sensitive to the power law fit to the linear, single-arm data, particularly for large f where the molecular weight of the

TABLE I

Static and Hydrodynamic Ratios of Star Polymers

f	g'	g'_θ	g	g_{rw}	g_θ	h	h_{rw}	h_θ
2	1.00	1.00	1.00	1.00	1.00	1.00	1.00	1.00
3	0.84	0.87	0.79	0.777	0.82	0.95	0.947	0.95
4	0.73	0.76	0.63	0.625	0.64	0.85	0.892	—
5	0.60	0.74	—	0.520	—	—	0.842	—
6	0.58	0.63	0.45	0.444	0.46	—	0.798	—
8	0.613	0.55	0.35	0.344	0.42	0.82	0.726	—
12	0.34	0.42	0.25	0.236	0.30	0.74	0.623	0.78
18	0.23	0.29	0.19	0.160	0.21	0.67	0.527	0.72
32	0.15	0.20	0.12	0.092	0.15	0.59	0.409	0.63
64	0.09	0.13	0.07	0.046	0.09	0.49	0.295	0.55
128	0.05	0.08	0.04	0.023	0.06	0.41	0.211	0.47
270	0.03	0.045	0.03	0.011	0.06	0.34	0.147	0.45

TABLE II

Static and Hydrodynamic Ratios for Star Polymers

	$[\eta]/[\eta]_a$		$\langle R_V \rangle / \langle R_V \rangle_a$		$\langle R_H \rangle / \langle R_H \rangle_a$		$\langle R_G^2 \rangle^{1/2} / \langle R_G^2 \rangle_a^{1/2}$		$\langle R_T \rangle / \langle R_T \rangle_a$
f	GSL	θ	GSL	θ	GSL	θ	GSL	θ	GSL
2	1.6	1.4	1.5	1.4	1.5	1.4	1.5	1.4	0.86
3	1.8	1.5	1.8	1.6	1.8	1.6	1.7	1.6	0.76
4	1.9	1.5	2.0	1.8	1.9	—	1.8	1.6	0.78
5	2.1	1.6	2.2	2.0	—	—	—	—	—
6	2.0	1.5	2.3	2.1	—	—	1.9	1.7	—
8	2.0	1.5	2.6	2.3	2.8	—	1.9	1.8	0.44
12	1.9	1.5	2.9	2.6	3.2	2.6	2.1	1.9	0.33
18	1.8	1.2	3.3	2.9	3.7	2.8	2.3	2.0	0.23
32	1.7	1.1	4.0	3.3	4.6	3.6	2.7	2.2	0.10
64	1.6	1.0	5.0	4.0	5.8	4.4	3.0	2.4	0.06
128	1.5	0.9	6.1	4.9	7.3	5.3	3.7	2.8	0.03
270	1.4	0.7	7.0	5.8	9.4	7.4	4.8	4.0	~0.01

star often exceeds the range where the linear chains have been measured. To produce Table I, we have used the power law fits to linear chain data presented in Fetters et al. [36]. Each functionality family was derived from the chlorosilane linking chemistry. Although valuable in their day the DVB-linked PS and PI star data are not included. These materials lack the near-monodisperse functionality of the chlorosilane materials and also are not members of *functionality families*, which, of course allow the development of power laws to describe the various parameters and which in turn lead to random errors being minimized. Furthermore, the subject polymers were based on polybutadiene, polyisoprene, and polystyrene. Although, none of these polymer systems covers the f range of 2–270 (where $f = 2$ represents the two-arm star), overlap in functionalities occurs and the data sets lead to coherent trends. Also, all three polymer families have been studied under Θ conditions as well as at the GSL. It should also be remembered that all three monomers polymerize in a disciplined fashion with retention of head group reactivity over an extended period of time. This gift from nature allows adequate post polymerization reaction time to achieve, within experimental error, uniform, high functionality stars in a controlled fashion.

In Table I, experimental results for g', g, and h for the GSL and for the Θ condition are presented and compared to the RW values for g_{rw} and h_{rw}. In Table II, results for the intrinsic viscosity $[\eta]$, as well as $\langle R_V \rangle$, $\langle R_H \rangle$, $\langle R_G^2 \rangle^{1/2}$, and $\langle R_T \rangle$, all normalized by their respective values for a single arm ($f = 1$), are presented for stars in the range f of 2–270. Results for the first four are given for the good solvent limit and for a Θ solvent. The RW predictions are in general accord with the corresponding g_Θ and h_Θ for small functionalities $f = 3$ to 6. However, for $f > 6$, the differences between the Θ condition values and those of the phantom chain RW model progressively increase, as seen in Figs. 3a and 3b. This dual behavior is a reflection of stretching to eliminate steric crowding of the arms. This stretching is reminiscent of that observed in polymer brushes attached to a substrate [7]. At the high functionality limit, $\langle R_G^2 \rangle^{1/2}$ for $f = 270$ does not follow the extrapolation from smaller f. This is presumably due to the fact that the backbone of the 270-arm star is more extended than that for $f \leq 128$. This leads to less crowding near the central core region of the star, resulting in an increase in $\langle R_G^2 \rangle$ [45]. While for small f there is no real difference in the size of a polymer molecule made of chains grafted to a small central core or a flexible backbone chain [46], this difference in architecture is important for large f. This is particularly true in the present case since for the largest 270-arm star produced to date, the arm molecular weight M_a is only four times the backbone. Somewhat surprising, and not explained, is that while $\langle R_G^2 \rangle^{1/2}$

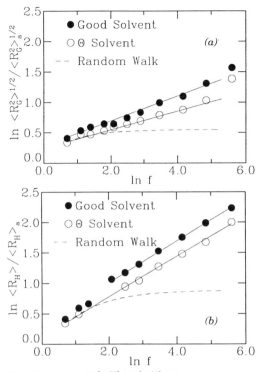

Figure 3. Radius of gyration $\langle R_G^2 \rangle^{1/2} / \langle R_G^2 \rangle_a^{1/2}$ (a) and hydrodynamic radius $\langle R_H \rangle / \langle R_H \rangle_a$ (b) versus f for both the good solvent and Θ solvent limit. The corresponding result for the random-walk model are shown as a dashed line.

does not follow the extrapolated curve, the hydrodynamic radius for the 270-arm star does. In Table III, the parameter for best fits and range of the fits are given for both the GSL and for Θ conditions to $K'f^{\beta'}$. Note that the respective results for the exponent β' are far from the predictions of the RW model and K' is a constant.

The use of radius ratios requires *transition* formulas for intercomparisons. These are defined as follows with the exponents found for the linear chains ($R_G \sim M_W^\nu$ and $[\eta] \sim M K_W^\alpha$) and g, h, and g' denoting experimental values:

$$\frac{\langle R_G^2 \rangle^{1/2}}{\langle R_G^2 \rangle_\alpha^{1/2}} = g^{1/2} f^\nu \qquad (11)$$

TABLE III
Coefficients Describing Functionality Dependence of Intrinsic Viscosity, Viscometric Radii,
Hydrodynamic Radii, and Radii of Gyration

Parameter	Good Solvent Limit			θ Conditions		
	K'	β'	f Range	K'	β'	f Range
g'	2.37	−0.789	5–270	2.22	−0.690	5–270
$[\eta]/[\eta]_a$	2.43	−0.0997	5–270	2.22	−0.190	5–270
$\langle R_V \rangle / \langle R_V \rangle_a$	1.34	0.298	5–270	1.30	0.275	5–270
h	1.38	−0.250	8–270	1.16	−0.174	2–270
$\langle R_H \rangle / \langle R_H \rangle_a$	1.38	0.346	8–270	1.16	0.326	2–270
g	1.76	−0.801	2–128	1.66	−0.694	2–128
$\langle R_G^2 \rangle^{1/2} / \langle R_G^2 \rangle_a^{1/2}$	1.33	0.206	2–128	1.29	0.154	2–128

$$\frac{\langle R_V \rangle}{\langle R_V \rangle_a} = (g')^{1/3} f^{(1+\alpha)/3} \tag{12}$$

$$\frac{\langle R_T \rangle}{\langle R_G^2 \rangle^{1/2}} = 1.10 \psi_F^{1/3} \tag{13}$$

$$\frac{\langle R_V \rangle}{\langle R_G^2 \rangle^{1/2}} = 5.41 \times 10^{-9} \phi_F^{1/3} \tag{14}$$

$$\frac{\langle R_T \rangle}{\langle R_V \rangle} = 0.856 \sigma_F^{1/3} \tag{15}$$

$$\frac{\langle R_V \rangle}{\langle R_H \rangle} = \frac{(\langle R_V \rangle / \langle R_H \rangle)_{\text{lin}} (g')^{1/3}}{h} \tag{16}$$

where the penetration function, ψ_F, is given by

$$\psi_F = \frac{A_2 M_W^2}{4 \pi^{3/2} N_A \langle R_G^2 \rangle^{3/2}} \tag{17}$$

The intrinsic viscosity–chain dimension relationships are expressed in terms of the Flory coefficient

$$\phi_F = \frac{[\eta] M_W}{\langle R_G^2 \rangle^{3/2}} \tag{18}$$

while the intrinsic viscosity–A_2 relationship can be expressed in terms of

$$\sigma_F = A_2 M_W [\eta]^{-1} \qquad (19)$$

For monodispersed spheres of equal density,

$$R_V = R_T = R_H = (\tfrac{5}{3})^{1/2} R_G \qquad (20)$$

Conventionally, dilute solution theories for the effects of branching on R_G, D_o, A_2, and $[\eta]$ are given in terms of parameters such as g, g', h, the Flory coefficients, and the penetration function. Our presentation has utilized the use of size ratios. For example, g is equivalent to $\langle R_G^2 \rangle / \langle R_G^2 \rangle_a$ and ψ_F parallels $\langle R_T \rangle / \langle R_G^2 \rangle^{1/2}$. The advantage of this approach is that, for example, the variation of $\langle R_T \rangle / \langle R_G^2 \rangle^{1/2}$ with f displays in an explicit fashion the approach of the star to the spherical structure (hard sphere) as f increases. In a similar fashion, the variation of $\langle R_G^2 \rangle^{1/2} / \langle R_G^2 \rangle_a^{1/2}$ with f directly demonstrates the swelling of the star chain dimensions as f increases. In Table IV values of four ratios are given as a function of functionality f: $\langle R_T \rangle / \langle R_G^2 \rangle^{1/2}$, $\langle R_H \rangle / \langle R_G^2 \rangle^{1/2}$, $\langle R_V \rangle / \langle R_G^2 \rangle^{1/2}$ and $\langle R_V \rangle / \langle R_H \rangle$. Following Roovers and Martin [44] we have used $\langle R_G^2 \rangle^{1/2}$ as the basis of comparison since it is the size parameter obtained directly from measurements whereas the other dimensions are based on the equivalent sphere model. Finally, in Table V, we present the

TABLE IV
Size Ratios for Linear and Star Polymers

f	$\langle R_T \rangle / \langle R_G^2 \rangle^{1/2}$	$\langle R_H \rangle / \langle R_G^2 \rangle^{1/2}$		$\langle R_V \rangle / \langle R_G^2 \rangle^{1/2}$		$\langle R_V \rangle / \langle R_H \rangle$	
		GSL	θ	GSL	θ	GSL	θ
2^a	0.68	0.71	0.79	0.77	0.83	1.08	1.09
3	0.77	0.76	0.80	0.78	0.86	1.04	1.08
4	0.89	0.75	—	0.81	0.94	1.09	—
6	0.94	—	—	0.88	1.06	—	—
8	1.00	0.99	—	0.99	1.02	0.99	—
12	1.12	1.05	1.07	1.06	1.04	1.01	1.06
18	1.24	1.12	1.18	1.11	1.21	1.08	1.03
32	1.28	1.23	1.19	1.26	1.25	1.03	1.04
64	1.40	1.28	1.36	1.31	1.39	1.04	1.04
128	1.41	1.34	1.40	1.36	1.42	1.03	1.03
270	1.22	1.32	1.30	1.32	1.23	1.00	0.95
Hard sphere	1.29	1.29	1.29	1.29	1.29	1.00	1.00

^{a} From data sets of Ref. 36.

TABLE V
Star Polymer Master Equations for $\langle R_G^2 \rangle^{1/2}(\text{Å})$

Sample	Condition	K	ν	β
PBd	GSL	0.172	0.609	−0.403
PBd	θ	0.489	0.5	−0.346
PI	GSL	0.168	0.610	−0.404
PI	θ	0.426	0.5	−0.346
PS	GSL	0.160	0.595	−0.389
PS	θ	0.340	0.5	−0.346
PIB	GSL	0.182	0.595	−0.389
PIB	θ	0.387	0.5	−0.346
PEP	GSL	0.221	0.600	−0.394
PEP	θ	0.491	0.5	−0.346

Note: $\langle R_G^2 \rangle^{1/2} = K M_W^\nu f^\beta$ for $f = 2, \ldots, 128$.

master equation for $\langle R_G^2 \rangle^{1/2}$ in the form,

$$\langle R_G^2 \rangle^{1/2} = K M_W^\nu f^\beta \tag{21}$$

for five polymer systems. See Appendix B for more details of the fitting and similar results for $[\eta]$ and $\langle R_H \rangle$. For this fitting, a linear chain is treated as a two-arm star. The first three are for PBd, PI, and PS, which are the most often studied star polymers. The other two, polyisobutylene (PIB) and *alt*-poly(ethylene-propylene) (PEP) have only been studied for linear chains (see Ref. 36 and Appendix B). Here M_W is the total molecular weight of the star and $\beta = \beta' - \nu$ (see Table III).

The intrinsic viscosity for stars, normalized by its value for a single arm, is shown in Fig. 4 as a function of f for both the Θ condition and the GSL. This ratios gradually increases for small f over a range of 2–5 for both solvent conditions. However, for larger values of f, $[\eta]/[\eta]_a$ decreases uniformly for f from 5 to 270. This decrease in $[\eta]/[\eta]_a$ is doubtlessly due to the inability of the hydrodynamic volume, which depends on $\langle R_G^2 \rangle$ and $\langle R_H \rangle$, to keep pace with the concurrent increase in molecular weight (f),

$$[\eta] \sim \langle R_G^2 \rangle \langle R_H \rangle (f M_a)^{-1} \tag{22}$$

The surprising result is that under Θ conditions, $[\eta] < [\eta]_a$ for $f \gtrsim 60$. This Θ condition behavior is not predicted by theory. Both the GSL and Θ condition behavior serve to emphasize the potential pitfalls that can be encountered in the use of dilute solution viscometry as the sole characterization tool for branched macromolecules.

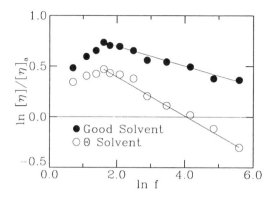

Figure 4. Intrinsic viscosity $[\eta]$, normalized by its value for a free chain of the same length as a single arm $[\eta]_a$ versus f for both the good solvent and Θ solvent limit.

From viscometry, one can also determine the Huggins coefficient k_H. The parameter k_H is sensitive to f; at constant f, an increase in arm molecular weight M_a leads to an increase in k_H as does an increase in f for constant M_a. The GSL and Θ solvent behavior of k_H reaches the hard sphere value $k_H = 0.99$ [47], for functionalities of order 50 and larger for large M_a [2, 21, 42]. The pattern is similar to that found for $\langle R_T \rangle / \langle R_G^2 \rangle^{1/2}$ and $\langle R_V \rangle / \langle R_H \rangle$, Table IV. An unambiguous evaluation of k_H, particularly under Θ conditions for large f is difficult to achieve since contributions from higher terms in Eq. (1) appear rather quickly when k_H is large. It is interesting to note that the DVB-linked PI stars yield k_H values in excess of the hard sphere limit under Θ conditions. The interplay between k_H and f for star-shaped species has not received theoretical scrutiny.

The final measure of a star that has been discussed in the literature is the thermodynamic radius $\langle R_T \rangle$, which is a measure of excluded volume and is determined by Eq. (4). The general behavior of $\langle R_T \rangle / \langle R_T \rangle_a$ is shown in Fig. 5.

IV. SCALING THEORY

Theoretically, to go beyond the RW model, many of the same techniques that work well for linear chains can also be applied to star polymers, including some exact results [48], scaling [8, 9, 49], renormalization group [50–52], and self-consistent minimization of the intermolecular free energy [53, 54]. Daoud and Cotton [8] and Birshtein and Zhulina [9]

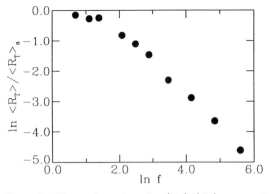

Figure 5. Thermodynamic radius $\langle R_T \rangle / \langle R_T \rangle_a$ versus f.

generalized the de Gennes scaling [55] model for linear polymers to star polymers. In this approach, the star consists of three regions, an inner meltlike extended core region, an intermediate region resembling a concentrated solution, and an outer semidilute region. Since the volume accessible to a given chain increases with the distance r from the center of the star, the monomer volume fraction $\rho(r)$ is expected to be a decreasing function of r. Each arm can be seen as a succession of growing spherical blobs as shown in Fig. 6. Within one blob each arm behaves as an isolated chain. At a given distance r from the center, a sphere of radius r is cut by f arms. The outer region of the star looks like a semidilute solution with a screening length $\xi(r)$, where $\xi(r)$ is a function of r and the number of arms f. Each blob contains monomers of a single chain. Since f blobs cover the sphere of radius r, the blob radius is

$$\xi(r) \simeq r f^{-1/2} \qquad\qquad (23)$$

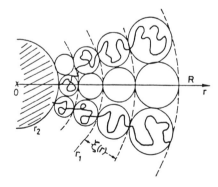

Figure 6. A representation of scaling model [8, 9] of a many armed star in which each arm is made of a succession of blobs of size $\xi(r)$ increasing from the center of the star to the outside. (From Ref. 8.)

$\xi(r)$ can also be related to the number $n(r)$ monomers in a blob and the excluded volume interaction v [55, 56]. Here, v takes into account the effective interaction between two monomers and is proportional to the monomer volume a^3 for an athermal solvent or at high temperature and is negative below the Θ temperature T_Θ, where v vanishes. The number $n(r)$ of monomers per blob of size $\xi(r)$ is the same as in a semidilute solution,

$$n(r) \simeq a\xi(r)^{1/\nu} \left(\frac{v}{a^3}\right)^{2\nu-1} \tag{24}$$

Here ν is the correlation length exponent, which is 0.588 for a good solvent in three dimensions [57] and $\frac{1}{2}$ for a Θ solvent. From this picture, one can easily verify that $\rho(r)$, the number density of monomers falls off as

$$\rho(r) \simeq \frac{n(r)}{\xi(r)^3} \simeq f^{(3\nu-1)/2\nu} \left(\frac{r}{a}\right)^{(1-3\nu)/\nu} \left(\frac{v}{a^3}\right)^{(1-2\nu)/\nu} \tag{25}$$

In a good solvent, $\rho(r) \simeq f^{0.65} r^{-1.30}$. Qualitatively, this scaling behavior is nicely illustrated in Fig. 7, in which we present a projection of a typical

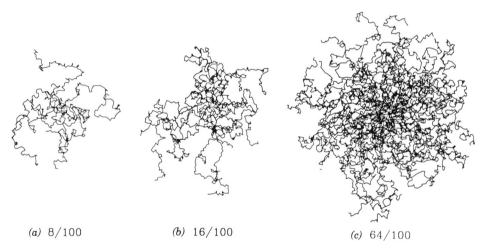

(a) 8/100   (b) 16/100   (c) 64/100

Figure 7. Projection of a typical configuration of a star with f values of 8, 16, and 64 with $N = 100$ monomers per arm in a good solvent. The pictures give an impression of the increasingly homogeneous density in the stars with f values of 16 and 64, while the 8-arm star obviously is governed much more by single-chain properties. These are from MD simulations of Ref. 60 with a Lennard-Jones interaction cutoff at $r_c = 2^{1/6}\sigma$ between nonbonded monomers.

configuration of a star polymer with $N = 100$ monomers per arm for $f = 8, 16, 64$. This simulation was carried out using the molecular dynamics method for monomers interacting with a Lennard-Jones interaction truncated at $r_c = 2^{1/6}\sigma$ [58–60]. This figure nicely shows the density falloff and corresponding increase in $\xi(r)$ with increasing r, as expected. Similar projections for stars in a Θ solvent are shown in Ref. 45.

This results for $\rho(r)$ explicitly assumes that the blob size $\xi(r)$ is larger than the size of a thermal blob $l_c \simeq a(v/a^3)^{-1}$ [55], where l_c is the measure of the distance over which the behavior of a chain is ideal. On longer length scales, a chain in a dilute solution is self-avoiding. For an athermal solvent $(v = a^3)$, the thermal blob is reduced to the size of a single monomer and the chain is swollen on all length scales. Therefore, Eq. (25) is valid only in the outer part of the star, in a region defined by $r > r_1$ where

$$r_1 \simeq af^{1/2}\left(\frac{v}{a^3}\right)^{-1} \tag{26}$$

For distances smaller than r_1, each arm can be pictured as a succession of growing ideal spherical blobs [8, 9]. In this case, the only difference with the above geometric considerations is that Eq. (23) is replaced by

$$\xi(r) \simeq an(r)^{1/2} \tag{27}$$

since now, within each blob, the arm behaves like an ideal chain. Thus for $r < r_1$, the monomer volume fraction is given by $\rho(r) \simeq (r/a)^{-1}f^{1/2}$. For still smaller distances $(r < r_2 \simeq af^{1/2})$, near the center, one reaches a central core region where $\rho(r) \simeq 1$.

Therefore, according to the scaling picture, a star with long enough arms consists of three regions. There is an inner meltlike core region $(r < r_2)$, an intermediate region $(r_2 < r < r_1)$ where the blobs are ideal, and an outer region $(r > r_1)$ where the blobs are swollen. In an athermal solvent $(v = a^3)$, the intermediate region disappears since $r_1 \simeq r_2$. Correspondingly, at the Θ temperature, the outer swollen region disappears and $\rho(r) \simeq (r/a)^{-1}f^{1/2}$.

The radius of the star can be deduced from the condition

$$Nfa^3 = \int_0^{R_0} d^3r\, \rho(r) \tag{28}$$

Using this result, it is easy to determine the outer radius of the star, R_0, within the scaling picture,

$$R_0 \simeq a N^{\nu} f^{(1-\nu)/2} \left(\frac{v}{a^3}\right)^{(2\nu-1)} \qquad N \gg f^{1/2} \left(\frac{v}{a^3}\right)^{-2} \qquad (29)$$

$$R_0 \simeq a N^{1/2} f^{1/4} \qquad f^{1/2} \left(\frac{v}{a^3}\right)^{-2} \gg N \gg f^{1/2} \qquad (30)$$

$$R_0 \simeq a N^{1/3} f^{1/3} \qquad f^{1/2} \gg N \qquad (31)$$

Thus in a good solvent, the mean-squared center-to-end distance $\langle R^2 \rangle$ and $\langle R_G^2 \rangle$ scale as $\langle R^2 \rangle \sim \langle R_G^2 \rangle \sim N^{1.18} f^{0.41}$. This result is in excellent agreement with the best fit of the experimental data to the scaling exponent for f, 0.41, presented in Table III. As the solvent quality is decreased and one approaches T_Θ, the range over which this result is valid decreases. At T_Θ, $\langle R^2 \rangle \sim N f^{1/2}$. Unlike the good solvent case, this prediction is significantly larger than the best fit to the experimental data, 0.31, presented in Table III. This lower experimental value, which is also found from computer simulations, see Section V.B, indicates that the scaling regime for a Θ solvent can only be reached for very large f. This last result is also valid in the intermediate region, though the crossover from the outer to the intermediate regime has not been observed. This is presumably due to the fact that r_1 is usually rather small and close to r_2, unless one studies stars with many long arms in the vicinity of the Θ point [61]. Note that these results for the star radius can also be obtained from a simple Flory-type free energy, except that the excluded correlation length exponent in Eq. (29) would be $\nu = \frac{3}{5}$ [8, 61].

All of the above results are for a star in a small molecular weight solvent. Raphaël et al. [61, 62] have generalized the scaling picture to the case of star dissolved in a melt of linear chains of length $P < N$, which are chemically identical to the star arms. They find that the monomer density varies as $\rho(r) \simeq 1$ in the melt regime for $r < a f^{1/2} P^{1/4}$ and $\rho(r) \simeq (r/a)^{-4/3} f^{2/3} P^{1/3}$ for $r > a f^{1/2} P^{1/4}$ (using the Flory value for $\nu = \frac{3}{5}$). Gay and Raphaël [63] have recently shown that this result is only valid for $f^2 > P$. In the opposite limit, $f^2 < P$, the monomer density varies as $\rho(r) = 1$ for $r < af$, $(r/a)^{-1} f$ for $af < r < aP/f$ and $(r/a)^{-4/3} f^{2/3} P^{1/3}$ for $r > aP/f$. Thus as P increases for fixed N, the star collapses and the melt chains are excluded from the star.

V. COMPUTER SIMULATIONS

Numerical simulations, both Monte Carlo (MC) and molecular dynamics (MD), have been used to test the scaling predictions and to determine other properties of a star polymer, including the static structure factor in the dilute limit. For a detailed overview of the various algorithms and methods used to study star polymers and other branched polymers see the review by Grest and Murat [45]. At present, it is not possible to simulate a melt or even a semidilute solution of many arm star polymers due to the long relaxation times. For few arm stars ($f \lesssim 12$) MC methods are clearly more efficient, while for large number of arms, MD methods work very well. For small f, the density of monomers of the star is low almost everywhere and static MC methods in which one generates the chains by constructing walks can be used [64–69]. Using this method, Batoulis and Kremer [68] were able to make very accurate estimates of the exponent $\gamma(f)$, which describes the dependence of the partition function $Z(N)$ on N, $Z(N) \sim N^{\gamma-1}$ for large N, in a good solvent. They also can determine very accurately $\rho(r)$ and $\langle R_G^2 \rangle$ for $f \leq 6$. Dynamic MC also works well in this limit, particularly if one invokes nonlocal moves, such as the pivot algorithm [70, 71]. However, as f increases, the interior becomes very dense and many of these methods fail or become inefficient. In this case, one can use either MD methods [72, 73] or a local stochastic MC method such as the bond fluctuation method on a lattice [74] or a simple off-lattice MC in which one attempts to move one monomer at a time. It is also possible to use nonlocal moves in the dilute, outer regions of the star and local moves near the interior, though this has not been done to the best of our knowledge.

A. Good Solvent

In a good solvent, the results from a number of simulations [58–60, 64–67, 70, 74–80] agree very well, though many of these are for small f. For small f, the tethered end of each arm can easily be attached to a single point. However, for large f, either the size of the central region must be increased or the maximum length of the bond between the central site and the first monomer must be increased [58]. Toral and Chakrabarti [81] have studied the crossover from star to brush as the radius of curvature of this sphere increases. In the dynamic methods, both MC and MD, the initial state of a star can be easily generated by growing chains from a point or a small sphere. Usually these simulations are initialized by simply letting the chains overlap in the initial configuration and equilibrating them until the excluded volume conditions are all satisfied. As an example, results for $\langle R_G^2 \rangle / \langle R_G^2 \rangle_a$ and $\langle R^2 \rangle / \langle R_G^2 \rangle_a$ are shown in Fig. 8. The off-lattice MC simulations of Freire et al. [76] and the MD simulations of Grest et al. [58, 60] are presented. Experimental results from Table II are shown as solid circles. Even though both simulations were carried out for coarse-grained models, in which no local bending

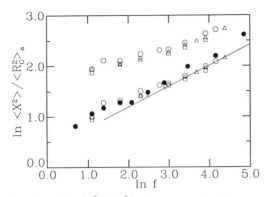

Figure 8. Radius of gyration $\langle R_G^2 \rangle / \langle R_G^2 \rangle_a$ (lower curve) and average squared end-to-end distance $\langle R^2 \rangle / \langle R_G^2 \rangle_a$ (upper curve) versus f for a good solvent. $\langle R_G^2 \rangle_a$ is the mean-squared radius of gyration for a single polymer chain. The solid circles are experimental data from Table II. The crosses are from the off-lattice MC simulations of Freire et al. [76] for $N = 49$ or 55. The other symbols are from MD simulations [58, 60] for monomers interacting with a purely repulsive Lennard-Jones interaction at $T = 1.2\varepsilon/k_B$ for (○) $N = 100$ and at $T = 4.0\varepsilon/k_B$ for $r_c = 2.5\sigma$ for $N = (\triangle)$ 50 and 100 (□). The solid line has slope of 0.41.

and torsion terms were included, the simulations describe the experimental results very well. As expected from the scaling theories, the global properties of the star are universal. The solid line in Fig. 8 indicates the predicted asymptotic power, 0.41. Note that the limiting scaling form is reached very early. Comparison of experimental results with renormalization group calculations are reviewed in Ref. 82. The density profile $\rho(r)$ scaled by $f^{0.65}$ is shown in Fig. 9 for four values of f ranging from 4 to 50 for $N = 100$ [60]. The measured slope is approximately 1.30 ± 0.03 for $f \geq 10$, in excellent agreement with the expected value 1.30. The rapid decay for large r is due to the finite chain length. As clearly seen, $\rho(r)$ scales with the number of arms f as predicted by scaling theory. This is in agreement with earlier simulations of Grest et al. [58] for $N = 50$ for a range of f and Batoulis and Kremer [67] for $3 \leq f \leq 6$. Using a self consistent field (SCF) theory, Dan and Tirrell [83] and Wijmans and Zhulina [84] found that $\rho(r)$ scaled as predicted by Eq. (25). As seen in Fig. 9, the scaling is valid even at very short distances from the center indicating that at least for these stars the simulations clearly exhibit the scaling predicted for a swollen star, and there is no need to consider the core regime. This core region is important for micellar stars [85] and for chains grafted onto a small colloidal particle [86].

·Because the interior of the star can become quite dense as f increases, one would expect that the free end of the chain is excluded from the core

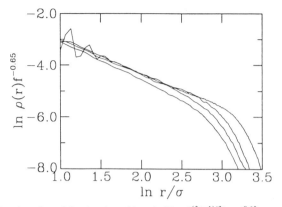

Figure 9. Log-log plot of the density $\rho(r)$ scaled by $f^{(3\nu-1)/2\nu} = f^{0.65}$ versus r/σ for stars with f values of 50, 20, 10, and 4 arms with $N = 100$ monomers per arm in a very good solvent. The larger f, the further the curves extend in r. These results are for $T = 1.2\varepsilon/k_B$ for a purely repulsive Lennard-Jones interaction truncated at $r_c = 2^{1/6}\sigma$. (From Ref. 60.)

region by simple steric effects. The expected width of the distribution $F(R)$ of center-to-end distances R can be estimated from the scaling model [8, 9], by considering the distance R of a chain confined in a narrow cone of opening $\theta = 2f^{-1/2}$, since a cone subtends a solid angle that is $1/f$ of the sphere. The end-to-end distance of such a chain fluctuates as through it were confined in a straight tube of diameter $R\theta$. The free energy of this chain, relative to an unconfined chain, is of order $k_B TR/(R\theta)$ or $k_B T$ per blob of size $R\theta$. Decreasing R to zero costs a further energy of this order. Thus the free energy associated with a small fluctuation ΔR of R away from its average value is of order $k_B TR/(R\theta)[\Delta R/R]^2$. Thus thermal fluctuations in ΔR, of energy $\sim k_B T$, are expected to be of order $\langle R\rangle\theta^{1/2} = 2^{1/2}\langle R\rangle f^{-1/4}$. For $f \gg 1$, $F(R)$ should approach a Gaussian shape of width $\Delta R \ll \langle R\rangle$. For small r, Ohno and Binder [49] found that in the scaling limit $F(R) \sim (R/\langle R\rangle)^{\theta(f)}$, where $\theta(f)$ is related to the exponent $\gamma(f)$. For large f, $\theta(f) \sim f^{1/2}$ in three dimensions [49], and $F(R)$ will appear to have an exclusion zone, although strictly speaking it does not. Li and Witten [87] used a variational approach to minimize the free energy with respect to the free end distribution and the stretching profiles of the polymer chains. Their results suggest a very large exclusion zone for the free ends and a different functional form for $F(R)$ for large f. The free ends would be restricted to the last 6% of the layer height and $F(R)$ would not be Gaussian. However, as seen in Fig. 10a, $F(R)$ for a 20- and a 50-arm star for $N = 100$ in a very good solvent are Gaussian as predicted by the scaling theory [60]. For smaller f, as shown in Fig. 10a for $f = 10$, there is

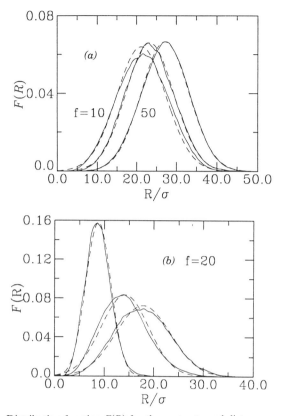

Figure 10.   Distribution function $F(R)$ for the center-to-end distances versus R for (a) a star with f values of 10, 20, and 50 arms in good solvent and (b) a star with $f = 20$ arms for T of $2.0\varepsilon/k_B$, $3.0\varepsilon/k_B$, and $4.0\varepsilon/k_B$ (from left to right) for $N = 100$ [60]. The results for (a) are for a purely repulsive Lennard-Jones potential at $T = 1.2\varepsilon/k_B$, while the results shown in (b) are for the potential truncated at $r_c = 2.5\sigma$. The solid line is the raw data, while the dashed line is a Gaussian with the same width and standard deviation as the data.

some deviation from the Gaussian, particularly for very small f, as the Gaussian overestimates the number of free ends near the origin. As the quality of the solvent decreases, $F(R)$ becomes somewhat nonsymmetric, with the center shifted to the center to that of the Gaussian [60]. This result is in agreement with the numerical SCF calculation of Wijmans and Zhulina [84] for polymers attached to the surface of a highly curved sphere.

While the center of the distribution $F(R)$ is clearly dependent on the number of arms, the absolute width depends only weakly on f [60]. The width depends mostly on the solvent quality and N. Thus the relative width decreases roughly as $1/\langle R \rangle \sim f^{-0.2}$ in the GSL. This behavior is consistent with the simple theoretical arguments [59], discussed above and with the variational calculation of Li and Witten [87]. They predict that the relative width of the end distribution does not decrease to zero but reaches a finite asymptote for large f and N of about 6%. Since the observed widths are much larger than this, stars with several times larger number of arms than those that have been used so far are necessary to approach the predicted relative width. It is also important to note that for a highly curved cylinder where they were able to also solve the SCF equations exactly, the size of the exclusion zone decreased significantly compared to the variational approach. In addition the variational approach appears to work less well as the dimension of the surface being tethered to decreases presumably because the strong stretching ansatz they use is not as applicable. Thus this lack of agreement is not too surprising and if the theory is correct, it applies only when the number of arms is very large. Finally, because many arms are attached to a central core, steric effects, which are not included in these calculations, must play an important role. Thus while the scaling theory is clearly an oversimplification, it describes the distribution of free ends very well for f in the range often studied experimentally.

From Fig. 3, we saw that at the high functionality limit, the 270-arm polymer did not follow the extrapolation from smaller f as expected. This was attributed to the fact that the 270-arm star had a more extended backbone than the other stars that were grafted to a small central core. To investigate this crossover in more detail, we present here some new MD simulation results comparing star polymers with f arms attached to a central core with the case where the arms are attached to a flexible backbone. The simulations are in a good solvent for monomers interacting with a Lennard-Jones interaction truncated at $r_c = 2.5\sigma$ [45, 60]. In Fig. 11, results for $\langle R_G^2 \rangle$ as a function of f for polymers with $N = 25$ and 50 monomers per arm. Results for a star polymer (solid circles) are compared to bottlebrush polymers with three different grafting densities. As expected, for small f the results are independent of the extension of the core, in agreement with the experimental results of Tsukahara et al. [46]. However, for larger f, $\langle R_G^2 \rangle$ increases significantly faster than a star. From Fig. 11, is it clear that the functionality f where the results deviate from each other clearly depends both on N and the branching density. These results are in qualitative agreement with the data for the 270-arm polymer shown in Fig. 3.

B. Θ and Poor Solvent

At the Θ point, the self-repulsion of the monomers, due to the excluded volume, is just compensated for by the interactions with the solvent.

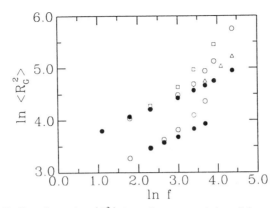

Figure 11. Radius of gyration $\langle R_G^2 \rangle$ for polymers consisting of f arms each of length $N = 50$ (upper set of points) and 25 attached to a ($\bullet$) central sphere or to a flexible line of density $\rho_l = (\square)$ 0.5, ($\bigcirc$) 1.0, and ($\triangle$) 2.0, where ρ_l is the number of branches per backbone monomer. The simulations were carried out under good solvent conditions at $T = 4.0\varepsilon/k_B$ [60].

While this tricritical point is well understood for linear chains, less is known about the properties of stars at T_Θ. Candau et al. [88] assumed that at T_Θ all the arms can interpenetrate each other completely. This means that all virial coefficients vanish and the size of the star is given by the Zimm–Stockmayer [37] equation, Eq. (8). While this is obviously an oversimplification, it turns out to work quite well as seen in Section III. As seen from Eq. (27), the Daoud and Cotton scaling argument gives $g_\Theta \sim f^{-1/2}$. However, since at T_Θ the second virial coefficient between a pair of arms vanishes, a third arm still gives a repulsive interaction [9]. Batoulis and Kremer [68] have argued that because of this effect, two, but not more than two, arms of the star can share a blob of diameter $\xi(r)$ at T_Θ, and one needs more arms than in a good solvent to observe the Daoud–Cotton scaling. They estimated that the RW results, Eq. (8), should cross over to the scaling result for $5 \le f \le 10$.

To study the importance of three body interactions on stars, Batoulis and Kremer [68] carried out high-precision MC simulations of linear and star polymers on the face-centered-cubic (fcc) lattice for $1 \le f \le 12$ and $N \le 900$ using the inverse restricted sampling method [89]. These simulations, extended earlier simulations on smaller stars by Mazur and McCrackin [64]. They found that T_Θ was independent of f for $N \to \infty$. This result was corroborated by Zifferer [77] for systems up to size $N = 3840$. Similar results were found experimentally for PI stars in dioxane [42] where the temperature at which the second virial coefficient A_2 vanishes for small f and N is lower than T_Θ for a corresponding linear chain but increases as N increased. In the limit of large M_W, T_Θ for the star is the same as for a linear chain.

Values for g_Θ obtained from simulations and experiment are compared to the RW result in Table VI. Lattice MC results [68], MD results on the bead–spring model [60], and experimental data from Table I are presented. Comparison of experimental and earlier MC results to renormalization group calculations are presented in Ref. 82. Since the Θ point is only known approximately for the off-lattice bead–spring model, the MD results are probably not as accurate as the lattice MC results of Ref. 68. Results from the off-lattice MC simulations of Freire et al. [76] for $f = 6$ and 12 also agree very well with the two simulations listed, while their results for $f = 18$ are somewhat too low. Zifferer [71] found $g_\Theta = 0.640$, 0.385, and 0.285 for $f = 4$, 8, and 12, respectively, using a pivot algorithm on the tetrahedral lattice. For small f, the RW result gives a reasonable approximation for g_Θ, though for larger f the RW result underestimates the size of the star. Note that all of these results are in contrast with those of Bruns and Carl [69] who interpret their MC results in terms of an f-dependent Θ point and, at least for $4 \leq f \leq 6$, at which g_Θ is within their numerical accuracy essentially the same as the RW result. In Fig. 12, $\langle R_G^2 \rangle / \langle R_G^2 \rangle_a$ is plotted versus f for the MC simulations of Batoulis and Kremer [68] and the MD simulations in which the monomers interact with a Lennard-Jones interaction, truncated at 2.5σ for $N = 50$ and 100 [60]. Experimental data from Table II are also shown in Fig. 12. Note that the experimental results agree very well with both the MC and MD ($N = 100$) results over the entire range of f. However, the slope (~ 0.32) is significantly less than the scaling prediction ($\frac{1}{2}$) for both the simulation

TABLE VI
Comparison of Simulation Results with Experiment for g_θ

f	RW	MC[a]	MD[b]	Experiment[c]
3	0.778	0.79	0.73	0.82
4	0.625	0.68	0.69	0.64
5	0.520	0.55	0.55	—
6	0.444	0.48	0.45	0.46
8	0.344	0.39	0.40	0.41
10	0.280	—	0.34	—
12	0.236	0.28	—	0.30
18	0.160	—	—	0.21
20	0.145	—	0.20	—

[a] Monte Carlo results of Batoulis and Kremer [68].
[b] Molecular dynamics results of Grest [60] for $T = T_\Theta = 3.0\varepsilon/k_B$ for $N = 100$.
[c] Results from Table I.

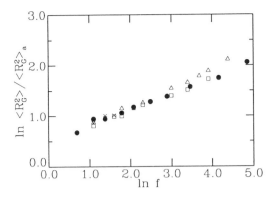

Figure 12. Radius of gyration $\langle R_G^2 \rangle / \langle R_G^2 \rangle_a$ versus f for a Θ solvent. The solid circles are experimental data from Table II. The crosses are from the MC simulations of Batoulis and Kremer [68] extrapolated to large N. The open triangles ($N = 50$) and squares ($N = 100$) are from MD simulations [60] in which nonbonded monomers interact with a Lennard-Jones interaction cutoff at $r_c = 2.5\sigma$ for $T = T_\Theta = 3.0\varepsilon/k_B$.

and experimental data, suggesting that scaling can only be reached for much larger values of f. This is in contrast to good solvent results were the data agree very well with the scaling predictions.

The monomer density $\rho(r)$ for a Θ solvent was found by Grest [60] to agree nicely with the scaling prediction, $f^{1/2}r^{-1}$. As in a good solvent the free ends are excluded from the center of the star. The distribution of free end for stars with $f = 20$ for a good, Θ, and poor solvent is shown in Fig. 10b. Note that the distributions are approximately symmetric, with a slight shift toward the center of the star. A Gaussian form is a pretty good fit, particularly for large f. However, for small f, it overestimates the number of free ends near the origin [60]. As the radius of curvature of the central core increases, the distributions are no longer described by a Gaussian.

C. Relaxation of Star Polymers

The scaling picture of a star can also be used to predict its dynamic relaxation processes [59]. There are at least three qualitative distinct relaxation processes for a star, which occur on different time scales and only weakly couple to each other. While all three of these mechanisms also occur for linear polymers, in a star they are easily separable.

First, the star relaxes via an overall shape fluctuation or elastic modes. This time scale is that of cooperative diffusion over the star size R. A second process is rotation diffusion of the object. For linear polymers, these first two have the same relaxation time, up to constants of order unity. However, for large f, the shape fluctuations relax progressively faster. Hence these two processes are expected to have the same dependence on N but a different dependence on f. The rotational diffusion is slower since it is not enhanced by the pressure within the star for large f, as the elastic modes are. The third process and by far the slowest is the disentangling of two or more interwined arms. Such a configuration can easily survive the shape or rotational relaxations. The larger f, the better one can distinguish these processes. If excluded volume interactions are ignored, as done by Zimm and Kilb [38] in the very first study of the dynamics of branched polymers, then these three different relaxation mechanisms could not have been distinguished.

The fastest process is the shape fluctuations, which can be measured directly by studying the fluctuations of the inertia tensor **M**, which is given by

$$M_{\alpha\beta} = \frac{1}{Nf} \sum_{i=1}^{fN} (\mathbf{r}_i - \mathbf{r}_{cm})_{\alpha} (\mathbf{r}_i - \mathbf{r}_{cm})_{\beta} \tag{32}$$

where $\mathbf{r}_{cm}$ is the position of the center of mass of the whole star. Defined this way, $\langle R_G^2 \rangle = \langle M_{xx} + M_{yy} + M_{zz} \rangle$. The typical shape fluctuations are then given by the time autocorrelation function of the elements of $\mathbf{M}, R_G^2, R_{Ga}^2$, or R^2 [58, 59, 74, 80]. For example, the autocorrelation function of $C_R(t)$ is defined by

$$C_R(t) = \frac{\langle R(t)R(0) \rangle - \langle R \rangle^2}{\langle R^2 \rangle - \langle R \rangle^2} \tag{33}$$

Results for $C_R(t)$ for stars in a good solvent are presented in Fig. 13 for a series of four stars with $N = 50$ [58]. The relaxation times, τ_{el} for the elastic modes, which are determined by the slope at long times on this semilog plot are essentially the same for all four values of f. Su et al. [74] found similar results for three- and four-arm stars using the bond fluctuation method. Ohno et al. [90] measured τ_{el} from the response of $\langle R_G^2 \rangle$ to a Kramers potential that describes the effect of shear flow in lowest order in the shear rate. Their MC simulation results using two different techniques also showed that τ_{el} decreased very slowly as f increased. This curious result that τ_{el} is nearly independent of f is in contrast with the independent strand model of Zimm and Kilb [38], which predicts that $\tau \propto f$ but can be explained using the scaling picture [58]. Consider the fluctuations in R, the center-to-end distance for a single arm. The fluctuations in the total length arise from independent fluctuations of order ξ within each blob. Thus R fluctuates by an amount of order $\xi(R/\xi)^2$. For a star, the largest length available is given by the size

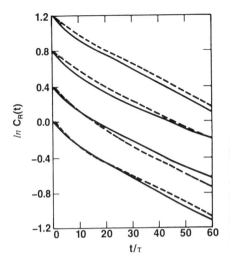

Figure 13. Autocorrelation function $C_R(t)$ for the end-to-center distance (solid lines) and single-arm radius of gyration (dashed lines) for four stars of arm length $N = 50$ with $f = 10, 20, 30$, and 50 arms in a good solvent. The curves have been displaced by 0.4 for clarity (f increases from bottom to top). Note that the relaxation times are essentially the same for all four values of f. (From Ref. 58.)

of the largest blob, which has a diameter $\xi_{max} \sim Rf^{-1/2}$. The bulk of the monomers are to be found in this largest blob [8, 9]. Thus the relative amplitude of the fluctuations in $\langle R^2 \rangle$ falls off as $\xi_{max}/R \sim f^{-1/2}$. The longest relaxation time τ_B of such a local fluctuation is given by the Rouse time of an isolated chain of $n_{B,max}$ monomers, $\tau_B \sim \xi_{max}^2 n_{B,max}$, where $n_{B,max} \sim \xi_{max}^{1/\nu}$. In the scaling picture, $n_{B,max} \sim Nf^{-1/2}$ and thus $\tau_B \sim (Nf^{-1/2})^{(1+2\nu)}$. This describes the initial stage of the local relaxation. In order to produce an overall shape fluctuation, a density fluctuation must diffuse a distance of the order of the diameter of the star, which is R. The diffusion constant for a semidilute solution is given by ξ^2 divided by the local relaxation time τ_B. Thus τ_{el} is of order

$$\tau_{el} \sim \tau_B \left(\frac{R}{\xi_{max}}\right)^2 \sim N^{1+2\nu} f^{1-(1+2\nu)/2} \tag{34}$$

In a good solvent, $\tau_{el} \sim N^{2.18} f^{-0.09}$, while at the T_Θ, $\tau_{el} \sim N^2 f^0$. Thus the dependence of τ_{el} on f is expected to be very weak, in agreement with the results shown in Fig. 13 for a good solvent for $N = 50$. Similar results for chains of length $N = 100$ were observed by Grest [60] for stars in both a good and a Θ solvent. The MC results of Ohno et al. [90] are in very excellent agreement with the predicted -0.09 dependence of exponent f in Eq. (34). Even for undiluted stars, Boese et al. [91] found using dielectric relaxation that the relaxation times of the center-to-end distance for *cis*-polyisoprene stars depended imperceptibly on f for $3 \leq f \leq 18$.

The next relaxation time is for the entire object to rotate or to move a distance comparable to its own diameter, since the same time is needed for the star to move its own distance or to make a complete rotation. For an assembly of Nf objects subjected to independent random forces as in the Rouse [92] model, the diffusion constant is given by $D \sim (Nf)^{-1}$. Within the diffusion time τ_D, the system moves a distance of about its own diameter:

$$\tau_D \sim \frac{R^2}{D} \sim NfN^{2\nu}f^{1-\nu} \sim N^{1+2\nu}f^{2-\nu} \qquad (35)$$

This time is smaller than for a linear chain of Nf monomers [55] but much larger than τ_{el}. Since D is very small, it is difficult to measure in a simulation and has not been done. The rotational diffusion time, which should be comparable to τ_D, can be ananalyzed by studying [59] the time autocorrelation function of the center-to-end vector $\mathbf{R}$ or the auto-correlation of the squares of the second-order spherical harmonics of the angles at which the principal axes of the ellipsoid are oriented with respect to a fixed-coordination system [74]. However, even for small f [74], there was no linear region in the semilog plots of $C(t)$ for the times that can presently be simulated, making it impossible at present to test Eq. (35).

Even when a star moves its own distance, the topological arrangement of two arms may be largely intact. A simple way to think about this is to imagine two interwined arms that can follow the shape fluctuations of the star without changing its topological character. Grest et al. [59] argued that this time should scale as

$$\tau_e \sim \exp(cf^{1/2}) \qquad (36)$$

where c is a constant. To measure this time, one needs to observe the relaxation of some quantity that survives both the shape fluctuations and rotation of the entire object but does not survive the disentangling fluctuations. One such sensitive quantity is the scalar product of two arbitrary center-to-end vectors: $\mathbf{R}_i \cdot \mathbf{R}_j$, where $\mathbf{R}_i$ is the vector from the vertex to the free end of the ith arm. This dot product is largest for nearby arms, which are the most likely to interwine. The product for two arms that were initially nearby relaxes to zero as these disentangle. Quantitively, one can define the entanglement correlation function $C_e(t)$ as

$$C_e(t) = \frac{1}{f(f-1)} \sum_{i=1}^{f} \sum_{j \neq i}^{f} \langle [\mathbf{R}_i(0) \cdot \mathbf{R}_j(0)][\mathbf{R}_i(t) \cdot \mathbf{R}_j(t)] \rangle \qquad (37)$$

This angular correlation can only decay after two arms that are strongly entangled have disentangled. Overall rotation of the star does not affect $C_e(t)$. Simulation results [59] for $C_e(t)$ clearly demonstrated a strong dependence of τ_e on f, but the decay was too slow to determine τ_e accurately.

The three relaxation times discussed above are for the Rouse [92] model, which applies to almost all the simulations on stars. However, experimentally hydrodynamic effects are important [93,94]. At present it is only possible to include solvent molecules explicitly in a simulation for very small stars ($Nf \sim 50$). Smit et al. [95] have studied a three-arm star with $N = 6$ in the presence of a solvent. While the introduction of hydrodynamic effects changes two of the relaxation times, there remain three distinct times for a star. In the Zimm [96] model, one can show [59] that $\tau_{el} \sim N^{3\nu} f^{(2-3\nu)/2}$ and $\tau_D \sim N^{3\nu} f^{3(1-\nu)/2}$. As in the Rouse case, τ_{el} and τ_D have the same N dependence but very different f dependencies. The prediction for τ_e remains unchanged. In a good solvent, $\tau_{el} \sim N^{1.77} f^{0.125}$ and $\tau_d \sim N^{1.77} f^{0.62}$. As studies of hydrodynamic effects on linear chains have only been feasible in the past few years [97, 98], it is not surprising that little has been done on simulating many arm stars with a solvent explicitly present. At present at least another order of magnitude in computer speed will be necessary for a serious study of the hydrodynamics of stars.

VI. SCATTERING METHODS

Experimentally, many properties of a star polymer can be found using scattering techniques. This is a useful way to determine not only the size of the star but also to learn about its internal structure. Light scattering [99, 100], small-angle x-ray scattering [101], and small-angle neutron scattering have been applied to examine structural aspects [93, 94, 100, 102–106] as well as some special features of star polymer interactions leading to ordering phenomena around the overlap concentration [106–108]. Quasi-elastic neutron scattering experiments using neutron spin echo (NSE) spectroscopy [13] have also been performed to study the collective and single-chain dynamics of the arms of the star polymers [93, 103].

Scattering techniques study the static structures of molecules and molecular organizations in the reciprocal space. For a given incident radiation of wavelength λ and scattering angle θ, the scattering intensity $I(\theta)$ is related to the spatial distribution of scattering density $\bar{\rho}(r)$ by [109]

$$I(\theta) = \int \bar{\rho}(\mathbf{r} - \mathbf{r}')\bar{\rho}(\mathbf{r}') \exp(\mathbf{Q} \cdot \mathbf{r}) \, d^3\mathbf{r} \, d^3\mathbf{r}' \tag{38}$$

Here $Q = 4\pi/\lambda \sin(\theta/2)$ is the scattering wave vector; $\bar{\rho}(\mathbf{r})$ is a material property that couples to the radiation. For light scattering, $\bar{\rho}(\mathbf{r})$ is the index of refraction (or the local polarizability) distribution function, for x-ray scattering, the electron density distribution function, and for neutron scattering, the nuclear scattering length density distribution. The choice of the radiation depends on the length scale and energy scale of interest: light scattering probes length scales in the 50-nm range and above, x-rays probes 0.1–50 nm; neutrons probe 0.01 nm to a few hundred nanometers. Modern instruments have achieved overlap in length scales for all three probes, but not yet in energy scales. However, current technology developments will allow the overlap of accessible energy (time) scales for quasi-elastic light scattering and NSE spectrometry in the near future.

Another difference between the various radiation sources is the intensity of the radiation. For instance, well-collimated monochromatic photons from a typical laser source can easily achieve a flux over 10^{22} photons $\text{cm}^{-2}\,\text{s}^{-1}$. Similar flux can also be generated in a synchrotron source for x-rays at much higher energies (kilo-electron-volts). On the other hand, the cold neutron flux is six or seven orders of magnitude lower even at the most powerful neutron source currently located in Grenoble, France. It should also be noted that the wavelength spread from a neutron source is generally much larger than that of light or x-ray: Typical values for $\Delta\lambda/\lambda$ from a neutron velocity selector are between 10 and 20%. However, since the absorption for neutron by the hydrocarbon systems is negligible, as is the case for most of the elements, this permits the usage of large volume samples as well as a broad range of choices for sample cells for high-temperature or high-pressure measurements. Furthermore, due to the largely different scattering properties of hydrogen and deuterium by partial deuteration, the scattering contrast for a neutron scattering experiment in general is orders of magnitude larger than for x-rays and light. Finally, radiation damage due to the cold neutrons is minimal, allowing long exposure times. Together these attributes serve to compensate for the low intensity of the neutron beams.

Theoretically, Benoit [110] in 1953 derived an analytic expression for the form factor $P(Q)$ for a star built from Gaussian chains. In Appendix A, we present a summary of this derivation. The influence of excluded volume interactions were incorporated by Alessandrini and Carignano [111]. A scaling form for $P(Q)$ within the scaling model of Daoud and Cotton [8] and Birshtein and Zhulina [9] was first discussed by Grest et al. [58]. As we will discuss below, the $I(Q)$ can be represented by the product $S_I(Q)P(Q)$. Here $S_I(Q)$ represents the interparticle structure factor determined by the interaction potential between the star molecules, while $P(Q)$ represents the form factor of the individual stars determined by the structure of the star. In a dilute solution, the interactions between stars can be neglected, $S_I(Q) = 1$ and the intensity $I(Q) \sim P(Q)$. Usually $P(Q)$ is determined by selectively labeling a small fraction of the stars so that $S_I(Q)$ can then be deduced from $I(Q)/P(Q)$. Note that in many papers on polymers the intra- and interparticle correlations are not separated and are simply referred to as the structure factor $S(Q)$, which is proportional to $I(Q)$.

A.   Small-Angle Neutron Scattering

Because neutron scattering is sensitive to isotope substitution (especially for the case of hydrogen and deuterium), it allows for labeling of specific part of the structure for study. This method, known as contrast variation, is unique to neutron scattering. As a result the small-angle neutron scattering (SANS) is found to be specially powerful for the study of hydrocarbon systems [112] such as polymers and micellar solutions. Even though all the scattering techniques share the same general principles, it is useful to develop the scattering formula for a particular probe. We shall chose SANS for the reasons just described.

The coherent scattering intensity $I(Q)$ can be defined as the coherent cross section, $d\Sigma(Q)/d\Omega$,

$$I(Q) = \frac{d\Sigma}{d\Omega}(Q) = \frac{\Delta\rho^2}{N_A} S(Q) \tag{39}$$

Here Ω is the solid angle into which the radiation is scattered and $S(Q)$ is the static scattering factor, and $\Delta\rho^2$ is the average contrast factor between the monomers and the solvent,

$$\Delta\rho^2 = \left(\frac{\Sigma b_P}{v_P} - \frac{\Sigma b_S}{v_S}\right)^2 \tag{40}$$

where Σb_P and Σb_S are the coherent scattering lengths, v_P and v_S the specific volumes of the monomer in the polymer and the solvent molecule, respectively. The coherent scattering length for hydrogen atom is -3.7×10^{-13} cm while the value for deuterium is 6.67×10^{-13} cm, so the scattering cross section (which is the square of the scattering length) of the solvents and parts of the polymer can be matched by judicial choice of selective deuterium hydrogen substitution. Because the scattering length of a hydrogen atom is negative (indicating the neutron wave changes phase upon scattering), it is possible to arrange for a partially deuterated polymer to be index matched to vacuum for neutron scattering.

For dilute polymer solutions, there is little interstar correlation, and the inverse scattering function can be expressed as [113]

$$\frac{\phi}{I(Q)} = \frac{1}{V_W P(Q)} + 2A_2\phi \tag{41}$$

Here V_W denotes the weight-average molar polymer volume and ϕ the monomer volume fraction. The magnitude of the second virial coefficient A_2 indicates the magnitude of deviations from van't Hoff's law for the osmotic pressure of ideal solutions. For a perfectly monodispersed system of stars, $V_W = Nfv_P$.

In order to determine the star form factor, $P(Q)$, it is important that the sample concentration be sufficiently dilute such that the interstar interaction term $A_2\phi$ can be neglected when compared with the first term on the right-hand side of Eq. (41). The higher the molecular weight (corresponding to a larger value of the V_W), the lower the concentration must be in order to satisfy this condition. A proper procedure to obtain the true form factor $P(Q)$ is to extrapolate the scattering curves to zero concentration via the classical Zimm plots [106].

Even though for most cases the single-chain form factor can be measured only at low concentrations (well below the overlap volume fraction, ϕ^*, a quantity we will define later), it is possible to obtain $P(Q)$ at finite concentrations using SANS by applying the experimental condition known as the zero-average contrast conditions [110, 114]. Under this condition, an equal volume blend of a deuterated and a protonated polymer is solubilized in a mixture of deuterated and protonated solvent having the same average scattering length density as that of the polymer blend:

$$\rho_0 = \frac{\phi_H \Sigma b_{S,H} + (1 - \phi_H)\Sigma b_{S,D}}{v_S} \tag{42}$$

where ϕ_H is the volume fraction of the protonated solvent. According to Eq. (39) the scattering cross section can be expressed as the sum of partial scattering functions:

$$\frac{d\Sigma}{d\Omega}(Q) = \frac{1}{N_A}(\Delta\rho_H^2 S_{HH} + \Delta\rho_D^2 S_{DD} + 2\Delta\rho_H \Delta\rho_D S_{DH}) \tag{43}$$

where the indices D and H refer to deuterated and protonated monomers, respectively.

For zero-average contrast conditions, $\Delta\rho_z = \Delta\rho_D = -\Delta\rho_H$, we can rewrite Eq. (43) as

$$\frac{d\Sigma}{d\Omega}(Q) = \frac{\Delta\rho_z^2}{N_A}(S_{HH} + S_{DD} + 2S_{DH}) \tag{44}$$

If we decompose of the partial scattering functions into a self (I) and a distinct (II) contribution according to

$$S_{ij} = S_i^I \delta_{ij} + S_{ij}^{II} \tag{45}$$

then assuming that the correlations between D chains, H chains, and H and D chains are all the same, the distinct contribution cancels. Assuming further that the form factors of both the H and D chains are identical, Eq. (45) simplifies to

$$\frac{d\Sigma}{d\Omega}(Q) = \frac{2\Delta\rho_z^2}{N_A} S_D^I \tag{46}$$

with S^I written in the normalized form

$$S_D^I = \frac{\phi}{2} V_w P(Q) \tag{47}$$

The coherent macroscopic scattering cross section is then directly related to the form factor at each concentration,

$$\frac{d\Sigma}{d\Omega}(Q) = \frac{\Delta\rho_z^2}{N_A} \phi V_w P(Q) \tag{48}$$

The importance of this scheme will become apparent when it is necessary to evaluate $P(Q)$ at moderate concentrations in order to evaluate interstar interactions by scattering techniques.

B.   Qualitative Features of the Form Factor

The scattering curves obtained from a star-branched polymer of high functionality is markedly different from that of a linear polymer. Figure 14 shows $P(Q)$ for a 128-arm PBd star in deuterated decane solvent [106]. Figure 15 shows the form factors for stars of 8, 18, 32, 64, and 128 arms. The intensities were scaled by the polymer volume and the number of monomers per arm, so all the high Q data coincide. As the functionality of the star increases, the center of the star becomes more crowded and the boundary of the star structure becomes more prominent.

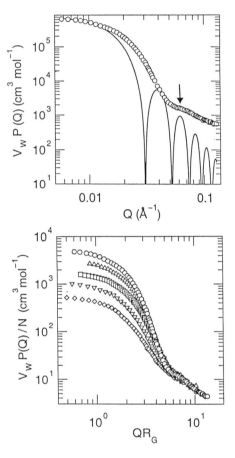

Figure 14. Form factor of the 128-arm PBd star (open circles) as compared to that of a hard sphere (solid curve) calculated by Eq. (50) [106].

Figure 15. Form factors obtained by SANS of stars of 8, 18, 32, 64, and 128 arms (from bottom to top). The data for 8- and 18-arm stars are for PI, and the larger arm stars are PBd. The intensities were scaled by the polymer volume and the number of monomers per arm. (From Ref. 106.)

The form factor reflects the changes in the structure and progressively deviates from that of the linear polymer.

This added structure in a star polymer compared to a linear chain comes from the fact that a star has three characteristic length scales: the radius R of the star, the correlation length ξ_{max} of the largest blob, and the monomer size σ. For $Q < R_G^{-1}$, the resolution of the probe is too coarse to distinguish any details of the star structure, the intensity is basically independent of Q, and the zero Q limit is determined by the star molecular weight. As QR_G approaches unity, there is still no detail structural features less than Q^{-1} that can be discerned. However, the interference effect between different parts of the star begins to set in and we enter the Guinier regime where the scattering is given by

$$P(Q) = \exp(-\tfrac{1}{3}Q^2 \langle R_G^2 \rangle) \qquad (49)$$

For $\xi_{\max}^{-1} < Q < \sigma^{-1}$, the scattering can be understood by covering the polymer with spheres of radius Q^{-1}. In the absence of strong correlations between the positions of the spheres, these scatter incoherently. The resulting scattering intensity $NfP(Q)$ is the number of spheres times the form factor of a single sphere. In the star the majority of these imaginary spheres are contained within blobs much larger than the spheres. The correlations between spheres are thus nearly the same as in a simple excluded volume of polymer. This gives a fractal scattering law $V_W P(Q) \sim Q^{-1/\nu}$, independent of f and N. This power law at high Q originates from the polymer self-avoiding random-walk structure within the blobs. Because of the limited number of data points one can measure in the appropriate Q range, an exact value cannot be obtained for ν. Power law fits at high-Q limit often yield higher values for ν than theoretically expected for a random polymer chain [45, 58, 93, 94]. This is due to the fact that in most previous studies, the chainlength of an arm are not large enough to see asymptotic behavior.

In between these two limits, $P(Q)$ must fall from about Nf to about $(\xi_{\max}/\sigma)^{1/\nu}$ or a factor of $f^{3/2}$. While the simplest way to connect these two regimes is by a power law decay Q^{-3}, the actual scattering is more complex [58]. In this regime, Q^{-1} is larger than the largest blob, and the polymer chain structure is invisible to the scattering. The dominant scattering is from the relatively sharp (see Fig. 14) outer boundary of the star, which gives rise to a faster decrease in the scattering envelope, Q^{-4}. For large f, oscillations superimposed on this Porod envelope should also be seen. It can be seen from Fig. 14 that the intensity dropped in a rather small Q range reflecting the fact that there exists a relatively sharp density cutoff in the neighborhood of $Q \approx R_G^{-1}$.

In the limit of very high functionality at finite arm molecular weight, the chains are expected to be so strongly stretched and tightly packed that the scattering curve will resemble that obtained from a solid sphere. A comparison of the scattering intensity obtained from a 128-arm PBd star and a hard sphere scattering curve (solid curve) is shown in Fig. 14. The sphere form factor was calculated according to

$$P(Q) = \left[\frac{3}{u_r^3} (\sin u_r - u_r \cos u_r) \right]^2 \tag{50}$$

where $u_r = QR$, and the sphere radius R was taken as $(\frac{5}{3})^{1/2} \langle R_G^2 \rangle^{1/2}$. The hard sphere form factor drops off beyond $Q \approx R_G^{-1}$ approaching Q^{-4} the Porod envelope for sharp interfaces. Figure 14 shows that $P(Q)$ of the star follows the sphere form factor covering about two of its oscillations before the scattering is dominated by the scattering from chains inside the blobs. It is also interesting to note that the form factors of the 128-arm stars exhibit a weak peak in the asymptotic range. The position of this peak coincides with the position of the third maximum in oscillation of the sphere form factor as marked by an arrow in Fig. 14. These features are absent in the scattering curves for stars with less than 18 arms as seen in Fig. 15. Thus we conclude that the high functionality stars assume a spherical shape with a relatively sharp density cutoff at the periphery of the star.

The scattering from star polymer melts is considerably simpler. Horton et al. [115] showed that the Benoit scattering function [110], Eq. (A13), gives a good fit to the SANS data for polyethylene (PE) star polymer melts. Figure 16 shows the least-square fit of the Benoit function to the experimental results, with $\langle R_G^2 \rangle$ and $P(0)$ as fitting parameters. The $\langle R_G^2 \rangle$ obtained from such fits show that the molecules are swollen with respect to a simple Gaussian conformation. Boothroyd and Ball [116] have shown that this data can be fit by a model without excluded volume interactions if one includes an impenetrable core that acts on all monomers. In a similar situation, Boothroyd et al. [117] has shown that the scattering for stars in a Θ solvent (PE star in toluene) were essentially similar to that of the observed in the melts, except that there was some additional swelling of the overall dimension.

The scattering curves for star polymers are often shown in a Kratky representation in which $I(Q)Q^2$ is plotted against Q as in Fig. 17. We find a peak in the curve near $Q \approx R_G^{-1}$, which becomes more pronounced with increasing f. This peak in a Kratky plot is a characteristic feature of star form factors that does not appear for linear polymers [93, 101, 103, 118]. From the peak position, Q_{max}, we can obtain the radii of gyration by using Benoit's approximation, Eq. (A13) in Appendix A, for a Gaussian star instead of the true form factor. From the first derivative of $Q^2 P(Q)$, the peak position was calculated to lie between $u_r = 1.1$ and $u_r = 1.3$

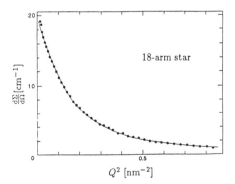

Figure 16. A least-square fit of the Benoit function (solid line) to SANS measurement for an 18-arm polyethylene star in the melt [115].

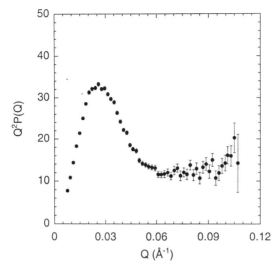

Figure 17.   Kratky plot of the form factor $P(Q)$ for an 18-arm PI star.

depending on functionality f [93]. Although this method is strictly valid only for a Θ solvent, the values obtained are close to those from the Zimm evaluation corrected according to Ullman [35, 119]. For stars for which the Zimm regime could not be accessed, it appears to be justified to take $\langle R_G^2 \rangle$ values from the Kratky evaluation [106].

Figure 18 shows a generalized Kratky plot of an 18-arm PI star form factor in decane, a good solvent [106]. The data are fitted by the Benoit function, Eq. (A13). The peak is well described by the fit. But significant deviations occur at higher Q, as expected. Here the theoretical curve

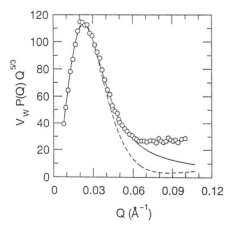

Figure 18.   Generalized Kratky representation of the form factor of an 18-arm PI star. The solid line represents a fit with the Gaussian model [Eq. (A13)]; the dashed line is fit to the renormalization group calculation [Eq. (51)]. (From Ref. 106.)

decreases continuously and does not follow the characteristic $Q^{-5/3}$ plateau since it does not take into account self-avoiding walk statistics for polymer chains in good solvents.

C. Theoretical Models for Form Factor

Excluded volume effects on the form factor has been included with renormalization group techniques by Alessandrini and Carignano [111]. They found a closed-form expression for $P(Q)$, which is quite complex. However, they have also determined a simplified formula for $P(Q)$ that can be used to fit to experimental data [111],

$$P(y) = f(y)\{1 + B_1[\exp(-B_2 y) - 1] + B_3 y + B_4 y^2\} \qquad (51)$$

Here

$$y = \frac{3}{4} \frac{fQ^2 \langle R_G^2 \rangle}{3f - 2}$$

and $f(y)$ is the Benoit function, Eq. (A13) with variable y. The parameter B_i depends on the functionality f. The dashed line in Fig. 16 represents a fit of an expression given by Eq. (51). For $f = 18$, $B_1 = 1.1$, $B_2 = 0.41$, $B_3 = 0.015$, and $B_4 = -2.0 \times 10^{-5}$. A comparison with the experimental data reveals an increasing deviation at $Q > Q_{max}$. The intensity of the theoretical curve decreases more rapidly and, thereby, overestimates the step between the peak maximum and the high Q plateau. Here the $Q^{-5/3}$ behavior is well established. However, the results show that both theories are limited in describing the experimentally obtained star form factors.

Using the scaling picture for a star, one can derive an approximate expression for the form factor. At length scales $r \sim \langle R_G^2 \rangle^{1/2}$ the star is described by its average monomer density distribution $\rho(r)$, while at scales $r < \xi_{max}$ the correlations within a single chain in a good solvent dominate. Thus, the overall pair correlation function may be written as

$$g(r) \sim \int \rho(r')\rho(r - r') d^3 r' \qquad r \sim \langle R_G^2 \rangle^{1/2}$$

$$g(r) \sim \sigma^{-1/\nu} r^{1/\nu - 3} \qquad r < \xi_{max} \qquad (52)$$

Recently Dozier et al. [105] presented an approximation to Eq. (52). In order to assure the correct behavior in the Guinier regime, they described the long-range correlations by a Gaussian giving rise to the proper radius of gyration. The short-range correlations between the monomer units were taken into account by the correlation function of a swollen chain including a cutoff function $\exp(-r/\xi)$, where ξ is an average blob size. Since both contributions are mainly on well-separated length scales, the total correlation function was written as the sum of two terms:

$$g(r) = c_1 \frac{\pi}{\langle R_G^2 \rangle^{1/2}} \exp\left(\frac{-3r^2}{4\langle R_G^2 \rangle}\right) + c_2 \exp\left(\frac{-r}{\xi}\right)\left(\frac{r}{\sigma}\right)^{1/\nu} r^{-3} \quad (53)$$

Here c_1 and c_2 are numerical constants. The scattering function is just taken as the Fourier transformation of the correlation function. From Eq. (53) we thus obtain

$$V_W P(Q) = V_W \exp\left(\frac{-Q^2 \langle R_G^2 \rangle}{3}\right) + \frac{4\pi\alpha}{Q\xi} \frac{\sin[\mu \tan^{-1}(Q\xi)]}{[1+(Q\xi)^2]^{\mu/2}} \Gamma(\mu) \quad (54)$$

where $\mu = 1/\nu - 1$, $\alpha \sim \upsilon_p(\xi/a)^{1/\nu} \sim \upsilon_p N f^{-1/2}$, and $\Gamma(\mu)$ is the gamma function. Note that according to Eq. (54) the ratio of $P(0)$ and the blob scattering obeys the $f^{3/2}$ law as discussed above. According to Dozier et al. [105] α, ξ, V_W, and $\langle R_G^2 \rangle$ are fitting parameters, and ν is set equal to the Flory value $\frac{3}{5}$. Using this value for ν one can obtain the best fits. We believe this might be an indication that the power law with $\nu = 0.67$ is partly due to the coherency effects arising from the superposition of scattering of different blobs. Since Eq. (54) contains no secondary maximum at higher values of Q, data points in the neighborhood of this feature were ignored so as not to be influenced by the secondary maximum in the experimental form factor for high functionality stars. The fits are generally very satisfactory. An example is displayed in Fig. 19 where the solid lines were calculated using Eq. (54) to fit the scattering data for stars with 8–128 arms [106].

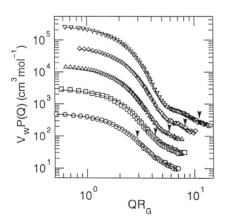

Figure 19. Comparison between the calculated form factor on a phenomenological approach [solid lines using Eq. (54)] and the measured form factors of stars with 8, 18, 32, 64, and 128 arms (offset by a multiplicative constant factor.) Arrows indicate the onset of the asymptotic regime according to $QR_G \sim f^{1/2}$ [106].

D. Comparison to Simulations

Another way to analyze the form factor is to compare the scattering data
with simulation results [58, 60]. Since these results are presented in terms
of $Q\sigma$, where σ is the segment length, we have scaled the simulation data
accordingly to compare to the SANS data. We estimated the segment
length from $\sigma = \sqrt{6}\langle R_G^2 \rangle^{1/2}/(nN)^{\nu}$, where n is the number of main-chain
bonds per monomer. For polybutadiene, $n = 3.86$. The intensity was
adjusted by multiplying a constant factor for all scattering curves such
that a comparison between simulated and measured stars of the same
number of arms becomes possible. The resulting plot is shown in Fig. 20.
In the simulations, $P(Q)$ is determined from Eq. (A9) for stars of length
$N = 100$ interacting with a purely repulsive Lennard-Jones interaction
[60]. Overall the agreement is remarkably good, particularly considering
that all four sets of data were scaled with the same segment length. For
large f, there is some deviations in detail. The difference mainly between
the secondary maximum for large f is probably due to polydispersity in
the experimental stars and also to the finite dispersion of the incident
wavelength. This is in contrast to the simulated stars in which the
dispersion of Q is zero. It is clear from this figure that the scattering
profile of the larger f experimental stars are smeared by a more fuzzy
edge than the simulated ones.

To get a closer look at the star structure, various partial structure

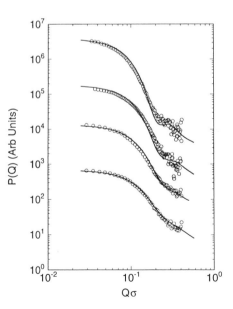

Figure 20. Form factor of stars of 8,
18, 32, and 64 arms (from the bottom to
top) as compared with the MD simulation results for stars of 8, 16, 30, and 64
arms each containing 100 repeat units; Q
is scaled by the segment length σ, which
was chosen to be 3.8 Å.

factors S_{11}, S_{22}, and S_{12} for a star with a deuterated core can also be
extracted from the three different contrasts conditions: one that matches
the shell (outer portion of the star), one that matches the core, and one
that matches the average scattering density of the whole star [94]. Here
S_{11} represents the partial structure factor given by the star core, S_{22}
represents the partial structure factor given by the star shell, and S_{12} is
the interference term. It turns out that these partial structure factors are
essential in our understanding of the dynamics of the star molecules.
Figure 21 shows S_{11} and S_{12} obtained for partially labeled PI stars in a
double-logarithmic plot [94]. The solid lines display the result of a MD
simulation [58] for a 10-arm star with 50 segments per arm, while the
dashed lines reproduce the Gaussian model proposed by Benoit and
Hadziioannou [120]. Qualitatively, the MD simulation result exhibits all
the essential features of the experimental data. In particular the intermediate intensity plateau of S_{22} is well reproduced; also the slightly
different asymptotic Q dependence of the two structure factors are
predicted. In comparison the Gaussian model (dashed lines) is virtually
featureless. However, quantitatively, the MD calculations and the experiment differ considerably. Most importantly this concerns the step height
of the steep low-Q increase of the partial structure factors. While the
experimental data exhibit roughly the same step height for both S_{11} and
S_{22}, the MD simulation predicts a larger step for S_{11} leading to an inverse
sequence of the intensity curves at high Q. Furthermore, the intermediate

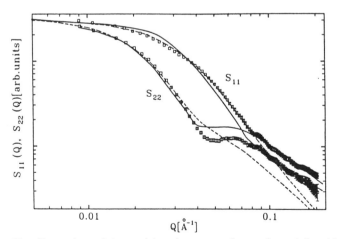

Figure 21. Comparison of the partial static structure factors S_{11} and S_{22} with the MD
simulation of a 10-arm 50-segment star (solid curve). The dashed lines represent the fits with
the Gaussian model [Eq. (A13)]. (From Ref. 106.)

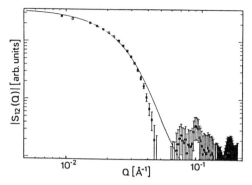

Figure 22. Partial structure factor S_{12} for a 12-arm PI star. The solid curve represents the fit with the Gaussian model. (From Ref. 94.)

plateau region of S_{22} is more narrow than predicted. Figure 22 displays the results for the interference term S_{12} in a double-logarithmic plot [94]. Since in the Q region $0.045 < Q < 0.06\,\text{Å}^{-1}$ S_{12} become negative, in Fig. 22 we plot the absolute values of S_{12}. Here the solid line reproduces the prediction of the Gaussian model. This quantity was not determined in the simulations. Again, we note that even though the Gaussian model fits are quite reasonable, the experimental data seem to exhibit considerably more structure than the Gaussian model predicts. The well-defined minimum at intermediate Q, where S_{12} becomes negative, indicates a well-developed core–shell structure of the star molecule.

Another interesting feature shows up when we compare the scattering curve of a fully labeled star to that of a center-labeled star. Figure 23 shows the structure factor [94] obtained from the fully labeled 12-arm PI star (total $M_W = 9.6 \times 10^4$) with the result of the MD simulation of a correspondingly labeled 10-arm 50-segment per arm star [58]. We again see the nearly perfect agreement. The radius of gyration of this star has been determined in the Guinier range using Eq. (49). In Fig. 23 we also show the structure factor of the star core, S_{11}, obtained from scattering patterns of a center-labeled PI star and scaled with the appropriate $\langle R_G^2 \rangle^{1/2}$'s to compare with the scattering curve of the fully labeled star. We see that the data sets do not match at higher Q values. The step height between the low-Q plateau and the asymptotic high-Q tail is different for the two cross sections. In addition the asymptotic Q dependence deviate from each other [$S_{11}(Q) \sim Q^{-1}$ for the star core structure and $S(Q) \sim Q^{-1.54}$ for the fully labeled star], indicating the chains are more stretched close to the star center. While the second result

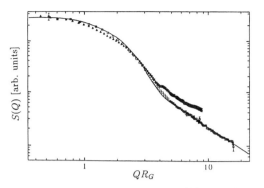

Figure 23. The structure factor obtained from fully ($\square$) labeled 12-arm PI stars (total $M_w = 9.6 \times 10^4$) as compared with the result of the MD simulation of a 10-arm 50-segment per arm star (solid line). The triangles represent the form factor S_{11} of core scattering from a center-labeled 12-arm star of the same molecular weight. (From Ref. 94.)

is not unexpected, the first may pose some questions on the scaling approach.

One can also study the properties of a single arm in the star by selective labeling. In Fig. 24 we display the SANS data for nominal one-arm labeled 12-arm PS stars. Assuming a Poisson distribution of the labeled arm among the stars with an average of one labeled arm per star, the scattering intensities should be dominated by stars with just one

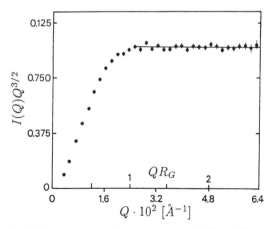

Figure 24. SANS data for nominally one-arm labeled 12-arm PS stars in a good solvent. (From Ref. 93.)

labeled arm. We demonstrate the asymptotic (large Q) behavior by plotting both $I(Q)Q^{3/2}$ versus Q. While the plot of $I(Q)Q^{3/2}$ versus Q achieves a high Q plateau, such a plateau is not reached for the scaling appropriate for a swollen coil, $I(Q)Q^{5/3}$ [93]. This excluded volume effect has been taken into account by a renormalization calculation to explain qualitatively the asymptotic behavior of the high-Q plateau for the single-arm static scattering form factor [111].

This apparently lower value for the inverse ν exponent (higher values for ν) indicates the fact that the star arms are more strongly stretched as compared to the linear polymer chain in a good solvent. This observation has been confirmed by MD simulations. Stretched-chain configurations with a Q scaling on the order of $Q^{-3/2}$ was also seen in the simulations.

For star polymers with the same number of arms but different arm molecular weights, the scattering curves appear to be all similar. In Fig. 25 the form factors of a series of $f = 32$-arm stars spanning a M_W range of about a factor of 7 are displayed [106]. Scaling the form factors obtained

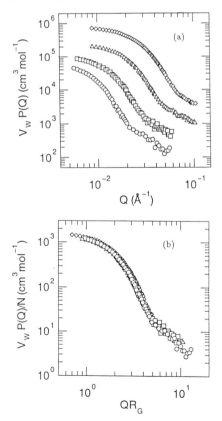

Figure 25. (a) Form factors for four 32-arm PBd stars with different molecular weights obtained by SANS [106]. (b) Same data scaled with R_G and M_W to show that the structure of the star polymer with the same functionality f but different M_W are self-similar. Arm molecular weights are ($\bigcirc$) 3.74×10^4, ($\square$) 1.94×10^4, ($\triangle$) 9.5×10^3, and ($\diamondsuit$) 5.4×10^3. (From Ref. 106.)

for different molecular weights with the corresponding values of $\langle R_G^2 \rangle^{1/2}$ and N allows the data set to collapse to a single master curve as shown in the lower panel of Fig. 25 [106]. Since the form factor is just the Fourier transform of the density distribution function, this scaling behavior implies that star polymers with different M_W but equal number of arms are self-similar in structure. Even though this behavior is expected for a linear polymer chain in solutions, it is by no means obvious that one would also find this scaling to be true for the highly functionalized star-branched chains. As the data in Fig. 25 shows that even for the lowest M_W star ($M_a = 5400$) no anomalies due to the relatively crowded center are visible.

VII.   CONCENTRATION DEPENDENCE

A.   Theory

In previous sections, we have considered the properties of star polymers in very dilute solutions. When the monomer concentration is increased, the same results should hold, except for correction terms due to finite concentrations, as long as the stars are far apart. However, for higher concentrations, above the overlap concentration c^*,

$$c^* \simeq \frac{Nf}{R^3} \tag{55}$$

the branches of different stars start to overlap and the scaling theory developed earlier has to be modified. In the semidilute range of concentration, $a^{-3} \gg c \gg c^*$, the properties of a star can be determined using scaling theory. Using the asymptotic expressions for the star radius, Eqs. (29)–(31), we find the following results for c^*,

$$c^* \simeq N^{1-3\nu} f^{(3\nu-1)/2} \left(\frac{v}{a^3}\right)^{-3(2\nu-1)} a^{-3} \qquad N \gg f^{1/2} \left(\frac{v}{a^3}\right)^{-2} \tag{56}$$

$$c^* \simeq N^{-1/2} f^{1/4} a^{-3} \qquad f^{1/2} \left(\frac{v}{a^3}\right)^{-2} \gg N \gg f^{1/2} \tag{57}$$

$$c^* \simeq a^{-3} \qquad f^{1/2} \gg N \tag{58}$$

Note that the last expression corresponds to the case when the stars reduce to their central core and do not overlap.

In the other two cases, the arms of the different stars interact. In the scaling picture, one imagines that the solution consists of a regime centered around the central core of the star where the star has the static properties of a single, star. Define the size of this region R_1. For distances larger than R_1, because of the overlap of the different stars, the interactions are screened and the size of the blob is controlled by the average concentration [8]. This concentration blob just corresponds to the size of the star blobs considered above a distance R_1 from the center of the star. In other words, the blobs are controlled by single-star conformations for $R < R_1$ and by the bulk concentration for distances $R > R_1$. Distance R_1 and the size of the concentration blob $\xi(c)$ can be determined by comparing the monomer concentration c to the local concentration $\rho(r)$ from center of a star.

Consider first the case where the branches are sufficiently long or the temperature sufficiently high $[N \gg f^{1/2}(v/a^3)^{-2}]$. Then the condition for the equality of the local and average concentration is given by

$$a^3 c \simeq \left(\frac{R_1}{a}\right)^{(1-3v)/v} f^{(3v-1)/2v} \left(\frac{v}{a^3}\right)^{(1-2v)/v}$$

Solving for R_1 gives

$$R_1 \simeq a(ca^3)^{v/(1-3v)} f^{1/2} \left(\frac{v}{a^3}\right)^{(2v-1)/(1-3v)} \tag{59}$$

For $v = 0.588$, $R_1 \simeq a(ca^3)^{-0.77} f^{1/2} v^{-0.23}$. The size of the concentration blob equals to that of a star blob for $r \sim R_1$,

$$\xi(c) \simeq R_1 f^{-1/2} \tag{60}$$

This gives $\xi(c) \simeq a(ca^3)^{-0.77}(v/a^3)^{-0.23}$. Note that this is the same law as for a solution of linear chains with the same concentration [55]. However, there is a basic difference between a solution of linear chains and stars, namely the extra length scale R_1. For distances less than R_1 from the center the star has a single star behavior. For larger distances the two solutions look the same. The above arguments are readily generalized to the case of a Θ solvent by using Eq. (57) for c^*. One can get the same result by setting $v = \frac{1}{2}$ in Eq. (59), $R_1 \simeq a(ca^3)^{-1} f^{1/2}$ and $\xi(c) \simeq a(ca^3)^{-1}$. Because a star in the semidilute consists of two regimes, scaling for the radius of the star is not expected to hold [8].

When f is large, Witten et al. [10] showed that for a solution of star polymers osmotic pressure Π increases rapidly in the vicinity of c^*. This crossover in the concentration dependence of Π leads to a macrocrystal ordering and a subsequent peak in the scattering intensity $I(Q)$ near $qR \sim 1$ for c in the vicinity of c^*. Consider a solution of star polymers in a very good solvent ($v = a^3$). At low concentrations, the dilute solution of stars has an osmotic pressure Π of $k_B T$ per star,

$$\Pi = \frac{k_B T c}{Nf} \qquad (61)$$

At $c^* \simeq NfR^{-3}$, Eq. (55), the osmotic pressure is of order $\Pi \simeq k_B T R^{-3}$. As discussed above, for concentrations far above c^* the solution of stars resemble an ordinary semidilute polymer solution. The bulk of the system is a uniform semidilute solution whose correlation length $\xi(c)$ is given by Eq. (60). The osmotic pressure in this regime is given by

$$\Pi \simeq k_B T \xi(c)^{-3} \qquad (62)$$

that is $k_B T$ per blob. This formula is valid for concentrations as low as c^*. When $c \simeq c^*$, $R_1 \simeq R[\xi(c^*) \simeq Rf^{-1/2}]$, the stars no longer overlap, the uniform concentration region disappears, and the above argument leading to Π is no longer valid. Thus the dilute and semidilute formulas for Π show a discrepancy at $c \simeq c^*$. The former gives $k_B T R^{-3}$, while the latter gives $k_B T \xi(c)^{-3} = k_B T R^{-3} f^{3/2}$. In order to pass from the dilute regime to the semidilute one, Witten et al. [10] point out that one must supply sufficient pressure to overcome the intrinsic osmotic pressure within each star. This osmotic pressure must increase by an amount of order $f^{3/2}$ as c increases by a factor of order unity, as shown in Fig. 26a. As f increases, so does the size of the jump.

This jump in Π implies a sharp decrease in the osmotic compressibility with increasing c in the same region. Thus at c^*, the "quasi-hard sphere"-like resistance to compressibility $(d\Pi/dc)^{-1}$ gives rise to a weaker neutron scattering amplitude in the forward direction and a peak in the total neutron scattering intensity $I(Q)$ for $Q \simeq R^{-1}$. For $Q = 0$, the scattering per monomer $I(0)$ is simply related to the osmotic compressibility $[c(\partial\Pi/\partial c)]^{-1}$ by $I(0) = k_B T (\partial\Pi/\partial c)^{-1}$. In the dilute limit, $\partial\Pi/\partial c \simeq \Pi/c = k_B T (Nf)^{-1}$. Thus for $Q \lesssim R^{-1}$, $I(Q)$ is of order Nf. Near c^*, Π increases by a factor of order $f^{3/2}$ as c increases by a factor of order unity, as seen above. Thus $\partial\Pi/\partial c \simeq \Pi/c^* f^{3/2} = k_B T (Nf)^{-1} f^{3/2}$. Accordingly $I(0)$ decreases by a factor of order $f^{3/2}$ as c increases near c^*. This decrease in $I(Q)$ occurs only for sufficiently small Q. For $Q \gg R^{-1}$, $I(Q)$ is virtually unaffected by concentration and should decrease with increasing Q. Thus $I(Q \simeq 1)$ is predicted by the scaling theory to be $f^{3/2}$ times larger than $I(0)$ for c near c^*, as illustrated in Fig. 26b.

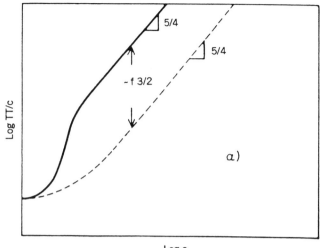

(a)

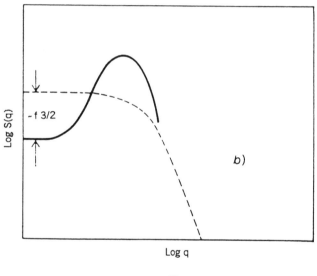

(b)

Figure 26. Predicted behavior of a star polymer solution. (a) Scaled osmotic pressure Π/c versus scaled concentration. Dashed curve is for linear polymers; solid curve is for stars. (b) Scattering intensity $I(q)$ versus q. Dashed curve is for the dilute limit; solid curve, $c \simeq c^*$. (From Ref. 10.)

B. Experiment

For star solutions at higher concentrations for which the interstar interaction cannot be taken into account by the second virial coefficient A_2 as in Eq. (39), we write the scattering intensity as

$$\frac{I(Q)}{\phi} = V_W P(Q) S_I(Q) = \frac{1}{\phi} \frac{N_A}{\Delta \rho^2} \frac{d\Sigma}{d\Omega}(Q) \tag{63}$$

relating the measured intensity to the form factor $P(Q)$ and the structure factor $S_I(Q)$ of the scattering system. Thereby, $S_I(Q)$ denotes the structure factor arising from the interstar interactions, which cause structure formation. Equation (63) presupposes that effects from the single-star configuration and the interstar structure can be well separated. It is expected that Eq. (63) is valid without complications below the overlap concentration ϕ^*. Beyond ϕ^* the stars begin to interpenetrate each other, and thereby the identity of each species is successively lost. In that concentration regime a straightforward application of Eq. (63) will become problematic.

Furthermore, in order to extract $S_I(Q)$ from Eq. (63) the concentration dependence of $P(Q)$ has to be known. Experimentally, this information can be gathered from a study of a star solution under zero-average contrast conditions [see Eqs. (42)–(48)]. As an example for such an experiment we display results taken from a mixture of identical hydrogenated and deuterated 18-arm PI stars. These stars were dissolved in a mixture of hydrogenated and deuterated methyl-cyclohexane. The zero-average contrast point was determined by contrast variation looking for the minimum of the scattered intensity. Under these conditions the experiments were performed [106].

As the volume fraction of the star polymer increases beyond the overlap volume fraction, ϕ^*, the outer rim of the stars will interpenetrate and form a semidilute solution identical to that of a linear polymer. However, near the star center, the structure of the star is preserved [8, 9]. Dozier et al. [105] have measured the form factors of a 0.05% labeled 12-arm PS star of 1.25×10^4 arm molecular weight in the presence of unlabeled identical PS stars at various volume fractions in a contrast-matched solvent. They found, as shown in Fig. 27, that the form factors apparently did not change very much in the range of total star volume fraction from $\phi = 0.01$ to $\phi = 0.10$. The highest values of ϕ were a factor of 2 beyond the nominal overlap volume fractions for this molecular weight. We expect that the form factor will eventually change at progressively higher values of ϕ. However, for ϕ twice that of ϕ^*, only

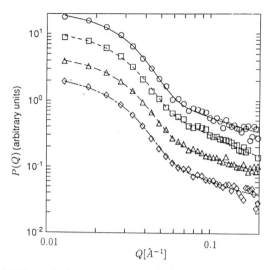

Figure 27. SANS result from constant matched blend solutions for four total star volume fractions: (○) $\phi = 0.01$, (□) 0.015, (△) 0.025, and (◇) 0.10. The volume fraction of the labeled star polymer is held at 0.005. (From Ref. 105.)

the outer quarter of the arm segments are expected to overlap between different stars, and the configuration of the labeled stars would not be expected to change much. Thus it is not surprising that for low-molecular-weight stars where ϕ^* is relatively large, as in the case studied by Dozier et al. [105] it was indeed observed that the form factors were virtually the same for all values of ϕ studied.

According to the scaling picture [8, 9], a semidilute solution of unlabeled star polymers resembles that of linear polymer solution containing regions of starlike coronas. For volume fractions ϕ much higher than ϕ^*, the ϕ dependence found for linear polymers [121] is expected to also hold for star polymers,

$$\langle R_G^2 \rangle^{1/2} \sim aN^{1/2} \left(\frac{\phi}{\phi^*} \right)^{-(\nu-1/2)/(3\nu-1)} \tag{64}$$

yielding $\langle R_G^2 \rangle^{1/2} \sim aN^{1/2}(\phi/\phi^*)^{1/8}$ for $\nu = \frac{3}{5}$. For ϕ close to ϕ^*, star structures are basically unaltered as discussed above. The scattering behavior is dominated by single stars. In this case the model gives a $\phi^{-3/4}$ dependence for $\langle R_G^2 \rangle^{1/2}$. To demonstrate the influence of ϕ on $\langle R_G^2 \rangle^{1/2}$ in Fig. 28, we plot $\langle R_G^2 \rangle^{1/2}$ for an 18-arm star as a function of ϕ [122]. While the size of the starlike region is constant up to a $\phi = 0.10$, which is

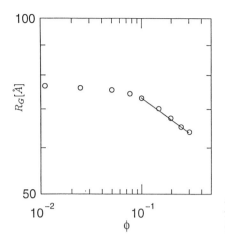

Figure 28. Radius of gyration of $f = 18$-arm stars as a function of monomer volume fraction ϕ. The solid line has a slope of $-\frac{1}{8}$. (From Ref. 106.)

ϕ^* [105], the shrinkage of the starlike region is clearly observed at higher ϕ. In accord with the scaling behavior for linear polymers, the slope in this semidilute region is -0.125. However, the plot does not reveal any sign of the $\phi^{-3/4}$ dependence in this volume fraction range.

Figure 29 presents Q-dependent intensities normalized to the volume fraction for 18-arm PI stars for different concentrations [108]. In this case, all of the stars are labeled. With increasing concentration at intermediate Q values, a peak develops indicating preferred interstar distances in this concentration range. The observation of such a peak is typical for all studies of star solutions at intermediate concentrations and shows directly the presence of ordering in such solutions. At low Q the intensity drops sharply, signifying the increasing osmotic pressure Π in the solution $[I^{-1}(0) \propto d\Pi/d\phi]$. At high Q the measured intensities at all concentrations approximately follow the same Q dependence resulting from the inside of the concentration blobs within the star. The SANS pattern for

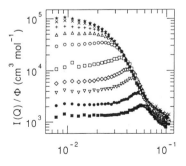

Figure 29. Normalized intensities $I(Q)$ versus Q for $f = 18$-arm PI stars for 10 values of the total star volume fraction ϕ (from top to bottom: $\phi =$ 0.002, 0.004, 0.01, 0.015, 0.025, 0.049, 0.074, 0.099, 0.146, and 0.192). (From Ref. 108.)

stars with higher functionality exhibit even more pronounced peak structures reflecting thereby the higher degree of order.

The structure factors were evaluated on the basis of Eq. (63); $P(Q, \phi)$ was approximated by the zero concentration form factors $P(Q)$, which is a good approximation below ϕ^* while beyond ϕ^* the shrinking of the star may affect the evaluated $S_I(Q)$. It was argued, however, that since the form factors are smooth the shrinking effects should mainly cause errors in the normalization of $S_I(Q)$ but not distort it strongly. Figure 30 displays such evaluated structure factors for the 64-arm star solutions [122]. From low to high concentration we observe the gradual buildup of the peak structure culminating at 0.103 where it nearly reaches 3. Calculating the overlap concentration from $\phi^* = V_W/(\frac{4}{3}\pi \langle R_G^2 \rangle^{3/2})$, a value of 0.11 is obtained. We note that beyond the first maximum no further peak structure develops. For higher concentration the height of the first peak in $S_I(Q)$ diminishes, seemingly in agreement with the prediction of Witten et al. [10] that beyond ϕ^* the mutual star interpenetration should diminish the ordering effects. However, one has to be aware that due to star interpenetration the basis of Eq. (63), namely the clear separation of form factor and structure factor contributions, is not fulfilled any more.

Figure 31 displays the structure factor $S_I(Q)$ for stars with different functionalities as a function of QR_G always at the overlap concentration

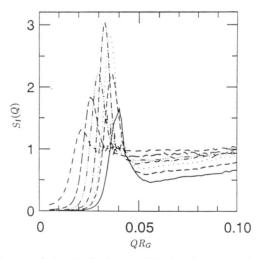

Figure 30. Structure factor for $f = 64$-arm PBd stars for seven values of the volume fraction ϕ (from left to right, $\phi = 0.028, 0.053, 0.078, 0.103, 0.13, 0.15,$ and 0.202). (From Ref. 122.)

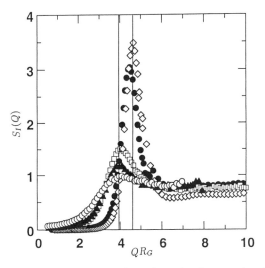

Figure 31. Structure factor at the overlap volume fraction as a function of $Q\langle R_G^2\rangle^{1/2}$ for ($\diamondsuit$) $f = 128$, ($\bullet$) 64, ($\square$) 32, ($\blacktriangle$) 18, and ($\bigcirc$) 8-arm stars. (From Ref. 122.)

ϕ^* demonstrating the buildup of structure with functionality [123]. While for the eight-arm star the peak in the structure factor is barely developed, with increasing f the peak height increases and the peak sharpens. Furthermore, the maxima of $S_I(Q)$ shift with increasing functionality from lower to higher QR_G. Thereby, stars with f values of 8 and 18 arms and the 64- and 128-arm stars appear to form two different groups with the $f = 32$ star in between. As we shall see this effect distinguishes liquidlike structure formation at low functionalities from crystalline star structures at high functionality.

In the liquid regime at f values of 8 and 18, the system can be described in terms of a colloidal approach, where the interparticle structure factor $S_I(Q)$ reflecting the pair correlation function is related to the interstar potential. Such calculations are performed on the basis of statistical mechanical procedures and in general require many approximations. The structure factors for the 18-arm stars were investigated in terms of two commonly applied models: (1) the hard-sphere approach in terms of the Percus–Yevick [124] approximation and (2) a soft Yukawa potential using the mean spherical approximation (MSA) by Hayter and Penfold [125]. The hard-sphere model is determined by one parameter, namely the effective diameter of the hard-sphere R_h. For interparticle distances larger than $2R_h$ the interaction energy is zero, while for distances shorter than $2R_h$ it becomes infinite. The soft potential Yukawa approach adds to the hard core a screened Coulomb tail,

$$V(r) = k_B T \frac{2\gamma R_h}{r} e^{-r/\lambda} \tag{65}$$

where γR_h describes the strength of the potential and λ the range of the soft potential interaction.

While the hard-sphere potential gives unsatisfying results below $\phi = \phi^*$, the soft Yukawa potential leads to a good description of the interparticle structure factors. Figure 32 presents a fit with the Yukawa potential to the $f = 18$ structure factors. We note that below ϕ^* the shown agreement can be obtained fixing both the screening length and the

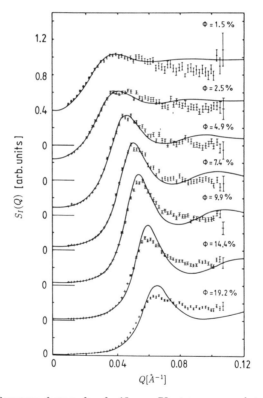

Figure 32. Structure factors for $f = 18$-arm PI stars compared to the calculated structure factor on the basis of a soft Yukawa potential using the mean spherical approximation according to Ref. 125. (From Ref. 122.)

potential strength. In each case the monomer concentration was kept to the experimental value. For the screening length a value of $\lambda = 25$ Å was found, which relates well to the outer blob diameter of $\xi \propto \langle R_G^2 \rangle^{1/2} f^{-1/2} \sim 20$ Å. Thus, the interaction range of the interacting stars correlates with the outer blob size. From the potential strength the interaction energy at $r = 2R_G$ can be calculated resulting in $V(2R_G) = 0.3k_BT$, which is of the order of the energy per blob. Thus, approaching the overlap concentration the soft potential model can consistently describe the interstar interactions revealing the appropriate structure factors. On the other hand, for concentrations larger than ϕ^* the Yukawa description breaks down, because it cannot handle the interstar interpenetration.

At higher functionality the colloidal approach for a liquidlike structure does not bear results. Using the Lindemann criterion, Witten et al. [10] have estimated that at a critical functionality of the order of $f_c \sim 100$ a transition to crystalline order should occur. It is also known that classical colloidal systems with a repulsive two-body interaction become crystalline, if the density is high enough. Furthermore, on the basis of computer simulations on colloidal systems with van der Waals interaction, Hansen and Verlet [126] conjectured that if the height of the first structure factor peak surpasses 2.85, a solidlike structure prevails. This criterion is clearly fulfilled for the 128-arm star, where $S_I(Q_{\max}) = 3.6$. But considering resolution smearing—the spectra in Fig. 31 have not been deconvoluted—the 64-arm star also falls into this category. Finally, the shift in $S_I(Q_{\max})$ from a lower value in QR_G for $f = 8, 18$ to a higher value for $f = 64, 128$ indicates a change of the type of ordering. Finally, recent shear experiments on 128-arm stars facilitated the orientation of the macrocrystalline grains and revealed a body-centered-cubic- (bbc) type structure [127].

As eluded to above, at low $Q(Q \sim 0)$ the scattered intensity $I(0)$ is inversely proportional to $d\Pi/d\phi$ or proportional to the osmotic compressibility $(\phi d\Pi/d\phi)^{-1}$, which is predicted to exhibit an anomalous decrease around $\phi \simeq \phi^*$. On the other hand at high concentrations the osmotic pressure of the interpenetrating star solutions should be indistinguishable from a semidilute solution of linear chains of the same concentration. This is demonstrated well by Fig. 33 displaying the osmotic pressure gradient obtained from light scattering at a series of PI stars with 8 and 18 arms and a corresponding linear PI chain [107]. As can be seen at high concentrations $\phi > 10^{-2}$, the osmotic pressure gradients for all samples collapse to a single curve determined by the osmotic pressure of the semidilute polymer solution. In this limit star and linear chain solutions cannot be distinguish.

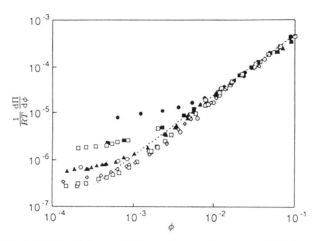

Figure 33. Variation of the gradient of the osmotic pressure $\partial\Pi/\partial\phi$ versus volume fraction ϕ for 8- and 18-arm stars and linear PI in semidilute solution. The dashed line corresponds to the first-order linear fit of the high-volume fraction part of the data. The slope is 1.30. Results are shown for ($\square$) linear chains ($M_W = 5.8 \times 10^5$), ($\bullet$) 8-arm stars $M_W = 1.3 \times 10^5$, ($\blacksquare$) 7.95×10^5, ($\blacktriangle$) 1.76×10^6, and 18-arm stars ($\bigcirc$) $M_W = 1.5 \times 10^6$, ($\diamondsuit$) 3.57×10^6, ($\square$) 6.8×10^6. (From Ref. 107.)

The predicted jump in the osmotic pressure gradient can be visualized qualitatively using the low-Q SANS data from the star solutions at different concentrations and functionalities. Figure 34 displays $d\Pi/d\phi$ as a function of ϕ/ϕ^* [123]. Thereby, the overlap concentration ϕ^* was taken from the measured second virial coefficient $\phi^* = (A_2 M_W)^{-1}$ and

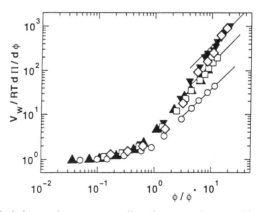

Figure 34. Scaled osmotic pressure gradient for star polymers with the functionalities ($\bigcirc$) $f = 8$, ($\blacktriangle$) $f = 18$, ($\square$) $f = 32$, ($\diamondsuit$) $f = 64$, and ($\blacktriangledown$) $f = 128$. The solid lines correspond to the expected scaling behavior of a semidilute polymer solution, $\phi^{1/(3\nu - 1)}$. (From Ref. 123.)

the pressure gradient was scaled by the molecular volume in order to bring together all curves at low ϕ/ϕ^*. While at low concentrations $d\Pi/d\phi$ remains nearly constant as expected, above ϕ^* a sudden increase of the osmotic pressure gradient is observed. At high concentrations the solid lines indicate the expected scaling behavior of a semidilute polymer solution $[(d\Pi/d\phi) \propto \phi^{1/(3\nu-1)}]$. According to theory, the jump in the osmotic pressure gradient should be proportional to $f^{3/2}$. While Fig. 34 clearly displays an increasing jump height of the osmotic pressure gradient with increasing functionality, quantitatively the $f^{3/2}$ dependence could not be verified. In particular, for the high functionality stars the observed increase of $d\Pi/d\phi$ appears to be too low. Possibly for these stars the conditions of a semidilute solution were not sufficiently reached. However Roovers et al. [128] have recently observed a steep rise in the osmotic modulus for PBd stars near the overlap concentration. Their results agree with the $f^{3/2}$ dependence predicted by Witten et al. [10] for the size of the jump in the osmotic pressure.

McConnell et al. [85] have investigated the ordering phenomena of concentrated diblock polymeric micellar aggregates in selective solvents. These spherical aggregates form very monodispersed structures. The aggregation number is about 100 or higher, resembling a high functionality star polymer. In this case, the solvent, decane, serves as a preferential solvent for PI. The micelles are comprised of a highly concentrated PS core surrounded by a diffuse corona of PI. McConnell et al. have found that for these "long-hair" micelles with large corona-thickness to core-radius ratios, the solution contained well-defined bcc crystalline phase as shown in Fig. 35. The ordering is similar to that found in concentrated star solutions. However, for the "short-hair" micelles, the prevailing structure is fcc as shown in Fig. 36. As expected for power law and Yukawa systems, as the length scale of the repulsion decreases, the disorder–order transition occurs at higher concentrations.

The order–disorder transition in starblock polymers has been investigated. These are star polymers in which each arm is an identical diblock copolymer. While many earlier reports [129] suggested that starblock polymers have an ordered bicontinuous double diamond structure, more recent work suggests that the bicontinuous phase has cubic symmetry [130]. For starblock polymers, the number of arms has an effect on the ordering transition, but the transition is symmetrical with respect to the composition [129, 131]. This differs from the behavior of mikto stars [12], in which the arms are made of distinct polymeric species. Well-characterized samples for this latter case have only recently be synthesized. While there has not been much work on these systems, what has been done [132, 133] suggests that this is an interesting case to explore in the future.

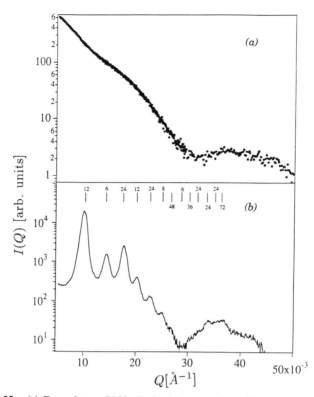

Figure 35. (a) Form factor $P(Q)$ obtained by x-ray for a PS–PI diblock micelle star with molecular weight of approximately 4.5×10^4 for each block in decane at 0.5 wt %. (b) Diffraction pattern for the same micelles at 12.5 wt %. The structure is bcc. The marks above the intensity profile indicate the location of the Bragg peaks for a bcc lattice with a lattice constant of 872 Å. (From Ref. 85.)

VIII. INTERNAL DYNAMICS OF STAR POLYMERS

In this part of the review we discuss the internal dynamics of star polymers, that is, the star relaxation in solution on length scales smaller than the size of the star. Such motional processes can uniquely be accessed by quasi-elastic neutron scattering and in particular by the NSE method. In the following we briefly outline the NSE method, display some basic features of quasi-elastic neutron scattering on polymer solutions, and thereafter present results on the collective relaxation and single-arm relaxation of the polymer star. Finally we discuss partial dynamic structure factors of the outer star corona and the star core in

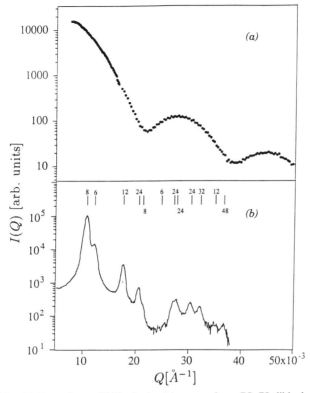

Figure 36. (a) Form factor $P(Q)$ obtained by x-ray for a PS–PI diblock micelle star with molecular weights of 4.6×10^4 and 2.0×10^4, respectively, in decane at 0.5 wt %. (b) Diffraction pattern for the same micelles at 22 wt %. The marks above the intensity profile indicate the locations of Bragg peaks for an fcc lattice with a lattice constant of 1007 Å. The structure is fcc. (From Ref. 85.)

terms of the random-phase approximation (RPA) and compare the theoretical approach with experimental results.

A. Neutron Spin Echo Method

Other than x-rays, where molecular or atomic resolution can be only achieved with kilo-electron-volt photons, as a consequence of their mass, neutrons with a de Broglie wavelength of angstroms possess thermal energies. For this reason neutrons are unique probes to access simultaneously the spatial and temporal behavior of condensed matter. While an x-ray photon reveals the position of a molecule, scattering from the neutron tells us where the atom is situated and in what direction and how fast it moves. In order to obtain information about the motional properties not only the direction of the scattered neutron but also its energy change during scattering has to be determined. This leads to the so-called double differential cross section, which for coherent scattering is

$$\frac{\partial^2 \Sigma}{\partial \Omega \, \partial \omega} = \frac{k_f}{k_i} \frac{\Delta \rho^2}{N_A} S(Q, \omega)$$ (66)

where k_f and k_i are the final and initial neutron wave vectors and $S(Q, \omega)$ measures the space–time Fourier transform of the density fluctuations in the system. This allows one to access the collective processes,

$$S(Q, \omega) = \frac{1}{2\pi\hbar(Nf)^2} \int_{-\infty}^{+\infty} e^{-i\omega t} \, dt \langle \bar{\rho}_Q(t) \bar{\rho}_{-Q}(0) \rangle$$

$$\bar{\rho}_Q(t) = \sum_j \exp[-i\mathbf{Q} \cdot \mathbf{r}_j(t)]$$ (67)

where $\bar{\rho}_Q$ is the Fourier component of the scattering density at wave vector Q. Here $\mathbf{r}_j(t)$ is the position of scatterer j at time t.

In conventional inelastic scattering techniques, the experiment takes place in two steps; first monochromatization of the incoming beam and thereafter the analysis of the scattered beam. The energy and momentum changes during scattering are then obtained by evaluating the appropriate difference from the initial measurements. For very slow motion, which is present in large-scale structure, very high energy resolution is needed. This requires that one select a very small primary energy interval from the comparatively weak spectrum of a neutron source. Therefore, such methods always have to fight with low neutron intensities.

The unique feature of NSE is its ability to determine energy changes of neutrons during scattering in the direct way [13]. NSE measures the neutron velocities of the incoming and scattered neutrons utilizing the Larmor precession of the neutron spin in an external guide field. Since the neutron spin vector acts like the hand of an internal clock attached to each neutron, which stores the results of the velocity measurement on the neutron itself, this measurement is performed for each neutron individually. Therefore, the incoming and outgoing velocities of each neutron can be compared directly and velocity difference measurements become possible. Thus, energy resolution and monochromatization of the primary beam or the proportional neutron intensity are decoupled and an energy-resolution of the order of $\Delta E/E \approx 10^{-5}$ can be achieved with an incident neutron spectrum of fairly broad bandwidth.

The basic experimental setup of a neutron spin echo spectrometer is

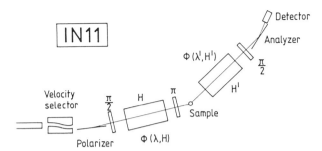

Figure 37.   Schematic sketch of the neutron spin echo spectrometer IN11 at the ILL Grenoble.

shown in Fig. 37. A velocity selector in the primary neutron beam selects a wavelength interval of about $\Delta\lambda/\lambda = 0.20$. The spectrometer offers primary and secondary neutron flight paths, where guide fields H and H' can be applied. At the beginning of the first flight path a supermirror polarizer polarizes the neutrons in the direction of propagation. A first $\frac{1}{2}\pi$ coil turns the neutron polarization into a direction perpendicular to the neutron momentum. From this well-defined initial condition the neutrons commence to precess in the applied guide field. Without the action of the π coil, which turns the neutron spin by 180° around a perpendicular axis, each neutron performs a phase angle $\Phi \propto \lambda \int H \, ds$. Since the wavelengths are distributed over a wide range, in front of the second $\frac{1}{2}\pi$ coil the phase angle will be different for each neutron and the beam will be depolarized.

A π coil positioned at half the value of the total field integral avoids this effects. On its way to the π coil the neutron may pass an angle $\Phi_1 = 2\pi n + \Delta\Phi_1(\lambda)$, where n is number of turns the spin undergoes. The action of the π coil transforms the angle $\Delta\Phi_1$ to $-\Delta\Phi_1$. In a symmetric setup where the both field integrals before and after the π coil are identical, the neutron spin turns by another phase angle $\Phi_2 = \Phi_1 = 2\pi n + \Delta\Phi_1$. The spin transformation at the π coil compensates the residual angles $\Delta\Phi_1$ and in front of the second $\pi/2$ coil the neutron spin points again into its original perpendicular direction, independent of its velocity. The second $\pi/2$ coils projects this perpendicular component of the polarization into the forward direction and then at the supermirror analyzer the total polarization is recovered. The experimental setup is spin focusing. As with nuclear magnetic resonance (NMR) spin echo methods, the phase is focused to its initial value in front of the second $\pi/2$ coil for each spin separately.

In a spin echo spectrometer the sample is positioned close to the π

coil. If the neutron energy is changed due to inelastic scattering at the sample, the neutron wavelength is modified from λ to $\lambda + \Delta\lambda$. Then the phase angles Φ_1 and Φ_2 do not compensate each other and the second $\pi/2$ coil projects only that component of the polarization that points into the original perpendicular direction. This component passes afterward through the analyzer. Apart from resolution corrections the final polarization P_f is then related to the initial polarization P_i by

$$P_f = P_i \int_{-\infty}^{+\infty} S(Q, \omega) \cos \omega t \, d\omega \qquad (68)$$

The scattering function $S(Q, \omega)$ is the probability for a certain momentum transfer $\hbar Q$ and energy change $\hbar\omega$ to occur during scattering. We have introduced the time variable $t \propto \lambda^3 H$. From Eq. (68) we see that NSE is a Fourier method and it essentially measures the real part of the intermediate scattering function $S(Q, t)$. The time variation in a spin echo scan is performed by changing the magnetic field H.

B. Quasielastic Scattering for Polymer Solutions: Zimm Model

The chain relaxation in a polymer solution is commonly described in terms of the Zimm model [96], which besides entropic and thermal forces considers also long-range hydrodynamic interactions between the chain segments mediated by the solvent. The model considers a Gaussian chain of N segments, the positions of which are denoted by a vector $\mathbf{a}_n$ along the chain. Then, in the limit of vanishing inertial forces the Langevin equation for segmental motion is

$$\zeta \dot{\mathbf{a}}_n + \frac{\partial^2}{\partial n^2} \mathbf{a}_n + \mathbf{K}_n = \mathbf{A}_n(\mathbf{t}) \qquad (69)$$

where ζ is a monomeric friction coefficient. The second derivative $\partial^2/\partial n^2$ originates from the entropic force resulting from the conformational chain entropy; $\mathbf{A}_n$ is the thermal random force, and

$$\mathbf{K}_n = \zeta \mathbf{v}(r_n) \qquad (70)$$

is the hydrodynamic force. The velocity field $\mathbf{v}(r_n)$ originates from the friction forces exhibited by all other segments on the solvent with ζ being the appropriate friction coefficient. A moving segment creates a back flow $\mathbf{v}$ in the solvent commonly described by the Oseen tensor $D_{ij} \propto 1/\eta_s |\mathbf{r}_i - \mathbf{r}_j|$. This leads to a long-range coupling between all segments. Equation

(69) has a spectrum of relaxation rates for the pth mode,

$$\tau_p^{-1} = \frac{12k_B T \pi^{1/2}}{(\langle R_G^2 \rangle^{3/2} \eta_s) p^{3/2}} \tag{71}$$

which only depend on the solvent viscosity η_s and the overall chain dimension $\langle R_G^2 \rangle$.

From the eigenvalues and eigenvectors of Eq. (69), the dynamic structure factor for Zimm relaxations can be calculated. Being a Fourier method NSE is sensitive to the intermediate scattering function $S(Q, t)$, which in turn is given by the pair correlation function

$$S(Q, t) = \frac{1}{(Nf)^2} \sum_{ij} \langle \exp[-i\mathbf{Q} \cdot \mathbf{r}_j(0)] \exp[i\mathbf{Q} \cdot \mathbf{r}_i(t)] \rangle \tag{72}$$

The brackets denote a thermal average. In the Gaussian approximation, Eq. (72) becomes

$$S(Q, t) = \frac{1}{(Nf)^2} \sum_{ij} \exp\{-\tfrac{1}{6}Q^2 \langle [r_i(0) - r_j(t)]^2 \rangle\} \tag{73}$$

In this case, Dubois-Violette and de Gennes [134] calculated explicitly the dynamic structure factor for the Zimm model. It turns out that $S(Q, t)$ is a function of a single scaling variable $x = (Q^2 l^2/3)[k_B T t/(6\sqrt{2}\pi\eta_s)]^{2/3}$. The characteristic decay rate at which the dynamic structure factor decays to $1/e$ is

$$\Gamma_z = \frac{Q^3 k_B T}{6\pi\eta_s} \tag{74}$$

This result is valid for the regime of internal relaxation modes. We note that no chain parameter enters Eq. (74). Furthermore, the Q^3 dependence distinguishes this relaxation distinctly from normal diffusive motion where $\Gamma \propto Q^2$.

C. Collective and Single-Star Relaxation

The dynamics of star molecules were first treated by Zimm and Kilb [38] and later on the basis of the Kirkwood diffusion equation by Burchard et al. [135] who calculated the first cumulant $\Gamma = -\{d[\ln S(Q, t)]/dt\}_{t=0}$. In this latter approach all pairwise interactions between segments of different arms of the star are also treated. The deviations from the relaxation behavior of a linear polymer are visualized most clearly if reduced cumulants Γ/Q^3 are plotted against the variable $u_r = \sqrt{Q^2 \langle R_G^2 \rangle f/(3f-2)}$ [see Appendix, Eqs. (A7) and (A13)]. Such a plot is displayed in the insert of Fig. 38. In this representation segmental diffusion of linear polymers ($f = 1$) is described by a line parallel to the abscissa. For small u_r the crossover to long-range diffusion $\Gamma/Q^3 \propto 1/Q$ is seen. Depending on functionality these calculations predict the occurrence of a minimum in Γ/Q^3 for Q values intermediate between translational diffusion and short-range segmental motion.

The NSE experiments were performed on linear polymers and 4-, 12- and 18-arm PI star molecules in dilute solution [103]. The neutron spin echo spectra were fitted to the Zimm scattering function. The relaxation rates obtained are presented in Fig. 38. The characteristic frequencies [see Eq. (74)] were reduced by Q^3 and are plotted versus the scaling

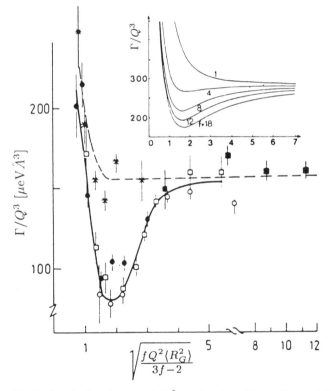

Figure 38. Reduced relaxation rates Γ/Q^3 as a function of the scaling variable u_r for PI stars. The solid line is a guide to the eye. The insert presents the calculated reduced first cumulants Γ/Q^3. Results are for (■) linear chains, (∗) 4-arm, (□) 12-arm, and (○●) 18-arm stars. (From Ref. 103.)

variable u_r using the experimental values for $\langle R_G^2 \rangle$ for the transformation from Q to u_r. At large u_r the reduced rates from all molecules fall on the same line parallel to the abscissa indicating the $\Gamma \propto Q^3$ behavior of the Zimm regime. Data for the four-arm star remain on this plateau until they bend up at $u_r \approx 1.5$. This crossover to an $\Gamma \propto Q^2$ behavior coincides with the change of lineshape to a single exponential decay and reflects the onset of translational diffusion. The stars of higher functionality exhibit a different behavior. Starting from the same level at large u_r, the reduced rates pass through a sharp minimum around $u_r = 1.5$ before they bend upward into the regime of translational diffusion. Thereby, all data points for different f and M_W appear to follow the same master curve. The full width at half-minimum (FWHM) amounts to about $\Delta u_r = 1.5$, its amplitude to 50% of the large u_r level.

The most important feature of the experimental result is the observation of a pronounced minimum in Γ/Q^3 as a function of u_r. It occurs in the same region where the static structure factor and its Kratky representation exhibits its maximum (see Fig. 18). Thus, apparently, the occurrence of the minimum in Γ/Q^3 is directly related to peculiarities of the star architecture. The peak structure in $Q^2 P(Q)$, which is related to a maximum in the monomer pair distribution function in the star, leads to a minimum in the reduced internal relaxation rate. In the theory of liquids such a phenomenon is well known under the term of de Gennes narrowing [136]. There, the peak in $S_I(Q)$ renormalizes the relaxation rate for density fluctuations. While in the liquid this is an effect involving different independent particles, here it occurs for the internal density fluctuations of one entity.

A comparison with Burchard et al.'s [135] first cumulant expansion shows qualitative agreement in particular with respect to the position of the minimum. Quantitatively, however, important differences are obvious. Both the sharpness as well as the amplitude of the phenomenon are underestimated in the calculations [$\Delta u_r = 1.5$ (experiment), $\Delta u_r = 2.5$ (theory); amplitude ($f = 18$), experiment, 50%; theory, 25%]. As we shall see in the following, these deviations mainly result from the inadequate structure factor for the star, which was assumed.

The fact that the minimum in Γ/Q^3 is due to the collective motion of the arms was confirmed by measurements on a 12-arm PS star where one single (on the average) arm was labeled [93]. Figure 39 compares the reduced relaxational rates Γ/Q^3 as a function of Q for the single labeled arm and a corresponding fully labeled 12-arm PS star. While the 12-arm fully labeled star exhibits a similar minimum in Γ/Q^3 as those observed on PI stars (see Fig. 38), the dynamics of a single labeled arm is very similar to that of a linear polymer in solution exhibiting no minimum,

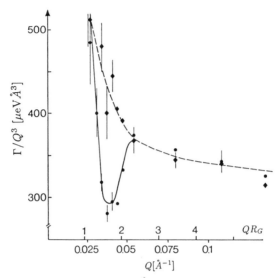

Figure 39. Reduced relaxation rate Γ/Q^3 for ($\blacklozenge$) the one-arm labeled star and ($\bullet$) the fully labelled star as a function of Q. The lines are guide to the eye. (From Ref. 93.)

crossing over directly from the high-Q Zimm regime to long-range center-of-mass diffusion.

D. Partial Dynamic Structure Factors and RPA Approach

To understand in more detail the relaxation properties of a star molecule and to test thoroughly the RPA approach, now we consider the dynamics of a partially labeled star where the star core is differently labeled than the outer parts of the star polymer [94]. In this case Eq. (72) has to be rewritten in terms of partial dynamics structure factors $S_{\alpha\beta}(Q, t)$, where α and β denote the core or shell, respectively,

$$S_{\alpha\beta}(Q, t) = \sum_{ij} \langle \exp[-i\mathbf{Q} \cdot \mathbf{r}_{i\alpha}(0)] \exp[i\mathbf{Q} \cdot \mathbf{r}_{j\beta}(t)] \rangle \qquad (75)$$

The vectors $\mathbf{r}_{i\alpha}$ and $\mathbf{r}_{j\beta}$ are the position vectors of the monomers with the label α or β, respectively. The double differential cross section (Eq. (66)) then is a sum over the partial structure factors weighted by the product of the appropriate scattering length $b_\alpha b_\beta$. In the framework of linear response theory and neglecting all memory effect in the calculation of the dynamic response, a simple equation of motion for the intermediate scattering function $S_{\alpha\beta}(Q, t)$ holds [137]

$$\frac{\partial}{\partial t} S_{\alpha\beta}(Q, t) + \sum_{\gamma} \Gamma_{\alpha\gamma} S_{\gamma\beta}(Q, t) = 0 \tag{76}$$

where Γ is the first cumulant matrix

$$\Gamma = \lim_{t \to 0} \frac{\partial}{\partial t} S(Q, t) S^{-1}(Q) \tag{77}$$

The matrix Γ can be related to the generalized mobility matrix $M(Q)$, which we will discuss later in some detail:

$$\Gamma = k_B T Q^2 M(Q) S^{-1}(Q) \tag{78}$$

The solution of Eq. (76) leads to two eigenvalues that describe the two relaxation modes of the system:

$$\Gamma_{1,2} = \Gamma_{av} \pm (\Gamma_{av}^2 - \Delta\Gamma)^{1/2} \tag{79}$$

where

$$\Gamma_{av} = \tfrac{1}{2}(\Gamma_{11} + \Gamma_{22})) \qquad \Delta\Gamma = \Gamma_{11}\Gamma_{22} - \Gamma_{12}\Gamma_{21}$$

The eigenvectors of Eq. (76) define the spectral weights of these relaxation modes in the dynamic structure function. The explicit solution is given in Ref. 94.

Following Benoit and Hadziioannou [120], the different static partial structure factors can be calculated independently of the Flory exponent ν as long as one assumes that the distance distribution between two arbitrary polymer segments i and j is Gaussian. Figures 21 and 22 compare such calculations with the partial structure factors S_{11}, S_{22}, and S_{12} obtained from SANS measurements on a 12-arm PI star with core–shell contrast.

In order to calculate the mobility matrix we have to consider the prevailing dynamics in a polymer solution, which is governed by the hydrodynamic interaction (Zimm model). According to Akcasu and Gurol [138] and Burchard [118] the generalized mobility $M(Q)$ can be expressed in terms of the Oseen diffusion tensor

$$M_{\alpha\beta} = \frac{1}{k_B T Q^2} \sum_{ij} \langle (Q D_{jk}^{\alpha\beta} Q \exp(-iQ r_{jk}^{\alpha\beta}) \rangle \tag{80}$$

with

$$D_{jk}^{\alpha\beta} = \frac{k_B T}{\zeta_j} \delta_{jk}^{\alpha\beta} + (E^{\alpha\beta} - \delta_{jk}^{\alpha\beta}) \frac{k_B T}{8\pi\eta_s r_{jk}^{\alpha\beta}} \left[E^{\alpha\beta} + \frac{r_{jk}^{\alpha\beta} r_{jk}^{\alpha\beta}}{(r_{jk}^{\alpha\beta})^2} \right]$$

The first term in Eq. (80) describes the so-called free draining ζ_i being the monomeric friction coefficient of monomer j. In dilute solution this contribution is negligible and will be omitted. The second term results from the hydrodynamic interaction between monomers j and k, which is determined by the strength of the back flow field over distances $r_{jk}^{\alpha\beta} = r_{i\alpha} - r_{j\beta}$ in a solvent with the viscosity η_s; E is the unity matrix. As Burchard et al. [139] have shown, Eq. (80) may be evaluated using a preaveraged form of the Oseen tensor and a correction term. With this approach for the preaveraged mobility matrix and the correction term, finally the following expressions are obtained:

$$M_{\alpha\beta}^{pre}(Q) = \frac{1}{3\pi^2\eta_s} \int_0^\infty S_{\alpha\beta}[(Q^2 + x^2)^{1/2}] \, dx$$

$$\Delta M_{\alpha\beta}(Q) = \frac{Q^2}{30\pi^2\eta_s} \int_0^\infty \frac{1}{x^2} \{S_{\alpha\beta}[(0.72Q^2)^{1/2}] - S_{\alpha\beta}[(0.72Q^2 + x^2)^{1/2}]\} \, dx$$

(81)

Equation (81) explicitly demonstrates that in the initial slope approximation the generalized mobility matrix can be expressed only in terms of integrals over the static structure factors. Equation (81) is valid as long as we assume a Gaussian distance distribution for the monomer distances $r_{ij}^{\alpha\beta}$.

Figure 40 displays the reduced relaxation rates $\Gamma_{1,2}/Q^3$ calculated for a Gaussian 12-arm star. In addition the effective $1/e$ decay rates of the dynamic structure factor for three different contrast situations are shown: (1) core contrast ($b_1 = 1, b_2 = 0$), (2) shell contrast ($b_1 = 0, b_2 = 1$), and (3) average contrast ($b_1 = -b_2$). Γ_2/Q^3 exhibits a minimum around $QR_G = 1$ and at low Q describes the translational diffusion of the whole star. This can be seen from the insert of Fig. 40, which displays the nonreduced eigenvalues as a function of QR_G [93]. At low QR_G, Γ_2 increases proportional to Q^2 exhibiting diffusive behavior. On the other hand Γ_1 reaches a finite value at low Q being an opticlike mode describing the relative motion of the star shell with respect to the core. At higher QR_G, the mode Γ_1 does not exhibit the minimum shown Γ_2. For very large QR_G, both modes converge describing the Zimm relaxation of the arms. Inspecting the results for the dynamic structure factors at different contrast conditions, we find that for core contrast basically the

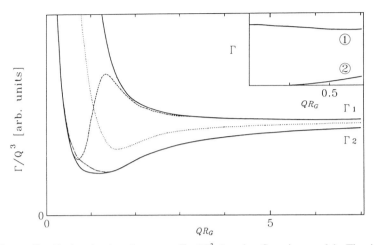

Figure 40. Reduced relaxation rates $\Gamma_{1,2}/Q^3$ for the Gaussian model. The insert presents the two rates in the low Q regime directly. Note that Γ_1 approaches a constant at low Q. The broken lines give the effective reduced relaxation rates for various contrast conditions. Dashed line: shell contrast; dashed dotted line: core contrast, dotted line: average contrast. (From Ref. 94.)

mode Γ_2 is observed while for shell contrast at higher QR_G the relaxation rate first follows Γ_1. For small QR_G a crossover to Γ_2 is observed. Thereby, the reduced relaxation rate passes through a maximum. Finally, for average contrast an intermediate relaxation rate is predicted that for low QR_G mainly picks up the optic mode.

Finally, we relate the general results stated in Eq. (78) to the observed apparent renormalization of the reduced relaxation rate with $Q^2S(Q)$ discussed above. In the Zimm regime, from Eq. (78) we have $\Gamma/Q^3 = Q^2M(Q)/[Q^3S(Q)]$. Since the Q dependence of the mobility $M(Q) \propto 1/Q$ for a Gaussian star polymer, this leads to the qualitatively observed relation

$$\frac{\Gamma}{Q^3} = \frac{\text{const}}{Q^2S(Q)} \tag{82}$$

valid for $QR_G > 1$. For large QR_G, the scattering functions from a Gaussian star leads to the scaling relation $S(Q) \propto Q^{-2}$ giving the expected scaling, $\Gamma \propto Q^3$.

Figure 41 displays NSE spectra taken from a dilute solution of a 12-arm PI diblock–copolymer star (inner part deuterated, outer part protonated in deuterated octane) under shell contrast [94]. The data are plotted logarithmically versus the Fourier time, the parameter being the

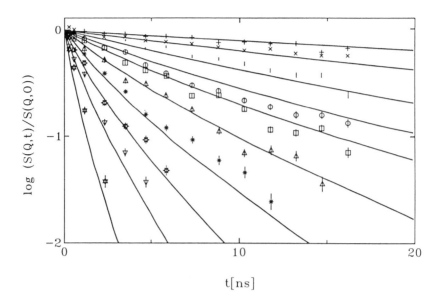

Figure 41. Neutron spin echo spectra obtained for a star with shell contrast. The solid lines are the results of a fit with the so-called structural model (see text) to the spectra using only data points for the first 4 ns. The data represent the following Q values: $(+)$ 0.026 Å^{-1}, $(\times)$ 0.032 Å^{-1}, $(|)$ 0.038 Å^{-1}, $(\bigcirc)$ 0.045 Å^{-1}, $(\square)$ 0.051 Å^{-1}, $(\triangle)$ 0.064 Å^{-1}, $(*)$ 0.076 Å^{-1}, $(*)$ 0.089 Å^{-1} (∇) 0.102 Å^{-1}, $(*)$ 0.127 Å^{-1}. (From Ref. 94.)

momentum transfer Q. With increasing Q we observe a strong increase of the star relaxation rate.

In the short time time regime ($t < 4$ ns) the data were analyzed in terms of the Zimm dynamic structure factor, which in this regime describes the experimental line shape very well. Figures 42 and 43 show the obtained reduced relaxation rates for core, shell, and intermediate contrast, which at low Q had to be corrected for the interstar structure factor $S_I(Q)$ [$\Gamma = \Gamma_{obs} S_I(Q)$] [94].

A comparison with the RPA approach was performed for two cases: (i) On the basis of Gaussian partial structure factors, where the values for the radii of gyration of the two blocks were taken from SANS, the reduced relaxation rates of the dynamic structure factors for the various contrasts were calculated with the adjustable parameter being the value of the Zimm rate. This lead to the dashed lines in Figs. 42 and 43. (ii) Alternatively, using the measured partial structure factors as an input (see Figs. 21 and 22) on the basis of Eq. (80) the dynamic response was

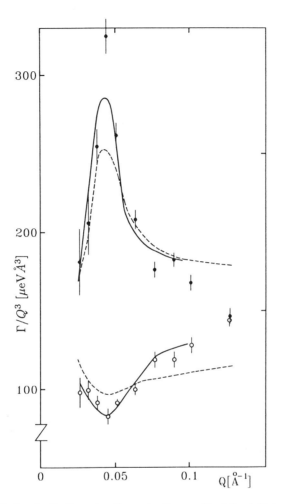

Figure 42. Reduced effective relaxation rates from a fit of the initial decay of the experimental spectra to the Zimm dynamic factor for (●) shell contrast and (○) core contrast. The dashed lines represent the Gaussian model; the solid lines are the result from the structural model. (From Ref. 94.)

evaluated again with the Zimm rate as adjustable parameter. This calculation gives the solid lines in Figs. 42 and 43 [94].

The Gaussian model provides a qualitative description of the experimental Q dependencies. The general features of the experimental results are present but are considerably less pronounced than observed.

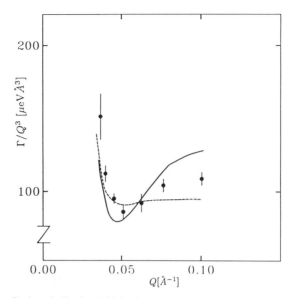

Figure 43. Reduced effective initial relaxation rates of the experimental spectral taken under average contrast conditions. Solid line: structural model; dashed line: Gaussian model. (From Ref. 94.)

Calculations in which the measured structure factors are used in Eq. (81) lead to superior results. These calculations reproduce the Q dependencies nearly quantitatively for all three partial dynamic structure factors. Thus, the RPA approach, which relates the initial star relaxation solely to structural properties of the star, is well confirmed.

While the calculated Q dependencies agree well with the observations, the results for the adjustable Zimm rates do not fit into this simple picture. In a polymer solution, where the hydrodynamic interaction dominates, the time scale of relaxation should be determined solely by the ratio of temperature and solvent viscosity [see Eq. (74)]. Here considerable deviations from the expected value of $T/\eta_s = 837° \, \text{K/cP}$ are evident, which also depend on the particular partial structure factor. For examples, a nearly 50% faster time scale is found for the shell relaxation compared to the core relaxation. This deviation has been related to the progressively increasing hydrodynamic screening if one moves from the rim to the center of the star [140]. Such screening effects are not accounted for in the calculations.

Up to now the discussion of the star relaxation has dealt with the short time behavior, which can be treated by RPA in a simple way. From Fig.

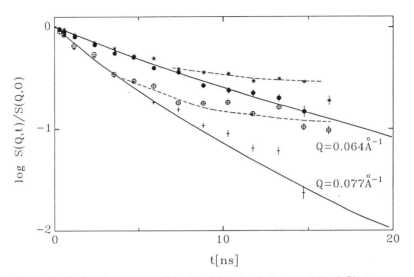

Figure 44. Relaxation spectra of the ($\bullet$+) fully labeled star and the ($*\oplus$) star core as two different Q values. The solid lines represent the result of a fit with the Zimm dynamic structure factor to the initial relaxation of the fully labeled star. The dashed lines are guides to the eye showing the retardation of the relaxation for the star core. (From Ref. 94.)

41 in which the relaxation spectra from the star shell are presented, we recognize that these spectra can also be well described at longer times by the Zimm dynamic structure factor. That this is not always the case is demonstrated in Fig. 44, which compares the NSE spectra of a fully labeled star with those of the star core of a corresponding star. While at short times the relaxation for both species proceeds in a similar way, at longer times the motion of the star core is strongly retarded compared to the full star. This effect may be related to the interarm entanglements in star polymers, which have been addressed above theoretically in Section V.C. [58].

IX. DIFFUSION AND VISCOSITY OF STAR POLYMERS

How a star molecule diffuses in a melt of other star polymers or a matrix of long, linear chains is an interesting question. The reader is probably familiar with the de Gennes–Edward's picture of how a linear chain moves by reptating like a snake [141]. However, a multiarm star with $f \geq 3$ cannot reptate in the same way, and other effects such as constraint release become important. If one assumes that in a topologically invariant medium only the fluctuations in path length are available to provide a long-range mobility, then diffusion only occurs when all but two arms of the star retract simultaneously to the branch point [142–146]. This leads to a strong exponential dependence of D on f and N,

$$D \sim \exp[-\alpha(f-2)N] \qquad (83)$$

The various theories differ only in the N dependence of the prefactor. Thus in a medium of long, linear chains, D for stars with long arms should be extremely small, on the order of the diffusion constant of the linear reptating species multiplied by a factor that decreases exponentially with both f and N. However as we see below, the dependence of D is much weaker than predicted by this simple model.

The tracer diffusion of star molecules in a fixed matrix of linear polymers has been studied by infrared microdensitometry [145], forced Rayleigh scattering [147], and forward recoil spectrometry [148, 149]. Klein and co-workers [145] studied the dynamics of symmetric three-arm PE stars (prepared by the saturation of the polybutadiene precursors) in linear PE. They also developed a theoretical model for the dynamics in terms of a general diffusive motion in an entropic potential field. This model predicted that D and the longest relaxation time τ_S scale as

$$D = \frac{D_1}{N} \exp(-\alpha N) \qquad (84)$$

and

$$\tau_S \sim N \exp(\alpha N) \qquad (85)$$

where α is a constant. For the case where linear chains served as the matrix, it was argued that for long enough arm length, the dynamics are dominated by tube-renewal effects. Figure 45 shows D versus $3N$ for the three-arm star diffusants. The variation of D with fN is initially more rapid than a power law. For $N < 500$, a power exponential relation emerges of the form given by Eq. (84), with $D_1 = 2.73 \times 10^{-6} \mathrm{cm}^2\,\mathrm{s}^{-1}$ and $\alpha = 8.4 \times 10^{-3}$. Alternatively, over the same N range, a good fit was obtainable via a pure exponential relation,

$$D = D_2 \exp(-\alpha N) \qquad (86)$$

with $D_2 = 3.9 \times 10^{-8} \mathrm{cm}^2\,\mathrm{s}^{-1}$ and $\alpha = 1.26 \times 10^{-2}$. For $N > 500$ the experimental data in Fig. 45 show a progressively attenuated dependence of D upon N. This was attributed to the enhanced rule of tube renewal effects. These results are in agreement with the findings of Antonietti and Sillescu for three-arm PS stars [147]. They also are in agreement with computer simulations on three-arm stars in a fixed matrix of obstacles by

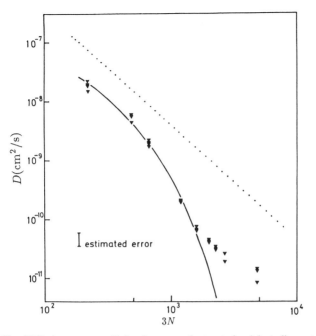

Figure 45. Diffusion constant D for three-arm deuterated polybutadiene stars diffusing in a high-density PE ($M_w = 1.6 \times 10^5$) melt at 176°C. The solid line is a least squared fit to Eq. (84). The dotted line shows D for linear chains of length N. (From Ref. 145.)

Needs and Edwards [150]. However, more recent simulations of Sikorski et al. [151] for three-arm stars in a linear, mobile matrix are best fit with a power law. However, the length of linear, matrix chains in this simulation were only about six entanglement lengths, much smaller than in the experimental studies. Therefore it is not surprising that the diffusion of the stars was faster than exponential. Simulations with long, linear chains comparable to those used experimentally are not feasible at present.

The dependence of D on functionality f was studied by Shull et al. [149]. Results for PS stars with $N = 530$ monomers power arm are shown in Fig. 46. The data show a much weaker dependence on f than Eq. (83), which for this value of N would have predicted a dependence of the form $D \sim \exp(-10.3f)$ (dashed line in Fig. 46). Instead the best fit to an exponential has the much weaker dependence,

$$D \sim \exp(-0.41f) \qquad (87)$$

The observed f dependence of D (Fig. 46) can be rationalized via two

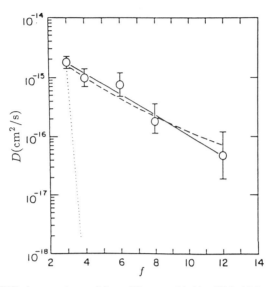

Figure 46.   Diffusion constants of f-arm PS stars with $N = 530$ in high-molecular-weight matrices at 178°C. The solid line represents the exponential dependence [Eq. (87)] and the dotted line represents the exponential dependence predicted by the simultaneous arm retraction model [Eq. (83)]. The dashed line represents the fit to Eq. (88), with $z = 6$ and a independent of f. (From Ref. 149.)

related viewpoints. The first [149] involves a branch point that can diffuse in a randomly spaced array of constraints each time a single arm retracts. In this case the jump frequency $\Gamma_S \sim f$. Thus, the observed dependence of D on f is controlled by the f dependence of the jump distance, a_S. The solid line through the data in Fig. 46 are consistent with $a_S \sim f^{-1.8}$ and $\Gamma_S \sim f$.

For a regular lattice of constraints, a_S does not depend on f. This leads to $D(f)$, which is proportional to Γ_S. Rubenstein [152] obtained an expression for $\Gamma_S(f)$ by assuming that the entanglement network around the star center defines gates through which the arms emanate. Thus, diffusive steps occur only when most of these gates are occupied by arms,

$$D \sim a_S^2 \frac{[1 + (z-1)(f-1)/2]z!f!}{(z+f-1)!} \tag{88}$$

where z denotes the number of gates. The data of Fig. 46 is best fitted (dashed line) by assuming $z = 6$ and a_S is independent of f. More recent work by Clarke and McLeish [153] also suggests that f only enters D as a prefactor. Obviously, the available data (Fig. 46) cannot distinguish between these related models.

The self-diffusion of three-arm PE stars has also been studied [154]. In contrast with earlier results for linear polymers, the self-diffusion of stars is much faster (20-fold) than tracter diffusion, when the latter occurs in a high-molecular-weight (effectively immobile) matrix. The difference was explained in terms of constraint release, a mechanism believed to dominate self-diffusion for entangled stars while conversely having little effect for entangled linear polymers (the matrix for tracer diffusion).

The initial rheological measurements for model stars was undertaken by Kraus and Gruver [155] for a series of three- and four-arm PBd stars. They found that the Newtonian viscosity η scaled in an exponential fashion with M_W; unlike the behavior for linear polymers, $\eta \sim M_W^{3.4}$ for chain lengths longer than the entanglement length. As a consequence of this exponential behavior, the melt viscosity for stars can either be higher or lower than a linear material of identical M_W.

Generally the rheological properties of stars are distinguished from those of linear chains by exhibiting a broader relaxation spectrum and a viscosity that increases exponentially with M_W. For functionalities ranging from 4 to ca. 33, the Newtonian viscosity has been found [156] to be independent of f while depending only on the arm molecular weight M_a. This observation [156] is in accord with that of Toporowski and Roovers [43] who found that 4- and 18-arm PBd stars with the same arm molecular weight had the same viscosity and longest relaxation times. However for larger f (32, 64, and 128) Roovers [157] observed that η is no longer independent of f.

Two factors need to be considered in comparing viscosities of star and linear polymers of the same M_W. One is the expected decrease in η due to smaller $\langle R_G^2 \rangle$, expressed by g [Eq. (5)], while the other is the expected enhancement of η due to the increase in the arm molecular weight. The reduction follows from

$$\eta = K \langle R_G^2 \rangle^a = K'[\eta]_\Theta^{2a} \tag{89}$$

where $a = 3.4$ in the entanglement regime. The enhancement of η is apparently caused by the slowing of the reptation event along the backbone contour as M_a increases. Under these conditions, η has been found to have the form

$$\eta \sim \left(\frac{M_a}{M_e}\right)^b \exp\left(\frac{\nu' M_a}{M_e}\right) \tag{90}$$

where M_e is the entanglement molecular weight. Since the pre-exponential factor in Eq. (90) has only a small influence on the viscosity, the exponent b is not determined from the experimental data. Instead, values of order 1 ($\frac{1}{2} \leq b \leq 2$) have been used [143, 156] and are consistent with theory.

The value of ν' can be related to the plateau modulus G_N. According to Doi and Kuzuu [143] and Pearson and Helfand [158],

$$\frac{15}{8} \frac{G_N M_W}{\rho RT} = \nu' \frac{M_a}{M_e} \qquad (91)$$

This leads to $\nu' = \frac{15}{8}$, which is larger than the experimental result, which is ~ 0.5. Ball and McLeish [159] explained this difference by noting that during the extended periods required for an arm to retract to its origin, the surrounding constraints (resulting from the star arms) are also retracting. This permits each arm to relax its orientation via tube renewal. This approach assumes that the value of M_e increases in inverse proportion to the number of unrelaxed arms. In an approximate fashion, they found that the value of ν' decreases by a factor of 3, thus reducing the value of ν' to that found experimentally [159]. For a more popular review of the relaxation of a star and other branched polymers see the recent article in *Physics World* by McLeish [160].

Stress relaxation for entangled polymers is achieved by a process of disorientations involving tube renewal and diffusion out of the tube. Klein [146] has suggested that three-arm stars can access a relaxation mechanism unavailable to stars with more arms. This involves one arm vacating its tube if the trifunctional branch point diffuses down one of the tubes. This motion requires one of the tubes to be occupied by $f - 1$ arms, an event of low probability for $f \geq 4$. Using an approximate potential field, Klein [146] predicted that the viscosity is 33% lower for a three-arm star relative to those with $f \geq 4$. This value is in reasonable accord with that found [157] experimentally for eight pairs of three- and four-arm stars with the same M_a for each pair. Hadjichristidis and Roovers [161] observed similar results in this comparison of three and four arm stars.

An important connection between relaxation in stars and the temperature dependence of their viscoelastic properties is available from Graessley [162]. This addressed the observation that in some polymer species the presence of long-chain branching leads to an increase in the temperature coefficient of the melt viscosity and departures from time–temperature superposition. In other species the temperature dependence is unchanged with branching.

TABLE VII
Temperature Dependence of Chain Dimensions

Polymer	$d \ln\langle R_G^2 \rangle / dT$ $10^3 \deg^{-1}$	Reference
Polyethylene	-1.1	165
alt-Poly(ethylene-propylene)	-1.1	166
Polybutadiene (1,4)	~ 0	167
Polyisoprene (1,4)	0.4	168
Polystyrene	~ 0	169

If stars relax via chain retraction, then that chain must access more compact configurations than experienced on average. The longest relaxation times should be associated with the more compact configurations since total disengagement requires that the chain retract a distance proportional to the length of an arm. Thus, this dictates that the retracted chains possess a larger fraction of gauche states. For the case in which the gauche confirmer has the higher energy the activation barrier for the longest time should increase in more or less a linear fashion with M_a. Proportionally less affected are the faster relaxation times.

This behavior is found for PE and PEP stars [162, 163] but not for PS, PBd, and PI stars [157, 162, 164]. Data for the temperature dependence of chain dimension, $d \ln\langle R_G^2 \rangle / dT$, reveals for the latter three polymers that the trans and gauche minima have essentially the same energy as seen in Table VII [165–169]. With little or no energy difference between contracted and expanded configurations, the temperature dependence for star and linear materials would be the same, a scenario observed for PS, PBd, and PI. For PE and PEP differences exist [163] between the temperature dependence of linear and star polymers. Those differences are accounted for by the negative temperature coefficients in Table VII.

In conclusion it is interesting to recall that while star diffusion exhibits a clear f dependency, the melt flow behavior of a star is independent of f for $f = 4$ to ca. 33. The interplay between diffusion and flow behavior, while well understood for linear chains, remains elusive for star-branched systems.

X. APPLICATIONS

The initial commercial application of a star copolymer was the K-resin of the Philips Petroleum Co. The synthetic route for this family of star polymers is outlined as follows [170]:

$$sec - BuLi + CH_2=CH\phi \rightarrow polystyryllithium(I)$$

$$I + sec - BuLi + CH_2=CH\phi \rightarrow bimodal\ polystyrllithium(II)$$

$$II + CH_2=CH—CH=CH_2 \rightarrow (PBd:1)PSLi + (PBd:2)PSLi$$

This diblock mixture contained about 75 wt % polystyrene.

This mixture is subsequently reacted with epoxidized linseed oil, a polyfunctional linking agent. The resultant multimodal product yields clear resins with modest impact properties. The usage of these resins are in products where optical clarity is necessary or desirable; e.g., packaging uses and disposable cups. The multimodal nature of the product is claimed to aid in processing.

The Shell Chemical Co. markets a hydrogenated polyisoprene star prepared by the DVB linking chemistry, Shellvis [171]. This product is used as a viscosity index modifier. The material exhibits excellent shear stability as a consequence of its near hard-sphere characteristics. The synthetic procedure has been given in Section II. The use of a commercial DVB–ethylbenzene mixture (DVB–PLi ≈ 3) at high temperature ($\sim70°C$) for the linking temperature is followed by the hydrogenation event. This leads to a high functionality alternating PEP star material.

Figure 47 summarizes the size exclusion chromatography (SEC) result of an authentic Shellvis material. The PEP star shows the primary peak and a lower molecular weight component that we assume to be an unlinked parent arm. The elution behavior of that component replicates that of a linear PEP with an $M_W = 3.5 \times 10^4$ g mol^{-1}. The PEP star was characterized by low-angle laser light scattering and viscometry. The data give a total molecular weight $M_W = 7.34 \times 10^5$ and $f = M_W/M_a = 21$.

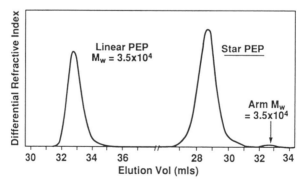

Figure 47. SEC profiles of linear PEP and DVB branched Shellvis star polymers.

Viscometric measurements in THF give $[\eta] = 0.92 \, \text{dL/g}$ and $k_H = 0.88$ [172].

The functionality of a star can be assayed in several different ways. The preferred route involves knowledge of the arm and star molecular weights, the ratio of which yields f. Alternative methods, though, do exist; knowledge of $\langle R_H \rangle (h)$, $\langle R_G^2 \rangle (g)$, $[\eta](g')$, and $\langle R_V \rangle$ can also be used. Perhaps the most simple method in the absence of the arm molecular weight involves knowledge of $[\eta]$ and the star M_W. This entails the calculation of $[\eta]$ for linear chains with the same M_W as that of the star. This facile operation is done via the use of the equation [173]

$$[\eta]_{\text{THF}}^{25°C} = 3.88 \times 10^{-4} M_W^{0.686} \qquad \text{dL g}^{-1} \tag{92}$$

where THF is tetrahydrofuran. For the GSL [168],

$$[\eta]_{\text{CYC}}^{25°C} = 2.73 \times 10^{-4} M_W^{0.744} \qquad \text{dL g}^{-1} \tag{93}$$

where CYC is cyclohexane.

The calculated value of 4.10 for linear PEP in THF yields a value of $g' = 0.224$ for the Shellvis star discussed above. This in turn yields, via the use of the g' GSL relation in Table III, a functionality f of 19.9. The corresponding equation, Table III, using the relation for $[\eta]/[\eta]_a$ yields $f = 19.1$. These values are slightly lower than the absolute value of 21; but within the confines of experimental error the agreement is satisfactory.

It should also be noted that the star polymers claimed in Ref. 28 were, in reality, coupled linear triblocks having little star-shaped polymer content. This is shown as follows. The values of $[\eta]$ and $[\eta]_a$ allow one, without knowledge of molecular weights, an estimate of star functionality. The power law expressions of Table III for $f \geq 5$ can be used to evaluate f for both homopolymer and block copolymer stars. For $f \geq 5$, the GSL expression is $[\eta]/[\eta]_a = 1.31 f^{0.284}$. Using these expressions reveals that the functionalities of the 'stars' of Ref. 28 range from 1.2 to ca. 3 with only one sample, out of 10, having an f around 3. The apparent high functionalities of 28 to 295 are precluded to the limited DVB–active center range of 2, which was used for all preparations but one. As has been found [174, 175], the ratio of 2 is inadequate for high yields of polydiene stars. Thus the claims of Ref. 28 that functionalities ranging from 3 to 10 were found are erroneous.

Star-branched diblock copolymers (polystyrene–butadiene) appear to have superior mechanical properties and processability relative to their linear analogues [31]. Materials with four arms and high polybutadiene content have been used in hot-melt and pressure-sensitive adhesives [176, 177].

The concept of star polymers containing reactive functional groups at the tips was first addressed by Schulz et al. [178]. This approach involves the use of "protected" initiators, for example, CH_3CH-$(OCH_2CH_3)ORLi$, where the initiator fragment at the end of a star arm can be hydrolyzed to yield a terminal hydroxy group. This approach yields a facile route to networks having crosslink sites of known functionality. This latter feature reflects the functionality of the star prior to network formation, which is brought about by use of a difunctional reagent which leads to chain extension. Star polymer networks exhibit mechanical properties that are equivalent to those of conventional sulfur-based vulcanizates.

Functionalized acrylic star polymers ($f \geq 5$) are claimed [179] to improve both the oxygen permeability of contact lenses while simultaneously enhancing the mechanical properties, that is, hardness and machinability. The functionalized stars, which have olefinic groups at the tips of the arms, are incorporated via copolymerization into the linear copolymer matrix. In this fashion a network is formed. An intriguing unanswered question is the mechanism by which the incorporation of the star polymer enhances oxygen permeability. Such an effect, if general, would have important ramifications in membrane and separation technologies.

In addition to the applications discussed above, star polymers have been considered for use in the coating industry [180, 181], as binders for toners for copying machines [182], and as resins for electrophotographic photoreceptors [183]. Star polymers and starburst dendrimers [19, 20, 184–186] also have important applications in semiconductor devices [187] and pharmaceutical and medical fields, such as in control release drug delivery [188].

We believe that in the near future there will be more applications taking advantage of the unique rheological properties of stars and that there will be an active investigation in the synthesis and characterization of mixed-arm star-branched systems [12, 132, 133], and selectively functionalized star molecules to form model networks.

APPENDIX A: DERIVATION OF $\langle R_G^2 \rangle$ AND $P(Q)$ FOR GAUSSIAN STAR

The mean-squared radius of gyration for a star polymer consisting of point masses, with a delta-function distribution of mass, is given by

$$\langle R_G^2 \rangle = \frac{1}{(Nf)^2} \sum_i \langle (\mathbf{r}_i - \mathbf{r}_{cm})^2 \rangle \tag{A1}$$

where $\mathbf{r}_{cm}$ is the center of mass of the star,

$$\mathbf{r}_{cm} = \frac{1}{Nf} \sum_j \mathbf{r}_j \qquad (A2)$$

The sums are over all Nf monomers of the star and $\langle \cdot \rangle$ stands for a configuration average. Substituting Eq. (A2) into (A1) gives

$$\langle R_G^2 \rangle = \frac{1}{2(Nf)^2} \sum_i \sum_j \langle (\mathbf{r}_i - \mathbf{r}_j)^2 \rangle \qquad (A3)$$

For a chain that obeys Gaussian statistics, the mean-squared distance between two chain elements i and j, depends only on the path length along the chain separating them,

$$\langle (\mathbf{r}_i - \mathbf{r}_j)^2 \rangle = |i - j| b^2 \qquad (A4)$$

where b is the bond length.

To determine $\langle R_G^2 \rangle$, select one special monomer, say the jth unit, as a reference point and all pairs of monomers of the same path length can be grouped together. The result can then be summed over all path lengths. This reference monomer can be thought of as the root of a tree. To get the total double sum, each monomer has to be selected as a reference point, and the result of the summation over the different reference points has to be added. This leads to the result,

$$\frac{2(Nf)^2 \langle R_G^2 \rangle}{b^2} = 1 + \sum_{j=1}^{Nf+1} \sum_{n=1}^{n_{max}} N_j(n) n \qquad (A5)$$

where $N_j(n)$ is the number of paths of length n for the jth monomer. This formula can be easily generated for any branched polymer without loops. For trees with the branching point as the root, the number of paths is $N(n) = 1$. However, this contribution turns out to be negligible for large N. For the jth monomer along a branch, there are two types of contributions, depending on whether the second monomer is on the same branch or not. For monomers on the same branch, there are two terms since $n_{max} = N - j(j)$ for monomer closer (further) from the branch point. The contribution to the sum from the jth monomer is

$$T_{rj} = 1 + \sum_{n=1}^{N-j} n + \sum_{n=1}^{j} n + (f-1) \sum_{n=1}^{N} (n+j)$$

$\langle R_G^2 \rangle$ can then be written as the sum over all trees:

$$\frac{2(Nf)^2 \langle R_G^2 \rangle}{b^2} = f \sum_{j=1}^{N} T_{rj} \tag{A6}$$

The summations in Eq. (A6) can easily be carried out with the result that the mean-squared radius of gyration of the whole star is related to that of one arm by [37]

$$\langle R_G^2 \rangle = (3f - 2) \langle R_G^2 \rangle_a / f \tag{A7}$$

where $\langle R_G^2 \rangle_a = \frac{1}{6} N b^2$.

The calculation for the form factor $P(Q)$ proceeds along similar lines. For a system of point masses, $P(Q)$ can be written as a sum over all monomers of the star,

$$P(Q) = \frac{1}{(Nf)^2} \sum_i \sum_j \langle \exp[i\mathbf{Q} \cdot (\mathbf{r}_i - \mathbf{r}_j)] \rangle \tag{A8}$$

Note that we have normalized the form factor such that $P(Q = 0) = 1$, which is consistent with the definition used in Section VI. However, in most theoretical and simulation papers, $P(Q)$ is normalized so that $P(Q = 0) = Nf$, the number of scatters. The difference comes in via the factor $V_w = Nf v_p$. While this formula is useful in deriving analytic closed-form formulas for $P(Q)$, for simulations it is computationally more efficient to use the equivalent form:

$$P(Q) = \frac{1}{(Nf)^2} \left\langle \left| \sum_i \exp(i\mathbf{Q} \cdot \mathbf{r}_i) \right|^2 \right\rangle \tag{A9}$$

For a star that obeys Gaussian statistics, $P(Q)$ can be determined analytically, since the exponential factor in Eq. (A9) can be rewritten as

$$\langle \exp[i\mathbf{Q} \cdot (\mathbf{r}_i - \mathbf{r}_j)] \rangle = \exp(-\tfrac{1}{6} b^2 Q^2 |i - j|) \tag{A10}$$

To determine $P(Q)$, one follows the same procedure as was outlined for $\langle R_G^2 \rangle$, by selecting one special monomer as a reference point and summing over all pairs of monomers of the same path length can be grouped together [118],

$$(Nf)^2 P(Q) = 1 + \sum_{j=1}^{Nf+1} \sum_{n=1}^{n_{max}} N_j(n) \exp(-\tfrac{1}{6} b^2 Q^2 n) \tag{A11}$$

For trees with the branching point as the root, $N(n) = 1$, and

$$T_b = 1 + f \sum_{n=1}^{N} \phi_Q^n$$

where $\phi_Q = \exp(-\frac{1}{6}b^2 Q^2)$. For the jth monomer along a branch, there are two types of contributions,

$$T_{rj} = 1 + \sum_{n=1}^{N-j} \phi_Q^n + \sum_{n=1}^{j} \phi_Q^n + (f-1) \sum_{n=1}^{N} \phi_Q^{n+j}$$

The form factor can then be written as the sum over all trees:

$$(Nf)^2 P(Q) = T_b + f \sum_{j=1}^{N} T_{rj} = T_b + fT_r \tag{A12}$$

The summations in Eq. (A12) can easily be carried out [110, 118]. For large N, T_b can be neglected. Also the relevant Q scale is such that $\phi_Q \simeq 1$ and $1 - \phi_Q \simeq \frac{1}{6}b^2 Q^2$. This leads to the following result for the form factor for regular stars,

$$P(Q) = \frac{2}{fu_r^4} \left[u_r^2 - [1 - \exp(-u_r^2)] + \frac{f-1}{2}[1 - \exp(-u_r^2)]^2 \right] \tag{A13}$$

where $u_r^2 = \frac{1}{6}Nb^2 Q^2 = Q^2 \langle R_G^2 \rangle_a$.

APPENDIX B

Power law fits for $\langle R_G^2 \rangle^{1/2}$ for five polymer systems were presented in Table V for both GSL and Θ conditions. These fits were obtained by combining the least-squares fits for linear chains [36] with those for the functionality dependence presented in Table III. This procedure was followed for all but the PEP stars, for which $\langle R_G^2 \rangle$ as a function of molecular weight has yet to be determined directly. In this case, experimental data for $\langle R_V \rangle$ was used in conjunction with the characteristic ratios for $\langle R_V \rangle / \langle R_G^2 \rangle^{1/2}$ presented in Table IV. The GSL results are based on Eq. (93), while the Θ solvent results are based on [189]

$$[\eta]_{2-\text{OCT}}^{21°C} = 8.22 \times 10^{-4} M_W^{0.5}$$

TABLE VIII
Star Polymer Master Equations for $[\eta]$ (dL/g)

Sample	Condition	K, $\times 10^3$	ν	β	Temperature (°C)	Solvent[a]	Reference
PBd	GSL	0.675	0.730	−0.789	35	Toluene	190
	θ	0.917	0.697	−0.789	25	Cyclohexane	36
		4.08	0.5	−0.690	26.5	Dioxane	36
PI	GSL	0.403	0.742	−0.789	34	Toluene	41
		0.467	0.733	−0.789	25	Cyclohexane	36
	θ	2.86	0.5	−0.690	34.5	Dioxane	36
PS	GSL	0.234	0.739	−0.789	30	Benzene	36
		0.220	0.734	−0.789	30	Toluene	36
		0.236	0.734	−0.789	30	THF	36
	θ	1.87	0.5	−0.690	34.5	Cyclohexane	36
PIB	GSL	0.287	0.751	−0.789	25	Cyclohexane	36
		0.474	0.670	−0.789	25	Toluene	191
	θ	2.44	0.5	−0.690	25	Benzene	36
PEP	GSL	0.647	0.744	−0.789	25	Cyclohexane	168
		0.920	0.686	−0.789	25	THF	193
	θ	4.51	0.5	−0.690	19	Benzene	192
PE	GSL	0.929	0.725	−0.789	130	TCB	193
	θ	6.94	0.5	−0.690	138	Dodecanol	193
PEE	GSL	0.195	0.760	−0.789	25	THF	194
	θ	1.82	0.5	−0.690	21	2-Octanol	189

Note: $[\eta] = K M_w^{\nu} f^{\beta}$.
[a] THF = tetrahydrofuran, TCB = trichlorobenzene.

TABLE IX
Star Polymer Master Equations for $\langle R_H \rangle (\text{Å})$

Sample	Condition	K	ν	β
PBd	GSL	0.206	0.570	−0.224
	θ	0.324	0.5	−0.168
PI	GSL	0.155	0.592	−0.246
	θ	0.289	0.5	−0.168
PS	GSL	0.164	0.575	−0.242
	θ	0.245	0.5	−0.168
PIB	GSL	0.186	0.573	−0.227
	θ	0.283	0.5	−0.168

Note: $\langle R_H \rangle = K M_w^\nu f^\beta$. GSL: $f = 6, \ldots, 270$; θ: $f = 2, \ldots, 270$.

The resulting expressions for linear PEP are

$$\langle R_G^2 \rangle^{1/2} = 0.166 M_W^{0.60} \quad \text{(GSL)}$$

$$\langle R_G^2 \rangle^{1/2} = 0.381 M_W^{0.5} \quad (\Theta)$$

We have also compiled master equations for $[\eta]$ and $\langle R_H \rangle$. These are presented in Tables VIII and IX for several systems.

ACKNOWLEDGMENT

We thank J. Roovers for a critical reading of the manuscript.

REFERENCES

1. B. J. Bauer and L. J. Fetters, *Rubber Chem. Tech.* **51**, 406 (1978).

2. J. Roovers, L.-L. Zhou, P. M. Toporowski, M. van der Zwan, H. Iatrou, and N. Hadjichristidis, *Macromolecules* **26**, 4324 (1993).

3. N. S. Davidson, L. J. Fetters, W. G. Funk, W. W. Graessley, and N. Hadjichristidis, *Macromolecules* **21**, 112 (1988).

4. G. Broze, R. Jerome, P. L. Teyssie, and C. Marco, *Macromolecules* **16**, 1771 (1983).

5. Z. Tuzar and P. Kratochvil, *Adv. Colloid Interface Sci.* **6**, 201 (1976); in *Surface and Colloid Science*, Vol. 15, E. Matijević, ed., Plenum, New York, 1993, p. 1.

6. L. J. M. Vagberg, K. A. Cogan, and A. P. Gast, *Macromolecules* **24**, 1670 (1991); K. A. Cogan, M. Capel, and A. P. Gast, *Macromolecules* **24**, 6512 (1991).

7. A. Halperin, M. Tirrell, and T. P. Lodge, *Adv. Polym. Sci.* **100**, 31 (1992).

8. M. Daoud and J. P. Cotton, *J. Phys. (Paris)* **43**, 531 (1982).

9. T. M. Birshtein and E. B. Zhulina, *Polymer* **25**, 1453 (1984); T. M. Birshtein, E. B. Zhulina, and O. V. Borisov, *Polymer* **27**, 1078 (1986).

10. T. A. Witten, P. A. Pincus, and M. E. Cates, *Europhys. Lett.* **2**, 137 (1986); T. A. Witten and P. Pincus, *Macromolecules* **19**, 2509 (1986).

11. N. Khasat, R. W. Pennisi, N. Hadjichristidis, and L. J. Fetters, *Macromolecules* **21**, 1100 (1988).

12. H. Iatrou and N. Hadjichristidis, *Macromolecules* **26**, 2479 (1993); **25**, 4649 (1992); H. Iatrou, E. Siakali-Kioulafa, N. Hadjichristidis, J. Roovers, and J. Mays, *J. Polym. Sci. Polym. Phys. Ed.* **33**, 1925 (1995).

13. F. Mezei in *Neutron Spin Echo*, F. Mezei, ed., Springer Verlag, Berlin, 1979.

14. M. Morton, T. E. Helminiak, S. D. Gadkary, and F. Bueche, *J. Polym. Sci.* **57**, 471 (1962).

15. R. P. Zelinski and C. F. Woffard, *J. Polym. Sci., Part A* **3**, 93 (1965).

16. J. Roovers and S. Bywater, *Macromolecules* **5**, 385 (1972); **7**, 443 (1974).

17. N. Hadjichristidis, A. Guyot, and L. J. Fetters, *Macromolecules* **11**, 668 (1978).

18. N. Hadjichristidis and L. J. Fetters, *Macromolecules* **13**, 191 (1980).

19. D. A. Tomalia, H. Baker, J. Dewald, M. Hall, G. Kallos, S. Martin, J. Roeck, J. Ryder, and P. Smith, *Macromolecules* **19**, 2466 (1986); D. A. Tomalia, V. Berry, M. Hall, and D. M. Hedstrand, *Macromolecules* **20**, 1167 (1987).

20. D. A. Tomalia, A. M. Naylor, and W. A. Goddard III, *Angew, Chem., Int. Ed. Engl.* **29**, 138 (1990).

21. L.-L. Zhou, N. Hadjichristidis, P. M. Toporowski, and J. Roovers, *Rubber Chem. Tech.* **65**, 303 (1992).

22. J. Roovers, P. Toporowski, and J. Martin, *Macromolecules* **22**, 1897 (1989).

23. L. L. Zhou and J. Roovers, *Macromolecules* **26**, 963 (1993).

24. L. J. Fetters, N. P. Balsara, J. S. Huang, H. S. Jeon, K. Almdal, and M. Y. Lin, *Macromolecules* **28**, 4996 (1995).

25. R. W. Pennisi and L. J. Fetters, *Macromolecules* **21**, 1094 (1988).

26. J. W. Mays, *Polym. Bull.* **23**, 247 (1990).

27. P. Jerome, P. Teyssie, and G. Huyuh-Ba, in *Anionic Polymerization: Kinetics, Mechanisms, and Synthesis*, Vol. 166, J. E. McGrath, ed., American Chemical Society, Washington, D.C., 1981, p. 211.

28. R. Milkovich, Canada Pat. 716,645 (Aug. 25, 1965).

29. J. G. Zilliox, P. Rempp, and J. Parrod, *J. Polym. Sci., C* **22**, 145 (1968).

30. D. J. Worsfold, J. G. Zilliox, and P. Rempp, *Can. J. Chem.* **47**, 3379 (1969).

31. L. K. Bi and L. J. Fetters, *Macromolecules* **8**, 90 (1975); **9**, 732 (1976).

32. L. J. Fetters and L. K. Bi, U.S. Pat. 3,985,830 (Oct. 12, 1976); British Pat. 1,502,800 (Mar. 1, 1978); Canadian Pat. 1,057,889 (Mar. 3, 1979).

33. D. J. Worsfold, *J. Polym. Sci., Part A1* **5**, 2783 (1967).

34. G. Quack, L. J. Fetters, N. Hadjichristidis, and R. N. Young, *Ind. Eng. Chem., Prod. Res. Div.* **19**, 587 (1980).

35. M. B. Huglin, ed., *Light Scattering from Polymer Solutions*, Academic, New York, 1972.

36. L. J. Fetters, N. Hadjichristidis, J. S. Lindner, and J. W. Mays, *J. Phys. Chem. Ref. Data* **23**, 619 (1994).

37. B. H. Zimm and W. H. Stockmayer, *J. Chem. Phys.* **17**, 1301 (1949).

38. B. H. Zimm and R. W. Kilb, *J. Polym. Sci.* **37**, 19 (1959).

39. W. Stockmayer and M. Fixman, *Ann. N.Y. Acad. Sci.* **57**, 334 (1953).

40. J. Roovers, *Polymer* **16**, 827 (1975); **20**, 843 (1979); J. Roovers and P. Toporowski, *J. Polym. Sci., Polym. Phys. Ed.* **18**, 1907 (1980).

41. N. Hadjichristidis and J. Roovers, *J. Polym. Sci., Polym. Phys. Ed.* **12**, 2521 (1974).

42. B. J. Bauer, L. J. Fetters, W. W. Graessley, N. Hadjichristidis, and G. F. Quack, *Macromolecules* **22**, 2337 (1989).

43. P. M. Toporowski and J. Roovers, *J. Polym. Sci., Polym. Chem. Ed.* **24**, 3009 (1986).

44. J. Roovers and J. E. Martin, *J. Polym. Sci., Polym. Phys. Ed.* **27**, 2513 (1989).

45. G. S. Grest and M. Murat, in *Monte Carlo and Molecular Dynamics Simulations in Polymer Science*, K. Binder, ed., Clarendon Press, Oxford, 1995.

46. Y. Tsukahara, K. Mizuno, A. Segawa, and Y. Yamashita, *Macromolecules* **22**, 1546 (1989).

47. G. K. Batchelor, *J. Fluid Mech.* **83**, 97 (1977).

48. K. De'Bell and T. Lookman, *Rev. Mod. Phys.* **65**, 87 (1993).

49. K. Ohno and K. Binder, *J. Chem. Phys.* **95**, 5444 (1991); **95**, 5459 (1991).

50. A. Miyake and K. F. Freed, *Macromolecules* **16**, 1228 (1983); **17**, 678 (1984); *J. F. Douglas and K. F. Freed, Macromolecules* **17**, 1854 (1984).

51. C. H. Vlahos and M. K. Kosmas, *Polymer* **25**, 1607 (1984).

52. B. Duplantier, *Phys. Rev. Lett.* **57**, 941 (1986); *Europhys. Lett.* **8**, 677 (1988); B. Duplantier and H. Saleur, *Phys. Rev. Lett.* **57**, 3179 (1986); **59**, 539 (1987).

53. G. Allegra and F. Ganazzoli, *Prog. Polym. Sci.* **16**, 463 (1991); G. Allegra, E. Colombo, and F. Ganazzoli, *Macromolecules* **26**, 330 (1993).

54. G. Allegra, M. De Vitis, and F. Ganazzoli, *Makromol. Chem., Theory Simul.* **2**, 829 (1993).

55. P. G. de Gennes, *Scaling Concepts in Polymer Physics*, Cornell University Press, Ithaca, NY, 1979.

56. P. Flory, *Statistical Mechanics of Chain Molecules*, Interscience, New York, 1969.

57. J. C. LeGuillou and J. Zinn-Justin, *Phys. Rev. B* **21**, 3976 (1980).

58. G. S. Grest, K. Kremer, and T. A. Witten, *Macromolecules* **20**, 1376 (1987).

59. G. S. Grest, K. Kremer, S. T. Milner, and T. A. Witten, *Macromolecules* **22**, 1904 (1989).

60. G. S. Grest, *Macromolecules* **27**, 3493 (1994).

61. E. Raphaël, P. Pincus, and G. H. Fredrickson, *Macromolecules* **26**, 1996 (1993).

62. M. Aubouy, G. H. Fredrickson, P. Pincus, and E. Raphaël, *Macromolecules* **28**, 2979 (1995).

63. C. Gay and E. Raphaël, private communication, 1995.

64. J. Mazur and F. McCrackin, *Macromolecules* **10**, 326 (1977).

65. J. E. G. Lipson, S. G. Whittington, M. K. Wilkinson, J. L. Martin, and D. S. Gaunt, *J. Phys. A* **18**, L469 (1985); M. K. Wilkinson, D. S. Gaunt, J. E. G. Lipson, and S. G. Whittington, *J. Phys. A* **19**, 789 (1986); S. G. Whittington, J. E. G. Lipson, M. K. Wilkinson, and D. S. Gaunt, *Macromolecules* **19**, 1241 (1986).

66. A. J. Barrett and D. L. Tremain, *Macromolecules* **20**, 1687 (1987).

67. J. Batoulis and K. Kremer, *Macromolecules* **22**, 4277 (1989).

68. J. Batoulis and K. Kremer, *Europhys. Lett.* **7**, 683 (1988).

69. W. Bruns and W. Carl, *Macromolecules* **24**, 209 (1991).

70. G. Zifferer, *Makromol. Chem.* **191**, 2717 (1990); **192**, 1555 (1991).

71. G. Zifferer, *Makroimol. Chem., Theory Simul.* **1**, 55 (1992); **2**, 653 (1993); *Macromol. Theory Simul.* **3**, 163 (1994).

72. M. P. Allen and D. J. Tildesley, *Computer Simulation of Liquids*, Clarendon Press, Oxford, 1987.

73. R. W. Hockney and J. W. Eastwood, *Computer Simulation Using Particles*, Adam Hilger, Bristol, 1988.

74. S.-J. Su, M. S. Denny, and J. Kovac, *Macromolecules* **24**, 917 (1991); S.-J. Su and J. Kovac, *J. Phys. Chem.* **96**, 3931 (1992).

75. A. Kolinski and A. Sikorski, *J. Polym. Sci., Polym. Chem. Ed.* **20**, 3147 (1982); **22**, 97 (1984).

76. J. J. Freire, J. Pla, A. Rey, and R. Prats, *Macromolecules* **19**, 452 (1986); J. J. Freire, A. Rey, and J. G. de la Torre, *Macromolecules*, **19**, 457 (1986); A. Rey, J. J. Freire, and J. G. de la Torre, *Macromolecules*, **20**, 342 (1987).

77. W. Mattice, *Macromolecules* **13**, 506 (1980).

78. B. H. Zimm, *Macromolecules* **17**, 795 (1984).

79. M. Bishop and J. H. R. Clarke, *J. Chem. Phys.* **90**, 6647 (1989).

80. A. Sikorski, *Makromol. Chem., Theory Simul.* **2**, 309 (1993).

81. R. Toral and A. Chakrabarti, *Phys. Rev. E* **47**, 4240 (1993).

82. J. F. Douglas, J. Roovers, and K. F. Freed, *Macromolecules* **23**, 4168 (1990).

83. N. Dan and M. Tirrell, *Macromolecules* **25**, 2890 (1992).

84. C. M. Wijmans and E. B. Zhulina, *Macromolecules* **26**, 7214 (1993).

85. G. A. McConnell, A. P. Gast, J. S. Huang, and S. D. Smith, *Phys. Rev. Lett.* **71**, 2102 (1993).

86. P. Tong, T. A. Witten, J. S. Huang, and L. J. Fetters, *J. Phys. France* **51**, 2813 (1990).

87. H. Li and T. A. Witten, *Macromolecules* **27**, 449 (1994).

88. F. Candau, P. Rempp, and H. Benoit, *Macromolecules* **5**, 627 (1972).

89. A. W. Rosenbluth and M. N. Rosenbluth, *J. Chem. Phys.* **23**, 356 (1955).

90. K. Ohno, M. Schulz, K. Binder, and H. L. Frisch, *J. Chem. Phys.* **101**, 4452 (1994).

91. D. Boese, F. Kremer, and L. J. Fetters, *Macromolecules* **23**, 1826 (1990).

92. P. E. Rouse, *J. Chem. Phys.* **21**, 1272 (1953).

93. D. Richter, B. Farago, J. S. Huang, L. J. Fetters, and B. Ewen, *Macromolecules* **22**, 468 (1989).

94. D. Richter, B. Farago, L. J. Fetters, J. S. Huang, and B. Ewen, *Macromolecules* **23**, 1845 (1990).

95. B. Smit, A. Van der Put, C. J. Peters, J. de Swaan Arons, and J. P. L. Michels, *J. Chem. Phys.* **88**, 3372 (1988).

96. B. H. Zimm, *J. Chem. Phys.* **24**, 269 (1954).

97. C. Pierleoni and J.-P. Ryckaert, *Phys. Rev. Lett.* **66**, 2992 (1991); *J. Chem. Phys.* **96**, 8539 (1992).

98. B. Dünweg and K. Kremer, *Phys. Rev. Lett.* **66**, 2996 (1991); *J. Chem. Phys.* **99**, 6983 (1993).

99. K. Huber, W. Burchard, and L. J. Fetters, *Macromolecules* **17**, 541 (1984).

100. K. Huber, S. Bantle, W. Burchard, and L. J. Fetters, *Macromolecules* **19**, 1404 (1986).

101. B. A. Khorramian and S. S. Stivala, *Polymer Commun.* **27**, 184 (1986).

102. K. Huber, W. Burchard, S. Bantle, and L. J. Fetters, *Polymer* **28**, 1990 (1987); **28**, 1997 (1987).

103. D. Richter, B. Stuhn, B. Ewen, and D. Nerger, *Phys. Rev. Lett.* **58**, 2462 (1987).

104. C. W. Lantman, W. J. MacKnight, A. R. Rennie, J. F. Tassin, and L. Monnerie, *Macromolecules* **23**, 836 (1990).

105. W. Dozier, J. S. Huang, and L. J. Fetters, *Macromolecules* **24**, 2810 (1991).

106. L. Willner, O. Jucknischke, D. Richter, J. Roovers, L.-L. Zhou, P. M. Toporowski, L. J. Fetters, J. S. Huang, M. Y. Lin, and N. Hadjichristidis, *Macromolecules* **27**, 3821 (1994).

107. M. Adam, L. J. Fetters, W. W. Graessley, and T. A. Witten, *Macromolecules* **24**, 2434 (1991).

108. L. Willner, O. Jucknischke, D. Richter, B. Farago, L. J. Fetters, and J. S. Huang, *Europhys. Lett.* **19**, 297 (1992).

109. See, for instance, O. Gratter and O. Kratky, *Small Angle X-Ray Scattering*, Academic, New York, 1982.

110. H. Benoit, *J. Polym. Sci.* **11**, 507 (1953).

111. J. L. Alessandrini and M. A. Carignano, *Macromolecules* **25**, 1157 (1992).

112. J. S. Higgins and H. C. Benoit, *Polymer and Neutron Scattering*, Clarendon Press, Oxford, 1994.

113. P. Flory, *Principles of Polymer Chemistry* Cornell University Press, Ithaca, NY, 1953.

114. T. Csiba, G. Jannick, D. Durand, R. Papoular, A. Lapp, L. Auvray, F. Boué, J. P. Cotton, and R. Borsali, *J. Phys. II France* **1**, 381 (1991).

115. J. C. Horton, G. L. Squires, A. T. Boothroyd, L. J. Fetters, A. R. Rennie, C. J. Glinka, and R. A. Robinson, *Macromolecules* **22**, 681 (1989).

116. A. T. Boothroyd and R. C. Ball, *Macromolecules* **23**, 1729 (1990).

117. A. T. Boothroyd, G. L. Squires, L. J. Fetters, A. R. Rennie, J. C. Horton, and A. M. B. G. de Vallêra, *Macromolecules* **22**, 3130 (1989).

118. W. Burchard, *Adv. Polym. Sci.* **48**, 1 (1983).

119. R. Ullman, *J. Polym. Sci., Polym. Phys. Ed.* **23**, 1477 (1985).

120. H. Benoit and G. Hadziioannou, *Macromolecules* **21**, 1449 (1988).

121. M. Daoud, J. P. Cotton, B. Farnoux, G. Jannick, G. Sarma, H. Benoit, R. Duplessix, G. Picot, and P. G. de Gennes, *Macromolecules* **8**, 804 (1975).

122. D. Richter, O. Jucknischke, L. Willner, L. J. Fetters, M. Lin, J. S. Huang, J. Roovers, P. M. Toporowski, and L. L. Zhou, *J. Phys. IV Suppl.* **3**, 3 (1993).

123. O. Jucknischke, Thesis, University of Münster, Germany, 1995.

124. J. K. Percus and G. J. Yevick, *Phys. Rev.* **110**, 1 (1958).

125. J. B. Hayter and J. Penfold, *Molec. Phys.* **42**, 106 (1981).

126. J. P. Hansen and L. Verlet, *Phys. Rev.* **184**, 151 (1969).

127. G. A. McConnell, M. Y. Lin, and A. P. Gast, *Macromolecules* **28**, 6754 (1995).

128. J. Roovers, P. M. Toporowski, and J. Douglas, *Macromolecules* **28**, 7064 (1995).

129. D. B. Alward, D. J. Kinning, E. L. Thomas, and L. J. Fetters, *Macromolecules* **19**, 215 (1986); D. J. Kinning, D. B. Alward, E. L. Thomas, L. J. Fetters, and D. L. Handlin, Jr., *Macromolecules* **19**, 1288 (1986).

130. D. A. Hajduk, P. E. Harper, S. M. Gruner, C. C. Honeker, E. L. Thomas, and L. J. Fetters, *Macromolecules* **28**, 2570 (1995).

131. T. Hashimoto, Y. Ijichi, and L. J. Fetters, *J. Chem. Phys.* **89**, 2463 (1989).

132. N. Hadjichristidis, H. Iatrou, S. K. Behal, J. J. Chludzinski, M. M. Disko, R. T. Garner, K. S. Liang, D. J. Lohse, and S. T. Milner, *Macromolecules* **26**, 5812 (1993).

133. S. T. Milner, *Macromolecules* **27**, 2333 (1994).

134. E. Dubois-Violette and P. G. de Gennes, *Physics (Long Island, NY)* **3**, 181 (1967).

135. W. Burchard, K. Kajiwara, D. Nerger, and W. H. Stockmayer, *Macromolecules* **17**, 222 (1984).

136. P. G. de Gennes, *Physica (Utrecht)* **25**, 825 (1959).

137. Z. A. Akcasu, M. Benmouna, and B. Hammouda, *J. Chem. Phys.* **80**, 2762 (1984).

138. Z. A. Akcasu and H. Gurol, *J. Polym. Sci., Polym. Phys. Ed.* **14**, 1 (1976).

139. W. Burchard, M. Schmidt, and W. H. Stockmayer, *Macromolecules* **13**, 580 (1980).

140. D. Richter, K. Binder, B. Ewen, and B. Stühn, *J. Chem. Phys.* **88**, 6618 (1984).

141. M. Doi and S. F. Edwards, *The Theory of Polymer Dynamics*, Clarendon Press, Oxford, 1986.

142. P. G. de Gennes, *J. Phys. France* **36**, 1199 (1975).

143. M. Doi and N. Kuzuu, *J. Polym. Sci., Polym. Lett.* **18**, 775 (1980).

144. W. W. Graessley, *Adv. Polym. Sci.* **47**, 67 (1982).

145. J. Klein, D. Fletcher, and L. J. Fetters, *J. Chem. Soc. Faraday Symp.* **18**, 159 (1983).

146. J. Klein, *Macromolecules* **19**, 105 (1986).

147. M. Antonietti and H. Sillescu, *Macromolecules* **19**, 798 (1986).

148. K. R. Shull, E. J. Kramer, G. Hadzioannou, M. Antoniette, and H. Sillescu, *Macromolecules* **21**, 2578 (1988).

149. K. R. Shull, E. J. Kramer, and L. J. Fetters, *Nature* **345**, 790 (1990).

150. R. J. Needs and S. F. Edwards, *Macromolecules* **16**, 1492 (1983).

151. A. Sikorski, A. Kolinski, and J. Skolnick, *Macromol. Theory Simul.* **3**, 715 (1994).

152. M. Rubinstein, *Phys. Rev. Lett.* **24**, 3023 (1986).

153. N. Clarke and T. C. B. McLeish, *Macromolecules* **26**, 5264 (1993).

154. C. R. Bartels, B. Crist, L. J. Fetters, and W. W. Graessley, *Macromolecules* **19**, 785 (1986).

155. G. Kraus and J. T. Gruver, *J. Polym. Sci., Part A* **3**, 105 (1965).

156. L. J. Fetters, A. D. Kiss, D. S. Pearson, G. F. Quack, and F. J. Vitus, *Macromolecules* **26**, 647 (1993); G. Quack and L. J. Fetters, *Polym. Prep.* **18**, 558 (1977).

157. J. Roovers, *Macromolecules* **24**, 5895 (1991); private communication, 1995.

158. D. S. Pearson and E. Helfand, *Macromolecules* **17**, 888 (1984).

159. R. C. Ball and T. C. B. McLeish, *Macromolecules* **22**, 1911 (1989).

160. T. C. B. McLeish, *Physics World* **8**(3), 32 (1995).

161. N. Hadjichristidis and J. Roovers, *Polymer* **20**, 1087 (1985).

162. W. W. Graessley, *Macromolecules* **15**, 1164 (1982).

163. J. M. Carella, J. T. Gotro, and W. W. Graessley, *Macromolecules* **19**, 659 (1986).

164. W. W. Graessley and J. Roovers, *Macromolecules* **12**, 959 (1979).

165. A. T. Boothroyd, A. R. Rennie, and C. B. Boothroyd, *Europhys. Lett.* **15**, 715 (1991).

166. A. Zirkel, D. Richter, W. Pyckhout-Hintzen, and L. J. Fetters, *Macromolecules* **25**, 954 (1992).

167. J. W. Mays, N. Hadjichristidis, W. W. Graessley, and L. J. Fetters, *J. Polym. Sci.*, *Polym. Phys. Ed.* **24**, 2553 (1986).

168. L. J. Fetters, unpublished.

169. A. T. Boothroyd, A. R. Rennie, and G. D. Wignall, *J. Chem. Phys.* **99**, 9135 (1993).

170. For a discussion of the K-resins see P. Dreyfuss, L. J. Fetters, and D. R. Hansen, *Rubber Chem. Tech.* **53**, 728 (1980).

171. R. J. Sutherland and R. B. Rhodes, U.S. Pat. 5,369,564 (Nov. 1, 1994).

172. N. Hadjichristidis, private communication.

173. J. T. Gotro and W. W. Graessley, *Macromolecules* **17**, 2767 (1984).

174. R. N. Young and L. J. Fetters, *Macromolecules* **11**, 899 (1978).

175. M. K. Martin, T. C. Ward, and J. E. McGrath, in *Anionic Polymerization: Kinetics, Mechanisms and Synthesis,* Vol. 166, J. E. McGrath, ed., American Chemical Society, Washington, D.C., 1981, p. 557.

176. O. L. Marrs and L. O. Edmunds, *Adhesives Age* **14**, 15 (1971).

177. O. L. Marrs, F. E. Naylor, and L. O. Edmunds, *J. Adhes.* **4**, 211 (1972).

178. D. N. Schulz, J. C. Sanda, and B. G. Willoughby in *Anionic Polymerization: Kinetics, Mechanisms and Synthesis,* Vol. 166, J. E. McGrath, ed., American Chemical Society, Washington, D.C., 1981, p. 427.

179. H. J. Spinelli, U.S. Pat. 5,019,628 (May 28, 1991); W. Anton, H. D. Coleman, and H. J. Spinelli, International Pat. WO 9,207,014 (1992).

180. J. A. Simms, *Proceedings of International Conference on Organic Coating Science Technology,* Organic Coating Science Technology, New Paltz, NY, 1992, p. 423.

181. J. C. Chen, European Pat. EP 389791 (1990).

182. M. Uchama, Japanese Pat. 9,919,196 (1994).

183. E. Kato and K. Ishii, Japanese Pat. 9,334,941 (1993).

184. P. G. de Gennes and H. Hervet, *J. Phys. Lett. France* **44**, L351 (1983).

185. M. L. Mansfield and L. I. Klushin, *J. Phys. Chem.* **96**, 3994 (1992); *Macromolecules* **26**, 4262 (1993).

186. B. I. Voit, *Acta. Polym.* **46**, 87 (1995).

187. H. Matsuka, Japanese Pat. JP 02189307 (1989).

188. N. A. Peppas and A. B. Argade, in *Controlled Release of Bioactive Materials*, T. J. Roseman and N. A. Peppas, eds., Controlled Release Society, Deerfield, IL, 1993, p. 143; N. A. Peppas, T. Nagai, and M. Miyajima, *Pharm. Tech. Jpn.* **10**, 611 (1994).

189. R. Hattam, S. Gauntlett, J. W. Mays, N. Hadjichristidis, R. N. Young, and L. J. Fetters, *Macromolecules* **24**, 6199 (1991).

190. J. Roovers, *Polym. J.* **18**, 153 (1986).

191. T. G. Fox and P. J. Flory, *J. Phys. Colloid Chem.* **53**, 197 (1949).

192. J. W. Mays and L. J. Fetters, *Macromolecules* **22**, 921 (1989).

193. H. Wagner, *J. Phys. Chem. Ref. Data* **14**, 611 (1985).

194. Z. Xu, N. Hadjichristidis, J. M. Carella, and L. J. Fetters, *Macromolecules* **16**, 925 (1983).

TETHERED POLYMER LAYERS

I. SZLEIFER AND M. A. CARIGNANO

Department of Chemistry
Purdue University
West Lafayette, Indiana

CONTENTS

I. INTRODUCTION

The behavior of polymer molecules when one of their ends is tethered to a surface or an interface is qualitatively different than that of chain

Advances in Chemical Physics, *Volume XCIV*, Edited by I. Prigogine and Stuart A. Rice.
ISBN 0-471-14324-3   © 1996 John Wiley & Sons, Inc.

molecules in bulk. The presence of a wall limits the configurational space of the chains and the two-dimensional anchoring makes the repulsions between neighboring chains different than that of polymers in bulk solutions. Consider, for example, the case in which the tethered polymer molecules are in contact with a low-molecular-weight good solvent; that is, the solvent molecules are of the size of a polymer segment and the effective interactions between the solvent and the polymer segments is attractive. In the limit of very low surface coverages, defined as the number of polymer chains per unit area of the tethering surface, the chains are isolated from each other and they form a structure called the *mushroom* regime, in which the polymer molecules are swollen within the constraints imposed by the presence of the wall (Fig. 1a). As the two-dimensional density of polymers is increased, the chain molecules try to avoid each other in order to have as much contact as possible with the solvent molecules. The only way that the chains can achieve this is by stretching out of the surface forming a layer, for moderate surface coverage this is called the *brush* regime (Fig. 1b). Clearly, there is also the interesting regime between these two limiting cases, and further one should expect dramatic changes in this behavior when the solvent is changed or if the surface has specific interactions with the segments of the polymer chain.

The interest in tethered polymer goes back to the middle of the century when it was found that the end grafting of polymer molecules to colloidal particles is a very effective way to prevent flocculation [1–11]. Namely, one can achieve colloidal stabilization by modifying the interactions between the large particles by the presence of a polymer layer [12]. Since then it has been found that tethered polymer layers can be very useful in many important applications, including the compatibility of bioimplants [13], the development of new adhesive materials [14,15], chromatographic devices [16], lubricants [17], the prevention of protein adsorption to biosurfaces [18], and drug delivery [19], among many others. Moreover, diblock copolymer molecules, that is, two polymer molecules formed by momomers of different chemical structure chemically bound at one of their ends, are the high-molecular-weight analog of short-chain amphiphilic molecules. Thus, they may be surface active and may also form a variety of phases, such as micellar and lamellar, among many others [20]. One can look at these aggregates as being formed by two tethered layers in which each block is anchored to an interface of opposite curvature (Fig. 2). Further, the molecular understanding of the organization of the layers, the mechanical properties, and the thermodynamic behavior of the modified surfaces can be used in the development of layers with desired properties.

Mushroom

Brush

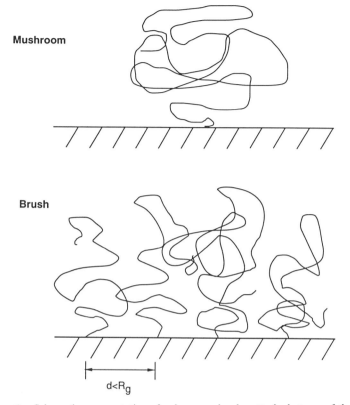

$d<R_g$

Figure 1. Schematic representation of polymer molecules attached at one of their ends to a planar surface in the good-solvent regime. The polymer molecules are represented by lines while the solvent molecules are not shown. (*a*) Isolated polymer molecule in the "mushroom" regime. The distance between neighboring molecules is much larger than the radius of gyration of the chain. The molecules are expanded coils very similar to those in bulk solution. (*b*) For large surface coverages, such that the distance between neighboring chains is much smaller than the radius of gyration of the chain, there is the formation of a polymer "brush." In this regime the chains stretch perpendicular to the surface to gain as much as possible contact with the solvent.

The study of tethered layers extends to many fields, including physics, chemistry, material science, and engineering. During the last two decades many scientists have investigated the behavior of tethered layers using experimental and theoretical methodologies. On the experimental side the measurements of the force–distance profiles have provided very valuable information on how the modified surfaces interact with each other [21–27]. Scattering techniques have been applied to elucidate the structure of the layers by providing the variation of the monomer density

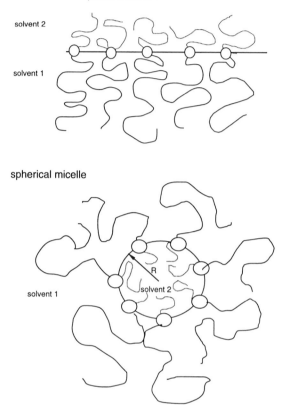

Figure 2. Schematic representation of diblock copolymers at the interface between two immiscible solvents. The circles between the blocks represent the chemical link between the two blocks. The planar interface can be part of a single layer of a multilamellar structure. The lower picture shows a cut of a spherical micelle. The radius of the interface in this case is measured from the center of the aggregate. It is positive for the blocks represented by the full line (outer blocks) and negative for the blocks represented by the dotted line (inner blocks).

as a function of the distance from the tethering surface and how this structural property changes with the quality of the solvent [28–43]. More recently the lateral interactions among the chains have been studied by measuring the reduction of the surface tension of an interface by the addition of diblock copolymer [44–48]. On the theoretical side there are scaling approaches [14,15,49–76], self-consistent field (SCF) calculations

[77–93], analytical solutions of the SCF equations for particular conditions [94–102], molecular approaches [103], and computer simulations including molecular dynamics (MD) [104–107] and Monte Carlo (MC) [108–122] methodologies.

There have been several excellent review articles covering some of the different aspects of the work mentioned above. Patel and Tirrell [123] wrote an extensive article on the force measurements in diblock copolymers adsorbed to surfaces. Halperin et al. [124] have compiled a comprehensive review that covers static and dynamic properties both experimentally and theoretically. On the theoretical side they mostly covered work carried out with the SCF theory and analytical approaches. Milner [125] has an article in which the analytical approaches are discussed in detail. More recently, Halperin [126] reviewed the application of the blob concept (scaling approach) to several problems related to chains tethered at interfaces. The simulation work has been extensively discussed by Grest and Murat [127].

The aim of this chapter is to review our understanding of the behavior of tethered polymer molecules concentrating on the coupling that exists between the molecular organization in the layer and the resulting thermodynamic behavior. This includes the work that has not been reviewed in the aforementioned review articles and the understanding that stems from the application of a molecular theoretical approach that enables us to look in detail at how the molecular organization of the polymer layers changes due to changes in thermodynamic variables (e.g., density and temperature) and due to changes in the molecular architecture of the chain molecules composing the layer. The range of applicability of the different theoretical approaches will be discussed in detail. Moreover, comparison with experimental observations and predictions of behaviors not yet found in measured systems will be presented.

Most of the work reviewed here is based on our own work on the development and application of the single-chain mean-field theory to the understanding of tethered polymer layers. Further, we concentrate our attention on monodisperse chain molecules of intermediate chain length in a sea of low-molecular-weight solvents. This range of chain lengths is interesting due to the fact that it cannot be described by the scaling and analytical theories, which are exact in the asymptotic limit of very long chain lengths, and also because such intermediate sizes of molecules find many applications in a variety of systems from biocompatibility [18, 19] to water clean-up [128]. Moreover, it will be shown that the use of the molecular theory enables the quantitative evaluation of the properties of the layers for all ranges of surface coverages.

II. THEORETICAL APPROACHES

The description of the theoretical methods will be on the simplest layer, that is, that containing a monodisperse sample of polymers attached to the surface. Throughout this chapter the generalizations necessary to treat other, more complex systems will be given.

The system of interest is composed of N_p polymer molecules attached at one of their ends to a surface of total area A and embedded in a sea of low-molecular-weight solvent (Fig. 3). The sea of solvent is actually a bath of solvent molecules at constant chemical potential μ_s. The surface coverage is defined as $\sigma = N_p/A$. We define z as the distance perpendicular to the surface and consider polymer molecules composed by n segments, each of volume v_p and bond length l, and the solvent volume is given by v_s.

In principle, the best way to solve this problem is by exact solution of the equation of motion of the molecules (MD) or by sampling of configuration space (MC). To do this, the interactions between the different units, polymer segments, and solvent molecules need to be defined and then the simulations can be run, thus obtaining the average properties of the system. This has been done in a variety of cases providing very valuable information about grafted polymer systems; for a recent review see ref. 127. The main advantages of computer simulations is that they provide the exact solution of the model chosen for the

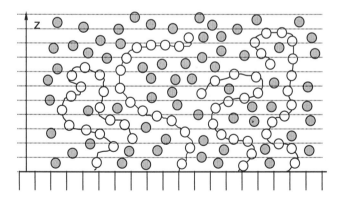

Figure 3. Schematic representation of a tethered polymer layer in a sea of low-molecular-weight solvent. The z axis is defined perpendicular to the surface, and the z layer is defined by the volume between z and $z + dz$.

molecules, and therefore they include information on the fluctuations in the system that in general are not taken into account in other theoretical descriptions. Thus, they also serve to check the validity of the approximations necessary to derive theoretical approaches. Furthermore, one can also study the average equilibrium properties of the system as well as the dynamic behavior. The main setbacks of this methodology are as follows:

1. They require a very large amount of computer time, and thus many systematic studies as a function of all the relevant variables in the problem are extremely time consuming.
2. The evaluation of thermodynamic information such as chemical potentials is very hard to obtain with a good degree of accuracy, in particular for intermediate to long chain lengths [129].

All the other theoretical methodologies require the use of approximations. One that is common to all the approaches is the incompressibility assumption that we describe next.

A. Volume-Filling Constraints or Incompressibility Assumption

The polymer solvent mixture in the tethered layers is at a liquidlike density. Most theoretical approaches assume that this liquid layer is incompressible, namely, that the volume accessible is filled by solvent molecules or polymer segments. In the case of chains attached to a surface or an interface the z direction is inhomogeneous; that is, the density of polymer as a function of the distance from the surface is not constant (see Fig. 3). In addition, it is assumed that in the planes parallel to the surface (xy) the system is homogeneous. Then, the assumption of incompressibility at all distances from the surface implies that

$$\langle \phi_p(z) \rangle + \langle \phi_s(z) \rangle = 1 \qquad 0 \le z \le \infty \qquad (1)$$

where $\langle \phi_p(z) \rangle$ and $\langle \phi_s(z) \rangle$ are the (ensemble) average volume fraction of polymer and solvent at layer z. The layer is defined as the volume occupied between z and $z + dz$.

We can write Eq. (1) in terms of the many chain probability distribution function of conformations, $P(\alpha_1, \alpha_2, \ldots, \alpha_{N_p})$, with α_i denoting the conformation of chain i, and the volume occupied by the chains at layer z, $[v_1(z; \alpha_1) + v_2(z; \alpha_2) + \cdots + v_{N_p}(z; \alpha_{N_p})] dz$, with $v_i(z; \alpha_i)$ being the volume that chain i in conformation α_i occupies at layer z. This yields

$$\frac{1}{A} \left(\sum_{\{\alpha_1, \alpha_2, \ldots, \alpha_{N_p}\}} P(\alpha_1, \alpha_2, \ldots, \alpha_{N_p})[v(z; \alpha_1) + v_2(z; \alpha_2) + \cdots + v_{N_p}(z; \alpha_{N_p})] \right.$$

$$\left. + N_s(z)v_s \right) = 1 \qquad 0 \leq z \leq \infty \tag{2}$$

where $N_s(z)\,dz$ is the number of solvent molecules at layer z.

Since all the polymers molecules are equivalent, we can write Eq. (2) in terms of the single-chain probability distribution function (pdf) in the form

$$\sigma \sum_{\{\alpha\}} P(\alpha)v_p(z; \alpha) + \frac{N_s(z)v_s}{A} = \sigma \langle v_p(z) \rangle + \langle \phi_s(z) \rangle = 1 \qquad 0 \leq z \leq \infty$$

$$\tag{3}$$

with $P(\alpha)$ denoting the probability of finding a chain in conformation α and the sum runs over all the possible configurations of a single chain.

The incompressibility assumption, or volume-filling constraint, is the way in which the repulsive interactions between segments (polymeric and/or solvent) are taken into account in the theories. In other words, no two segments can occupy the same volume due to the repulsive interaction between them. This is clearly an approximate way to consider these repulsions, but as it will be shown below, it seems to take into account these interactions in an appropriate way.

The next step in the derivation of a theory is to find the polymer and solvent density profiles that fulfill the volume constraint equations and that also properly describe the behavior of the chain molecules.

B. Single-Chain Mean-Field Theory

The single-chain mean-field [103,130–132] (SCMF) theory is based on looking at a central chain with all its intramolecular interactions taken into account in an exact way, while the interactions of this chain with the other polymer and solvent molecules are taken within a mean-field approximation. This is schematically shown in Fig. 4. The mean-field interactions are functions of the polymer density, and this density depends upon the distance from the surface due to the inhomogeneous profile in the z direction. Therefore, the probability of a given chain configuration will depend upon its intramolecular interactions and its distribution of segments that will determine, together with the density profile, the intermolecular interactions. This approach enables one to

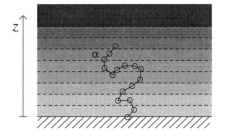

Figure 4. Schematic representation of the "central" chain in the mean-field of the surrounding polymer and solvent molecules. The different gray levels represent the different mean fields with which the central molecule interacts. The variation of the mean field is due to the inhomogeneous density of polymer and solvent as a function of z. See Fig. 3.

look at how the same set of configurations changes the relative weight of the different conformations depending upon the thermodynamic control variables, that is, surface coverage and temperature (quality of solvent). This provides insightful information on the coupling that exists between the chain conformations and the thermodynamic behavior of the layer. Namely, the polymer density profile is given by the average properties of the chain [see, e.g., Eq. (3)]. Thus, there is a coupling between the average chain configurations and the probability of a chain to have a given distribution of segments, that is, to be in a given conformation. The thermodynamic properties of the layers are determined by the average properties of the chains, and here is where the coupling between the macroscopic behavior of the layer and the microscopic characteristics of the chain molecules explicitly manifest itself.

This theoretical approach was originally developed to treat short surfactant molecules in amphiphilic aggregates [133–136], and it was later generalized for longer chains in the presence of solvent molecules [103]. One of the main advantages of the theory is that it explicitly treats the chain configurations [103], and thus the theory can be applied for any chain architecture. Moreover, since the theory requires as input the set of single-chain configurations, this approach is exact in the very low surface coverage limit. As has been shown [130,132], with increased density the predictions of the theory are in quantitative agreement with computer simulations, providing thus a very good approach applicable at all densities. The question that arises, then, is how an approach in which the intermolecular interactions are considered within a mean-field framework, but includes the exact configurations at the single-chain level, can be quantitative for both conformational and thermodynamic properties. This will become clear below once the second derivation of the pdf is shown. A discussion of the limitations of the approach will then follow.

1. Minimization of Free Energy

We are interested in finding the pdf of chain conformations $P(\alpha)$ from which the polymer density profile and all other average conformational and thermodynamic properties can be obtained and the solvent density profile. The simplest way to do this is by explicitly writing the free-energy density of the layer, free energy per unit area, in terms of the pdf and the solvent density profile and obtain their functional form by minimization subject to the volume-filling constraints [Eqs. (3)].

To write the free energy, we have to keep in mind that the inter-molecular repulsive interactions are already taken into account through the incompressibility constraint. The intramolecular repulsions are taken through the set of single-chain configurations; that is, a set of self-avoiding walks is used.

The free energy will have the following terms:

1. The translational (two-dimensional) entropy of the polymer chains. This term will be absent if the chains are grafted to the surface. However, since it is of the form $\ln \sigma$, the functional form of the pdf and the solvent density profile are unaffected whether it is present or not.

2. The conformational entropy of the chains. This term is of the form $-\Sigma_{\{\alpha\}} P(\alpha) \ln P(\alpha)$.

3. The internal energy of the chains, $\langle \varepsilon_{\text{int}} \rangle$. This term is exactly taken into account and it contains two contributions, one arising from the intramolecular attractive interactions and the other arising from the bare interactions of the polymer monomers with the tethering surface.

4. The intermolecular (mean-field) attractions between the chain segments. It is important at this point to recall that since we are considering an incompressible system there is only one interaction parameter that determines the behavior of the system. We chose to model it by the effective attractions between the chains since in this way it provides a direct way to visualize the effect of the quality of the solvent to the distribution of conformations of chain molecules. This term has the form $\frac{1}{2}\sigma \int \int \chi(|z - z'|)\langle v_p(z)\rangle\langle \phi_p(z')\rangle \, dz \, dz'$, with $\chi(|z - z'|)$ being the average interaction parameter between a segment at z and the "mean field" at z' (see the Appendix).

5. The last contribution arises from the translational entropy of the solvent molecules and their chemical potential. These have the form $\int \langle \phi_s(z)\rangle(\ln \langle \phi_s(z)\rangle + \mu_s) \, dz$, which is integrated over the whole layer to account for the different contributions arising from the variation of the solvent density as a function of z.

Summing all these contributions, we obtain, for the free energy density,

$$
\beta \frac{f}{A} = \sigma \sum_{\{\alpha\}} P(\alpha) \left[\ln P(\alpha) + \beta \varepsilon_{\text{int}}(\alpha) + \beta \frac{1}{2} \int \int \chi(|z - z'|) v_p(z; \alpha) \right.
$$

$$
\left. \times \left(\sigma \sum_{\{\gamma\}} P(\gamma) v_p(z', \gamma) \right) dz \, dz' \right]
$$

$$
+ \sigma \ln \sigma + \int \langle \phi_s(z) \rangle (\ln \langle \phi_s(z) \rangle + \mu_s) \, dz \tag{4}
$$

The terms proportional to the surface coverage, σ, correspond to the free energy per chain molecule, while the other term is the solvent contribution.

We can now minimize Eq. (4) subject to the volume-filling constraints [Eqs. (3)]. Introducing the set of Lagrange multipliers $\beta\pi(z)$, we obtain, by straightforward minimization for the pdf of chain conformations,

$$
P(\alpha) = \frac{1}{q} \exp \left[-\beta \left(\int \pi(z) v_p(z; \alpha) \, dz + \varepsilon_{\text{int}}(\alpha) + \int \int \chi(|z - z'|) v_p(z; \alpha) \right. \right.
$$

$$
\left. \left. \times [1 - \langle \phi_s(z') \rangle] \, dz \, dz' \right) \right] \tag{5}
$$

where we have used the constraint equation to replace the volume fraction of polymer by that of the solvent and q is the normalization constant that ensures $\Sigma_{\{\alpha\}} P(\alpha) = 1$.

From the solvent density profile we obtain

$$
\langle \phi_s(z) \rangle = \exp[-\beta(\pi(z) v_s - \mu_s)] \tag{6}
$$

Now we have explicit closed-form expressions for the pdf and the solvent density profile. The only unknown are the set of Lagrange multipliers $\{\pi(z)\}$. The way to determine them is by introducing expressions (5) and (6) into the constraint equations (3), which provides for a set of equations in which the only unknowns are the Lagrange multipliers. The input necessary to solve these equations are the surface coverage, the temperature (quality of solvent), the solvent chemical potential, and the set of single-chain conformations; see the Appendix for details on how the calculations are carried out.

It is interesting to see what is the physical meaning of the Lagrange multipliers. This can be seen from Eq. (6), which shows that these quantities are the osmotic pressure necessary to have a constant solvent chemical throughout the layer. Further, since $\{\pi(z)\}$ are the Lagrange multipliers conjugated to the packing constraints, they represent the average repulsive energy by the chains from the neighboring polymer and solvent molecules. A more thorough discussion of this point is presented in the next section, where the pdf is derived by considering the many molecular partition function.

2. Expansion of Partition Function

The derivation of the pdf by minimization of the free energy of the system provides a simple way to visualize what are the relevant contributions that determine the behavior of the layers. However, it is not easy to see what are the approximations involved in the derivation of the pdf and the solvent density profile. This is best obtained by determining these quantities from the expansion of the total partition function of the system.

To keep the discussion as simple as possible, we will consider the case of an athermal polymer–solvent mixture. This is a good representation of the good-solvent regime in which effectively the polymer segments like to be in the environment of the solvent molecules. We denote by $Q(N_p, N_s, V, T)$ the canonical partition function of a system containing N_p polymer molecules tethered at one of their ends to a surface or interface, and there are N_s solvent molecules in a total volume V at temperature T. The probability of finding the chains in configuration $\alpha_1, \alpha_2, \ldots, \alpha_{N_p}$ is given by

$$
P(\alpha_1, \alpha_2, \ldots, \alpha_{N_p})
$$

$$
= \frac{\int \cdots \int \exp[-\beta U(\alpha_1, \alpha_2, \ldots, \alpha_{N_p}, r_1, \ldots, r_{N_s})]\, dr_1 \cdots dr_{N_s}}{Q(N_p, N_s, V, T)} \tag{7}
$$

where r_i denotes the position of solvent molecule i and the integrals are performed over all possible configurations of the solvent molecules. Here, $U(\alpha_1, \alpha_2, \ldots, \alpha_{N_p}, r_1, \ldots, r_{N_s})$ is the total interaction energy between the polymer and solvent molecules.

The probability of finding, without the loss of generality, chain 1 in configuration α is given by the sum over all the possible configurations of the other $N_p - 1$ polymer molecules of the total pdf given by Eq. (7).

Namely,

$$
P(\alpha) = \frac{\Sigma_{\{\alpha_2,\ldots,\alpha_{N_p}\}} \int \cdots \int \exp[-\beta U(\alpha_1, \alpha_2, \ldots, \alpha_{N_p}, r_1, \ldots, r_{N_s})] \, dr_1 \cdots dr_{N_s}}{Q(N_p, N_s, V, T)}
$$

(8)

Now we separate the interaction energy into two terms, one containing the interactions of the central chain in conformation α with all the other polymer and solvent molecules. We denote this quantity $u(\alpha; \underline{R})$, where $\underline{R}$ is a vector that represents the coordinates of all other polymer and solvent molecules. The second term includes the interactions between the other $N_p - 1$ polymer molecules and N_s solvent molecules, and it is denoted $U(\alpha_2, \ldots, \alpha_{N_p}, r_1, \ldots, r_{N_s})$. Then the pdf of a single chain can be written as

$$
P(\alpha) = \frac{\Sigma_{\{\alpha_2,\ldots,\alpha_{N_p}\}} \int \cdots \int \exp[-\beta U(\alpha_2, \ldots, \alpha_{N_p}, r_1, \ldots, r_{N_s})] \exp[-\beta u(\alpha; \underline{R})] \, dr_1 \cdots dr_{N_s}}{Q(N_p, N_s, V, T)}
$$

(9)

Since we are considering only excluded-volume interactions, we can see that in the numerator in Eq. (9) the Boltzmann factor of the central chain is 1 if the central chain does not overlap with any other molecule or zero if it does. Thus, effectively the central chain in conformation α reduces the volume available to all the other molecules by that taken by its segments (Fig. 5). The numerator can be interpreted as the partition function of a system with $N_p - 1$ polymer molecules, and with N_s solvent molecules that have available the volume $\int [A(z) - v_p(z; \alpha)] \, dz$, that is, the total volume $[V = \int A(z) \, dz]$ minus the volume taken by the chain in conformation α. Thus,

$$
P(\alpha) = \frac{Q(N_p - 1, N_s, \int [A(z) - v_p(z; \alpha)] \, dz, T)}{Q(N_p, N_s, V, T)}
$$

(10)

Since the volume taken by one polymer molecule is microscopic and $A(z) \, dz$ is macroscopic, we can expand the pdf in Eq. (10) in terms of the volumes, $v_p(z; \alpha) \, dz$, and also in the number of polymer molecules to

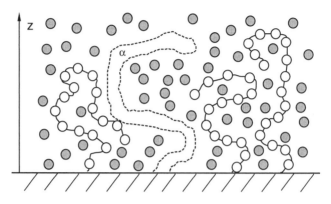

Figure 5. Schematic representation of the system whose partition function is $Q(N_p - 1, N_s, \int [A(z) - v_p(z; a)] \, dz, T)$, that is, a partition function with $N_p - 1$ polymer molecules in which the volume occupied by the central chain in conformation α is not accessible to the other polymer and solvent molecules. See Eq. (10).

obtain

$$\ln P(\alpha) = -\left(\frac{\partial \ln Q}{\partial N_p}\right)_{N_s, V, T} - \int \left(\frac{\partial \ln Q}{\partial A(z)}\right)_{A(z' \neq z), N_p, N_s, T} v_p(z; \alpha) \, dz$$

$$(11)$$

where higher order terms are not taken into account because they are negligible in the thermodynamic limit [134]. Rearranging the pdf, we get

$$P(\alpha) = \frac{1}{q} \exp\left[-\beta \int \pi(z) v_p(z; \alpha) \, dz \right] \qquad (12)$$

where we have defined the quantities

$$\beta \pi(z) = \left(\frac{\partial \ln Q}{\partial A(z)}\right)_{A(z' \neq z), N_p, N_s, T} \qquad (13)$$

and the single-chain partition function is given by $q = \exp(-\beta \mu_p)$. As can be seen, the pdf obtained from the expansion of the partition function is identical to that obtained by minimizing the free energy. Further, the meaning of the quantities $\pi(z)$ is transparent from this derivation. Namely, they represent (minus) the change in the free energy of the layer when the volume at z is increased. Therefore, they are the lateral pressures responsible for stretching the chains.

For the solvent volume fraction we can follow the same lines as those used in the derivation of Eq. (12), and again we will obtain the same expression as derived from the direct minimization of the free energy with $\pi(z)$ defined in Eq. (13).

The main advantage of this more length derivation is that it shows that the form of the pdf is not a mean-field result but can be obtained by proper steps from the partition function. Moreover, it shows that the only approximation in the *single-chain* pdf is the replacement of the chain configuration by the volume the conformation occupies. As discussed in detail elsewhere [134], this is a very good assumption that will further be confirmed in the sections below.

It is important to realize that the main approximation enters in the way the lateral pressures are calculated. This is in reality where the mean-field approximation is used, namely, the use of the packing constraints as the source to determine $\pi(z)$. In the cases discussed here, that is, polymer–solvent mixtures, the approximation can be seen by noting that in reality the packing constraints, with the use of the solvent volume fraction derived above, are a way to include the "equation of state" into the problem. Further, it has been shown that one can also treat compressible systems with this theory [137], but this is beyond the scope of this chapter.

One of the main advantages of the theory is that the expressions obtained are independent of the chain model used for the polymer molecules. Therefore, one can treat any type of chains, including free space and lattice. Further, the self-consistent field theory described below has been shown [103] to be a particular case in which the chains are generated by the Green functions of a random walk in the presence of a field; see discussion below.

3. Thermodynamic Quantities

The theory just described gives simple expressions to determine the thermodynamic behavior of the polymer solvent layer. For example, we can write the free energy per unit area of the system by replacing the expression of the single-chain pdf [Eq. (5)] and that of the solvent density profile [Eq. (6)] into Eq. (4) to obtain

$$\beta \frac{f}{A} = \sigma \ln \sigma - \sigma \ln q - \int \pi(z) \, dz$$

$$- \frac{1}{2} \int \int \chi(|z - z'|) \langle \phi_p(z) \rangle \langle \phi_p(z') \rangle \, dz \, dz'$$

$$+ \int \mu_s \langle \phi_s(z) \rangle \, dz \tag{14}$$

where the packing constraints [Eqs. (3)] have been used.

All other thermodynamic quantities can be obtained by taking the derivatives of the free energy. For example, the lateral osmotic pressure of the layer is given by $\Pi_\alpha = -(\partial f/\partial A)_{N_p, N_s, T}$, which, using Eq. (14), yields

$$\beta\Pi_a = \sigma + \int \pi(z)\, dz - \sigma n + \frac{1}{2} \int \int \chi(|z - z'|)$$

$$\times \langle \phi_p(z) \rangle \langle \phi_p(z') \rangle \, dz\, dz' \qquad (15)$$

This is an important quantity that can be directly measured experimentally. For example, in the case of diblock copolymers adsorbed to a fluid–fluid interface, the lateral pressure is the reduction of the surface tension induced by the presence of the adsorbed polymers.

Another quantity that we will refer to below is the chemical potential of the tethered polymers. This quantity determines the amount of polymer that will be adsorbed to a surface in the case of adsorption of end-functionalized polymers. The chemical potential of the polymer chains is given by $\mu_p = (\partial f/\partial N_p)_{N_s, a, T}$, which, from Eq. (14), gives

$$\beta\mu_p = \ln \sigma - \ln q \qquad (16)$$

The terms σ in Eq. (15) and $\ln \sigma$ in Eq. (16) represent the translational contribution of the chains to the pressure and chemical potentials, respectively. These terms will be absent when the polymer molecules are grafted to the surface. Note that the configurational contribution to μ_p obtained here is the same as obtained from the expansion of the partition function.

In Section VI we will describe how to determine elastic constants for polymer layers by taking derivatives of the free energy. These derivatives can then be easily calculated from the knowledge of a single-layer geometry.

C. Self-Consistent Field Theory

The SCF theory was originally developed by Edwards [138] to treat bulk polymer systems and later generalized by Dolan and Edwards [92, 93] to treat polymer molecules in inhomogeneous environments. Since then, the SCF theory has been applied in a variety of problems related to tethered polymer chains. Scheutjens and Fleer [85, 86] have developed the approach for lattice chains. The basic idea of this approach is to solve for the Green function of a random walk in the presence of the mean field imposed by the repulsive and attractive interactions. This mean field is a function of the density, which in turn is determined, self-consistently, by the average value obtained from the Green functions.

The SCF theory can be derived also from the SCMF presented above. The way to do that is to replace the sum over chain conformations by the appropriate product of distribution of segments, that is, the Green functions. The advantage of presenting the SCF in this way is that then the thermodynamic functions can be obtained in a straightforward way as described in the previous section. However, it has to be kept in mind that the replacement of the pdf by the Green functions implies that now the chain molecules can no longer be treated as whole units; for a detailed discussion see the Appendix of ref. 103 and ref. 139.

The main advantage of the SCF approach over the SCMF theory is that since the chains are obtained as random walks on an external field, one can perform calculations for longer chain lengths than those computationally feasible from the SCMF theory. However, the reference state for the SCF approach is a Gaussian chain, which is not the appropriate state for polymers in good solvents, while in the SCMF theory the reference state is the exact single chain in solvent. The effect of this on predictions from both theories will be discussed in detail below.

D. Analytical Approaches

The self-consistent equations of the previous section have an analytical solution, for a particular limit, usually referred to as the *classical limit*. This regime is obtained, in good solvents, for very long chains (formally when $n \rightarrow \infty$) when the surface is relatively high and the chains are stretched. In this limit, based on ideas of Semenov [140] for diblock copolymers, Milner et al. [94, 95] and independently Zhulina et al. [100] have found that the density profile of the grafted chains is parabolic. Using the parabolic profile Milner et al., and Zhulina et al. derived scaling relations for the structure and thermodynamic properties of the layers. A very good review of the basic ideas and the main results obtained by the analytical SCF approach has been published by Milner [125]. Further, the analytical approach was generalized for a variety of different systems [96–99, 101, 102, 141–143].

The other main analytical approach is based on the seminal work of Alexander [49] in which the behavior of the grafted layers was obtained by assuming (also in the limit of highly stretched chains) that the density profile is a step function. In this case one can explicitly write the free energy of the layer by balancing the stretching entropy with the inter-action term. The scaling predictions obtained by this approach are very similar to those of the analytical SCF approach even though the latter provides better predictions for the structure of the layer. A recent review of the application of such theoretical ideas has been written by Halperin [126].

The main results from both approaches in the good-solvent regime concern the scaling behavior of the thickness of the grafted layer, h. This quantity depends upon the chain length and the surface coverage as

$$h \propto n\sigma^{1/3} \qquad (17)$$

while the free energy of the chains is found to have the scaling form

$$f \propto n\sigma^{2/3} \qquad (18)$$

The validity of these predictions as well as others obtained from the analytical approaches will be discussed below. The main advantage of the analytical approaches is that they provide very simple expressions from which the physical behavior of the layers can be understood from the main contributions to the free energy. Namely, they provide physical insight into what are the factors determining the behavior of the system.

The theoretical study of polymer systems in general and those of interest here in particular, by the proper combination of computer simulations, molecular theories, self-consistent field approaches, and analytical theories, provide complete understanding of the behavior of macromolecular fluids. The different methodologies complement each other in their range of applicability and ability to be applied in different regimes of interest.

III.   TETHERED POLYMERS IN GOOD-SOLVENT REGIME

In this section we describe the behavior of tethered polymers in the good-solvent regime, namely, in the regime in which the effective interactions between polymer monomers and solvent are more favorable than those between polymer segments among themselves. In other words, in this regime the polymer monomer–polymer monomer second virial coefficient is positive.

Most of the results that will be shown were obtained by the SCMF theory. This approach has been shown to provide predictions that quantitatively agree with MD and MC simulations for conformational properties [132] and the few available calculations for thermodynamic quantities [130]. In some cases comparisons with simulations will be shown as well as with other theoretical approaches and experimental observations.

The predictions of the SCMF theory presented in this chapter have been carried out using the rotational isomeric state model [144] for the chain molecules in free space, unless otherwise stated. The details of the calculations are presented in the Appendix, and for the chain model the reader is referred elsewhere [130, 144]. The unit length is taken to be the bond length l; thus area is measured as l^2, and the free energies are measured in units of the thermal energy kT.

A.   Conformational Behavior of Linear Chains

One of the properties that characterizes the structure of the layer is the variation of the polymer volume fraction (density) as a function of the distance from the tethering surface. These density profiles are a function of the grafting density. In the good-solvent regime we expect the isolated chain to be a swollen chain with the same characteristics as that of a polymer chain in very dilute solution. This was shown by computer simulations [145] in which it was found that a grafted single chain is different from the bulk single chain, as far as its scaling behavior, only in the so-called enhancement exponent [146]. Namely, the number of self-avoiding walks (SAWs) is given by

$$N_{SAW} = C z_{eff}^n n^{\gamma - 1} \tag{19}$$

where γ is the enhancement exponent, $\gamma_{graf} = 0.695$ while $\gamma_{bulk} = 7/6$; z_{eff} is the effective coordination number of the lattice, which is the same for grafted and bulk chains; and C is a constant. The exponent describing the variation of the radius of gyration with chain length, both parallel and perpendicular to the walls, is the same as that of the bulk chain.

As the surface coverage increases, the repulsions between the chains start to play a role and the polymer molecules stretch perpendicular to the surface. The analytical SCF theory [94, 95, 100] predicts that for moderately high surface coverage the profile is parabolic, while the theory of Alexander [49] assumes that it is a step function. Further, Shim and Cates [141] have extended the analytical SCF approach by considering finite extensibility of the chain, and they found that the density profile is flatter than parabolic for very high surface coverages.

Figure 6 shows the average number of polymer segments as a function of the distance from the grafting surface from the limit of isolated chains to high surface coverages as predicted by the SCMF theory. The volume fraction is obtained by multiplying the average number of segments by the surface coverage. The figure shows the qualitative changes of the shape of the chain profile as the surface coverage increases. The free chain, that is, a chain with the boundary conditions imposed by the

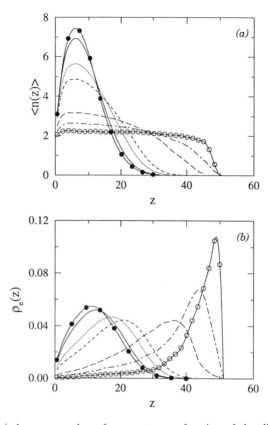

Figure 6. (*a*) Average number of segments as a function of the distance from the tethering surface for chains with $n = 100$ in the good-solvent regime. (*b*) The probability of finding the free end of the chain as a function of the distance from the tethering surface. The distance is measured in units of bond length. Full line with filled symbols: free chain; full line: $\sigma = 0.001$; dotted line: $\sigma = 0.005$; dashed line: $\sigma = 0.01$; long-dashed line: $\sigma = 0.05$; dot-dashed line: $\sigma = 0.1$; and full line with open symbols: $\sigma = 0.2$.

surface but surrounded only by solvent molecules, has a profile that shows that the chain swells in the three directions. The presence of the neighboring molecules is felt even for very low surface coverages, where it can be observed that the chains start to stretch perpendicular to the surface for average distances between molecules much larger than the average lateral extension of the chains (see Fig. 8 on page 187). In the intermediate range of surface coverages the profiles are more or less parabolic beside a small depletion layer near the surface and an exponential-like tail at the end of the layer. For very large surface coverages the profile becomes more and more steplike due to the very strong repulsions between the chain molecules, which are now tethered to the surface at average distances much smaller than the lateral dimensions of the free chains.

The figure also shows the distribution of the free ends for a variety of surface coverages. The shape of the end profiles clearly reflects the stretching of the chains as the surface coverage increases. There is a qualitative change in these profiles from the mushroom regime to the highly stretched brush regime. A more detailed analysis of these distributions will be presented below.

The question that arises next is whether the density profile reflects the average shape of the polymer molecules. To show that this is indeed the case, in Fig. 7 we show the variation of the z-dependent lateral radius of gyration, defined by

$$
\langle R_{g,xy}^2(z)\rangle = \tfrac{1}{2}[\langle R_{g,x}^2(z)\rangle + \langle R_{g,y}^2(z)\rangle]
$$

$$
= \frac{1}{2n}\sum_{\{\alpha\}} P(\alpha)\left(\sum_{i=1}^{n} [x_i(z;\alpha) - x_{cm}(z;\alpha)]^2\right.
$$

$$
\left. + [y_i(z;\alpha) - y_{cm}(z;\alpha)]^2\right) \tag{20}
$$

where cm denotes center of mass. Even though the density profiles do not quantitatively scale with the lateral radius of gyration, the qualitative

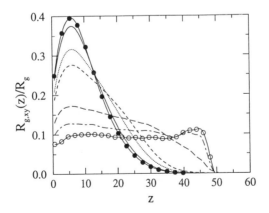

Figure 7. The z-dependent lateral radius of gyration, scaled by the bulk radius of gyration, as a function of the distance from the tethering surface for chains with $n = 100$ in the good-solvent regime. Lines and symbols as in Fig. 6.

changes of the shape of the molecules can be well understood from any of these quantities.

More interestingly, the z-dependent radii of gyration provide a better picture of the arrangement of the layers by replotting them in a form that includes the neighboring molecules. To visualize the relevant length scales of the tethered layers, we will scale all the distances by the bulk radius of gyration of the chains. This is clearly the only relevant length scale in the limit of very low surface coverages since the chains are isolated from each other. Thus, we define the reduced surface coverage as $\sigma^* = \sigma \pi R_g^2$, which provides the number of chains per unit area, with the area measured in terms of the cross-sectional area of a bulk isolated chain. For large values of σ^*, that is, in the so-called brush regime, the bulk radius of gyration can no longer be the relevant length scale. The analytical approaches consider that in this regime the only relevant length scale is the distance between grafting points. This is expected since the chains are highly stretched and therefore the chains do not resemble at all the isolated chains. Actually, they are described as a chain of stretched "blobs" each of size $\sigma^{-1/2}$ [49].

When considering two regimes (mushroom and brush), this description is very useful since it provides a simple picture. However, the questions that arise are what the crossover regime is and how broad it is. This is particularly important when trying to interpret experimental observations in terms of analytical theories that have been designed for a particular range of surface coverages.

Let us first look at the shape of the chain molecules for a variety of reduced surface coverages. Figure 8 shows the z-dependent lateral radius of gyration as a function of the distance from the surface scaled by the bulk R_g. In the very low surface coverage regime, $\sigma^* = 0.5$, the chains are already feeling each other and they start to stretch. However, the average distance between tethering points is large enough so that the chains are in a mushroom-like structure. Increasing the surface coverage to five times the cross-sectional area of the bulk molecule results in a more stretched average structure in which the chains also compress in the direction parallel to the surface. Furthermore, the chains almost overlap with each other on average. The two other curves show much larger surface coverages in which the distances between grafting points are $0.28R_g$ and $0.22R_g$. These are well in the brush regime and the chains are highly stretched. There is also a considerable amount of overlap between the chains. However, the lateral spreading of the chains is not found to follow the predictions of de Gennes [52] for the lateral dimensions of a brush.

An important point that will be discussed throughout this chapter is

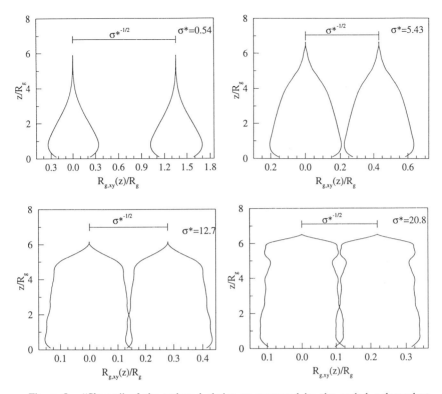

Figure 8. "Shapes" of the tethered chains as measured by the scaled z-dependent radius of gyration. Two neighboring molecules are shown; the distance between the tethering points is $\sigma^{*-1/2}$.

that in general most of the predictions obtained from the SCMF theory have been done for chains of $n \leq 250$. In this range of chain lengths it turns out that scaling predictions are not appropriate due to the finite chain length used in the calculations. This is also true for the computer simulation results reviewed in ref. 127. However, these approaches shed light on the behavior of chains of intermediate chain length that are realistic polymer molecules, in particular for biological applications [19]. Further, the SCMF theory predictions are in rather good agreement with experimental observations for quite long (real polymeric) chain lengths [147]. Therefore, it is important to emphasize in which ranges of chain length and surface coverage one should expect the analytical (scaling) predictions to apply and in which they should not.

A better way to visualize the mushroom-to-brush transition is by looking at the height of the tethered layer as a function of surface coverage. A way to quantify this property is by calculating the first moment of the density profile, namely,

$$\langle z \rangle = \frac{\int \phi(z) z \, dz}{\int \phi(z) \, dz} \tag{21}$$

where we will define the height of the brush by $h = 2\langle z \rangle$. This quantity, scaled by the bulk radius of gyration, is shown in Fig. 9 as a function of the reduced surface coverage in a log-log representation. While the choice of R_g as a length scale may not be the most appropriate for all the surface coverage regimes, experimental observations (to be shown below) suggest that this is a proper choice for the range of measured heights.

The scaling of the height is not perfect; that is, the different results do not perfectly collapse into a universal curve. However, they provide a good way to discuss the data. At very low σ^* the height of the layer changes very slowly, showing the mushroom regime. At $\sigma^* \simeq 1$ the height start to increase sharply with surface coverage. However, the brush regime seems to start for the chain lengths shown here at $\sigma^* > 5$. The full line represents $h \propto \sigma^{*1/4}$, while the dashed line corresponds to the analytical predictions $h \propto \sigma^{*1/3}$.

The power law found to fit the calculations is not exactly the one predicted by the analytical approaches and reflects the fact that the chain lengths used in the calculations presented here are not very long.

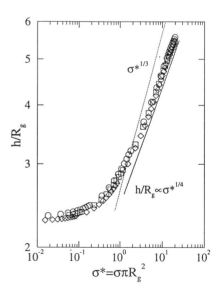

Figure 9. Log-log representation of the height of the layer, scaled by the bulk radius of gyration, as a function of the scaled surface coverage. Circles $n = 100$, squares $n = 70$, and diamond $n = 50$. The full line represents the best fit to the data in the "brush" regime, while the dotted line represents the prediction of the analytical approaches.

However, it is a good representation of the brush regime, which is the main purpose of the discussion at this point.

The layer height is a quantity that is experimentally accessible. Kent et al. [45–48] have measured the density profiles and the height of a layer of polystyrene blocks as a function of the surface coverage using neutron reflectivity. The layer was formed by spreading polystyrene–polydimethylsiloxane (PS–PDMS) diblock copolymers at the ethyl benzoate–air interface.

The results of the measurements for a variety of surface coverages are shown in Fig. 10. The experimental results (filled symbols) seem to suggest that R_g is a relevant length scale for all the surface coverages. However, the theoretical predictions (open symbols) are in excellent quantitative agreement with the experimental observations, but as discussed above, this form of plot does not necessarily suggest that the only length scale in the problem is the bulk radius of gyration. It is important to emphasize also that the experimental observations cover the range of reduced surface coverages from the mushroom into brush regime. Therefore, they cannot be interpreted with the analytical brush approaches since they are applicable only in the brush regime.

It is interesting to note that the experimental observations include chain lengths that have a high molecular weight and the predictions of the SCMF theory are very good, even though these were obtained for chain lengths up to $n = 100$. Similar good agreement has been obtained using

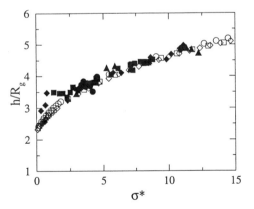

Figure 10. Same as in Fig. 9 but in a linear representation. The open symbols are the values calculated with the SCMF theory as in Fig. 9. The filled symbols belong to the experimental observation of Kent et al. [45, 46] for four different molecular weights: 4–30 (circles), 4.5–60 (squares), 15–175 (diamonds), and 20–338 (triangles). The first number is the molecular weight of the PDMS block while the second is that of the tethering PS block, both numbers should be multiplied by $10^3 \, \text{g mol}^{-1}$.

numerical SCF calculations [148]; however, the bulk radius of gyration had to be taken from experiments since the SCF approach considers a random walk, with $R_g \propto n^{0.5}$, as the bulk reference state. However, the stretching of the chains is properly taken into account by the SCF theory even though the whole scaled curve cannot be obtained without introducing information that the approach cannot predict.

Another way to visualize the differences in the behavior of the chain molecules as the surface coverage changes is by looking at the distribution of segments along the chain. This is also an important quantity if one is interested in chemical reactions of functionalized groups in different positions along the chain with reactants in the solvent or near the tethered surface/interface. The distributions are shown in Fig. 11 for a variety of segments along the chain at two different surface coverages. The lower surface coverage corresponds to the brush-to-mushroom transition region. Here the chains are slightly stretched and the segments span throughout the layer. In the case of very high surface coverage, well in the stretch brush region, the distribution of segments is highly localized even though all the segments (including the free end) have a finite probability to be found in any region they can reach; that is, the longest elongation segment i can reach is il, where l is the bond length.

To summarize this section, we have shown that the structure of the tethered polymer layers is well described by the mushroom picture in the low surface coverage and the brush (and stretched brush) in the high surface coverage regime. The transition region between these two structures is quite broad. The predictions from the analytical SCF approaches and scaling theories have been confirmed in the literature, in particular for the scaling behavior of the height of the layer and the parabolic-like profile predicted by the analytical SCF approach, both for the brush regime [125]. It is important to remember that these predictions are for the limit of very long chains, while for intermediate chain lengths more detailed analysis is needed to properly describe the behavior of the layers.

The numerical SCF approach provides very similar density profiles and segment distributions than those shown above [149]. Further, this approach has the advantage that it can be used for relatively long chain lengths. However, since the intramolecular excluded-volume interactions are taken into account within the same mean-field approximation as the intermolecular interactions, the predictions of the SCF approach in the mushroom regime and in the crossover region are not adequate. This results from the fact that in good solvents the intramolecular excluded-volume interactions are dominant for low to intermediate σ. Thus, the statistics of the chains are those of SAWs. These factors are naturally

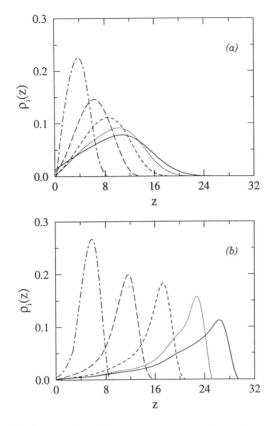

Figure 11. Distribution of individual segments along the chain as a function of the distance from the surface for chains with $n = 50$ in the good-solvent regime. (a) $\sigma = 0.03$ and (b) $\sigma = 0.20$; $i = 10$ (dotted–dashed line), $i = 20$ (long dashed line), $i = 30$ (dashed line), $i = 40$ (dotted line), and $i = 50$ (solid line). The segment number is counted from the one grafted to the surface, thus $i = 50$ is the free end of the chain.

included in the SCMF approach since the single-chain statistics are exactly taken into account. Further, the intermolecular repulsions seem to be properly taken into account as confirmed by the excellent quantitative agreement that can be achieved with computer simulations, as shown in ref. 132. The fact that the single-chain statistics are exactly taken into account also enables the calculation of the lateral dimensions of the chains, which in the SCF approaches remain Gaussian at all surfaces coverages. The main limitation of the SCMF theory is the need to have a set of single-chain configurations. Thus, only chains of a few hundred segments can be properly calculated. The proper combination of all these approaches, including computer simulations, is required to obtain a very good description of the structure of the layers in all the surface coverages and molecular weights regimes. We will see that some of the limitations of the different approaches seem to be more pronounced in the determination of thermodynamic properties.

B.   Thermodynamic Properties

The different conformational behavior of the chain molecules as a function of surface coverage manifests itself in the thermodynamic properties. The relevant quantities for a tethered layer are the lateral pressure and the polymer's chemical potential. The former measures the average repulsions between the chain molecules parallel to the tethering surface [recall that we are considering the good solvent (athermal) limit.] The chemical potential is the work required to add a polymer chain to the layer, and thus it is the quantity that will determine the amount of polymer tethered to the surface when in equilibrium with a bath of end-functionalized polymers in bulk solution.

The analytical approaches predict that in the good-solvent regime the pressure scales as $\Pi_a \propto n\sigma^x$, with x values of $\frac{5}{3}$ and $\frac{11}{6}$ for the SCF and scaling theories, respectively. Recent studies have shown that this result does not agree with the full-scale simulations and that $x = 2.5$ seems to better describe the MD results [107]. We have compared the predictions from the SCMF theory with the MD simulations. As can be seen from Fig. 12, there is quantitative agreement between the two for the whole range of surface coverages and different chain lengths. These results are for the good-solvent regime but at finite (above Θ) temperatures. In ref.

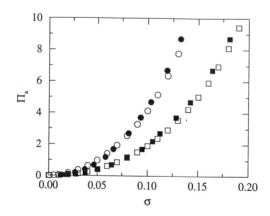

Figure 12.   Lateral pressure as a function of the surface coverage for chains with $n = 50$ (squares) and $n = 100$ (circles). The open symbols are the predictions of the SCMF theory while the filled symbols are the results of MD simulations by Grest [107].

130 a thorough discussion of the reasons for the disagreement between the MD and the SCMF theory and analytical predictions are presented. The main conclusion is that the chain lengths in the MD simulations and the SCMF theory are too low to be in the asymptotic regime where the analytical predictions are expected to be valid. Further, it was concluded that a single scaling exponent, as suggested by Grest [107], cannot predict the behavior of the pressures due to the fact that for these chain lengths there are contributions from many virial coefficients as the surface coverage is varied.

To see if for longer chain lengths there will be a regime of surface coverages for which the analytical predictions are confirmed, Martin and Wang [139] have used the SCMF theory in conjunction with the matrix of DiMarzio and Rubin [150] to study the behavior of chains up to $n = 1000$. This approach is in reality equivalent to the SCF approach as discussed in Section II.C and ref. 103. By looking at the pressure as a function of surface coverage in a log-log representation, they found that, for $n = 200$, 1000, there is a considerable regime for which $\Pi_\alpha \propto \sigma^{5/3}$. A more careful check of the scaling can be obtained by looking at $\Pi_a / n\sigma^{5/3}$ as a function of $n\sigma^{5/3}$. This is shown in Fig. 13 together with the log-log representation from ref. 139. The log-log plot seems to indicate almost two orders of magnitude in surface coverage where the scaling predictions are verified. The lower graph, on the other hand, shows that there seems to be a region of scaling only for $n = 1000$. The different conclusions that can be drawn from the different ways in which the data are represented shows that, in order to obtain, or check, scaling predictions, the log-log representation may be misleading and a further check is important to ensure the reliability of the conclusions.

The question that arises is why the pressures are not properly predicted by the analytical approaches, while it is known that the analytical SCF theory properly predicts the behavior of the density profiles, in the appropriate limit. This point was discussed in detail in ref. 130 and also by Martin and Wang [139]. One finds that, even for relatively short chains, the shape of the density profiles are more or less parabolic in quite a wide range of surface coverages. However, when this profile is used to calculate the pressures within the second virial coefficient approximation, the results are valid only in the asymptotic regime.

The lack of agreement with the analytical approaches opens the question of what, if any, is the relevant scaling variable needed to describe the pressure in tethered polymer layers. There are two alternative choices that one can use [147]. Both of them are based on arguments valid in the low surface coverage regime. The first one is based on the fact that the free energy of a single chain scales (up to logarithmic

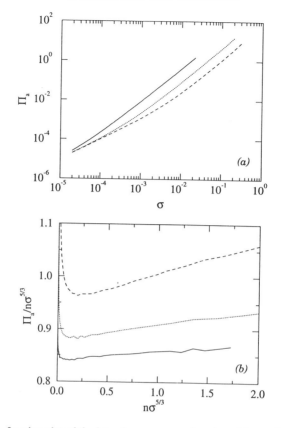

Figure 13. Log-log plot of the lateral pressure as a function of the surface coverage for chains of length $n = 50$ dashed line, $n = 200$ dotted line, and $n = 1000$ full line. The lower graph shows the same data in scaled form. The longest chain length seems to show a plateau region, but the data for $n = 200$ does not; see text. The data is from Martin and Wang [139].

corrections) linear with molecular weight. Therefore, the pressure must also be linear in n as $\sigma \to 0$. Further, the first term in the pressure must be quadratic in surface coverage. Recall that we are considering the pressure beyond the ideal term. Thus, we have $\Pi_a \propto n\sigma^2$, and then it seems natural to consider $y = \sigma n^{0.5}$ as the scaling variable. Figure 14 shows the predictions of the SCMF theory with this scaling variable. It is clear that up to $y = 1.5$ the data collapse in one universal curve. However, we have not found in this regime that the data could be described by a single scaling exponent. This is due to the fact that the results scale for a wide range of values of surface coverages for which simple second virial coefficient arguments do not hold. The problem with

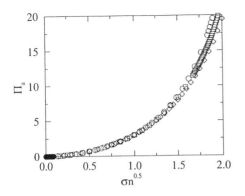

Figure 14. Lateral pressure as a function of the surface coverage scaled by the square root of the chain length. The symbols represent: circles $n = 50$, squares $n = 70$, and diamond $n = 100$. Note the good scaling of the data up to $\sigma n^{0.5} = 1.5$.

this argument is that while the data follows a universal curve, a change in the chain model provides a different curve.

The second argument goes back to Flory's idea [151] for bulk polymer solutions in the good-solvent regime. Flory argued that the second virial coefficient of chain molecules is equivalent to that of two hard spheres with a radius determined by the size of the polymer chain (i.e., R_g). The reason is that when two polymers 'collide,'' several monomers come in contact; if the energy per monomer contact is kT, then the repulsion is very strong. We have shown above that at low surface coverages the size of the polymer chains in the planes parallel to the surface is proportional to the bulk R_g; therefore generalizing Flory's argument to two dimensions we expect the second virial coefficient to be that of the equivalent "hard disk" with a radius given by the size of the polymer, again the bulk R_g. Then, $\Pi_a \propto R_g^2 \sigma^2$. Further, since we have seen that σ^* is the proper scaled surface coverage at low and intermediate two-dimensional densities, we expect $\Pi_a R_g^2 \propto \sigma^{*2}$ to be adequate at small σ^*. Figure 15 shows the scaled pressure as a function of the reduced surface coverage for a variety of chain lengths. It is interesting to note that the data have off-lattice calculations and lattice results and all follow the same universal curve up to $\sigma^* \simeq 6$, that is, in the whole mushroom regime and in the transition region. In the "brush" and stretched-brush regimes the pressure shows a different dependence for each molecular weight; however, as mentioned above, it does not seem to follow the analytical predictions for this quantity.

The advantage of the Flory argument is that it becomes completely independent of the chain model and, thus, at least in the regime of low and intermediate surface coverages, can be compared with experimental observations. This has been recently done [147] with the measurements of Kent and co-workers, and good agreement was found with the ex-

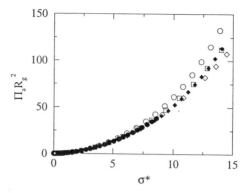

Figure 15. Lateral pressure scaled by the bulk radius of gyration square as a function of the scaled surface coverage, $\sigma^* = \sigma \pi R_g^2$. Symbols as in Fig. 3.9; open symbols off-lattice chains, filled symbols are for cubic lattice chains.

perimental data in the regime where $\Pi_a R_g^2$ scales with σ^*. It also confirms that even beyond the mushroom regime, and up to the beginning of the brush regime, the bulk radius of gyration is a relevant length scale. It is important to consider this due to the broadness of the transition region between the mushroom and brush regimes.

The other important thermodynamic quantity to completely character-ize the tethered layers is the chain's chemical potential. We first discuss the predictions of the different approaches for the chemical potential and later show how they can be used to predict adsorption isotherms.

Shaffer [121] has calculated the chemical potentials of chain molecules using MC simulations in which the chains are modeled by the bond fluctuation model [152]. Shaffer used a log-log representation of the data and concluded that even for $n = 10$ the scaling predictions seem to be confirmed. Figure 16 shows a comparison of the MC simulations with the SCMF theory. The predictions of the theory were obtained with a set of single chains with the same bond fluctuation model as used in the simulations. There is excellent quantitative agreement between the two calculations. The lower graph shows the predictions of the SCMF theory but now in the form $\mu/n\sigma^{5/6}$ versus $n\sigma^{5/6}$. If the scaling predictions were correct, this representation should show a plateau region. The fact that there is no scaling is the same as discussed above for the pressures, and again it shows that in order to check scaling predictions, special care has to be taken and one must not use only log-log representations to obtain the apparent exponents.

The numerical SCF approach provides reliable information on the structure of the tethered layer, in particular for the brush regime. Here

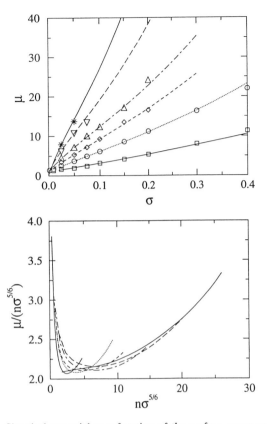

Figure 16. Chemical potential as a function of the surface coverage for lattice chains (bond fluctuation model). The symbols are the results of MC simulations by Shaffer [121] while the lines are the SCMF theory predictions. Squares (full line) $n = 10$, circles (dotted line) $n = 20$, diamond (dashed line) $n = 30$, triangle up (dot–dashed line) $n = 40$, triangle down (long dashed line) $n = 60$, and stars (upper full line) $n = 80$. The lower figure shows the SCMF calculations in scaled form. If the analytical predictions were correct for these short-chain length, a plateau region would appear.

we compare the chemical potentials predicted by the SCF approach and those of the SCMF theory. Since the SCMF predictions are in quantitative agreement with the MC calculations, the comparison between the two theories provides a test of the SCF approach. This is shown in Fig. 17 for two different molecular weights. The variation of the chemical potential with surface coverage is different at low surface coverages, while at intermediate density the slope of the predictions from the two theories is similar. The reason for the discrepancy is due to the fact that the SCF approach considers the chains to be random walks. In the low

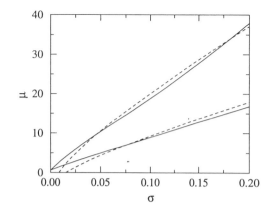

Figure 17. Chemical potential as a function of surface coverage, for the same chain model as those of Fig. 16, according to the SCMF theory (full lines) and SCF theory (dashed line). The upper two curves are for $n = 60$ while the lower ones are for $n = 30$. Note the discrepancy between the two predictions at the lowest surface coverages.

surface coverage limit the increase in surface coverage induces a slight stretching of the chains. As can be seen from the SCMF calculations, this effect does not carry too much free-energy cost due to the fact that even at zero surface coverage the chains statistics are those of self-avoiding chains. However, to stretch a random walk costs much more free energy.

C. Adsorption from Solution

In all the cases described so far the surface coverage was considered as a thermodynamic variable that can be controlled. In general, the amount of polymer tethered to the surface is governed by the conditions from which the layer is prepared, namely, the type of functionalized end group that interacts with the surface, the bulk polymer concentration, and the temperature (or quality of solvent). There are very few experimental studies of how the surface coverage depends upon the variation of the above-mentioned quantities, and in general once the sample is prepared, the surface coverage is obtained by direct or indirect measurements.

The most systematic study of the adsorption of end-functionalized polymers to surfaces was carried out by Auroy et al. [38]. In this study they looked at the amount of polydimethylsiloxane (PDMS) grafted to a silica surface as a function of the bulk polymer concentration. The grafting is irreversible as it is obtained from the chemical reaction of OH end groups with the silica surface that is believed to result in a bond with strength of the order of several hundred kT. On the other hand, the studies of Klein and co-workers [22, 24] were done for zwitterionic end-functionalized polystyrene (PS). It is believed that the zwitterion has an adsorption energy with the mica surfaces of the order of 6–8 kT. These studies were carried out for a variety of chain lengths of PS, but the bulk concentration was not varied systematically. A more recent study of adsorption by Perahia et al. [28] looked at cases in which the monomers of the polymers have an attractive interaction with the surface and, therefore, there is the additional competition between end anchoring and polymer adsorption, which is beyond the scope of this chapter.

The first systematic theoretical study of the adsorption isotherms was developed by Ligoure and Leibler [153]. They calculated the amount of surface coverage as a function of the bulk density using a generalization of the analytical SCF approach. This approach is expected to be valid in the brush regime for very long chain lengths.

As discussed above in relation to the experimental observations of Kent et al. [45–48] and as mentioned in ref. 147, many of the experimental studies carried out from dilute polymer solution are probably not in the brush regime but in the crossover between mushroom and brush. Further, it was shown that the SCMF theory provides good quantitative information on the thermodynamic properties for all the surface coverage regimes [130, 132], and therefore one would expect that it will also provide valuable information on the adsorption isotherms.

The equilibrium amount of polymer adsorbed by the end-functionalized group is obtained by equating the chemical potential of the polymers in bulk and at the surface. The chemical potential of the tethered chains is obtained by adding to Eq. (16) the contribution from the interaction of the end-functionalized group with the surface, ε_s. The bulk chemical potential has to be calculated within the same theoretical approach as that of the tethered chains. However, the SCMF for bulk polymers will be appropriate only in the dilute-solution case. At higher bulk concentrations isotropic chain–chain repulsions have to be taken into account. This can, in principle, be done by including correlation terms in the pdf of the chain conformations [Eq. (5)]. However, it is rather a difficult task, and therefore we concentrate out attention on the case of adsorption from dilute solution, for which the chemical potential is given by

$$\mu_{\text{bulk}}(\phi_{\text{bulk}}) = \ln \frac{\phi_{\text{bulk}}}{n} - \ln(1 - \phi_{\text{bulk}})^n - \ln q_{\text{bulk}} \qquad (22)$$

where ϕ_{bulk} is the bulk polymer volume fraction and q_{bulk} is the single-chain partition function of the polymer in bulk solvent. In the good-solvent regime, for dilute solutions, this last quantity is the number of self-avoiding conformations of the bulk chain.

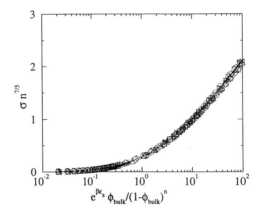

Figure 18. Universal adsorption isotherm as obtained from the SCMF theory. There are 16 different curves corresponding to four chain lengths, $n = 30$, 50, 70, and 100. For each chain length four different adsorption energies have been used: $\beta\varepsilon_s = -3$, -7, -10, and -15.

Figure 18 shows the results of calculations obtained for a variety of chain length and different adsorption energies for the functionalized end group. The adsorption isotherms show a universal behavior in the range of surface coverages and bulk concentrations calculated. It is found that $\sigma n^{7/5}$ is a universal function of $x = e^{\beta\varepsilon_s}\phi_{bulk}/(1 - \phi_{bulk})^n$. While the variable x follows naturally from the bulk chemical potential [Eq. (22)] and the Boltzmann factor of the adsorption energy added to the tethered chain's chemical potential, it is not clear why $\sigma n^{7/5}$ should be a universal measure of the amount adsorbed. Further, as can be seen from the figure, the dependency of the scaled surface coverage on x is rather complicated.

The main feature from the isotherm is that it clearly shows that to obtain a high concentration of polymer tethered to the surface the functionalized end group needs to have a very strong attractive interaction with the surface, in particular for the dilute bulk solutions that we are treating here. This is a straightforward reflection of the loss of conformational entropy induced by the packing of the chains at the surface as it is manifested in the chemical potentials (see Fig. 16). This may explain why most of the experimental observations in end-functionalized tethered layers from dilute solutions result in a surface coverage that is just at the onset of the brush regime.

It is interesting to note that while the estimates for the interactions between the end-functionalized zwitterion and the mica surface is 6–8 kT [26], the adsorption isotherm shown in Fig. 18 predicts much higher anchoring energies necessary to reach the experimental surface coverage. While a direct comparison with experiments in this case is hard to do, the calculations are more in line with the energies obtained from the measurements of Kent and co-workers [45] who suggest that the anchoring energy may be an order of magnitude larger than previously estimated.

A more direct comparison between the predictions of the theory and experimental observations can be done with the data of Auroy et al. [38]. In this case the grafting is obtained by chemical reaction of the end group. Therefore, the adsorption is irreversible. Assuming that the amount of polymer grafted is determined by the equality of chemical potentials in the bulk and the surface, as suggested by the experimentally found dependency of the amount of polymer grafted on the bulk polymer density, one can calculate the adsorption isotherms. The chemical potential of the grafted chains *does not* include the two-dimensional translational term, $\ln \sigma$, and thus it is given by $\beta\mu_p = -\ln q + \beta\varepsilon_s$, while the bulk polymer chemical potential is again given by Eq. (22). The experimental observations were carried out for PDMS with molecular weight (MW) of $145{,}000 \, \text{gr mol}^{-1}$. Since the calculations cannot be done with the SCMF theory for such high molecular weight, we have used $n = 100$ and determined the length scale of the problem by equating the bulk radius of gyration of the PDMS to that of the model chain. This provides an equivalent bond length $l = 7.4 \, \text{Å}$. Using this bond length and $\beta\varepsilon_s = -20$, Fig. 19 shows the comparison between the experimental observations and the calculated isotherm. Note that the calculated curve is only up to $\phi_{\text{bulk}} = 0.1$ because of the limitation of the bulk chemical potential as calculated from the SCMF theory. The agreement is very

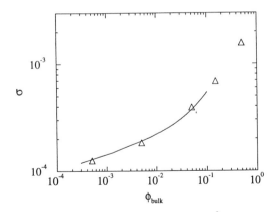

Figure 19. Adsorption isotherm, surface coverage in $1/\text{Å}^2$ vs. bulk polymer volume fraction. The symbols are the experimental measurements of Auroy et al. [38] and the line is the prediction of the SCMF theory. See text.

good; however, in order for the predictions of the theory to be useful, one needs to determine what is the equivalent adsorption energy in the experimental system. This can be obtained by scaling the experimental data and the theoretical predictions. Further experimental observations are needed in order to do this.

The experimental observations of Auroy et al. are the only ones in which very large surface coverages are obtained. There are reasons for this. One is that the grafting is obtained by chemical reaction, not adsorption. The second, which seems to be more important, is the fact that for high surface coverages the adsorption is done from semidilute and concentrated bulk solutions.

D. Changing Chain Architecture

The discussion so far has concentrated on cases in which the polymer molecules are linear, flexible chains. An important question in the design of tethered layers is how changes in the architecture of the chains will affect the molecular organization and the thermodynamic behavior of the modified surface.

There are very few studies of the effects of molecular structure on the tethered layer for low-molecular-weight solvents. On nematic solvents, however, Halperin and Williams have studied the behavior of liquid crystalline polymers under a variety of conditions [154–161]. Gersappe and co-workers [162] have considered chain molecules in which the free ends are functionalized to have attractions between them. They found that the proper choice of these functionalized groups induces microphase separation at the end of the layer. This is an interesting example of tailoring the properties of the layers by chemical modification of the chain molecules. Very recently, Zhulina and Vilgis [163] have looked at the behavior of tethered layers composed of molecules with backbone and many branches, comblike polymers, and starlike polymers as a function of the solvent quality. They have found that the layer exhibits a rather rich phase behavior, as compared to linear chains. For example, they have found that comblike polymers tethered to a planar surface will show a scaling behavior for the layer thickness as $h \propto \sigma^{1/3}$ for solvents ranging from good to Θ, while for linear chains the same scaling is found for good solvents, but Θ solvents show $h \propto \sigma^{1/2}$.

To show the dramatic changes in the molecular organization of the tethered layer upon changes in the molecular architecture, we have chosen to treat two different classes of polymer chain architectures using the SCMF theory. The first one is that of three branches of polymers [131]. These molecules are composed of two symmetric free branches and one different branch tethered to the surface (see Fig. 20). The second

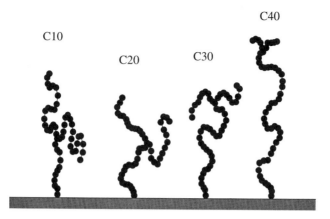

Figure 20. Schematic representation of three branch chains with a branch of length i tethered to the surface and two free branches of length $(50 - i)/2$. The chains are called Ci as indicated in the figure.

class is that of chain molecules, where a portion of the chain is rigid while the other segments are fully flexible (see Fig. 21). This class of chain molecules is of the same structure as those molecules forming liquid crystalline phases in bulk. Note that we are not calling these chains liquid crystalline polymers, due to the short chain length considered here.

Consider the case of a polymer chain with a short branch tethered to the surface and two free branches that are longer than the surface-

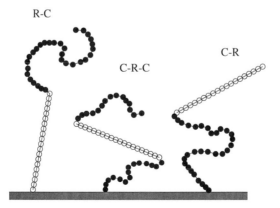

Figure 21. Schematic representation of chains with flexible and rigid blocks. For all the chains the rod block has 20 segments and the total chain length is $n = 50$. The names used in the text are as noted in the figure.

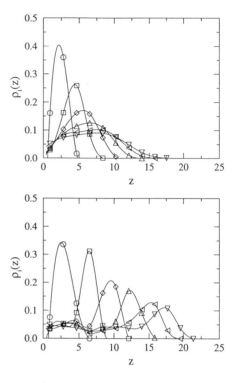

Figure 22. Distribution of individual segments for chains with three branches, $C10$. The upper graph is for $\sigma = 0.01$ while the lower one is for $\sigma = 0.15$: circles $i = 5$, squares $i = 10$, diamond $i = 15$, triangle up $i = 20$, triangle left $i = 25$, and triangle down $i = 30$. Note that the distributions for $i > 10$ are for the two equivalent segments of the two symmetric free branches.

attached branch. The question that arises is how the segments are distributed as a function of the surface coverage. This is shown in Fig. 22 for the $C10$ chain. At low surface coverages the segments are distributed more or less as in the linear chain (see Fig. 11); that is, they span as much as possible in a mushroom-like configuration. At the higher surface coverage the chains stretch perpendicular to the surface but in a rather different way than the linear chains. The distributions for segments 5 and 10, the later being the branching points, show the much more stretched average configuration of the grafted branch as compared to the linear chain at the same surface coverage. The maximum of the distribution for $i = 10$ is found at $z = 6.8l$ for the branched chain and at $z = 5.8l$ for the linear one. Note that the maximum distance that this segment can achieve is $z_{max} = 8.3l$. This implies that the branch grafted to the surface is very highly stretched. The segments of the two free branches show an interesting behavior at the high surface coverage. While the distribution shows a high probability of finding these segments far from the surface, the probability of finding them near the surface is much higher than the equivalent segment in the linear chain.

This behavior is also seen in the chains with a larger grafted branch and shorter free branches, as shown in Fig. 23, but to a much smaller extent. Looking in detail at other configurational properties, one finds [131] that in the case of a short tethered branch at high surface coverages the best organization of the chain is such that the short branch stretches as much as possible out of the surface. The configurations in which one of the free branches is toward the surface and the other is toward the solvent are those that have the largest probability. This can be understood with the following simple argument. The free-energy cost of stretching the shorter tethered branch is lower than the free-energy gain of the longer free branches. Further, the stretching of the short branch leaves volume available to one of the free branches near the surface. Therefore, the

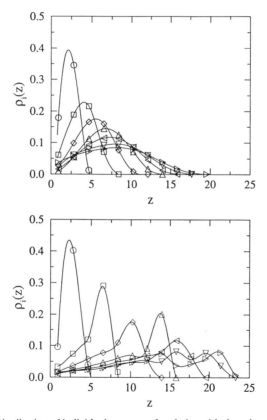

Figure 23. Distribution of individual segments for chains with three branches, $C20$. The upper graph is for $\sigma = 0.01$ while the lower one is for $\sigma = 0.15$; circles $i = 5$, squares $i = 10$, diamond $i = 15$, triangle up $i = 20$, triangle left $i = 25$, triangle down $i = 30$, and triangle right $i = 35$. Note that the distributions for $i > 20$ are for the two equivalent segments of the two symmetric free branches.

conformations for which one of the free branches is toward the surface and the other toward the solvent enhance their probability very much.

One can ask the question of when it will be more favorable for one of the free branches to fold toward the surface as a function of the ratio of size of the tethered branch and the size of the free branches. Marques [164] has derived a criterion based on the simple Alexander picture [49]. Consider a branched chain with a branch of n_a segments grafted on the surface and two free branches each with n_b segments. The idea is to compare the free energy of the system in which the two free branches are stretched toward the solvent as compared with that of a chain with one of the free branches toward the surface and the other toward the solvent.

In the cases of the two free branches stretched toward the solvent we can treat the system as forming two brushes, one of thickness h_a formed by the grafted branch and the second formed by the two b chains that are tethered at the end of the a brush. According to the Alexander approach, the minimized free energy of the brush is given by

$$F = \frac{3}{2^{5/3}} n\sigma^{5/3} \tag{23}$$

Therefore, the free energy of the brush with the two branches up is $F_{up} = F(n_a, \sigma) + F(n_b, 2\sigma)$, the 2 appears due to the fact that the brush is formed by the two b branches. Defining $x = n_a/(n_a + n_b)$ and using Eq. (23), we obtain

$$F_{up} = \frac{3}{2^{5/3}} (n_a + n_b)\sigma^{5/3}[x + 2^{5/3}(1 - x)] \tag{24}$$

Now consider the case in which one of the b branches folds toward the surface. Then we have a brush with $F(n_b, \sigma)$ formed by the b branch that is toward the solvent and we have to determine the free energy of the "folded" brush. Within the Alexander picture the free energy of this brush is

$$F = \frac{1}{2} h^2\sigma\left(\frac{1}{n_a} + \frac{1}{n_b}\right) + \frac{1}{2} (n_a + n_b)^2 \frac{\sigma^2}{h} \tag{25}$$

where the first term is the elastic energy and the second represents the interaction energy. Minimizing Eq. (25) with respect to h and adding the free energy of the other b brush, we obtain for the "down" configuration

$$F_{down} = \frac{3}{2^{5/3}} (n_a + n_b)\sigma^{5/3}\left[\frac{1}{[x(1 - x)]^{1/3}} + (1 - x)\right] \tag{26}$$

The optimal configuration of the brush will be the one with the minimal free energy. Therefore, when $F_{down} < F_{up}$, the down configuration will be the equilibrium one. Using Eqs. (26) and (24) one finds that for fixed surface coverage the condition is

$$\frac{1}{[x(1-x)]^{1/3}} + (1-x) - x - 2^{5/3}(1-x) < 0 \qquad (27)$$

which is fulfilled for $0.145 < x < 0.5$. This is in agreement with our results for the $C10$ chains for which $x = \frac{1}{3}$. It is interesting to note that the simple Marques picture also predicts a lower bound for x. This is necessary because in the limit of a very small grafted branch the stretching of the short branch does not allow enough free space for one of the longer free branches to fold back toward the surface.

This is one of the many examples in which the simple analytical picture provides bounds and a physically simple intuitive picture that is very helpful in understanding the behavior of the chain molecules. Further, it provides guidelines to which regimes will be more interesting to look at experimentally and by more sophisticated (numerically intensive) theoretical methods such as the SCMF approach.

We have seen how a change in the molecular architecture of the chains can dramatically change the molecular organization of the layer. In particular, for the $C10$ chains it was found that one can increase the probability of finding a free end at the surface, even at large surface coverages. The question that arises is how the thermodynamic properties of the layers are modified due to the different molecular organization.

Figure 24 shows the chemical potential and the lateral pressure of the layers composed of chains of different chemical architecture as a function of the surface coverage. The molecule $C40$, not shown, has a chemical potential and lateral pressure almost identical to the same molecular weight linear chain [131]. This is actually true for all the conformational properties as well. The main difference is that there are two free ends in the branched chain. Thus, if one is interested in very large concentrations of free ends close to the solvent, to be used, for example, in a chemical reaction with a solvent-miscible reactant, the short branches provide a better alternative than the linear chain, the reason being that preparing a layer at a given surface coverage will require almost the same conditions for the linear and the $C40$ chain. On the other hand, the chains with a shorter tethered branch show a higher chemical potential and a higher lateral pressure than the linear chains for fixed surface coverage. The reason is that the branched chains are more bulky and effectively shorter (see Fig. 20), and since all the description is for the good-solvent regime,

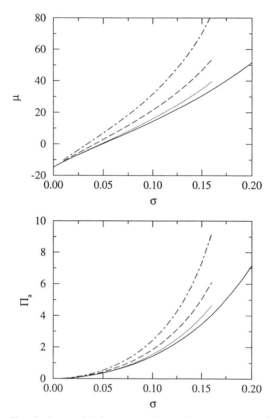

Figure 24. Chemical potential (upper graph) and lateral pressure (lower graph) as a function of the surface coverage for branched chains, all with $n = 50$. Full line is the linear chain, dotted line $C30$, long dashed line $C20$, and dot-dashed line $C10$.

a bulkier environment implies higher repulsions. Therefore, it will be harder to prepare a high surface coverage layer of these branched chains.

An interesting question that arises in connection with colloidal stabilization is what are the changes in the interactions between two tethered surfaces due to the different molecular architectures. This is shown in Fig. 25 for the different chains, all at a fixed, relatively high surface coverage. The shorter grafted branch results in a sharper repulsive interaction. This is the result of having the same amount of polymer segments in a much shorter layer, due to the different chain architectures. The repulsions between the layers results in an interesting rearrangement of the molecules that has been discussed in detail in ref. 131.

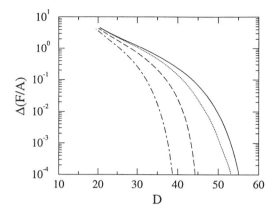

Figure 25. Interaction between two tethered layers as a function of the distance between the surfaces for linear and branched chains. The lines are as in Fig. 24.

We see that changing the chain architecture one can tune the range and strength of the interactions between grafted layers. One can think of additional chemical modifications that can result in an attractive interaction followed by a repulsive one. Actually, for linear chains changes in the quality of the solvent have been shown to induce a range of distance where the interactions become attractive [123]. Branched chains, in which the branches are made of different monomers, may lead to a way to tune range and strength of interactions, depending upon the length and number of branches as well as the affinity of the solvent to each of the different monomers. While this type of molecule has been synthesized [165], the predictions for the branched molecules are yet to be studied experimentally.

A different molecular architecture is that of chain molecules with portions of the chains that are rigid. It is well known that rigidlike molecules form liquid crystalline phases in bulk, and the isotropic–nematic transition in bulk can be described as a purely entropic effect [166]. However, if these molecules are tethered to a surface, it has been shown that there cannot be a first-order transition for purely repulsive systems [167], that is, in a good-solvent environment. Here we are interested in looking at what are the effects on the molecular organization of the rigid rods when they are chemically bound to flexible chains. To this end, we will consider the molecules shown schematically in Fig. 21, which all have 50 segments, and how the surface coverage changes the properties of the layer.

We start by looking at the structure of a single-chain molecule at the surface, that is, the mushroom regime. Figure 26 shows the distribution of

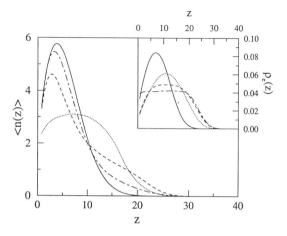

Figure 26. Average number of segments as a function of the distance from the surface for free chains of different architectures. Full line is for linear chains, dot–dashed line is C–R, dashed line is C–R–C, and dotted line is R–C. The inset shows the distribution of the free ends of the chains.

segments as a function of the distance from the tethering point for the three different rod-flexible chains and for comparison a fully flexible chain of the same length. The chain with the rod at the end of the tethered flexible branch shows a distribution similar in shape to that of the fully flexible chain but is more stretched due to the rod structure of its last twenty segments. The $C-R-C$ chain shows three distinct regimes, each corresponding to the different flexibility of the portion of the chain. The most different profile is found for the chain that has its rodlike portion attached to the surface. This chain shows the most "mushroom"-like configuration, due to the inability of the chain to have a large number of segments near the surface. This results from the fact that when the rod is parallel to the surface the flexible block has to be stretched out of the surface due to the boundary conditions. However, when the rod is perpendicular to the surface, the flexible part can still fold back toward the surface. This results in a wide distribution as compared to the other chain architectures.

The inset in the figure shows the distribution of the free end as a function of the distance from the surface for the different molecular architectures. The distributions change continuously from that of a fully flexible chain, in which the end segment spans along the layer with a bell-like shape, to that of the rod as the free end, which shows a constant distribution almost at all distances from the surface. The constant distribution is what one will also obtain for a rodlike object in which all the angles are equally probable.

Now we look at the influence that increasing the surface coverage has on the molecular organization of the layer. Fig. 27 shows the density profile for the four different chains at a relatively high surface coverage. While the linear chain shows a parabolic-like shape, the presence of the rod block manifests itself in the distribution of the other layers. The $C-R$ chain shows two regimes of densities. The first is very similar to that of the fully flexible chain and corresponds mostly to the flexible segments of the chains. The second regime is that which contains mostly the rodlike portion of the chain, and the shape of the profile is a direct result of the structure of the rod. The other two profiles show an almost constant density in the region that the rod is present while the portions of the flexible part of the chain show the typical brushlike shape.

A more detailed picture of the molecular organization can be obtained by looking at the distribution of individual segments along the layer. Figure 28 shows the distribution of selected segments for the three molecular architectures for the same surface coverage as Fig. 27. In all the cases the segments are highly localized due to the large surface coverage that induces chain stretching. However, the distribution of the segments is rather different for the different chains, particularly in how the stretching occurs. For example, the chain with rod block as the free end shows that its free-end segment has zero probability to be found up to $z = 12.5l$. A similar dead zone is found for segments 40 and 30. This shows that when this particular chain stretches, the rod block cannot turn itself toward the surface due to the strong repulsions that would result

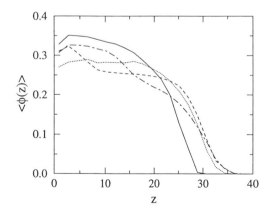

Figure 27. Density profiles for chains of different architectures all at $\sigma = 0.15$. Lines are as in Fig. 26.

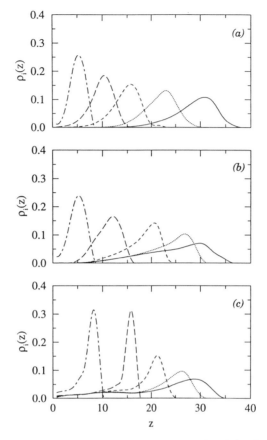

Figure 28. Distribution of individual segments along the chains for : (a) C–R, (b) C–R–C, and (c) R–C, all at $\sigma = 0.15$. The lines are for the segments number as in Fig. 11.

from the flexible, grafted block. Note the very dramatic effect that surface coverage has on the probability of the different chain configurations, by comparing the end distribution of this chain in the case that no neighbors are present (inset of Fig. 26), where the end distribution is flat starting from the surface to the high surface coverage with a large dead end zone.

The case of the chain with the rod block between the two flexible ones also shows a dead zone for almost all its segments, $i > 15$, but this region is smaller than in the previous case. The reason is that the rod block orients preferentially perpendicular to the surface (see Fig. 29 below and discussion thereafter), and thus the free flexible end block is not long enough to bend and reach the surface. The chain with the rod grafted to the surface shows a very sharp distribution of the segments of the rods, indicating a preferential perpendicular orientation, while the free end, which is part of the flexible block, shows a distribution with two maxima. One is toward the solvent (i.e., highly stretched), and the second is at about $z = 10l$. This second maximum results from the folding back of the flexible block toward the surface.

The density and distribution profiles suggest that the rod portion of the molecules is ordering mostly perpendicular to the surface. To look at the ordering in a more quantitative way, we define the order parameter of a segment by

$$\langle P_2(\cos\theta_i)\rangle = \tfrac{3}{2}\langle \cos^2\theta_i\rangle - \tfrac{1}{2} \tag{28}$$

where θ_i is the angle between the vector that links segment $i-1$ and segment $i+1$ and the normal to the surface. Thus, we have $P_2 = 0, 1, -\tfrac{1}{2}$ for a completely random, perfectly perpendicular, and perfectly parallel distribution of angles. Therefore, if the average order parameter is positive, the segment is preferentially oriented perpendicular to the surface.

Figure 29 shows the order parameter as a function of the segment number along the chain for a moderately high surface coverage. Clearly, the rodlike blocks have the same orientation for all their segments. The orientation of the rod may have important implications for the optical properties of the layers and can be measured by spectroscopic techniques. The results illustrate that while the orientation of the flexible segments of the chains is preferentially perpendicular to the surface, the segments are

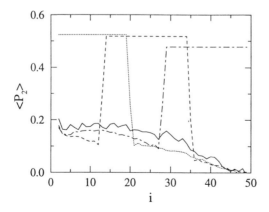

Figure 29.   Bond order parameter as a function of the segment number for $\sigma = 0.15$. Lines are as in Fig. 26.

not very highly oriented and the same segment number, if it is flexible, has almost the same orientation in all the molecular architectures including the fully flexible one. However, the rod blocks show a very large orientation that at this relatively high surface coverage is almost independent of the location of the rod along the chain.

The reason for the high ordering of the rod blocks as compared to the flexible ones is that whenever the rod feels repulsions from the neighboring molecules the only way to avoid them is by orienting perpendicular to the surface. On the other hand, the flexible segments can find many ways to rearrange, even though this would result in some stretching, that does not result in predominantly perpendicular orientations because this will carry with it too much conformational entropy.

The question that arises now is how does the orientation of the rod block change as a function of the surface coverage for the different molecular architectures. This is shown in Fig. 30. In the limit of isolated chains ($\sigma = 0$) the molecules with the rod block tethered to the surface show a high orientation due to the fact that if the rod tends to orient parallel to the surface there is not enough space for the flexible block attached to the end of the rod to visit its conformational phase space. A rather different behavior is found for the two molecules in which the rod is the midblock or the end block. In both cases the rod has a preferential orientation that is parallel to the surface, that is, negative order parameter. In the case of an isolated rod tethered to a surface one finds that the random orientation results in $\langle P_2 \rangle = 0$. Therefore, the parallel orientation of the molecules with rod and flexible blocks must be the result of

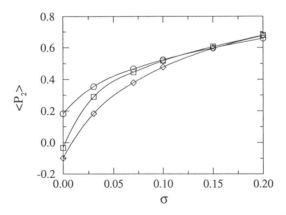

Figure 30. Order parameter of the rigid portion of the chains as a function of the surface coverage. Circles R–C, squares C–R–C, and diamonds C–R.

the best way the flexible blocks can realize their many different conformations.

There is a sharp rise of the order parameter of the rods with increasing surface coverage, resulting in a predominantly perpendicular orientation with the surface. Note that the sharpest increase results for the molecules in which the rod blocks are more parallel at $\sigma \to 0$. At very high surface coverages the three molecular architectures show almost the same degree of ordering of the rod blocks regardless of their position along the chain. It is important to note that the orientation of the rods is even higher than that found in nematic phases of liquid crystals [168]; thus the layers will have interesting anisotropic optical properties.

Now that the molecular organization of the different molecular architectures is understood, we turn to the thermodynamic properties. Figure 31 shows the chemical potential and the lateral pressures of the

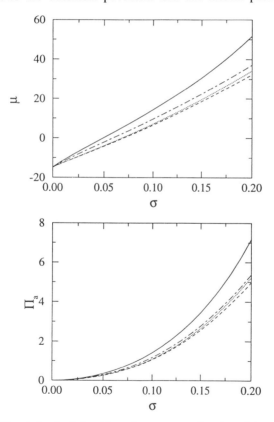

Figure 31. Chemical potential and lateral pressure as a function of the surface coverage for different chain architectures. Lines are as in Fig. 26.

different chains as a function of the surface coverage. The main feature is that for the three molecules with a rod block the chemical potential and the lateral pressure are lower than the corresponding fully flexible chain. This can be understood from the molecular organization of the layers. We have seen that the rod block orients preferentially parallel to the surface. This ordering brings with it a lowering of the orientational entropy. However, this entropic loss is much lower than what is required to stretch the same number of segments when they are fully flexible, resulting in a much lower free energy of packing. Thus, one should expect to be able to tether molecules at larger surface coverages when a portion of the chain is a rod as compared to the fully flexible chain.

IV. VARYING SOLVENT QUALITY

The conformational statistics of isolated bulk chains in solvent are a function of the quality of the solvent, that is, the temperature. In the limit of very high temperatures, or the good-solvent regime, polymer chains have self-avoiding walk statistics. In this case the chains are said to be swollen by the solvent, because there is an effective repulsion between the monomers of the chains. As the temperature is lowered, one reaches the Θ temperature, defined as the temperature where the polymer–polymer second virial coefficient vanishes. In this case the chains have Gaussian statistics due to the effective cancellation of the attractive and repulsive interactions. For temperatures below Θ, the "poor" solvent regime, the chains form compact "collapsed" structures in order to avoid contact with the solvent as much as possible. In bulk polymer solutions there is a phase separation for temperatures below Θ into two phases, one that is almost pure solvent and the other that contains the polymer solution. We are interested to see how the changes in temperature, or quality of solvent, affect the behavior of polymer molecules when one of the ends of the polymer is tethered to a surface.

As discussed above, bulk polymer solutions at temperatures below Θ tend to phase separate. In the case of tethered polymers there are two possible cases that need to be considered. One is when the chains are grafted by a chemical bond between the surface and the polymer end group. This is the case we will denote grafted. The second is when the chains are adsorbed to the surface by a finite interaction between the functionalized end group and the surface. In this case the chains have two-dimensional translations on the surface, and thus we will call them mobile chains. It is clear that the phase behavior in the two cases will be rather different, the reason being that in the case of grafted chains the system is frustrated. Namely, in the poor solvent regime the system cannot undergo a macroscopic phase separation due to the chemical bond between the chain molecule and the surface. Therefore, the tendency of the molecules to avoid the solvent cannot be completely fulfilled, and one could expect the layer to be unstable to lateral fluctuations, resulting in the formation of spatial domains of large concentration of polymer segments while others will be almost pure solvent. For the mobile chains macroscopic (two-dimensional) phase separation is possible due to the ability of the molecules to freely diffuse along the surface.

Halperin [169] studied the scaling properties of collapsed layers ($T \ll \Theta$), and he found that in the mushroom regime the collapse of the polymer molecules is very similar to that of the bulk polymer solutions. However, at high surface coverages the collapse is "weak"; namely, while the layer is thinner than in the good or Θ regimes, the scaling of the thickness is still $h \propto n$. Zhulina et al. [101] studied the behavior of the tethered layers as a function of the temperature using the analytical SCF theory. Shim and Cates [141] also looked at the behavior of the brushes as a function of temperature using the finite extensibility approach for the chains. Both studies found that at fixed surface coverage the density profiles become more steplike as the temperature decreases.

Ross and Pincus (RP) [170] looked at the stability of the collapsed brush, in the high surface coverage regime, by combining the Alexander approach with the random-phase approximation (RPA) and found that the brushes are stable to lateral fluctuations. Later, Yeung, Balazs, and Jasnow (YBJ) [171] looked at the same problem by combining the SCF approach with the RPA and found that there is a regime of surface coverages for which the homogeneous layer is unstable and there is a formation of cluster on the surface. This seemingly contradiction between the results of RP and YBJ was resolved by Tang and Szleifer (TS) [172] who showed that extending the RP approach to the low surface coverage regime one finds that at very low surface coverages mushrooms are stable followed by cluster and at higher surface coverage the homogeneous layer becomes stable (see Fig. 32). The scaling predictions of TS are in agreement with those of YBJ. Further, MC [117, 122] and MD [107] simulations have also shown these different regimes. Very recently, Soga et al. [173] presented simulations of the Edwards Hamiltonian also showing the formation of clusters at intermediate surface coverages followed by homogeneous layers for high σ. The phase boundary between the two regimes are in agreement with the predictions of YBJ and TS. Recent experimental observations using atomic force microscopy have observed for the first time the formation of the clusters [174]. For very

Mushroom

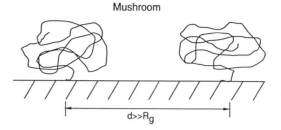

Cluster

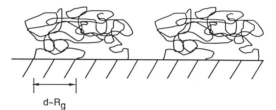

Layer (homogeneous)

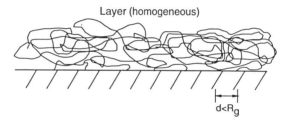

Figure 32. Schematic representation of the mushroom, cluster, and layer regimes for grafted polymer molecules in poor solvent. In the mushroom regime the chains are isolated collapsed coils. When the distance between grafting points is of the order of R_g, the homogeneous layer may not be stable and the formation of clusters is possible. For high enough surface coverages the chains are crowded enough that the homogeneous layer becomes stable.

low temperatures Williams [63], using the scaling approach, predicts the formation of stable surface micelles.

A. Conformational Properties

To understand the behavior of the layers as the quality of the solvent is reduced, it is useful to look at the molecular organization. Figure 33 shows the distribution of segments along the chain at a relative low surface coverage for Θ and a poor solvent. In the case of $T = \Theta$ one observes that the distribution of segments is very similar to that of the

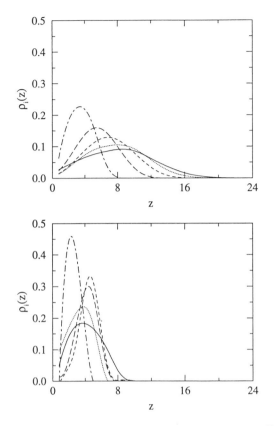

Figure 33. Distribution of individual segments as a function of the distance from the surface for low surface coverage, $\sigma = 0.03$, for two different temperatures. Upper graph is for $T = \Theta$, and the lower one is for a poor solvent, $T = 0.38\Theta$. Lines are as in Fig. 11.

good solvent, but a little less stretched. This is because there is the additional attractions between the monomers of the chain. However, on average the repulsions and attractions cancel each other. In the poor solvent there is a complete collapse of the chain, in order to avoid as much as possible contact of the polymer monomers with the solvent. This is clearly a situation in which the chains are losing much conformational freedom. For relatively large surface coverages the chains must stretch at all temperatures due to the large crowding near the surface. This effect results in stretched average configurations at all temperatures; however, the higher the temperature, the larger the amount of stretching.

One can better visualize the different behaviors at low and large surface coverages for the different temperatures by looking at the lateral radius of gyration as a function of the distance from the surface [see Eq. (20)]. This is shown in Fig. 34, where all the distances have been scaled by the bulk radius of gyration of the chain in the good-solvent regime. In the low surface coverage, as the temperature decreases, the chains compress toward the surface. At Θ the chains show a larger lateral extension than the good-solvent molecules. However, the most dramatic effect is seen for the poor-solvent chains, which show a highly collapsed configuration and at the same time a large lateral extension in order to reach the neighboring chains so as to minimize solvent contact and gain in polymer–polymer interactions. Note that there is overlap between the shapes of neighboring chains. At high surface coverages, the chains are highly stretched at all temperatures with a larger degree of stretching the higher the temperature. However, in this limit, while the layer is shorter,

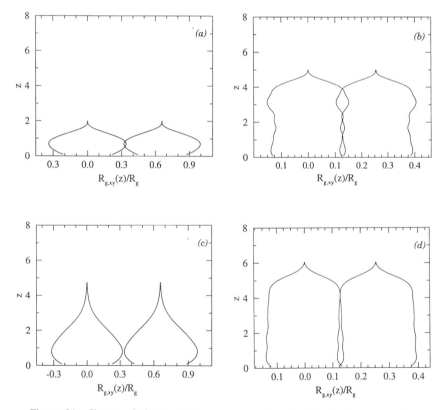

Figure 34. Shapes of the molecules, represented as in Fig. 8, for poor solvent, $T = 0.38\Theta$, for two different surface coverages, (a) $\sigma = 0.03$, (b) $\sigma = 0.20$. Figures (c) and (d) correspond to the same surface coverages as (a) and (b), respectively, but for Θ solvent.

there is not a qualitative difference in the average shapes of the chains, as is the case in the low-density limit.

The compression of the chains in the low surface coverage regime can be measured by the layer thickness. This is shown in Fig. 35 for different temperatures as a function of the surface coverage. For each surface coverage the average thickness of the film decreases with decreasing temperature. What is interesting to note is that for $T < \Theta$ there is a minimum in the layer thickness as a function of surface coverage. The minimum is in the region close to $\sigma^* = 1$, that is, where the coils in the good-solvent regime start overlap. However, as was shown before, in the poor solvent case the chains stretch laterally in order to gain more contact with neighboring polymers, in particular when the distance between tethering points is very large. This results in more "pancake"-like configurations than the isolated poor-solvent mushroom, that is, a minimum in film thickness. This minimum has also been predicted by the scaling theory developed by Tang et al. [175]. In this approach the intramolecular and intermolecular interactions are separated, and both are treated within a scaling approach. The separation of the two contributions to the interactions enables study of the coupling between the dimensions of the chains, parallel and perpendicular to the surface, and the thermodynamic state of the layer.

In general the behavior of tethered layers is discussed in terms of the density profiles. For the cases just discussed this is shown in Fig. 36. While the shape of the profiles show the collapse of the chains, the detailed molecular organization enables one to have a much better understanding of the behavior of the chain molecules. These profiles are in qualitative agreement with the recent experimental observations of Karim et al. [176].

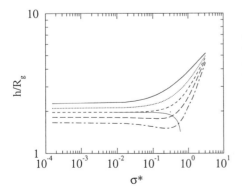

Figure 35. Height of the layer, scaled by the bulk radius of gyration of the chain in good solvent, as a function of the scaled surface coverage for a variety of qualities of solvents. Full line $T = 1.33\Theta$, dotted line $T = 1.14\Theta$, dashed line $T = \Theta$, long dashed line $T = 0.76\Theta$, and dot-dashed line $T = 0.57\Theta$. The thin line marks the onset of instability, that is, it corresponds, for each T, to the surface coverage at which the isothermal compressibility of the grafted chains is zero.

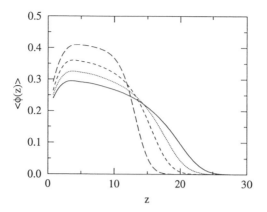

Figure 36. Density profiles for chains with $n = 50$ at $\sigma = 0.1$ for different solvent qualities. Full line $T = 1.5\Theta$, dotted line $T = \Theta$, dashed line $T = 0.76\Theta$, and long dashed line $T = 0.57\Theta$.

B. Phase Behavior

The calculations from the SCMF theory assume that the layer is homogeneous under all conditions. Thus, the calculated thermodynamic properties are not necessarily the ones that correspond to the true equilibrium system, as would be the case if clusters form for grafted polymers. Figure 37 shows the pressure–area isotherms for a variety of temperatures below Θ in the case of grafted chains. The pressure decreases as the area increases (surface coverage decreases) reaching

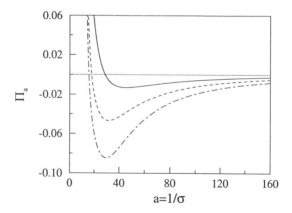

Figure 37. Pressure isotherms for grafted chains at temperatures below Θ. The dotted line marks $\Pi_a = 0$. Full line $T = 0.76\Theta$, dashed line $T = 0.61\Theta$, and dot–dashed line $T = 0.57\Theta$.

negative values up to a point where there is a minimum, and then for even larger areas the pressure increases again. For areas per chain larger than the one that corresponds to the minimum pressure the isothermal compressibility would be negative, indicating an unstable system. One would expect that in this regime there is the possibility of cluster formation. Indeed, this is the regime where YBJ and TS predict cluster formation.

The physical origin of the phase diagram of grafted polymers in poor solvents can be visualized by looking at the structure of the homogeneous (not always the stable) case, as shown in Fig. 4.4. The light line shows the locus of points in which the pressure isotherms show a minimum. In the very low surface coverage regime, the chains are isolated mushrooms that even upon large lateral stretching cannot reach the neighboring chains and therefore are the only possible stable states. As the surface coverage increases, the homogeneous layers show large lateral extensions (see Fig. 34), resulting in a decreasing brush height. In this regime, the distance between grafting points is not too large, and therefore the chains will form clusters with regions of high polymer concentration that will reduce the amount of unfavourable contacts with the solvent, as predicted by YBJ and TS. For high surface coverages the chains are crowded enough that there is no need to form laterally inhomogeneous layers to have a favorable interaction energy (Fig. 34b), again in agreement with the predictions of RP, YBJ, and TS.

In the case of mobile chains the picture is rather different due to the fact that the translation of the chains enables them to macroscopically phase separate. Also the addition of the term σ to the pressure warrants that at low enough surface coverage (very large area per molecule), the pressure will approach zero from the positive side of the pressure. Then, the isotherms are expected to show a van der Waals loop characteristic of a first-order phase transition. The isotherms are shown in Fig. 38. The first thing to note is that while in the grafted chains case there is a region of negative isothermal compressibility for all $T \le \Theta$, the critical temperature appears at much lower temperatures.

The phase diagrams for two different chain lengths are shown in Fig. 39, where both the bimodal and the spinodal are shown. This type of phase diagram was also obtained by Wang and co-workers [177, 178] and by Cantor and McIlroy [179] for shorter chains to explain the phase behavior of short amphiphilic monolayers. The phase separation between a very dilute polymer phase and a more concentrated one is actually very similar to that of bulk polymer solutions. The scaling variables σn and $n^{1/2}(T/\Theta - 1)$ have been derived by Tang et al. [175]. It is interesting to note that Tang et al. have also analyzed the critical exponents for both

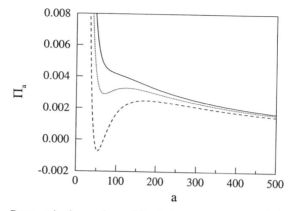

Figure 38. Pressure isotherms for mobile chains at temperatures below Θ. Full line $T = 0.76\Theta$, dotted line $T = 0.73\Theta$, and dashed line $T = 0.70\Theta$. Note the van der Waals loop for the two lowest temperatures.

the temperature and the degree of polymerization. The results obtained show that the scaling behavior of mobile tethered layers is different than that of two-dimensional polymeric systems, due to the ability of the tethered polymers to stretch in the third dimension.

V. TETHERED POLYMERS ON CURVED SURFACES

Many of the applications of tethered polymer layers are in cases in which the surface or interface is curved [180]. For example, in the case of

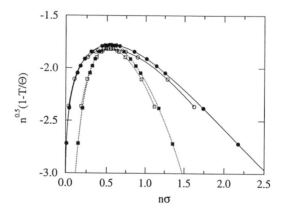

Figure 39. The bimodal (full lines) and spinodal (dotted lines) of the mobile chains in scaled variables $(1 - T/\Theta)n^{0.5}$ as a function of $n\sigma$, for chains with $n = 50$ filled symbols and $n = 30$ open symbols. The scaling variables are the ones predicted in ref. 175.

colloidal stabilization chain molecules are end grafted to a spherical surface. Also, diblock copolymer micelles can be thought of as two tethered layers with opposite curvature. The understanding of the properties of tethered layers on planar surfaces does not directly translate into the understanding of the properties of the system in the case that the surface or interface is not planar. A limiting example can be chains tethered to a point, that is, star polymers [181–183] in which the boundary conditions imposed by the "surface" are completely different than those of a planar surface. Further, the volume available to the molecules as a function of the distance from the surface changes with curvature, and thus one should expect important changes in the molecular organization and thermodynamic behavior as a function of curvature.

Another important point to recognize in many practical uses of curved tethered surfaces or interfaces is that many of the systems studied experimentally correspond to relatively short polymers. For example, the micellar behavior of $(C_2H_4O)_n(C_3H_6O)_m(C_2H_4O)_n$ (PEO–PPO–PEO), known by the commercial name of Pluronic, has been widely studied for chains in which the number of segments per block ranges from a few units to 150 [128, 184]. In this regime of molecular sizes we have shown that the thermodynamic properties of chains tethered to planar surfaces are not properly given by the analytical and scaling predictions. Therefore, it is important to understand the behavior of these "polymers" with a molecular approach in order to be able to make quantitative predictions of the thermodynamic properties that determine the phase behavior of the systems.

A. Conformational Properties

The analytical solution of the SCF approach for curved surfaces requires the existence of a dead zone for the end segments of the chains [142]. The existence of such a region has been investigated by MC [111] and MD [185] simulations as well as by numerical SCF [186, 187] approaches. The picture that emerges is that for long enough chains and intermediate to high surface coverages that region exists and the predictions from the SCF theory seem to appropriately describe the structure of the polymer layer.

In this section we concentrate on the special features observed for intermediate chain length, that is, those of relevance for the above-mentioned PEO–PPO–PEO polymers. We want to see what is the effect of changing the geometry of the surface on the molecular organization of the layer. To this end, Fig. 40 shows the density profiles, predicted by the SCMF theory (for the generalization to curved surfaces, see the Appendix), for a variety of radii of spheres where the chains are tethered. All

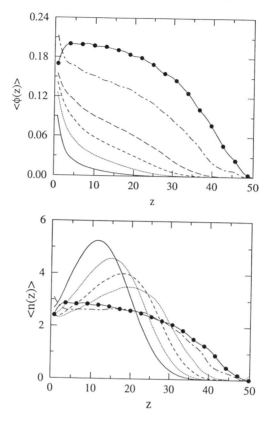

Figure 40. Density profiles (upper graph) for chains tethered at spherical surfaces for a variety of radii. z is the radial distance from the tethering surface. The full line with filled circles is for a planar film, dot–dashed line $R = 100l$, long dashed line $R = 20l$, dashed line $R = 10l$, dotted line $R = 5l$, and full line $R = 2l$. The lower graph shows the segments distributions but normalized as explained in the text. All curves correspond to chains with $n = 100$ and $\sigma = 0.07$.

the curves correspond to the same number of molecules per unit area. The profiles for spherical surfaces are normalized such that

$$\int_R^\infty \langle \phi(r) \rangle 4\pi r^2 \, dr = \sigma n \qquad (29)$$

where R is the radius of the sphere. For the flat surface $\int_0^\infty \langle \phi(z) \rangle \, dz = \sigma n$. The picture that emerges is that as the radius of the surface gets smaller the density profile, for fixed surface coverage, shows a sharp fall-off as a function of the distance from the surface. This simply reflects the fact that the volume available to the polymer segments in spherical surfaces increases as $(r/R)^2$, resulting in more available volume for a fixed number of chains per unit area. Even though this is a direct result of the effect of the geometry, it is not clear from the density profiles what are the changes in the molecular organization of the chains. The lower figure shows the number of segments per chain as a function of the distance from the surface, such that $\int_R^\infty \langle n(r) \rangle \, dr = n$. The segment profiles show that as the radius increases, for fixed surface coverage, the chains are more stretched due to the larger crowding induced by the smaller amount of increase of volume as a function of r. Note that for radius $R/l = 100$ the segment profile is almost identical to that of the linear chain, showing a comparable amount of stretching. However, the density profile is rather different due to the different volume dependence on the distance from the surface in the two cases. (For flat surfaces the volume available as a function of the distance from the surface is constant.)

The distribution of free ends for different surface radii is shown in Fig. 41. The behavior obtained shows the different degree of stretching of the chains for the various radii of the surface. Also there seems to be a dead zone for the free ends. We find that this zone, as it was found in MD simulations [185], results purely from a steric effect, in particular at high surface coverages and small radii. The predictions from the SCMF theory have been compared with MD and MC simulations of grafted chains on spherical and cylindrical surfaces, and quantitative agreement was found for all the different radii geometries and chain lengths for which simulations are available [132].

Numerical SCF calculations by Dan and Tirrell [186] and Wijmans and

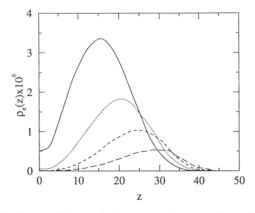

Figure 41. Distribution of free ends for some of the cases shown in Fig. 40.

Zhulina [187] provide the same qualitative behavior for the density profiles and the end distributions on spherical surfaces.

The behavior of the conformational properties for chains end grafted to cylindrical surfaces, not shown here, shows intermediate behavior between that of planar and spherical surfaces [132, 185].

As mentioned above, one of the important applications of tethered polymers on curved surfaces is the formation of micellar aggregates by block copolymers [20, 188, 189]. In these cases the system has at the same time chains tethered in one surface but they occupy different sides of the interface (see Fig. 2). Thus, for one of the blocks the radius is positive while for the other it is negative (see Fig. 2 for the definitions of the sign of the radius of the interface). Clearly, a negative radius imposes stringent conditions on the packing of the chains. To see this effect more quantitatively, Fig. 42 shows the density profiles and the segment distributions for chains tethered to a planar surface as well as in surfaces with positive and negative curvature. The shapes of the density profiles are very similar for the three curvatures, only the magnitude of the density changes due to the difference in volume available as a function of the distance from the surface for the three different geometries. The stretching of the chains, however, is different as seen in the segment distributions. While the chains tethered at a surface with negative radius are less stretched than the other two surfaces, these chains feel very strong repulsions due to the fact that as the distance from the surface increases the volume available decreases rather sharply. This effect causes the chains to be in very unfavorable states from the conformational point of view, as will be reflected in the thermodynamic behavior.

It is interesting to note that combination of the density profiles and segment distributions provide a very helpful tool in understanding the effects of the surface geometry to the packing of the chains. Looking at only one of them may prove to be slightly misleading due to the effect that the geometry has on changing the appearance of the profiles.

B. Thermodynamic Behavior

The central thermodynamic quantity in tethered polymers in curved surfaces is the polymer's chemical potential. As discussed in reference to planar surfaces, this is the quantity that determines the amount of polymer molecules tethered on the surface. Further, in the case of block copolymers self-assembly into micellar aggregates, the chemical potential will determine the size and shape distribution of the aggregates [190]. Clearly, the different molecular organizations of the chains tethered on curved surfaces as a function of the radius of the surface or interface will result in different chemical potentials. For example, we have seen that for

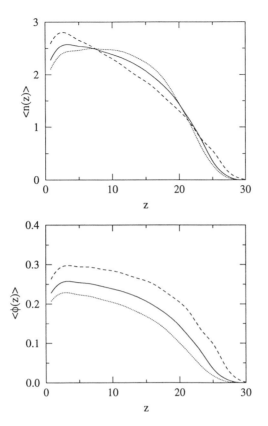

Figure 42. Average segment distribution (upper graph) for chains with $n = 50$ and $\sigma = 0.1$ for three different geometries. The full line is for a planar surface, the dashed line is for $R = -100l$, and the dotted line is for $R = 100l$. The lower graph shows the density profiles for the same three cases.

a fixed surface coverage and positive radii, the larger the radius of the surface, the more stretched the chains are. Thus, one would expect a higher chemical potential for larger radii. The results shown below are a direct manifestation of this effect and how they compete in the formation of aggregates.

Consider first the case in which the chains are grafted to a curved surface. The chemical potential as a function of the surface coverage is shown in Fig. 43 for spherical and cylindrical surfaces for a variety of curvatures. As expected from the stretching of the chains, the smaller the radius, the lower the chemical potential for all surface coverages. For negative radii we see that since the chains have less available volume as a function of the distance from the grafting surface the chemical potential

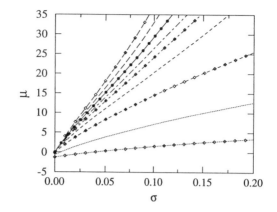

Figure 43. Chemical potential as a function of surface coverage for chains grafted at surfaces of different geometries. Lines with open symbols are for spherical surfaces while lines are for cylindrical surfaces. The full line with filled circles correspond to a planar surface. Dotted lines $R = 2l$, dashed lines $R = 20l$, dot–dashed lines $R = 100l$, and long dashed lines $R = -100l$.

for all σ is even higher than in the planar case. Also note that the chemical potential of the chains grafted on spherical surfaces, with positive curvature, is lower than the equivalent radius in a cylindrical geometry, the reason being that the volume accessible to the molecules increases as r/R for cylinders and $(r/R)^2$ for spheres. Therefore, much more volume to span the chains is accessible in the spherical case. We find that there is a marked variation of the chemical potential with surface coverage even for very low values of σ. This implies that, even though chains grafted on a spherical surface are relatively "free" to search almost all of their conformational space, the presence of a few neighbors already limits the conformational phase space increasing the chemical potential. This effect is important in determining the adsorption isotherms of the chains.

The behavior of the chemical potentials suggests that the amount of polymer tethered to a surface by a functionalized end group from solution will be larger the smaller the radius of the surface. Figure 44 shows the amount of polymer per unit area that is end adsorbed from a bulk solution of fixed polymer concentration as a function of the curvature of the surface. There are two sets of curves corresponding to spherical and cylindrical surfaces. In each group three different adsorption interactions of the functionalized end group are considered. The following trends emerge from the adsorption isotherms. In all cases the adsorption increases as the radius of curvature of the surface decreases. The

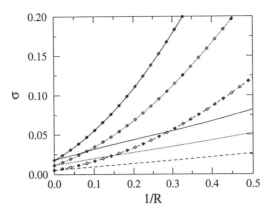

Figure 44. Equilibrium adsorption as a function of the curvature of the surface for spherical (lines with symbols) and cylindrical (lines) geometries. All the cases are for $\phi_{bulk} = 0.01$, and the adsorption energy of the functionalized end group is full lines $\varepsilon_s = -8\,\text{kT}$, dotted lines $\varepsilon_s = -6\,\text{kT}$, and dashed lines $\varepsilon_s = -4\,\text{kT}$.

adsorption in spherical surfaces is much higher than the equivalent curve for cylindrical geometry. The higher the energy of adsorption of the functionalized end group, the higher the amount of polymer end tethered to the surface. All of these trends are a direct manifestation of the variations of the chemical potentials of the tethered chains discussed in relation with Fig. 43.

We have found that for planar surfaces there is a universal adsorption isotherm (see Fig. 18 and discussion thereafter). In the case of curved geometries we could not find a universal behavior as a function of the radii of the surfaces. Note, however, that we are looking at very large changes in curvature from planar to very small spheres. Thus, one should not expect a universal behavior to cover this wide range of variation in surface geometry.

Understanding of the size and shape distribution of micelles requires consideration of the total free energy of the solution. In general, one treats these systems with the mass action model [191] in which the information needed for each aggregate's size and shape is the chemical potential of the chains in that aggregate as compared to the free polymer in solution. Here we want to show how different block copolymers have a preferential geometry by looking only at the chemical potential of the chains in the cases in which the systems are composed by two solvents, each good for one of the blocks but infinitely poor for the other. Thus, we assume that the polymers assemble at the interface between the two solvents but there is no protrusion of the blocks into the poor-solvent side of the interface. Clearly, this is an oversimplification of the system, but it is considered here in order to show the effect of curvature on the optimal aggregate geometry.

Linse [192–195] has looked at the properties of micelles formed by Pluronic molecules using the lattice SCF approach. Further, Hurter et al. [196] have also studied the solubilization abilities of Pluronic micelles with the lattice SCF theory. In both cases good agreement with experimental observations was obtained. However, in these cases the micelles are formed in a single solvent.

We consider two simple cases to show the effects of opposite curvature in the case of diblock copolymers. Figure 45a shows the chemical potential of symmetric diblock copolymer molecules at the interface between two immiscible solvents. The lower curve corresponds to the case of a planar interface while the upper curve is for spherical aggregates of radius $R = 100l$. For all surface coverages the planar interface is preferred over the curved one for these symmetric molecules. The reason is that even though the blocks that are at the positive side of the interface lower the chemical potentials, the block for which the interface has negative curvature loses more free energy than the gain obtained by the other block. Therefore, in terms of the optimal packing of the chain molecules symmetric diblocks will have a preference to planar interfaces, that is, lamellar phases.

The chemical potential of the chains depends on the interfacial geometry, the surface coverage, and chain length, as shown in Figs. 43 and 16. Further, we have seen that the planar interface is the most stable for symmetric molecules due to the fact that the loss of free energy of the block facing the negative curvature interface is larger than the gain of the block in the positive side. This suggests that by changing the chain lengths of the blocks one may tune the optimal size and shape of the aggregate. For example, consider a case in which the polymer is formed by blocks of different sizes and compare the chemical potentials of the diblock molecule for different interfacial geometries. This is shown in Fig. 45b for a planar interface and a spherical aggregate. The curve in the middle corresponds to a planar interface. The one that shows the highest chemical potentials for all surface coverage corresponds to diblocks in a spherical aggregate in which the larger block is at the interface with negative curvature while for the short one the curvature is positive. The lower chemical potential for all σ is for a polymer chain in which the short block is at a negative curvature interface while the long one has a positive curvature interface. In this particular example we see that the curved geometry, with the appropriate arrangement of the blocks in the aggregate, is more favorable than the planar interface, the reason being

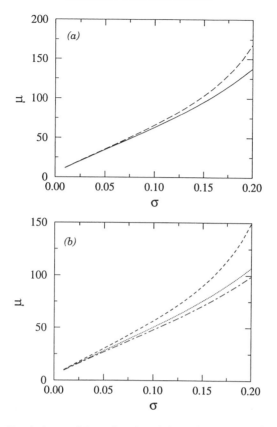

Figure 45. Chemical potential as a function of the surface coverage for $n_1 - n_2$ diblock copolymers: (a) symmetric diblocks $n_1 = n_2 = 50$ at a planar interface (full line) and at a spherical interface with $R = 100l$ (long dashed line). (b) asymmetric diblock with $n_1 = 50$ and $n_2 = 30$ at a planar interface (dotted line), an interface with positive radius, $R = 100l$, for the long block (dot–dashed line), and negative curvature for the short block. The dashed line corresponds to an interface with the same positive radius for the short block (dashed line) and negative curvature for the long one.

that the gain in conformational freedom by the long block, by being at a positive curvature interface, is larger than the loss of the short block.

The example just given shows that manipulations of the size of the blocks in the molecules can lead to preferential aggregate curvature. This statement will be discussed in a more quantitative and systematic way in the next section, which deals with elastic properties of diblock copolymers at liquid–liquid interfaces.

VI. ELASTIC PROPERTIES OF DIBLOCK COPOLYMER FILMS AT LIQUID–LIQUID INTERFACES

Diblock copolymers can be considered to be the high-molecular-weight analog of short surfactant molecules. Namely, they are formed by two blocks that may have different affinities to the surrounding solvent. The molecules form aggregates that can be of various shapes and sizes, depending upon the particular properties of the molecules, the solvent or solvents, and the thermodynamic conditions, that is, density and temperature, in order for the two blocks of the molecules to be in an environment that has favorable interactions. In this section we concentrate our attention on one particular case that corresponds to diblock copolymers at the interface between two immiscible solvents such that each solvent is good only for one of the blocks but is poor for the other. This can be thought of as the equivalent case of surfactant molecules at the water–oil interface, in which by addition of the amphiphilic molecules one can achieve solubilization of water in oil (or oil in water).

The first thing to consider is the amount of polymer chains at the liquid–liquid interface. Since the system under consideration is self-assembled, the area per molecule is not a thermodynamic variable but is determined by the optimal free energy of the system. To keep the discussion at a simple level, we consider here only the case in which the solvents are infinitely poor for one of the blocks. Namely, the diblock copolymers have no solubility in either of the solvents and therefore all of the chains are at the liquid–liquid interface. For this particular case, the surface coverage of the diblock molecules is obtained by minimizing the free energy of the interface in the presence of the polymer.

The free energy of the interface can be written as

$$F = \gamma_b A + F_0 \tag{30}$$

where γ_b is the bare liquid–liquid surface tension, A is the total surface of the interface, and F_0 is the total free energy of the diblock copolymer solvent mixture, which can be obtained from Eq. (14). The equilibrium amount of polymer at the interface is obtained from

$$\left(\frac{\partial f}{\partial a}\right)_{N_p, \beta} = 0 = \gamma_b - \Pi_a \tag{31}$$

where the athermal limit is considered for each block in its "good" solvent, and Eq. (14) is used. Note that in this approximation, that is, that the diblocks are insoluble in both solvents, the equilibrium surface coverage corresponds to zero surface tension of the diblock modified interface. Here we have used the free energy per chain, $f = F/N_p$, and the area per molecule, $a = A/N_p$.

The equilibrium surface coverage as a function of the bare liquid–liquid surface tension for a variety of diblock copolymers is shown in Fig. 6.1. The surface coverage is scaled by the square root of the total length of the chains, and it is found that the resulting curve is universal. This is a direct reflection of the behavior of the lateral pressures as discussed above (see Fig. 3.9). The figure shows that to minimize the free energy in the case of relatively high bare surface tension not too many diblock copolymers are needed. This is a consequence of having small local repulsions, but due to the size of the polymer, it results in a relatively high pressure opposing the bare liquid–liquid surface tension [see Eq. (15)].

To see how the results of Fig. 46 translate to measurable quantities, one needs to consider that the units of γ_b are given by kT/l^2, where l is the segment size. The water–oil surface tension is on the order of $\gamma_{o/w} = 0.12\,kT/\text{Å}^2$, which taking $l = 5\,\text{Å}$ corresponds to $\gamma_b = 3$. For this value of the bare surface tension one has $\sigma_e(n_1 + n_2)^{0.5} = 1$. Recalling that σ_e is in units of $1/l^2$ a chain with a total number length of 100 segments has an area per molecules $a_e = 1/\sigma_e = 250\,\text{Å}^2$, which is a reasonable number for this diblock chain length. In this way one can use the results of the "isotherm" to obtain a reasonable estimate of the equilibrium area per molecule for different diblocks at the interface of different immiscible solvents. It should be stressed that this estimate will be good only in the

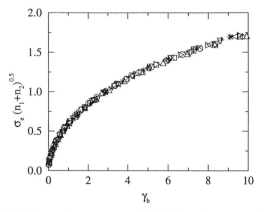

Figure 46. Equilibrium surface coverage of diblock copolymers, scaled by the square root of the total chain length, as a function of the bare surface tension of two immiscible solvents. The symbols correspond to $n_1 - n_2$: 30–30 circles, 30–50 squares, 30–70 diamond, 30–100 triangle up, 50–50 triangle left, 50–70 triangle down, 50–100 triangle right, 70–70 plus, 70–100X, and 100–100 stars.

cases that the solubility of the diblock in the solvents is very low; otherwise a more complete calculation that considers the equilibrium of the diblock chains in the solvents and the interface is necessary. The question that arises now is what will be the optimal geometry of the interface formed. One can consider the total free energy of the interface to include, in a phenomenological description, terms associated with changes in the area per molecule and those associated with variations of the curvature, defined as $c = 1/R$, of the interface. If one is interested in small deviations from some reference state, one can write the phenomenological change of the free energy per molecule up to quadratic order in the form

$$\delta f = \tfrac{1}{2}\lambda(a - a_e)^2 + \tfrac{1}{2}K(c_1 + c_2 - c_0)^2 + \bar{K}c_1 c_2 \tag{32}$$

where it is assumed that the reference state, with $a = a_e$, corresponds to the area per molecule that minimizes the free energy and hence there is no linear term in area. The two curvature terms arise from the fact that a surface has two independent curvatures, here c_1 and c_2 and hence two independent "normal modes". The quantity c_0 is called the spontaneous curvature and is a measure of the asymmetry of the diblock at the interface. The two constants K and $\bar{K}$ are called the bending and saddle-splay constants, respectively. The curvature expansion was first proposed by Helfrich [197] for the cases of interfaces at constant area per molecule. As discussed by Wang and Safran [198], the constants defined here are related to the Helfrich constants by $k = K/a_e$ and $\bar{k} = \bar{K}/a_e$, where the lowercase constants are the ones defined by Helfrich.

The phenomenological constants are clearly related to the derivatives of the free energy. We can write a general expansion of the free energy around the planar film at the equilibrium surface coverage in the form

$$\delta f = \frac{1}{2}\frac{\partial^2 f}{\partial a^2}(a - a_e)^2 + \frac{\partial^2 f}{\partial a\, \partial c_+}(a - a_e)c_+ + \frac{1}{2}\frac{\partial^2 f}{\partial c_+^2}c_+^2 + \frac{1}{2}\frac{\partial^2 f}{\partial c_-^2}c_-^2$$

$$\tag{33}$$

where we have used $c_+ = c_1 + c_2$ and $c_- = c_1 - c_2$ as the two independent curvatures for convenience and there are no linear terms in c_- due to the symmetry of the free energy [199]. Further, all the derivatives are evaluated at a_e; $c_+ = c_- = 0$. By comparing Eqs. (32) and (33) one can directly relate the phenomenological constants to the derivatives of the free energy. First, note that the phenomenological expression does not include a cross term between area and curvature. This implies that

$\partial^2 f/\partial a\, \partial c_+ = 0$. This relation defines the surface of inextension, which is the surface in which the area and the curvature have to be measured. This is an important consideration because in reality the layer is not a two-dimensional surface. Therefore, one can measure the curvature and area of the interface at any convenient choice provided that care is taken in including all the appropriate terms in the free energy. Since we find it more convenient to measure everything in terms of the surface of inextension, the cross term must vanish. This is a valid choice only in the cases in which the free energy is quadratic; otherwise the surface of inextension cannot be easily defined.

The reason for this choice is that it enables us to have a clear physical picture of what are the changes in the system because upon curvature deformation the area of the surface of inextension does not change. Then, one can analyze the changes of the free energy of the layer by looking independently at the curvature and area deformations. Another choice may have been the liquid–liquid interface, as was done by Wang and Safran [198], who properly took into account the additional contributions to the elastic constants arising from the different positions of the surface of inextension.

The elastic constants from Eq. (32) using Eq. (33) are

$$\lambda = \left(\frac{\partial^2 f}{\partial a^2}\right)_{a=a_e, c_+ = c_- = 0} \tag{34}$$

$$K = \left(\frac{\partial^2 f}{\partial c_+^2}\right)_{a=a_e, c_+ = c_- = 0} + \left(\frac{\partial^2 f}{\partial c_-^2}\right)_{a=a_e, c_+ = c_- = 0} \tag{35}$$

$$\bar{K} = -2\left(\frac{\partial^2 f}{\partial c_-^2}\right)_{a=a_e, c_+ = c_- = 0} \tag{36}$$

$$Kc_0 = -\left(\frac{\partial f}{\partial c_+}\right)_{a=a_e, c_+ = c_- = 0} \tag{37}$$

To obtain the elastic constants, we can use the free-energy expression from the SCMF theory and take the appropriate derivatives. This yields the expressions [199, 200]

$$\lambda = -\int_{-\infty}^{\infty} \left(\frac{\partial \Pi(z)}{\partial a}\right)_{a_e, 0, 0} dz \,, \tag{38}$$

$$K = -a_e \int_{-\infty}^{\infty} \left(\frac{\partial \Pi(z)}{\partial c_+}\right)_{a_e, 0, 0} z\, dz \tag{39}$$

$$\bar{K} = -a_e \int_{-\infty}^{\infty} \Pi(z)z^2 \, dz + \gamma_b a_e \xi^2 \tag{40}$$

$$Kc_0 = a_e \int_{-\infty}^{\infty} \Pi(z)z \, dz - \gamma_b a_e \xi \tag{41}$$

where we have used the "generalized" lateral pressure $\Pi(z) = e^{-\pi(z)} - 1 + \pi(z)$ for convenience. The origin of the z axis must be taken at the surface of inextension, which is determined by equating $(\partial^2 f / \partial a \, \partial c_+)_{a_e,0,0}$ to zero, and the limits of the integral are taken as infinity, but in practice $\Pi(z) \neq 0$ only in the region of the interface where there is polymer. The quantity ξ is the distance from the surface of inextension to the solvent–solvent interface. Note that $\bar{K}$ and c_0 depend upon the bare surface tension, but λ and K do not.

All the elastic constants are given by moments of the generalized pressure and its derivatives. Further, the pressures and their derivatives should be evaluated at the planar configuration. The derivatives are determined by differentiating the volume constraints [Eq. (3)] with respect to the desired variable. The result (see ref. 199) is a set of linear integral equations for the derivatives of the form

$$\int_{-\infty}^{\infty} \left(\frac{\partial \pi(z')}{\partial a} \right)_{a_e,0,0} [\langle n(z) \rangle \langle n(z') \rangle - \langle n(z)n(z') \rangle] \, dz'$$

$$- \left(\frac{\partial \pi(z)}{\partial a} \right)_{a_e,0,0} a_e e^{-\pi(z)} + e^{-\pi(z)} - 1 = 0 \qquad -\infty < z < \infty \tag{42}$$

From the derivatives of $\pi(z)$ the variation of the generalized pressures is straightforward. The term in brackets is related to the intramolecular density–density correlation function, and it has to be evaluated at the planar film. Therefore, all that is needed to obtain the elastic constants is the planar film. Clearly, the elastic constants could also be calculated by determining the free energy of spheres and cylinders and variations of the area per molecule in the planar film. However, this requires lots of computational effort, and further the expressions just presented enable us to understand the elastic constants and their variations by just looking at the pressure profiles in the planar film.

Cantor [201] calculated the bending elastic constant of diblock co-polymer at liquid–liquid interfaces using a mean-field approach. He concentrated on determining the free-energy cost of taking the monolayer from a planar geometry to that of sphere. Wang and Safran have carried out the most detailed calculations of the elastic constants using the analytical SCF approach. In both cases, the results should be valid in the asymptotic regime. Since we have seen that the pressures for intermediate chain length are not properly given by the analytical approaches, the elastic constants, which are moments of the pressures and their derivatives, will show the same discrepancies. Therefore, detailed comparisons between the SCMF calculations and the analytical approaches are not attempted here. Further, the results that correspond to reasonable bare surface tensions, up to $\beta\gamma_b l^2 = 10$, correspond to the regime where the chains are in the mushroom-to-brush transition region.

The variation of the stretching constant as a function of the bare surface tension is shown in Fig. 47. The stretching constant has been scaled by the square root of the total chain length of the diblock. This scaling results from the variation of σ_e shown in Fig. 46. As can be seen, the scaled stretching constant is a universal function of the bare surface tension. Further, we have found that $\lambda(n_1 + n_2)^{0.5} \propto \gamma_b^{1.6}$ over the range shown. It is not clear to us what is the origin of this exponent, and we are currently studying this behavior. What is clear is the origin of the overall dependence of the stretching constant on surface tension and chain length. For fixed γ_b the increase in molecular weight results in a decrease of the free-energy cost of taking the layer out of its equilibrium area per molecule, the reason being that at fixed bare surface tension the equilibrium area per molecule increases with increasing chain length. On the other hand, at a fixed molecular weight of the diblock an increase of γ_b results in a decrease of a_e; that is, the chains are more stretched, and therefore the restoring force to equilibrium is larger.

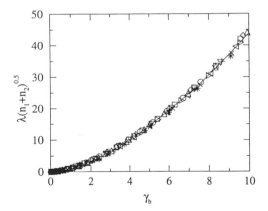

Figure 47. Stretching constant, scaled by the square root of the total chain length, as a function of the bare surface tension. Symbols as in Fig. 46. The line corresponds to $\lambda(n_1 + n_2)^{0.5} \propto \gamma_b^{1.6}$.

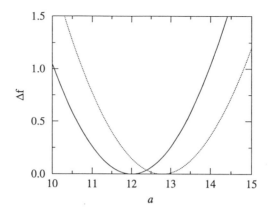

Figure 48. Free energy change upon stretching the interface from its equilibrium configuration at $\gamma_b = 3\,kT/l^2$ for 50–100 (full line) and 70–100 (dotted line) diblock copolymers.

Figure 48 shows the change in free energy upon stretching for two different molecular weights at a fixed bare surface tension to illustrate the energy scales for stretching. The larger diblock has a larger equilibrium area per molecule, as discussed above, but in the particular case shown the stretching constants are not very different. In both cases the change in area necessary to change the free energy by kT per molecule is on the order of $2l^2$. Taking $l = 5$ Å we have a change of 50 Å^2. This seems to be a very large change in area, resulting in a small change of free energy per molecule. However, it is due to the fact that at this bare surface tension, corresponding to the experimental water–oil tension, the chains are not highly stretched, implying an interface that can undergo very large area fluctuations. Experimentally, it has been found that very short molecules with 6–12 segments are enough to form microemulsions with ultralow surface tensions [202]. Thus, long chains can achieve the same reduction of the surface tension at much larger values of a_e, resulting in a more "elastic" interface.

The bending constant K can be understood as measuring the free-energy cost of forming a cylindrical interface. It is also this constant that is responsible for the amplitude of the fluctuations of the interface. Figure 49 shows the variation of the bending constant with the bare surface tension for a variety of diblock chains. We have found that the variation of the constants with total chain length are in line with the predictions of Wang and Safran [198], even though the calculations presented here are not in the proper scaling regime. However, the dependence on surface tension is rather different. Both elastic constants can be measured

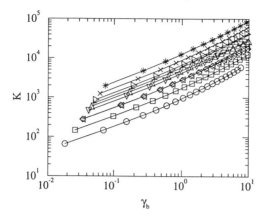

Figure 49. Log-log representation of the bending constant as a function of the bare surface tension. Symbols as in Fig. 46.

experimentally, but so far only measurements with short surfactants have been reported (see, e.g. ref. 203).

As in the case of stretching constants, increasing the bare surface tension at fixed chain length results in an increase of the bending constants. As γ_b increases, the change in the lateral pressures by changing the curvature increases too, due to the higher degree of stretching of the polymer molecules. Further, on the same physical ground, at fixed bare surface tension the bending constants increase with increasing overall chain length. Note that the two diblocks that have the same overall molecular weight have very similar bending constants, in particular for low surface tensions.

Figure 50 shows $-\bar{K}$ as a function of the bare surface tension. For all the cases shown the saddle-splay constant is negative, showing that saddle deformations always cost in free energy. However, the behavior of $\bar{K}$ is not as simple as that of the bending constant. For asymmetric diblocks there is a minimum in the saddle-splay constant for high enough bare surface tensions. The reason is the competing contributions of the chain lateral pressures and the bare surface tension [see Eq. (38)]. The more asymmetric the diblock, the larger the distance of the surface of inextension to the interface, that is the larger is ξ due to the fact that this quantity is determined by mechanical balance achieved by zero torque, and the larger the positive contribution to $\bar{K}$, and one may achieve the condition of positive values in which a saddle deformation (e.g., bicontinuous phases) is the equilibrium structure.

Consider now the case in which the chains are symmetric and therefore $c_0 = 0$. The quadratic free energy [Eq. (32)] predicts that the planar film

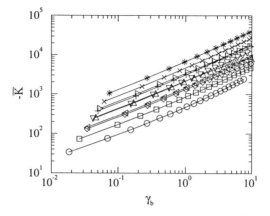

Figure 50. Log-log representation of the negative of the saddle-play constant as a function of the bare surface tension. Symbols as in Fig. 46.

will be stable provided that $K + \bar{K}/2 > 0$. Therefore, if $\bar{K}$ is negative but much larger in absolute value than K, one can make the planar film formed by symmetric molecules unstable and spherical droplets will form. (Higher than quadratic terms in the free energy prevent it from decreasing without bound, and these will also be needed to determine the optimal radius for $c_0 = 0$.) While this has not yet been observed for monolayers with $c_0 = 0$, it has been found in bilayers in which stable vesicle formation is possible [199, 204–206].

In these examples we can see how the proper choice of the blocks forming the polymer can be used to obtain a desired structure. We now will show some examples in which this can be achieved by looking at the equilibrium curvature of the layer, or the parameter that has been predicted to determine the shape of droplets in microemulsion phases [207].

The phenomenological free energy [Eq. (32)] predicts that the structure of the equilibrium layer (i.e., minimal f) is the one having $a = a_e$ and a spherical shape with curvature $c_- = 0$, and

$$c_{+,\text{eq}} = c_{\text{eq}} = \frac{Kc_0}{K + \bar{K}/2} \tag{43}$$

which for a symmetric diblock will give $c_{\text{eq}} = 0$ because $c_0 = 0$ and the planar structure will be the equilibrium one, provided $K + \bar{K}/2 > 0$.

The equilibrium curvature is a function of the asymmetry of the diblock and the bare surface tension through the bending and saddle-splay constants. Figure 51a shows the equilibrium radius of curvature,

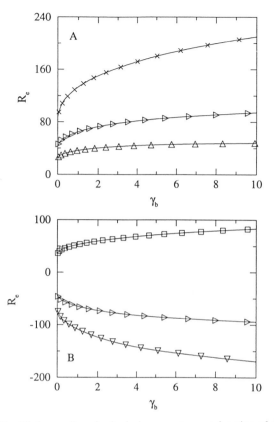

Figure 51. Equilibrium radius of spherical aggregates as a function of the bare surface tension for (a) $100 - n_2$ chains with $n_2 = 30$ triangles up, 50 triangles right, and 70 X. (b) $50 - n_2$ chains with $n_2 = 30$ squares, 100 triangle right, and 70 triangle down. Note that the sign of the radius is defined with respect to the n_1 block.

$R_{eq} = 2/c_{eq}$, as a function of the bare surface tension for diblock copolymers in which one block, the longer one, has a fixed length and the short block is of varying chain length. The results show that the radius of the equilibrium sphere increases with the bare surface tension. Further, as the asymmetry of the diblock increases, the radius becomes smaller. This reflects the tendency of the long chain to have the smallest positive radii in order to gain as much conformational freedom as possible. However, the other block will also prefer a positive radius, and the resulting equilibrium radius will be the compromise between these two opposing effects.

The interplay between the sizes of the two blocks in determining the optimal spherical droplet size can be seen in Fig. 51b, where the sign of the radius is measured with respect to the fixed length block. One can observe how the predicted droplets have $n_1 = 50$ blocks on the inside for the sphere when the other block is longer, but it goes to the outer side when the length of the second block is shorter.

An interesting result from Fig. 51 is that the size of the droplets for the particular chain lengths studied here varies from $30l$ to over $200l$, which (taking $l = 5$ Å) are of the same order of magnitude as water-in-oil (and oil-in-water) droplets in microemulsions in which short surfactant molecules are at the interface. It may be interesting to see if one can obtain stable microemulsions with the much longer chain length predicted here. Clearly, the properties of the interfaces will be different than those of the corresponding short surfactant system due to different character of the interface at constant γ_b as the chain length increases. The longer the chains, the more distributed the same overall stress is, resulting in rather different mechanical properties, that is, elasticity.

Another important quantity in the determination of the phase behavior of microemulsions that arises from the phenomenological treatment of the interfaces is the ratio $K/\bar{K}$ [207]. The analytical approaches that have been used to study the elastic properties of diblock copolymers predict that this ratio is independent of γ_b [198]. However, one can see, by analyzing the results in Figs. 49 and 50 and as has been shown for short surfactant molecules [199], that the variation of K and $\bar{K}$ with surface tension is not the same. The ratio is shown in Fig. 52 for a variety of diblock chain lengths. For very small bare surface tensions there is a maximum in this quantity. However, those surface tensions are very small, and in reality the chains are in the mushroom regime so that

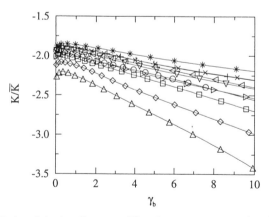

Figure 52. Ratio of the bending to saddle-splay constants as a function of the bare surface tension. Symbols as in Fig. 46.

probably no aggregates may form. The most important aspect of these results is that the ratio is not constant, and thus one may tune the shape of the aggregates formed by proper choice of the diblock length and asymmetry depending upon the surface tension of the solvent–solvent interface. It should be noted that this ratio is more than a factor of 4 as compared with the analytical predictions of Wang and Safran [198]. The large differences between the two predictions may lead to different stable structures of the aggregates. However, we are not aware of experimental evidence that may shed light on which of the predictions is more reliable, even though as discussed above, the analytical predictions are going to be valid in the asymptotic limit and only in the brush regime, while the SCMF calculations are more suitable for intermediate chain length and are applicable for all surface coverages.

VII. SUMMARY

In this chapter we have reviewed some of the aspects of the thermo-dynamic behavior and molecular organization of chain molecules tethered to surfaces or at interfaces. The systems considered are only those in which the chain molecules are embedded in a sea of low-molecular-weight solvent. Most of the studies presented here are based on calculations using the single-chain mean-field theory. This theoretical approach is based on looking at a central chain exactly; that is, all the intramolecular interactions as well as the interactions with the surface are exactly taken into account, and the intermolecular interactions are taken within a mean-field approximation. The mean field of the polymer and solvent is inhomogeneous due to the confinement of the end of the chains to the surface, and it is determined by the average properties of the central chain. The average density profiles are determined by assuming a local incompressibility, namely, that the volume accessible to the solvent and polymer chains as a function of the distance from the surface is completely filled. In reality this represents the inclusion of the complete equation of state, which is necessary in order to close the necessary equations to be able to perform any calculation.

The central quantity of the theory is the probability distribution function of chain conformations, which is a function of the surface coverage, the quality of the solvent, and the surface geometry. The pdf determines the relative weight of each different single-chain conformation under different conditions. Therefore, once the single-chain set of conformations is generated, one can calculate the properties of the system under all the different conditions by solving a set of nonlinear coupled equations to determine the "repulsion" field, or the local lateral interactions among the molecules. In this way one can follow how different conformations change their weight in the sample as the different conditions are studied. The computationally cost step is the generation of single-chain conformations, but once this is obtained, the calculations become simple (see Appendix).

The theory is exact in the limit of very low surface coverages, since the single-chain conformations are exactly taken into account. It turns out that for all surface coverages, geometries of the surface, and solvent quality, the predictions of the theory are in quantitative agreement with full-scale simulations. This implies that the mean-field approximation, used in determining the lateral pressures, is of excellent quality. Therefore, one can argue that in all these systems fluctuations parallel to the surface beyond the size of a chain are not very important. Otherwise the theory should fail to predict the behavior of these systems as compared with simulations.

The theory is applicable for chain lengths in the range of short to intermediate (up to a few hundred segments). For longer chain lengths the sampling of the single-chain conformations is too costly computationally, and therefore the approach is impractical. In its regime of applicability the main advantage of the theory is that it can be applied to any chain architecture and chain model. Therefore, one can study how different molecular structures result in different properties of the layers. This is also one of the main differences with the numerical self-consistent field theories, which are based on looking at a random walk in the presence of an external field. Therefore, in the SCF approach the intramolecular and intermolecular interactions are taken within the same mean-field approximation. As has been shown, the main drawback of this approach is that predictions for the chain chemical potential do not seem to be correct in the mushroom regime and in the mushroom-to-brush transition region. This is due to the fact that the work required to stretch a real chain, which is always self-avoiding, is not properly accounted for. However, the structure of the layer as measured by the density profile is properly predicted by the numerical SCF approaches, particularly when the chains are relatively stretched. The advantage of the SCF approach is that since it considers a random walk one can perform calculations on longer chain lengths than with the SCMF theory. However, changes in molecular architecture, while possible, are not straightforward with the SCF approach. More importantly, in the limit of small surface coverages the SCF theory treats the chains as isolated random walks and thus do not properly describe the whole regime of surface coverages.

The regime of chain length that can be covered by the SCMF theory is of practical applicability from surfactant molecules to chain molecules used for many biological applications, such as PEO molecules that have been used for drug delivery and for prevention of protein adsorption to biosurfaces [215].

The behavior of chain molecules tethered to a surface in the good-solvent regime show a mushroom-like structure at very low surface coverages. As the two-dimensional polymer density is increased, there is a broad region in which the chains undergo a "continuous" transition from mushroom to the so-called brush regime. These findings explain recent experimental observations in which the density profiles and height of the layers were studied by neutron reflectivity. It is important to emphasize that, as observed in the experiments and predicted by the SCMF theory, the radius of gyration of the chains is the relevant length scale of the problem in the experimental regime of surface coverages, not the distance between tethering points, as suggested by the analytical approaches. However, the reason is not a failure of the analytical approaches but the fact that most of the experimental observations are in the mushroom-to-brush transition region. This example shows that in order to interpret experimental observations with theoretical approaches one must be certain that the comparisons are made for measurements and predictions that correspond to the same regime; otherwise the comparisons are meaningless.

In the good-solvent regime the increase of surface coverage induces a stretching of the chains perpendicular to the surface. The distribution of segments becomes narrower and more sharply peaked (localized) as the surface coverage increases. The stretching of the chains is induced by the effective repulsive interactions among the polymer segments for this solvent quality. This repulsion can be measured by the lateral pressures, which show a sharp increase as the surface coverage increases. For intermediate chain lengths it is found that the lateral pressures multiplied by the bulk radius of gyration of the chains is a universal function of the reduced surface coverage ($\sigma^* = \sigma \pi R_g^2$) for values of σ^* up to the onset of the brush regime. However, the scaling function is not given by a single exponent. This can be explained by noting that the second virial description of the interactions is valid over a very narrow regime of surface coverage for the chain lengths presented here. Therefore, there is a constant increase of the number of virial coefficients that must be included.

The large increase of the lateral pressure is also reflected in the chemical potential of the chains. This quantity determines the equilibrium surface coverage of the chains. The work required to bring a chain to the surface in the brush regime is very high, and therefore the adsorption isotherms show that in dilute solution, even for high adsorption interactions of the functionalized end groups, the layers will form in the brush-to-mushroom region predominantly, as observed experimentally. Further, in dilute bulk solution the adsorption isotherm is predicted to be universal in the scaled variables $\sigma n^{7/5}$ versus $e^{\beta \varepsilon_s} \phi_{bulk}/(1 - \phi_{bulk})^n$. The molecular organization of chain molecules of different chain architectures has been shown to dramatically change the properties of the layers. Here, the behavior of chains with three branches and molecules with flexible and rigid portions was described. The main conclusion from these studies is that the proper choice of the chain chemistry can be used to tune several properties, such as interactions between grafted layers, the distribution of free ends, and the optical properties of the layers. In parallel, the study of the thermodynamic behavior of the different chain molecules enables us to predict whether the preparation of these layers is feasible in the regime where the chains show the behavior of interest.

In the poor-solvent regime the chain molecules are organized so that they avoid as much as possible contact with the solvent molecules. In the case that the polymers are chemically grafted to the surface, in order for the chains to have enough contact with other polymer chains, they form clustered structures in the intermediate surface coverage regime. For low enough surface coverage the chains are collapsed mushrooms with the same characteristics of bulk chains in dilute (poor-solvent) solutions. At high surface coverage homogeneous layers become stable due to the fact that the distance between chains is small enough so that many segment–segment contacts are possible without the need of microphase separation. In the case of molecules with lateral mobility the chains undergo a two-dimensional phase separation with one phase almost pure solvent and the other phase more concentrated in polymer.

The behavior of the chain molecules tethered on curved surfaces in the good-solvent regime is mainly determined by the different volumes accessible to the chains as a function of the distance from the tethering surface. Therefore, for fixed surface coverage, the chains are more stretched when grafted on planar surfaces as compared to cylindrical and spherical surfaces. Thus, the chemical potential of the chains is lower in spherical surfaces. Further, the smaller the radius of the surfaces, the lower the chemical potential. As a result, the amount of polymer per unit area that can be end adsorbed from solution is larger the smaller the radius of the spherical surface. It is found that for all surface geometries the chemical potential is a very sensitive function of the surface coverage, indicating that a proper estimate of this quantity is necessary in order to determine the adsorption isotherms.

The equilibrium surface coverage of diblock copolymers at the interface between two immiscible solvents that are selectively good for one of the blocks but infinitely poor for the other was shown. It has been found that $\sigma_e(n_1 + n_2)^{0.5}$ is a universal function of the bare solvent–solvent surface tension. For the equilibrium structure of the layers the stretching and bending constants of the fluid layer can be expressed in terms of moments of the lateral pressure profiles and their derivatives. The calculation of all the elastic constants require only the knowledge of one interfacial geometry. The expressions found provide a very useful tool to determine these material properties of the copolymer layers since other theoretical approaches require the determination of the free energy as a function of curvature, which is computationally very demanding.

As is the case for all other thermodynamic properties, it is found that the analytical predictions based on the SCF theory do not show the same surface tension dependence of most elastic properties as those obtained from the SCMF approach. The reason, again, is the finite chain length considered in the SCMF theory as compared to the asymptotic regime that is studied with the analytical approaches. For the chain lengths presented here, one finds that by the proper choice of the diblock copolymer molecule, one can tune the sizes of the aggregates formed and the phase behavior. We are not aware of microemulsion phases formed by diblock copolymers in low-molecular-weight solvents (e.g., water and oil). The amount of polymer predicted to be needed to obtain zero effective surface tension is relatively low. Further, the properties of the interface (i.e., elastic constants) are calculated to be easier to tune than the low-molecular-weight surfactant counterparts. It will be interesting to see if this type of interface can be formed experimentally.

The use of the single-chain mean-field theory to treat tethered polymer layers of intermediate chain lengths shows that the specific chemistry of the chain molecules plays an important role in determining the behavior of these systems. Moreover, it also shows that the fact that polymers are composed by segments connected to each other and constrained by the interactions among them has important consequences in the behavior of these complex fluids. While many properties and different behaviors of these systems can be studied by simple and physically insightful analytical approaches, the complement that can be given by combining molecular theories with analytical approaches enables us to learn about the behavior of the polymer layers at all length scales. This understanding at all length scales is very important because it will allow us to bridge the gap between the chemistry "manufacturing" of the molecules and the desired behavior of the polymeric layers. Further, it provides information that is very hard to access with experimental measurements, helping thus to complement the understanding that results from direct observation of these systems. The study of the behavior of polymeric systems at all length scales (i.e., universal and non-universal properties) also provides very important information in bulk homogeneous polymeric fluids [208]. In this chapter we have considered only the cases of monodisperse, nonpolar chain molecules in low-molecular-weight solvents. The behavior of these simple systems is rich and complex. Several other related systems have been studied such as polymeric solvents [80, 90, 139], polyelectrolyte layers [83, 209–211], and mixtures of tethered polymers [98, 212–214], among many others. The number of variables that can be manipulated in these systems is very large, and their understanding provides information on the basic physical chemistry underlying macromolecular systems. Further, it is also the hope that the microscopic understanding of these many different polymeric layers will provide guidelines on how to tailor surfaces with desired properties.

APPENDIX

In this appendix we present some details regarding the implementation of the SCMF theory to a specific example. Consider monodisperse polymer chains grafted on a planar surface and embedded in a low-molecular-weight solvent at a finite temperature T. Here, n is the degree of polymerization, l is the length of the chain segment, v_0 is the volume of the monomer, and the volume of the solvent molecules is also taken as v_0.

As was explained in Section II.B, the SCMF theory gives an integral equation for the anisotropic osmotic pressure $\pi(z)$. This equation is actually the volume-filling constraint Eq. (3) with the introduction of the probability distribution function of chain conformations $P(\alpha)$ given by Eq. (5) and the density of solvent molecules given by Eq. (6). Explicitly it reads

$$\sigma \sum_{\{\alpha\}} P(\alpha) v_p(z, \alpha) + \exp[-\beta\pi(z)v_0] = 1 \qquad 0 \le z \le \infty \qquad (44)$$

where

$$P(\alpha) = \frac{1}{q} \exp\left\{ -\beta\left[\int \pi(z)v_p(z;\alpha)\,dz + \varepsilon_{\text{int}}(\alpha) \right.\right.$$
$$\left.\left. + \int\int \chi(|z - z'|)\frac{v_p(z;\alpha)}{v_0}(1 - e^{-[\beta\pi(z')v_0]})\,dz\,dz' \right] \right\} \qquad (45)$$

Here, $v_p(z, \alpha)\,dz$ is the volume that the chain in conformation α occupies at layer z and q is the normalization constant. In principle, it is possible to solve this equation for $\pi(z)$ in terms of a given set of functions $v_p(z, \alpha)\,dz$ and $\varepsilon_{int}(\alpha)$, together with the mean-field interaction parameter $\chi(|z - z'|)$. However, this is impossible for practical applications.

To implement the theory, we discretize the space in layers of finite thickness δ. Then, the integrals over the variable z are converted into sums over the different finite layers. It is useful at this point to define a discrete variable $z_i = i\delta$, and the term "layer i" will refer to the volume between z_{i-1} and z_i. The number M of layers necessary should be sufficiently large to account properly for the more extended chain. In any case the election $M = ln/\delta$ warrants a proper counting.

Any chain model may be used to provide the inputs for Eq. (44). For example, we have applied the theory for continuous space chains generated with the rotational isomeric state (RIS) model, bead–spring chains, bead–rod chains, and chains on a variety of lattices. In the case of lattice chains, the discretization is already imposed by the lattice and there is no need to define the thickness δ, the volume v_0, and the segment length l.

To generate a complete set of single-chain conformations is a formidable task (and practically impossible) for chains with more than 20 segments with any chain model. However, we can get excellent results by taking a set of chains generated following a simple sampling technique. The size of this set depends on the length of the chain. However, in most of our application we have found that sets of 10^6–10^7 chain conformations provide excellent results. Note that the set of single-chain configurations used in the calculations is composed by self-avoiding chains that also considers the boundary conditions imposed by the surface. Namely, the intramolecular and surface repulsive interactions are taken into account through the set of single-chain conformations.

In the discrete space description, the volumes $v_p(z, \alpha)\,dz$ are integrated over each finite layer to give

$$\int_{z_{i-1}}^{z_i} v_p(z, \alpha)\,dz = n_\alpha(i)v_0 \tag{46}$$

where $n_\alpha(i)$ is the number of polymer segments that a chain in conformation α has in the layer i. In the same way, Eq. (4) is integrated over each layer, yielding

$$[\sigma l^2] \sum_{\{\alpha\}} P(\alpha)n_\alpha(i)\frac{v_0}{l^2\delta} + e^{-[\beta\pi(i)v_0]} = 1 \qquad 1 \le i \le M \tag{47}$$

where $\pi(i)$ is the value of the osmotic pressure in the layer i. Following the same discretization rules, the probability $P(\alpha)$ has the form

$$P(\alpha) = \frac{1}{q} \exp\left(-\sum_{i=1}^{M} [\beta\pi(i)v_0]n_\alpha(i) - \beta\varepsilon_{\text{int}}(\alpha) \right.$$

$$\left. - \sum_{i=1}^{M} \sum_{j=1}^{M} \beta\chi_{i,j} n_\alpha(i)(1 - e^{-[\beta\pi(j)v_0]}) \right) \tag{48}$$

where $\chi_{i,j}$ is the discretized interaction parameter between a segment in layer i and the mean field in layer j; see below.

Introducing Eq. (48) into Eq. (47) and rearranging using the definition of the normalization constant q, we get

$$\sum_{\{\alpha\}} \exp\left(-\sum_{j=1}^{M} [\beta\pi(j)v_0]n_\alpha(j) - \beta\varepsilon_{\text{int}}(\alpha) - \sum_{j=1}^{M} \sum_{k=1}^{M} \beta\chi_{j,k} n_\alpha(j)(1 - e^{-[\beta\pi(k)v_0]}) \right)$$

$$\times \left([\sigma l^2]n_\alpha(i)\frac{v_0}{l^2\delta} + e^{-[\beta\pi(i)v_0]} - 1 \right) = 0 \qquad i = 1, \ldots, M$$

$$\tag{49}$$

which constitutes a set of M nonlinear coupled equations for the discrete pressure $\pi(i)$. This set of equations is the equivalent in the discrete representation of the integral equation for the function $\pi(z)$. The inputs in Eq. (49) are the set of single-chain conformations through the numbers $n_\alpha(i)$ and the internal energy $\varepsilon_{\text{int}}(\alpha)$ (which includes segment–surface and the intramolecular attractions), the grafting density, and the temperature. Note that the grafting density is in units of l^2 and the osmotic pressure in units of βv_0.

To model the interactions in the system we have taken an effective potential between polymer segments. This potential $u(r)$ has a hard core (see Section II.B) and an attractive tail. The intermolecular hard-core repulsions are taken into account by the packing constraints. To calculate the values of the intermolecular attractive interaction parameters $\chi_{i,j}$, consider a monomer at the center of the layer i and integrate the potential over the volume of the layer j. Introducing cylindrical coordinates, it results in

$$\chi_{i,j} = 2\pi \int_{z_{(j-i)-1}}^{z_{j-1}} dz \int_0^\infty d\rho \, u[(\rho^2 + z^2)^{0.5}] \tag{50}$$

Note that this expression depends on the difference $i - j$. In the calculation of $\chi_{i,i}$, the integral is over the same layer the monomer is placed but excluding the volume of the interacting segment i:

$$\chi_{i,i} = 2\pi \int_{-\delta/2}^{\delta/2} dz \int_{(z^2+l^2)^{0.5}}^{\infty} d\rho\, u[(\rho^2 + z^2)^{0.5}] \qquad (51)$$

In our example we have chosen $\delta > 1$. If a more detailed study of the fine structure of the layer is desired, it can be done by using a much smaller value of δ. In this case, the volume of the segment contributes to more than a single layer and care should be taken in the calculation of the interaction parameters.

Let us now consider the case of a cylindrical or spherical interface and look at the differences with respect to the planar film. The volume-filling constraint should account for the change in the area per molecule as a function of the distance to the surface, $r - R$, R being the radius of curvature of the surface,

$$N_p \sum_{\{\alpha\}} P(\alpha)v_p(r, \alpha)\, dr + \exp[-\beta\pi(r)v_0]a_0 G(r)\, dr = a_0 G(r)\, dr \qquad r > R$$

$$(52)$$

The geometric factor $G(r)$ is equal to r/R (r^2/R^2) for a cylindrical (spherical) interface. The discretized version of Eq. (52), after rearrangement, reads

$$[\sigma l^2]\langle n(i)\rangle[\Delta G(i)]^{-1}\frac{v_0}{\delta l^2} + \exp[-\beta\pi(i)v_0] = 1 \qquad i = 1, \ldots, M \quad (53)$$

where

$$\Delta G(i) = \frac{1}{\delta}\int_{r_{i-1}}^{r_i} G(r)\, dr \qquad (54)$$

For the athermal case, the discrete $P(\alpha)$ is the same as in the planar film, with only the first term in the exponential. Then, the set of nonlinear coupled equations to be solved is given as

$$\sum_{\{\alpha\}} \exp\left(-\sum_{j=1}^{M} [\beta\pi(j)v_0]n_\alpha(j)\right)$$
$$\times \left[[\sigma l^2]n_\alpha(i)[\Delta G(i)]^{-1}\frac{v_0}{l^2\delta} + e^{-[\beta\pi(i)v_0]} - 1 \right] = 0 \qquad i = 1, \ldots, M$$

$$(55)$$

This set of equations reduces to Eq. (49) in the case of a planar system $[G(i) = 1]$ under athermal conditions $[\varepsilon_{int}(\alpha) = 0, \chi_{i,j} = 0]$.

ACKNOWLEDGMENTS

We thank our collaborator Hai Tang, with whom part of the work described here was done. We thank G. Grest, S. Shaffer, A. Chakrabarti, and S. Kumar for providing us with the simulations results, Zheng-Gang Wang for the calculations of the pressure and for many enlightening discussions, and M. Kent for providing us with raw experimental data. We also thank C. Marques for sending us his derivation on the branched chains prior to publication. This work was partially supported by a grant from Shell Research B.V. (Amsterdam).

REFERENCES

1. M. van der Waarden, *J. Colloid Sci.* **5**, 317 (1950).
2. M. van der Waarden, *J. Colloid Sci.* **6**, 443 (1951).
3. E. L. Mackor, *J. Colloid Sci.* **6**, 492 (1951).
4. E. L. Mackor and J. H. van der Waals, *J. Colloid Sci.* **7**, 535 (1952).
5. E. J. Clayfield and E. C. Lumb, *J. Colloid Interface Sci.* **22**, 269 (1966).
6. E. J. Clayfield and E. C. Lumb, *J. Colloid Interface Sci.* **22**, 285 (1966).
7. D. J. Meier, *J. Phys. Chem.* **71**, 1861 (1967).
8. R. H. Ottewill in *Nonionic Surfactants*, Vol. 1, *Surfactant Science Series*, M. J. Schick and F. M. Foekes, eds., Marcel Dekker, New York, 1967, pp. 627–682.
9. J. Lyklema, *Adv. Colloid Interface Sci.* **2**, 65 (1968).
10. F. T. Hesselink, *J. Phys. Chem.* **73**, 3488 (1969).
11. F. T. Hesselink, *J. Phys. Chem.* **75**, 65 (1971).
12. D. H. Napper, *Polymeric Stabilization of Colloid Dispersions*, Academic Press, New York, 1983.
13. R. S. Ward, *IEEE Eng. Med. Biol. Mag.* **6**, 22 (1989).
14. E. Raphaël and P. G. de Gennes, *J. Phys. Chem.* **96**, 4002 (1992).
15. H. Ji and P. G. de Gennes, *Macromolecules* **26**, 520 (1993).
16. J. H. van Zanten, *Macromolecules* **27**, 6797 (1994).
17. J.-F. Joanny, *Langmuir* **8**, 989 (1992).
18. M. Amiji and K. Park, *J. Biomater. Sci. Polymer Ed.* **4**, 217 (1993).
19. R. J. Lee and P. S. Low, *J. Biol. Chem.* **269**, 3198 (1994).
20. Z. Tuzar and P. Kratochvíl, *Surf. Coll. Sci.* **15**, 1 (1993).
21. J. Klein, *J. Chem. Soc., Faraday Trans. 1* **79**, 99 (1983).
22. H. J. Taunton, C. Toprakcioglu, L. J. Fetters, and J. Klein, *Nature* **332**, 712 (1988).
23. H. J. Taunton, C. Toprakcioglu, and J. Klein, *Macromolecules* **21**, 3333 (1988).
24. H. J. Taunton, C. Toprakcioglu, L. J. Fetters, and J. Klein, *Macromolecules* **23**, 571 (1990).

25. J. Klein, D. Perahia, and S. Walburg, *Nature* **352**, 143 (1991).

26. J. Klein, Y. Kamiyama, H. Yoshizawa, J. N. Israelachvili, L. Fetters, and P. Pincus, *Macromolecules* **25**, 2062 (1992).

27. G. Hadziioannou, S. Patel, S. Granick, and M. Tirrell, *J. Am. Chem. Soc.* **108**, 2869 (1986).

28. D. Perahia, D. G. Wiesler, S. K. Satija, L. J. Fetters, S. K. Sinha, and S. T. Milner, *Phys. Rev. Lett.* **72**, 100 (1994).

29. J. Field, C. Toprakcioglu, R. Ball, H. Stanley, L. Dai, W. Barford, J. Penfold, G. Smith, and W. Hamilton, *Macromolecules* **25**, 434 (1992).

30. T. Cosgrave, T. G. Heath, J. S. Phipps, and R. M. Richardson, *Macromolecules* **24**, 94 (1991).

31. T. Cosgrove, T. Heath, K. Ryan, and B. van Lent, *Polym. Commun.* **28**, 64 (1987).

32. T. Cosgrove and K. Ryan, *Langmuir* **6**, 136 (1990).

33. T. Cosgrove, *Mac. Reports* **A29**, 125 (1992).

34. T. Cosgrove, T. G. Heath, K. Ruan, and T. L. Crowley, *Macromolecules* **20**, 2879 (1987).

35. P. Auroy, L. Auvray, and L. Leger, *Phys. Rev. Lett.* **66**, 719 (1991).

36. L. Auvray and J. Cotton, *Macromolecules* **20**, 202 (1987).

37. P. Auroy, L. Auvray, and L. Leger, *Macromolecules* **24**, 2523 (1991).

38. P. Auroy, L. Auvray, and L. Leger, *Macromolecules* **24**, 5158 (1991).

39. P. Auroy and L. Auvray, *Molecules* **25**, 4134 (1992).

40. P. Auroy, Y. Mir, and L. Auvray, *Phys. Rev. Lett.* **69**, 93 (1992).

41. P. Auroy and L. Auvray, *J. Phys. II France* **3**, 227 (1993).

42. L. Auvray and P. G. de Gennes, *Europhys. Lett.* **2**, 647 (1986).

43. L. Auvray and P. Auroy, Neutron, *X-Ray and Light Scattering*, P. Linder and Th. Zemb, Eds., Elsevier Science Publishers, 1991, p. 199.

44. S. Granick and J. Herz, *Macromolecules* **18**, 460 (1985).

45. M. S. Kent, L. T. Lee, B. J. Factor, F. Rondelez, and G. S. Smith, *J. Chem. Phys.* **103**, 2320 (1995).

46. M. S. Kent, L. T. Lee, B. J. Factor, F. Rondelez, and G. Smith, *J. Phys. IV, Colloque C8* **3**, 49 (1993).

47. M. S. Kent, L.-T. Lee, B. Farnoux, and F. Rondelez, *Macromolecules* **25**, 6240 (1992).

48. B. J. Factor, L.-T. Lee, M. S. Kent, and F. Rondelez, *Phys. Rev. E* **48**, 2354 (1993).

49. S. Alexander, *J. Phys. (Paris)* **38**, 983 (1977).

50. S. Alexander, *J. Phys. (Paris)* **38**, 977 (1977).

51. P. G. de Gennes, *J. Phys. (Paris)* **37**, 1445 (1976).

52. P. G. de Gennes, *Macromolecules* **13**, 1069 (1980).

53. E. Raphaël and P.-G. de Gennes, *C. R. Acad. Sci. Paris* **315**, 937 (1992).

54. M. Auboy and E. Raphaël, *J. Phys. II France* **3**, 443 (1993).

55. E. Raphaël, P. Pincus, and G. H. Fredrickson, *Macromolecules* **26**, 1996 (1993).

56. E. Raphaël and P. G. de Gennes, *Makromol. Chem. Macromol. Symp.* **62**, 1 (1992).

57. E. Raphaël and P. G. de Gennes, *Physica A* **177**, 294 (1991).

58. M. Auboy, J.-M. di Meglio, and E. Raphaël, *Europhys. Lett.* **24**, 87 (1993).

59. F. Brochard-Wyart, P.-G. de Gennes, and P. Pincus, *C. R. Acad. Sci. Paris, Serie II* **134**, 873 (1992).

60. F. Brochard-Wyart and P.-G. de Gennes, *C. R. Acad. Sci. Paris, Serie II* **137**, 13 (1993).

61. F. Brochard and P. G. de Gennes, *Langmuir* **8**, 3033 (1992).

62. M. Auboy, F. Brochard-Wyart, and E. Raphaël, *Macromolecules* **26**, 5885 (1993).

63. D. R. M. Williams, *J. Phys. II France* **3**, 1313 (1993).

64. D. R. M. Williams and P. A. Pincus, *Europhys. Lett.* **24**, 29 (1993).

65. E. M. Sevick and D. R. M. Williams, *Macromolecules* **27**, 5285 (1994).

66. D. R. M. Williams, *Macromolecules* **26**, 5096 (1993).

67. D. R. M. Williams, *Langmuir* **9**, 2215 (1993).

68. D. R. M. Williams, *Macromolecules* **26**, (1993).

69. D. R. M. Williams, *Macromolecules* **26**, 6667 (1993).

70. A. Halperin and D. R. M. Williams, *Europhys. Lett.* **20**, 601 (1992).

71. P.-Y. Lai and A. Halperin, *Macromolecules* **25**, 6693 (1992).

72. P.-Y. Lai and A. Halperin, *Macromolecules* **25**, 4981 (1991).

73. A. Halperin, *Macromolecules* **20**, 2943 (1987).

74. A. Halperin, *Macromolecules* **22**, 3806(1989).

75. A. Halperin, *J. Phys. France* **49**, 131 (1988).

76. T. P. Lodge and G. H. Fredrickson, *Macromolecules* **25**, 5643 (1992).

77. S. S. Patel, unpublished preprint.

78. K. R. Shull, *J. Chem. Phys.* **94**, 5723 (1991).

79. C. M. Wijmans, J. M. H. M. Scheutjens, and E. B. Zhulina, *Macrmolecules* **25**, 2657 (1992).

80. C. M. Wijmans, E. B. Zhulina, and G. J. Fleer, *Macromolecules* **27**, 3238 (1994).

81. R. Israëls, F. A. M. Leermakers, G. J. Fleer, and E. B. Zhulina, *Macromolecules* **27**, 3249 (1994).

82. R. Israëls, F. A. M. Leermakers, and G. J. Fleer, *Macromolecules* **27**, 3087 (1994).

83. E. B. Zhulina, R. Israels, and G. J. Fleer, *Colloids Surface A* **86**, 11 (1994).

84. P.-Y. Lai and E. B. Zhulina, *Macromolecules* **25**, 5201 (1992).

85. J. M. H. M. Scheutjens and G. J. Fleer, *J. Phys. Chem.* **83**, 619 (1979).

86. J. M. H. M. Scheutjens and G. J. Fleer, *J. Phys. Chem.* **84**, 178 (1980).

87. G. J. Fleer and J. M. H. M. Scheutjens, *Croatica Chem. Acta* **60**, 477 (1987).

88. J. M. H. M. Scheutjens and G. J. Fleer, *Macromolecules* **18**, 1882 (1985).

89. T. Cosgrove, T. Heath, B. van Lent, F. Leermakers, and J. Scheutjens, *Macromolecules* **20**, 1692 (1987).

90. B. van Lent, R. Israels, J. M. H. M. Scheutjens, and G. J. Fleer, *J. Colloid Interface Sci.* **137**, 380 (1990).

91. M. Muthukumar and J.-S. Ho, *Macromolecules* **22**, 965 (1989).

92. A. K. Dolan and S. F. Edwards, *Proc. R. Soc. London, Ser. A* **337**, 509 (1974).

93. A. K. Dolan and S. F. Edwards, *Proc. R. Soc. London, Ser. A* **343**, 427 (1975).

94. S. T. Milner, T. A. Witten, and M. E. Cates, *Europhys. Lett.* **5**, 413 (1988).

95. S. T. Milner, T. A. Witten, and M. E. Cates, *Macromolecules* **21**, 2610 (1988).

96. S. T. Milner, *Europhys. Lett.* **7**, 695 (1988).

97. S. T. Milner, Z.-G. Wang, and T. A. Witten, *Macromolecules* **22**, 489 (1989).

98. S. T. Milner, T. A. Witten, and M. E. Cates, *Macomolecules* **22**, 853 (1989).

99. S. T. Milner and T. A. Witten, *Macromolecules* **25**, 5495 (1992).

100. E. B. Zhulina, O. V. Borisov, and V. A. Priamitsyn, *J. Colloid Interface Sci.* **137**, 495 (1990).

101. E. B. Zhulina, O. V. Borisov, V. A. Pryamitsyn, and T. Birshtein, *Macromolecules* **24**, 140 (1991).

102. Y. V. Lyatskaya, F. A. M. Leermakers, G. J. Fleer, E. B. Zhulina, and T. M. Birshtein, *Macromolecules*, **28**, 3562 (1995).

103. M. A. Carignano and I. Szleifer, *J. Chem. Phys.* **98**, 5006 (1993).

104. M. Murat and G. S. Grest, *Phys. Rev. Lett.* **63**, 1074 (1989).

105. M. Murat and G. S. Grest, *Macromolecules* **22**, 4054 (1989).

106. G. S. Grest and M. Murat, *Macromolecules* **26**, 3108 (1993).

107. G. S. Grest, *Macromolecules* **27**, 418 (1994).

108. J. Harris and S. A. Rice, *J. Chem. Phys.* **88**, 1298 (1988).

109. A. Chakrabarti and R. Toral, *Macromolecules* **23**, 2016 (1990).

110. A. Chakrabarti, P. Nelson, and R. Toral, *Phys. Rev. A* **46**, 4930 (1992).

111. R. Toral and A. Chakrabarti, *Phys. Rev. E* **47**, 4240 (1993).

112. A. Chakrabarti, P. Nelson, and R. Toral, *J. Chem. Phys.* **100**, 748 (1994).

113. R. Toral, A. Chakrabarti, and R. Dickman, *Phys. Rev. E* **50**, 343 (1994).

114. R. Dickman and D. C. Hong, *J. Chem. Phys.* **95**, 4650 (1991).

115. R. Dickman and P. E. Anderson, *J. Chem. Phys.* **99**, 3112 (1993).

116. P.-Y. Lai and K. Binder, *J. Chem. Phys.* **95**, 9288 (1991).

117. P.-Y. Lai and K. Binder, *J. Chem. Phys.* **97**, 586 (1992).

118. P.-Y. Lai, *Computational Polym. Sci.* **2**, 157 (1992).

119. P.-Y. Lai, *J. Chem. Phys.* **100**, 3351 (1994).

120. M. Laradji, H. Guo, and M. J. Zuckermann, *Phys. Rev. E* **49**, 3199 (1994).

121. J. S. Shaffer, *Phys. Rev. E* **50**, 683 (1994).

122. J. D. Weinhold and S. Kumar, *J. Chem. Phys.* **101**, 4312 (1994).

123. S. S. Patel and M. Tirrell, *Annu. Rev. Phys. Chem.* **40**, 597 (1989).

124. A, Halperin, M. Tirrell, and T. P. Lodge, *Adv. Pol. Sci.* **100**, 31 (1991).

125. S. Milner, *Science* **251**, 905 (1991).

126. A. Halperin in *Soft Order in Physical Systems*, Vol. 323 of *Series B: Physics*, Y. Rabin and R. Bruisma, eds., Plenum Press, NATO ASI Series, New York and London, 1994, pp. 33–56.

127. G. S. Grest and M. Murat, *Monte Carlo and Molecular Dynamics Simulations in Polymer Science*, K. Binder, ed., Clarendon Press, Oxford, 1994.

128. P. N. Hurter and T. A. Hatton, *Langmuir* **8**, 1291 (1992).

129. S. K. Kumar, I. Szleifer, and A. Z. Panagiotopoulos, *Phys. Rev. Lett.* **66**, 2935 (1991).

130. M. A. Carignano and I. Szleifer, *J. Chem. Phys.* **100**, 3210 (1994).

131. M. A. Carignano and I. Szleifer, *Macromolecules* **27**, 702 (1994).

132. M. A. Carignano and I. Szleifer, *J. Chem. Phys.*, **102**, 8662 (1995).

133. A. Ben-Shaul, I. Szleifer, and W. M. Gelbart, *Proc. Natl. Acad. Sci., U.S.A.* **81**, 4601 (1984).

134. A. Ben-Shaul, I. Szleifer, and W. M. Gelbart, *J. Chem. Phys.* **83**, 3597 (1985).

135. I. Szleifer, A. Ben-Shaul, and W. M. Gelbart, *J. Chem. Phys.* **83**, 3612 (1985).

136. I. Szleifer, A. Ben-Shaul, and W. M. Gelbart, *J. Phys. Chem.* **94**, 5081 (1990).

137. M. A. Carignano and I. Szleifer, *Europhys. Lett.*, **30**, 525 (1995).

138. S. F. Edwards, *Proc. Phys. Soc. (London)* **85**, 613 (1965).

139. J. I. Martin and Z.-G. Wang, *J. Phys. Chem.* **99**, 2833 (1995).

140. A. N. Semenov, *Sov. Phys. JETP* **61**, 733 (1985).

141. D. F. K. Shim and M. E. Cates, *J. Phys. France* **50**, 3535 (1980).

142. R. C. Ball, J. F. Marko, S. T. Milner, and T. A. Witten, *Macromolecules* **24**, 693 (1991).

143. E. Zhulina, O. Borisov, and L. Brombacher, *Macromolecules* **24**, 4679 (1991).

144. P. Flory, *Statistical Mechanics of Chain Molecules*, Oxford University Press, New York, 1988.

145. E. Eisenriegler, K. Kremer, and K. Binder, *J. Chem. Phys.* **77**, 6296 (1982).

146. P. de Gennes, *Scaling Concepts in Polymer Physics*, Cornell University Press, Ithaca, NY, 1991.

147. M. A. Carignano and I. Szleifer, *Macromolecules* **28**, (1995).

148. R. Baranowski and M. Whitmore, *J. Chem. Phys.* **103**, 2343 (1995).

149. S. Milner, *J. Chem. Soc. Faraday Trans.* **86**, 1349 (1990).

150. E. DiMarzio and R. Rubin, *J. Chem. Phys.* **55**, 4318 (1971).

151. P. Flory, *Principles of Polymer Chemistry*, Cornell University Press, Ithaca, NY, 1953.

152. I. Carmesin and K. Kremer, *Macromolecules* **21**, 2819 (1988).

153. C. Ligoure and L. Leibler, *J. Phys. France* **51**, 1313 (1990).

154. D. R. M. Williams and A. Halperin, *Macromolecules* **26**, 4209 (1993).

155. D. R. M. Williams and A. Halperin, *Phys. Rev. Lett.* **71**, 1557 (1993).

156. D. R. M. Williams and A. Halperin, *Macromolecules* **26**, 2025 (1993).

157. A. Halperin and D. R. M. Williams, *Europhys. Lett.* **21**, 575 (1993).

158. A. Halperin and D. R. M. Williams, *Phys. Rev. E* **49**, 986 (1994).

159. D. R. M. Williams and A. Halperin, *Europhys. Lett.* **19**, 693 (1992).

160. D. R. M. Williams, *Phys. Rev. E* **49**, 1811 (1994).

161. D. R. M. Williams and G. H. Fredrickson, *Macromolecules* **25**, 3561 (1992).

162. D. Gersappe, M. Fasolka, A. C. Balazs, and S. H. Jacobson, *J. Chem. Phys.* **100**, 9170 (1994).

163. E. B. Zhulina and T. A. Vilgis, *Macromolecules* **28**, 1009 (1995).

164. C. Marques, private communication.

165. H. Iatrou and N. Hadjichristidis, *Macromolecules* **25**, 4649 (1992).

166. L. Onsager, *Ann. N.Y. Acad. Sci.* **51**, 627 (1949).

167. Z.-Y. Chen, J. Talbot, W. M. Gelbart, and A. Ben-Shaul, *Phys. Rev. Lett.* **61**, 1376 (1988).

168. D. Frenkel in *Liquids, Freezing and Galss Transition*, Vol. 2, J. P. Hansen, D. Levesque, and J. Zinn-Justin, eds., North-Holland, Amsterdam, 1991.

169. A. Halperin, *J. Phys. France* **49**, 547 (1988).

170. R. S. Ross and P. Pincus, *Europhys. Lett.* **19**, 79 (1992).

171. C. Yeung, A. C. Balazs, and D. Jasnow, *Macromolecules* **26**, 1914 (1993).

172. H. Tang and I. Szleifer, *Europhys. Lett.* **28**, 19 (1994).

173. K. G. Soga, H. Guo, and M. J. Zuckermann, *Europhys. Lett.* **28**, 531 (1995).

174. W. Zhao, G. Krausch, M. H. Rafailovich, and J. Sokolov, *Macromolecules* **27**, 2933 (1994).

175. H. Tang, M. A. Carignano, and I. Szleifer, *J. Chem. Phys.* **102**, 3404 (1995).

176. A. Karim, S. K. Satija, J. F. Douglas, J. F. Ankner, and L. J. Fetters, *Phys. Rev. Lett.* **73**, 3407 (1995).

177. Z.-G. Wang and S. A. Rice, *J. Chem. Phys.* **88**, 1290 (1988).

178. S. Shin, Z.-G. Wang, and S. A. Rice, *J. Chem. Phys.* **92**, 1427 (1990).

179. R. S. Cantor and P. M. McIlroy, *J. Chem. Phys.* **90**, 4423 (1989).

180. T. A. Witten and P. A. Pincus, *Macromolecules* **19**, 2509 (1986).

181. M. Daoud and J. P. Cotton, *J. Phys. (France)* **43**, 531 (1982).

182. G. S. Grest, K. Kremer, and T. A. Witten, *Macromolecules* **20**, 1376 (1987).

183. D. Richter, B. Stün, B. Ewen, and D. Nerger, *Phys. Rev. Lett.* **58**, 2462 (1987).

184. G. Wu and B. Chu, *Macromolecules* **27**, 1766 (1994).

185. M. Murat and G. S. Grest, *Macromolecules* **24**, 704 (1991).

186. N. Dan and M. Tirrell, *Macromolecules* **25**, 2890 (1992).

187. C. M. Wijmans and E. B. Zhulina, *Macromolecules* **26**, 7214 (1993).

188. M. R. Munch and A. P. Gast, *Macromolecules* **21**, 1360 (1988).

189. M. R. Munch and A. P. Gast, *Molecules* **21**, 1366 (1988).

190. J. Israelachvili, *Intermolecular and Surface Forces*, Academic Press, London, 1991.

191. A. I. Rusanov, *Adv. Coll. Interface Sci.* **45**, 1 (1993).

192. P. Linse and M. Malmsten, *Macromolecules* **25**, 5434 (1992).

193. P. Linse, *Macromolecules* **26**, 4437 (1993).

194. P. Linse, *Macromolecules* **27**, 2685 (1994).

195. P. Linse, *Macromolecules* **27**, 6404 (1994).

196. P. N. Hurter, J. M. H. M. Scheutjens, and T. A. Hatton, *Macromolecules* **26**, 5592 (1993).

197. W. Helfrich, *Z. Naturforsch. Teil C* **28**, 693 (1973).

198. Z.-G. Wang and S. A. Safran, *J. Chem. Phys.* **94**, 679 (1991).

199. I. Szleifer, D. Kramer, A. Ben-Shaul, W. M. Gelbart, and S. A. Safran, *J. Chem. Phys.* **92**, 6800 (1990).

200. M. A. Carignano and I. Szleifer, to be published.

201. R. S. Cantor, *Macromolecules* **14**, 1186 (1981).

202. *Physics of Amphiphiles: Micelles, Vesicles and Microemulsions*, V. Degiorgio and M. Corti, eds., North-Holland, Amsterdam, 1985.

203. H. Kellay, B. P. Binks, Y. Hendrikx, L. T. Lee, and J. Meunier, *Adv. Colloid Interface Sci.* **49**, 85 (1994).

204. E. W. Kaler, A. K. Murthy, B. E. Rodriguez, and J. A. N. Zasadzinski, *Sci* **245**, 1371 (1989).

205. S. A. Safran, P. Pincus, and D. Andelman, *Science* **248**, 354 (1990). G. Porte and C. Ligoure, *J. Chem. Phys.* **102**, 4290 (1995).

206. Z.-G. Wang, *Macromolecules* **25**, 3702 (1992).

207. Z.-G. Wang and S. A. Safran, *Europhys. Lett.* **11**, 425 (1990).

208. K. S. Schweizer, *Adv. Chem. Phys*, this volume.

209. E. B. Zhulina, O. V. Borisov, and T. M. Birshtein, *J. Phys. II* **2**, 63 (1992).

210. O. V. Borisov, T. M. Birshtein, and E. B. Zhulina, *J. Phys. II* **1**, 521 (1991).

211. R. Ross and P. Pincus, *Macromolecules* **25**, 2177 (1992).

212. J. Marko and T. Witten, *Phys. Rev. Lett.* **66**, 1541 (1991).

213. S. Dhoot, H. Watanabe, and M. Tirrell, *Colloids Surfaces A* **86**, 47 (1994).

214. N. Dan and M. Tirrell, *Macromolecules* **26**, 6467 (1993).

215. Poly (ethylene glycol) Chemistry: Biotechnical and Biomedical Applications, J. Milton Harris, eds., Plenum Press, New York, 1992.

LIVING POLYMERS

SANDRA C. GREER

Department of Chemical Engineering, University of Maryland at College Park, College Park, Maryland

CONTENTS

Advances in Chemical Physics, *Volume XCIV*, Edited by I. Prigogine and Stuart A. Rice.
ISBN 0-471-14324-3 © 1996 John Wiley & Sons, Inc.

I. INTRODUCTION

A. General Reaction Mechanism

There are three main steps in addition/chain polymerization reactions. First, there is the *initiation* step: the activation of a monomer molecule, M, by an initiator molecule, I, to form an active species, M*:

$$M + I \rightleftarrows M^* \tag{1}$$

Then there is the *propagation* of the active species into a polymer by the addition of monomer:

$$M^* + M \rightleftarrows M_2^* \tag{2}$$

or, in general,

$$M_n^* + M \rightleftarrows M_{n+1}^* \tag{3}$$

where the subscript n indicates the number of monomers in the active polymer. We note that the propagation process is, in principle, reversible; given time, the active polymers can depolymerize as well as polymerize. The final step in a polymerization is the *termination* step, in which an active polymer molecule becomes deactivated by some process, usually reacting with some other species, T, and producing some other product, T′:

$$M_{n+1}^* + T \rightleftarrows M_{n+1} + T' \tag{4}$$

after which the "dead" polymer no longer polymerizes or depolymerizes.

We focus here on those cases in which the termination step is avoided, and thus the activated polymers remain "alive" and continue to polymerize and depolymerize; the term *living polymer* is applied to such cases [1]. We are particularly interested in the development of full thermodynamic equilibrium such that the active polymer is in dynamic equilibrium with its monomer; the term *equilibrium polymerization* is applied in this case.

A crucial feature of the systems we consider here is that the propagation of activated monomer into polymer will only occur below a temperature particular to each monomer, the *ceiling temperature* [2]. The thermodynamic requirement for propagation to proceed is for the Gibbs free energy of propagation to be negative: When the Gibbs free energy is positive, there is no propagation; when the Gibbs free energy passes through zero and becomes negative, propagation commences [3]. Such a transition will occur only if the enthalpy of propagation ΔH_p and the entropy of propagation ΔS_p have the same sign. If ΔH_p is positive and ΔS_p is negative, then propagation never occurs. If ΔH_p is negative and ΔS_p is positive, then propagation always occurs. If both ΔH_p and ΔS_p are positive, then propagation occurs only at temperatures above a "floor temperature"; the polymerization of liquid sulfur is an example of this case, and there has been recent theoretical [4–12] and experimental [13–15] work on the polymerization transition in sulfur.

We consider here the cases for which both ΔH_p and ΔS_p are negative and thus for which propagation occurs only at temperatures below a ceiling temperature T_p. A great many monomers fall into this category. Some monomers and their ceiling temperatures in the pure state are shown in Table I [where quantities preceded by the approximate symbol were calculated from the relation $T_p^0 = \Delta H_p^0 / \Delta S_p^0$; see Eq. (5) below].

While the initiation step does not determine T_p, this step does allow for interesting variety and complexity in the particularities of the reaction mechanism. The initiation can be (1) by formation of a cation as the active polymer (*cationic polymerization*) or (2) by formation of an anion as the active polymer (*anionic polymerization*). Some initiators form active polymers with only one active site, whereas others form more than one active site [3, 22].

Let us now consider the *microscopic* nature of a solution of living polymers in a state of equilibrium polymerization. Above T_p, the solution consists of activated monomeric species in a solution of nonactivated monomer molecules and solvent molecules. As the temperature is lowered to below T_p, each activated monomer starts to react with other monomer molecules to form activated polymer molecules. Initially at each temperature, we expect the monomers to add to all active sites equally and thus to form a nearly monodisperse solution of living

TABLE I
Thermodynamic Parameters of Some Common Monomers

Monomer	T_p^0 (K)	ΔH_p^0 (kJ mol^{-1})	ΔS_p^0 (J mol^{-1} K)
Styrene	~710	-74.9 ± 0.3[a]	-104 ± 1[a]
α-Methylstyrene	321 ± 10[b]	-35 ± 1[c]	-110 ± 1[b]
			-105 ± 1[d]
Methyl methacrylate	~482	-56.1 ± 0.1[e]	-116 ± 1[e]
Tetrahydrofuran	358 ± 2[f]	-19.2 ± 0.1[f]	~ -53.6[f]

[a] From ref. 16.
[b] From ref. 17.
[c] From ref. 18.
[d] From ref. 19.
[e] From ref. 20.
[f] From ref. 21.

polymers. However, if time is allowed for depolymerization to set in and for equilibrium to be reached, we expect a broad distribution of the molecular weights of the living polymer molecules [23]. The living polymer molecules may not all have the same tacticity [24]. Depending on the choice of initiator, there can be anionic or cationic ions as counterions to the living polymeric ions. These counterions can be totally free ions, they can be separated from the polymeric ion by one solvent molecule, or they can be tightly attached to the polymeric ion to form an ion pair [3]. The ion pairs can form intermolecular and intramolecular Coulombic associations in nonpolar solvents [3]. If the temperature is lowered further, the average molecular weight of the living polymer and the fraction of the initial monomer that has been converted to polymer will both increase. If the temperature is then increased, the average molecular weight of the living polymer and the fraction of monomer that has been converted to polymer will both decrease.

Unlike micelles [25] and some self-assembling biological macromolecules [26], living polymers are held together by covalent carbon-to-carbon bonds. Therefore, they do not break in the middle of the polymer chain: They polymerize and depolymerize only from the active sites. Neither do two living polymer molecules combine together to make a larger living polymer molecule; indeed, the presence of active sites with like charges may cause them to repel one another. For the same reason, a living polymer molecule will not react intramolecularly to form a ring polymer. On the other hand, the presence of the charged active sites and their counterions makes possible intramolecular and intermolecular Coulombic associations in solvents of low dielectric constant [3]. These features are different for micelles and also for the radical polymerization of liquid sulfur [27], where, indeed, there is a dynamic equilibrium with the monomer species but for which fragmentation and combination are possible.

B. History

Szwarc [3] and Morawetz [28] have reviewed the early history of living polymerization. Ziegler and Baehr in 1928 [29] and Abkin and Medvedev [30] in 1936 established experimentally the existence of active, living polymer ends in the polymerization of butadiene, initiated by sodium. In 1940, Flory [31] predicted the molecular weight distribution to be a narrow, Poisson distribution in the case of living polymerization without depolymerization. In 1948, Dainton and Ivin [2] recognized the thermodynamic significance of the ceiling temperature, and Beaman [32] realized the possibility of anionic polymerization. In 1956, Szwarc [1] introduced the term "living polymer" and with collaborators [33] began an extensive set of investigations culminating in Szwarc's 1968 book [3], which is to this day an encyclopedic guide to the subject. This book has been augmented by a recent monograph by Szwarc and Van Beylen [22].

The techniques of cationic and anionic living polymerization have been very important in the synthesis of novel polymer molecules [34]. The presence of active sites on the living polymer permits the manipulation of the structure of the growing polymer by the addition of a different monomer or monomers to make copolymers and by the termination of the active sites in such a way as to create useful functional groups on the monomer.

The history of the theoretical modeling of living polymers will be reviewed as each experimental aspect is discussed.

C. Issues

Bronowski [35], in a discussion of the nature of scientific truth, finds a wonderful metaphor in the experiences of climbers in the Hamalayan mountains. Mountain climbers hire native guides who live in the villages around a given mountain. The natives from a given village know only the face of the mountain that they see from their village. They have never seen the other faces of the mountain and cannot construct a model of the mountain as a whole. A model of the whole mountain requires a knowledge of all its faces and of how the faces fit together. Bronowski wrote, "we know a thing only by mapping and joining our experiences of its aspects." Organic chemists, polymer chemists, and polymer engineers have all studied living polymers, and each group has much to contribute about its own "face of the mountain." We see different faces, and we have different tools to illuminate different facets of that face.

Synthetic chemists have been interested in terminating a living polymerization at its early, *nonequilibrium* stage in order to achieve a narrow distribution of molecular weights. In physical chemistry and chemical physics, we are interested in the nature of the living polymer system *at equilibrium*, because only then can we apply statistical mechanical models. We are also interested in the *approach to equilibrium*: The chemical kinetics and the transport properties that determine how equilibrium is attained. In addition to macroscopic thermodynamic and transport properties, we are interested in the microscopic structure of the living polymer solutions: the distribution of the molecular weights of the polymer molecules, the conformations of the polymer chains, the entanglements of chains, and the diffusion of monomer and polymer molecules. We need to put all this information together for a full understanding of the nature of living polymer solutions. Our contributions to the map of the mountain will then help synthetic chemists and engineers do a better job in scaling their parts of the mountain.

II. REACTION PARAMETERS

A. Thermodynamic Parameters

The important thermodynamic parameters for living polymerization are the enthalpy and entropy of initiation [Eq. (1)] and the entropy and enthalpy of propagation [Eqs. (2) and (3)]. For the purposes of this discussion, we will take the initiation to be complete and irreversible, and thus the initiation parameters will not be considered. The propagation parameters are, however, important to our discussion. Table I shows these parameters for several monomers. We note that we take as the standard state the pure liquid monomer at one atmosphere, converting to liquid polymer. The quantities are given per mole of monomer converted to polymer, regardless of the molecular weight of the polymer. Then in the case of the pure monomer,

$$T_p^0 = \frac{\Delta H_p^0}{\Delta S_p^0} \tag{5}$$

When a solvent is added to the monomer, T_p decreases as the concentration of monomer decreases. Here, T_p as a function of the weight fraction (ϕ_m^0) or mole fraction (x_m^0) of *initial* monomer is the "polymerization line" for a given monomer, for which Dainton and Ivin [2] derived an equation. The equilibrium constant for polymerization, K_p, in which a polymer molecule of k monomers combines with another monomer molecule to make a polymer molecule of $k + 1$ monomers, is $a_{k+1}/a_k a_m$, where "a" is the activity of each species. If $a_k \approx a_{k+1}$, then $K_p \approx 1/a_m$. If the solution is ideal and the standard state is the pure monomer, then $K_p \approx x_m^{-1} \approx \exp[(\Delta S_p^0/R) - (\Delta H_p^0/RT)]$, where x_m is the mole fraction of monomer at equilibrium. Then $T = \Delta H_p^0/(\Delta S_p^0 + R \ln x_m)$ for any T. At T_p, $x_m = x_m^0$, so the polymerization line is

$$T_p = \frac{\Delta H_p^0}{\Delta S_p^0 + R \ln x_m^0} . \tag{6}$$

In fact, if the solution is ideal, x_m depends only on T, so a measure of x_m at some T_i is a measure of x_m^0 in the case $T_p = T_i$. Note that Eq. (6) assumes that solution is ideal and that $a_k \approx a_{k+1}$.

It is invariably assumed, in the above equations, in tabulations of ΔH_p^0

and ΔS_p^0, and in using these parameters in various models of living polymerization, that there is no dependence of thermodynamic quantities on molecular weight. This assumption may be especially suspect near T_p, where the polymer molecular weight is small and the addition of each monomer may have a significant effect. Roberts and Jessup [18] measured the heat of combustion of poly(α-methylstyrene) at various molecular weights, and from those measurements calculated ΔH_p^0 as a function of molecular weight; their values are plotted in Fig. 1. Note that these values are for *terminated* polymer; there may be a still stronger effect in *living* polymer because of the presence of ionic sites, where Coulombic interactions enter. Figure 1 shows that ΔH_p^0 becomes less negative as the molecular weight increases. In this discussion, while noting the probable dependence of thermodynamic quantities on molecular weight, we will still use the "asymptotic" values obtained at relatively high molecular weights.

The assumption of Eq. (6) of an *ideal* solution means that the polymerization line is not expected to depend on the particular solvent used. There is only one published polymerization line that includes

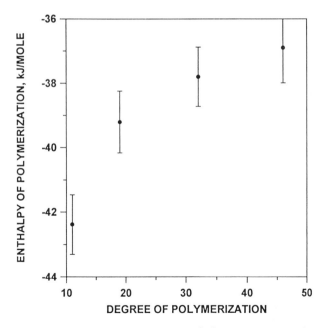

Figure 1. Measurements by Roberts and Jessup [18] of the enthalpy of propagation, ΔH_p° (at 298 K), of poly(α-methylstyrene) as a function of degree of polymerization (number of monomers per polymer). Error bars indicate one standard deviation.

various solvents and initiators [17]. That polymerization line is shown in Fig. 2 for living poly(α-methylstyrene) and will be discussed further below. Within the experimental error of a degree or two, the line does not depend on the solvent, even though these are ionic solutions and must be nonideal. Figure 2 also shows that, within experimental error, the polymerization line does not depend on the choice of initiator.

B. Solvent Effects

As discussed above, the nature of the solvent does not very much affect the polymerization line. However, the phase diagram of the living polymer in solvent does develop interesting new features if the polymer is not entirely soluble in the solvent (a "poor" solvent as opposed to a "good" solvent [36]). We will discuss these cases in Section IV.

Solvent properties also have profound effects on the chemical kinetics

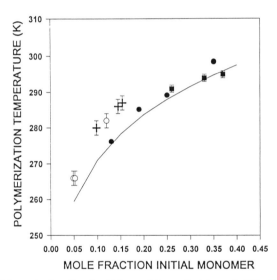

Figure 2. Polymerization line for living poly(α-methylstyrene) [17]. Region of phase diagram above line consists of monomer and initiated dimer (see text) in solvent; region below line consists of living polymer and monomer in chemical equilibrium in solvent. Determinations of T_p as function of mole fraction of monomer are from various physical measurements, with various solvents and various initiators: ($\blacksquare$) mass density measurements in tetrahydrofuran with sodium naphthalide initiator [17]; ($\bigcirc$) small-angle neutron scattering in tetrahydrofuran-d_8 with sodium naphthalide initiator [63]; (+) extent of polymerization measurements in tetrahydrofuran with sodium naphthalide initiator and cesium naphthalide initiator (point at mole fraction 0.098) [60]; ($\bullet$) shear viscosity in methylcyclohexane with n-butyllithium initiator [76]. Error bars correspond to uncertainties as given in papers (99% confidence intervals). Line corresponds to Eq. (6) with parameters set at the values in Table I.

of living polymers. The living ionic ends and their counterions can be closely associated ion pairs, solvent-separated ion pairs, or entirely free ions, depending the nature of the solvent [3]. The rate constants for propagation and depropagation will be very different for each case.

C. Initiator Effects

The initiation process [Eq. (1)] will occur at all temperatures, not just below the polymerization temperature. The situation is simplified if the initiation is allowed to proceed above T_p and can thus be taken to be complete and irreversible thereafter.

The nature of the initiator is still of interest because this choice can alter the reaction mechanism. Some initiators create living polymers with one active site, and some create two active sites. For a particular initiator, the choice of the counterion will affect the chemical kinetics and can affect the propensity for intramolecular association to form rings, perhaps altering the ultimate molecular distribution [3]. Some examples of such cases will be discussed below. We refer the reader to Szwarc [3], Szwarc and Van Beylen [22], and books on polymerization [37] for further details about the initiation step.

III. SPECIFIC EXAMPLE: α-METHYLSTYRENE

We will focus here on one particular living polymer: living poly(α-methylstyrene). This system has the advantages of having a polymerization line at experimentally convenient temperatures (Fig. 2) and of having been studied extensively by organic and polymer chemists. It has the disadvantage that there is a strong tendency to produce stable oligomers above T_p [38, 39].

A. Mechanisms

We will consider two basic mechanisms for the living anionic polymerization of α-methylstyrene: (1) using an alkali metal naphthalide initiator to produce a living linear polymer with both ends active and (2) using an alkyl lithium initiator to produce a living linear polymer with one end active.

The mechanism using an alkali metal naphthalide initiator is shown in Fig. 3 [33]. Metallic sodium reacts with naphthalene to make sodium naphthalide, which is green in solution. The sodium naphthalide reacts with α-methylstyrene to form a radical ion, which immediately dimerizes. The presence of the living ends turns the solution from green to a deep red. The dimer is then the species that propagates to the polymer, as

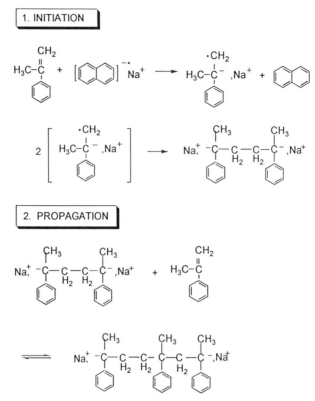

Figure 3. Reaction mechanism using alkali metal naphthalide initiator to produce living poly(α-methylstyrene) with both ends active.

shown. The formation of the dimer is "head to head", but the propagation itself is "head to tail".

The mechanism using an alkyl lithium initiator is shown in Fig. 4 [40]. The *n*-butyllithium reacts with the α-methylstyrene to make a species with one active end. The initiator molecule is incorporated into the activated monomer. The propagation is straightforward, as shown. The complications with this kind of initiator come from the tendency of the alkyl lithium molecules to agglomerate in a nonpolar solvent, an effect that can be avoided by the addition of a small amount of a polar solvent [41].

B. Polymerization Line and Thermodynamic Parameters

The polymerization line for living poly(α-methylstyrene) has been presented in Fig. 2. The data shown are all from the author's laboratory and

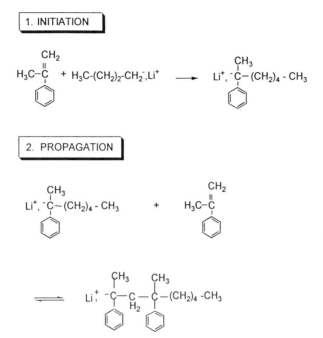

Figure 4. Reaction mechanism using alkyl lithium initiator to produce living poly(α-methylstyrene) with one end active.

were determined by the change in some physical property as a function of temperature [17]. Prior literature data lie about 10 K higher; we assume this difference is because most of the prior determinations of $T_p(x_m^0)$ were made by analysis of the concentration of residual monomer, which (as discussed above) can be complicated in the region above T_p by the presence of oligomers.

In Fig. 2 we also show the line corresponding to Eq. (6) with the parameters given in Table I for α-methylstyrene. The choice of these parameters is described below. Given that Eq. (6) assumes that the solution is ideal and that the parameters are independent of molecular weight, the agreement of the line with the measurements is not bad.

The measurements of the heat of polymerization of poly(α-methylstyrene) by Roberts and Jessup [18] (shown in Fig. 1) extrapolated to "infinite" molecular weight give $\Delta H_p^0 = -35 \pm 1 \, \text{kJ mol}^{-1}$, as given in Table I. The ΔS_p^0 is harder to measure [42]; we give in Table I a value of $-110 \pm 1 \, \text{J mol}^{-1} \text{K}^{-1}$ as determined by fitting the polymerization line to Eq. (6) and a value of $-105 \pm 1 \, \text{J mol}^{-1} \text{K}^{-1}$ as determined from fitting small-angle neutron scattering to a mean-field theory (see below).

IV. THERMODYNAMICS

That living, equilibrium polymerization commences on cooling through a polymerization or ceiling temperature and reverses on heating back through that temperature is reminiscent of a phase transition. The statistical mechanics of phase transitions has bloomed gloriously in the last two decades and is at our disposal to apply to such a phenomenon.

A. Statistical Mechanical Models

Here we will give a brief overview of the theory of polymerization transitions and then present the relevant theoretical equations as they are used.

1. Mean Field

Pivotal theoretical (and concomitant experimental) work on the polymerization transition was done by Tobolsky, Eisenberg, and their collaborators in the period 1959–1962 [43–47]. They derived various thermodynamic properties of these systems by starting from the equations for the reaction equilibrium constants. Wheeler et al. [4, 5] have demonstrated that the theories of Tobolsky and Eisenberg correspond to what we now call "mean-field" models of *second-order* phase transitions. In this view, the polymerization line is to be seen as a line of second-order phase transitions. Oosawa et al. [26, 48] did similar calculations for the equilibrium polymerization of biopolymers but were viewing the polymerization transition as a *first-order* transition, analogous to crystallization. It was Wheeler et al. [4] who first recognized that the polymerization transition is second order.

2. Non–Mean Field

Wheeler et al. [4] further recognized that the relationship already noted by de Gennes [49] and by de Cloizeaux [50] between polymer statistics and critical phenomena could be extended to provide a non–mean-field theory for the polymerization transition. They noted that, in the language of renormalization theory where n is the dimension of the order parameter, pure flexible living linear polymers will be in the $n \to 0$ universality class [51], and pure living ring polymers will be in the $n = 1$ universality class [8, 52, 53]. If the polymers are in a solvent, then flexible linear living polymers will be in the *dilute* $n \to 0$ universality class [11, 51, 54], and living ring polymers will be in the *dilute* $n = 1$ universality class [11].

B. Experimental Measurements

A major problem in the experimental study of equilibrium polymerization has been the preparation of samples that will remain "alive", that is, in which there are no impurities to cause termination. The major concerns are air and water: Water terminates the polymer molecules by proton transfer [3]; the oxygen in air terminates the polymer molecules by extracting electrons to make radicals, which then combine together [1]. The procedures for achieving clean samples have been discussed in a classical paper by Fetters [55] and in a recent paper by Ndoni et al. [56].

The contribution of the author and collaborators (especially Andrews) has been to note that the stability of living polymer solutions can be improved if the solution of monomer, initiator, and solvent can be prepared and sealed *above* the polymerization or ceiling temperature, so that the initiator acts as a "scavenger" to rid the solution of impurities while in the process of forming the initiated species. Some initiator molecules and some initiated monomers will be lost to impurities, but no polymers will be terminated, because none have yet been formed. Then the initiated species will form living polymer when the temperature is lowered below T_p. The result is a very stable living polymer solution: Samples remain alive for months, even years. When the living polymer is terminated and analyzed by gel permeation chromatography, the result indicates only one polymer product, the molecular weight of which is within 10% of that expected from the initiator concentration [57, 58]. This procedure also assures the validity of the assumption noted above that the initiation process is not an issue in the polymerization process.

1. Extent of Polymerization

The extent of polymerization $\phi(T)$, the fraction of initial monomer converted to polymer, as a function of temperature by a living polymer system is a primary prediction of the theories. Note that here ϕ is *not* a volume fraction of polymer, but a fraction of monomer reacted to make living polymer. Mean-field predictions were made by Tobolsky and Eisenberg [47] for pure living poly(α-methylstyrene). Non–mean-field predictions were made for pure living poly(α-methylstyrene) and for pure living poly(tetrahydrofuran) by Kennedy and Wheeler [51]. Experimental data exist for living poly(α-methylstyrene) in solution in tetrahydrofuran, but not for the pure monomer. Indeed, the polymerizing pure α-methylstyrene is very viscous and difficult to work with [59].

The mean-field theoretical prediction for $\phi(T)$ depends on whether the reaction mechanism is that of Fig. 3, where the propagating species is a bifunctional dimer, or that of Fig. 4, where the propagating species is a monofunctional activated monomer. If the propagating species is a bifunctional dimer, then the mean-field prediction for the mole fraction of unreacted monomer is [45, 60]

$$x_m(T) = x_m^0 - \frac{(x_i/2)[2 - K_p(T)x_m(T)]}{1 - K_p(T)x_m(T)} \qquad (7)$$

where $x_m(T)$ is the mole fraction of residual (unpolymerized) monomer at equilibrium at T, x_m^0 is the mole fraction of initial monomer, x_i is the mole fraction of initiator, and K_p is the equilibrium constant for the polymerization of monomer. From Eq. (7) and the conservation relation,

$$x_m^p(T) = x_m^0 - x_m(T) \qquad (8)$$

where $x_m^p(T)$ is the mole fraction of monomer molecules incorporated into polymer molecules at equilibrium at T. Equation (7) is a quadratic equation for $x_m(T)$. Thus

$$\phi(T) = \frac{x_m^p(T)}{x_m^0} \qquad (9)$$

where x_m^0 is known from the experimental preparation of the sample and $x_m^p(T)$ is calculated from Eq. (8) using Eq. (7). We define

$$r = \frac{N_i}{N_m^0} = \frac{x_i}{x_m^0} \qquad (10)$$

where N represents the number of molecules of a given species (subscript i for initiator and subscript m for monomer) in the system, and thus

$$x_i = rx_m^0 \qquad (11)$$

The equilibrium constant $K_p(T)$ can be calculated from

$$K_p(T) = \exp[(-\Delta H_p^0 + T \, \Delta S_p^0)/RT] \qquad (12)$$

where R is the gas constant.

If the propagating species is a monofunctional activated monomer (see Fig. 4), then instead of Eq. (7), the appropriate equation is [51]

$$x_m(T) = x_m^0 - \frac{x_i}{1 - K_p(T)x_m} \qquad (13)$$

and Eqs. (8)–(12) are used again.

Kennedy and Wheeler [51] have pointed out that for the case in which the dimer is the propagating species (Fig. 3), there may be a distribution of locations of the initial dimer within the polymer molecules and that this feature may negate the correspondence to the non–mean-field $n \to 0$ magnet model. We note this possibility, but proceed to test the non–mean-field dilute $n \to 0$ magnet model because (a) no alternative non–mean-field model is available and (b) we can thereby test the importance of the dimer feature.

The non–mean-field model has not yet been solved exactly, but it has been solved with the approximation that there is no interaction between the monomer and the solvent [54]. The important equations are

$$\frac{K_p^{c0}(1 - \zeta)}{K_p(T)} - 1 = r_p(1 - b^2\theta^2) \tag{14}$$

$$x_b = \tfrac{1}{2}am_0\gamma r_p^{1-\alpha}\theta^2 \tag{15}$$

$$x_p = \tfrac{1}{2}am_0 r_p^{2-\alpha}\theta^2(1 - \theta^2) \tag{16}$$

$$x_s = \left[\frac{\zeta}{1 + \zeta}\right](1 - x_b - x_p) \tag{17}$$

which are the same as Eqs. (5) and (12) in Ref. 7, except that we use a different notation. The equilibrium constant K_p^{c0} refers to that for the pure monomer polymerizing at its T_p, which is calculated from Eq. (12) by putting $T = T_p^0$ for pure monomer. The parameter $\zeta(T)$ is the activity of the solvent relative to monomer and is to be evaluated. The parameters $r_p(T)$ and $\theta(T)$ are parametric variables [4] indicating the proximity to the "critical point". The quantities $x_b(T)$, x_p, and x_s are the mole fractions of monomer-to-monomer bonds, polymer molecules, and solvent molecules, respectively. The constants α and γ are the critical exponents [4]. The constants a, m_0, and b are constants in the parametric representation [4, 51].

This non–mean-field model is solved numerically at each temperature by a computer program as follows. First we pick a trial value of θ. By setting $x_p = \tfrac{1}{2}x_i$ [see Eqs. (10) and (11)], we can substitute x_p and the trial θ into Eq. (16) and solve for r_p. Having r_p, we can calculate $x_b(T)$ using Eq. (15). Setting $x_s = 1 - x_m^0$, we can substitute $x_b(T)$ and x_p into Eq. (17) to obtain ζ. Now we can test for a true solution by testing to see if the two sides of Eq. (14) are equal using Eq. (11) for $K_p(T)$. The program varies θ until a solution is obtained within a specified accuracy.

When we have a solution within the desired accuracy, we can calculate

$$\phi(T) = \frac{x_b(T) + x_p}{x_m^0} \tag{18}$$

The predictions of the mean-field theory [Eqs. (7)–(9)] and the non–mean-field dilute $n \to 0$ magnet model [Eqs. (14)–(18)] for $\phi(T)$ for living poly(α-methylstyrene) by the mechanism shown in Fig. 3 were recently tested by Das et al. [60]. The measurements were made by preparing a number of identical samples of monomer an initiator in solvent at temperatures above T_p, then cooling each sample to a temperature below T_p, keeping it at that temperature until the equilibrium extent of reaction was attained. Then each sample was deliberately terminated, the residual free monomer concentration determined by gas chromatographic analysis, and $\phi(T)$ calculated from that measurement. The theories and the experiment are compared in Fig. 5. The theoretical predictions both use the parameters in Table I. The only truly free parameter is T_p^0, used in the non–mean-field theory. Both theories describe the data qualitatively. The non–mean-field dilute $n \to 0$ magnet model is a quantitatively better description.

The same conclusion is reached from a comparison of data from the literature [21, 61, 62] on the extent of polymerization of pure poly(tetrahydrofuran) [51, 60].

2. Mass Density

The mass density of a living polymer system in the temperature range near the polymerization transition is an indirect test of the theoretical models because additional assumptions must be made to calculate the density from the extent of polymerization. Kennedy and Wheeler [9] analyzed this issue for polymerizing sulfur in 1983, and Zheng and Greer [17] applied that analysis to new density measurements for living poly(α-methylstyrene) initiated by sodium naphthalide in tetrahydrofuran in 1992.

Kennedy and Wheeler [9] calculated the mass density as a function of temperature by assuming linear thermal expansions for the pure solvent, the pure monomer, and the pure polymer and then assuming the ideality of the solution, so that the density of the solution is calculated as a sum of the contributions of its components weighted by the weight fraction of each. Thus the study of the mass density is an indirect study of the extent of polymerization, that is, the weight fraction of polymer. On the other hand, the mass density can be measured much more precisely than can the extent of polymerization (0.004 vs. 1%).

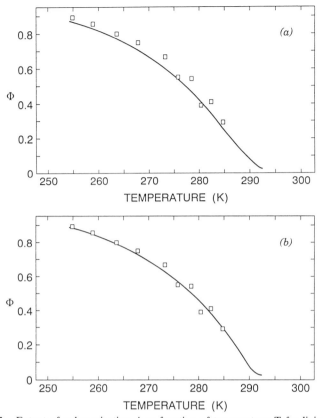

Figure 5. Extent of polymerization ϕ as function of temperature T for living poly(α-methylstyrene) as initiated by sodium naphthalene in solvent tetrahydrofuran [60]. Note that ϕ is not volume fraction, but is fraction of initial monomer converted to polymer. Sample had initial mole fraction of monomer $x_m^0 = 0.15378 \pm 0.00003$ and ratio of mole fraction initiator to mole fraction monomer $r = 0.0044 \pm 0.0001$. Data are compared to (a) mean-field theory and (b) the non–mean-field, dilute $n \to 0$ theory.

Zheng and Greer [17] measured the mass density by dilatometry, which has the advantages of high precision and an easily sealed cell but has the disadvantage that the measurements can only be made by heating runs. The latter is in conflict with the need (see above) to keep the sample above T_p until measurements are made. Moreover, these were the very first published data on these systems from the Greer laboratory and could suffer from the lack of experience of the experimentalists; in addition, the measurements showed evidence of drift with time. Thus, while these are the only published measurements of the mass density of a living polymer system, we consider them with reservations.

When Zheng and Greer published their density data, the dilute $n \to 0$ magnet model was not yet available [54], so they compared their measurements to the only available non–mean-field model, the $n \to 0$ magnet model for pure polymer [51]. In addition, at that time it was not realized that the mean-field equation of state [Eqs. (7) and (13) above] depends on the reaction mechanism [45, 63]. Zheng and Greer [17] concluded that while both models could describe the data qualitatively, the non–mean-field (nondilute) $n \to 0$ magnet model could provide a more accurate description of the data, but only if the value of T_p used in the model were different from the experimental value.

We now compare the data of Zheng and Greer to the proper mean-field model using Eq. (7) and to the appropriate non–mean-field dilute $n \to 0$ magnet model using Eqs. (14)–(18). We use all the same parameters as the original paper, except as follows. For the mean-field model, we first test the model using the value $\Delta S_p^0 = -105 \, \text{J mol}^{-1} \text{K}^{-1}$ determined from the neutron scattering studies [63] and used successfully to describe the extent of polymerization (see above) [60]. Figure 6a shows that the resulting prediction (solid line) of the mean-field theory does not describe the data. If ΔS_p^0 is instead set at $-109.5 \, \text{J mol}^{-1} \text{K}^{-1}$, then the mean-field theory describes the data very well, as shown in Fig. 6b. We show only sample 4 of Zheng and Greer [17], but the results are similar for their other two samples.

Similarly, the important parameter in the dilute $n \to 0$ magnet theory is the polymerization temperature of the pure monomer, T_p^0, which was found to be $334 \pm 2 \, \text{K}$ from the extent of polymerization [60] and $329 \pm 6 \, \text{K}$ from the small-angle neutron scattering [63]. If we use $T_p^0 = 334 \, \text{K}$ to predict the mass density, then we get the result shown by the dotted line in Fig. 6a, a line that does not describe the data. If $T_p^0 = 319 \, \text{K}$, then the theory describes the data, as shown in Fig. 6b. Again, we show only sample 4 of Zheng and Greer [17], but the results are similar for their other two samples.

Thus the values of the thermodynamic parameters that best describe other properties of this system do not work well for the mass density. This could be either because the data are distorted by sample problems or because the theoretical prediction of the mass density requires several other assumptions. We conclude that the measurements of mass density do not differentiate between the two theories.

3. Concentration Susceptibility

The concentration susceptibility, $(\partial x_s / \partial \Delta)_{P,T}$, where Δ is the difference in chemical potentials between the solvent and the monomer, can be obtained experimentally from small-angle neutron scattering (SANS)

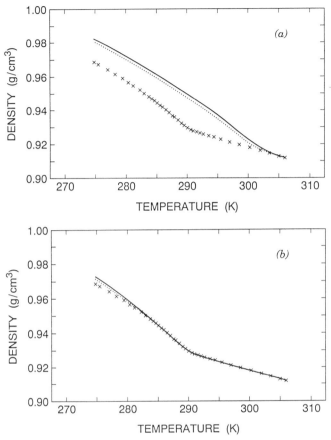

Figure 6. Mass density as function of temperature T for living poly(α-methylstyrene) as initiated by sodium naphthalene in solvent tetrahydrofuran. Crosses are data for sample 4 of Zheng and Greer [17], which had initial mole fraction of monomer $x_m^0 = 0.26$ and ratio of mole fraction initiator to mole fraction monomer $r = 0.0018 \pm 0.0001$. (a) Mean-field theory [Eq. (7)] with $\Delta S_p^0 = -105\,\mathrm{J\,mol^{-1}K^{-1}}$ (solid line); non–mean-field dilute $n \to 0$ theory with $T_p^0 = 334\,\mathrm{K}$ (dotted line). (b) Mean-field theory [Eq. (7)] with $\Delta S_p^0 = -109.5\,\mathrm{J\,mol^{-1}K^{-1}}$ (solid line); non–mean-field dilute $n \to 0$ theory with $T_p^0 = 319\,\mathrm{K}$ (dotted line).

[64]. The mean-field theory for the concentration susceptibility for a living polymer solution was presented by Andrews et al. [63], following the work of Boué et al. [65] for polymerizing sulfur. The structure function $S(q, T)$ at the scattering vector $q = 0$ is directly proportional to the scattering intensity and to the concentration susceptibility and is given

in the *mean-field theory* by

$$S(0, T)^{-1} = -\frac{4T_1}{T} + \frac{1}{x_m^0(1 - x_m^0)}$$

$$- \frac{K_p(T)[K_p(T)m + h]^3}{\{1 + \frac{1}{2}[K_p(T)m + h]^2\}\{[K_p(T)m + \frac{1}{2}h][K_p(T)m + h]^2 + h\}}$$

(19)

where the new symbols and mapping are the external magnetic field $h = 2x_p/m = x_i/m$ and the square of the magnetization $m^2 = 2x_m(x_m^0 - x_m)$, with x_m given by Eq. (7) above and where there is one new parameter, T_1, which can be considered to be the mean-field temperature of the upper critical solution point of the solvent–monomer solution.

The non–mean-field $n \to 0$ magnet theory for the concentration susceptibility in the case of no monomer–solvent interaction (i.e., $T_1 = 0$) was presented by Wheeler and Pfeuty [54] and is the same theory discussed above in connection with Eqs. (14)–(17). The expression for $S(0, T)$ for the *non–mean-field model* is given as Eq. (13) of Wheeler and Pfeuty:

$$S(0, T) = \frac{x_s}{1 + \zeta} + \left(\frac{\zeta}{1 + \zeta}\right)^2 \frac{(1 + \zeta)K_p^{c0}}{K_p} (\frac{1}{2}am_0\gamma)r_p^{-\alpha}\theta^2 F(\theta^2) \quad (20)$$

using

$$F(\theta^2) = \frac{2\beta\delta(1 - \theta^2) - (1 - \alpha)(1 - 3\theta^2)}{(1 - b^2\theta^2)(1 - 3\theta^2) + 2\beta\delta b^2\theta^2(1 - \theta^2)} \quad (21)$$

where r_p and θ are again the parametric variables and other variables are as defined above.

The SANS measurements have recently been made on poly(α-methylstyrene) initiated by sodium naphthalide in the deuterated solvent tetrahydrofuran-d_8 [63, 66]. We note that the exact nature of the solvent was not made clear in the original publications: The solvent was in fact tetrahydrofuran, which had been "hydrogenated" with deuterium. Figure 7 shows the measurements of the intensity at $q = 0$ (which is just proportional to $S(0, t)$ by a contrast factor) for the sample 91-1 of Andrews et al. [63] for two cooling runs (*A* and *B*). Sample 91-1 was characterized by $x_m^0 = 0.12 \pm 0.01$, $r = (3.9 \pm 0.5) \times 10^3$, and $T_p = 282 \pm 2$ K.

Also shown in Fig. 7 are predictions for the mean-field and non–mean-

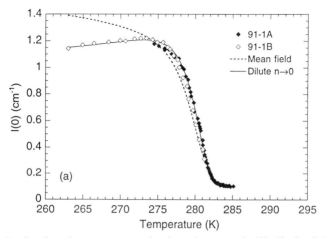

Figure 7. Small-angle neutron scattering intensity at $q = 0$, $I(0, T)$, for living poly(α-methylstyrene) initiated by sodium naphthalide in tetrahydrofuran-d_8 [63]. Symbols are experimental values of $I(0, T)$ for sample 91-1, where A and B refer to initial and second cooling runs. Sample 91-1 characterized by $x_m^0 = 0.12 \pm 0.01$, $r = (3.9 \pm 0.5) \times 10^{-3}$, and $T_p = 282 \pm 2$ K. Theoretical prediction for mean-field theory [Eq. (19)] using parameters discussed in text shown as dashed line. Theoretical prediction for the non–mean-field, dilute $n \to 0$ magnet theory [Eq. (20)] using parameters discussed in text shown as solid line.

field theories. For the mean-field theory, the various parameters (ΔH_p^v, ΔS_p^0, x_m^0, and r) were varied only within their uncertainties to obtain the best fit, except for T_1, which was the only true "free" parameter. The best fit value was $T_1 = 37 \pm 5$ K. The predictions were not very sensitive to changes in T_1 or r. However, the fits were quite sensitive to changes in ΔH_p^0 [a 1% change in ΔH_p^0 changes $I(0, T)$ by a factor of 2–3] and quite sensitive to changes in ΔS_p^0 [a 1% change in ΔS_p^0 changes $I(0, T)$ by a factor of about 2]. The mean-field theory clearly predicts the correct qualitative behavior for $I(0, T)$, but it also clearly is not quantitatively accurate. In particular, it fails to predict the maximum in $I(0, T)$ below T_p at sufficiently low r, as is seen for sample 91-1 in Fig. 7.

For the non–mean-field theory, the various parameters (ΔH_p^0, x_m^0, T_p^0, and r) were again varied only within their uncertainties to obtain the best fit. The best fit value for T_p^0 was 329 ± 6 K. The fits were not sensitive to changes in r or ΔH_p^0, but a 1% change in T_p^0 changed $I(0, T)$ by 10%. Figure 7 shows that the non–mean-field dilute $n \to 0$ magnetic model clearly gives better agreement with the data for $I(0, T)$ than does the mean-field theory. In particular, the dilute $n \to 0$ magnet model correctly predicts the maximum in $I(0, T)$ below T_p for sample 91-1, a feature not predicted by the mean-field theory [54].

4. Phase Diagram in Poor Solvent

Figure 2 is the phase diagram for a living polymer dissolved in a "good" solvent, in which the monomer and the living polymer are completely soluble [36]. If, however, the solvent is a "poor" solvent for the polymer and the solubility of the polymer in the solvent is limited, then the phase diagram becomes more interesting: We can have the two kinds of phase transitions, equilibrium polymerization and liquid–liquid phase separation, intersecting one another. The mean-field theory for calculating such a phase diagram has been developed, but no non–mean-field theory is yet available. The mean-field treatment of these phase diagrams has its origins 30 years ago in Scott's combination [67] of the treatment of Tobolsky and Eisenberg [68] of equilibrium polymerization with the Flory–Huggins theory [69–71] of polymer solutions to predict the phase diagrams of sulfur polymerizing in solution. In 1983, Kennedy and Wheeler [51] applied essentially the same model, but, in the language of the "mean field limit of the dilute $n \rightarrow 0$ magnet model", to living polymer solutions.

The nature of the phase diagram of a living polymer in a poor solvent is fundamentally dependent on the parameter that (in comparing the partition functions) maps to the external field h in the analogous magnet model. Four cases are possible [11]:

(I) The equilibrium constant for initiation, K_i, maps to $\frac{1}{2}h^2$. If K_i is close enough to zero, then we expect the polymerization line to meet the liquid–liquid coexistence at its critical point and to form a *symmetric tricritical point* with a critical exponent $\beta = 1$ [72, 73]. A K_i near zero corresponds to a very low concentration of activated monomer, which in turn corresponds to the creation of very large polymers.

(II) If K_i is *not zero* but is a *constant* over the phase diagram, then we expect the polymerization line to meet the liquid–liquid coexistence curve, generally at some point other than its critical point, in which case the critical point will be an "ordinary" one with critical exponent $\beta = 0.32$. If the polymerization line should meet exactly the liquid–liquid critical point, the result will be a *nonsymmetrical tricritical point* with a critical exponent $\beta = 0.25$. We note that K_i is a thermodynamic *field* rather than a thermodynamic density, and thus a field and not a density is held constant over the phase diagram [74].

(III) It can be that the quantity held constant over the phase diagram is instead the concentration x_i of initiators, a thermodynamic *density* that maps to $\frac{1}{2}hm$, where m is the magnetization. Such a case will still have a nonsymmetric tricritical point, but the phase separation curve is not expected to coincide with the coexistence curve, and the critical point is not expected to be at the apex of the phase separation curve [75].

(IV) It can be that the quantity held constant over the phase diagram is not x_i, but $r = x_i/x_m^0$, the ratio of the mole fraction of initiators to the mole fraction of initial monomers, again a density and not a field. The resulting phase diagram will be topologically similar to that for case III, but different in detail [76].

Case I will be very difficult to realize experimentally: The initiator itself must be in a state of chemical equilibrium with a very small equilibrium constant. Case II requires that the initiation step have a constant, if finite, equilibrium constant; this case will be hard to realize because the temperature, and thus the equilibrium constant, will vary across the phase diagram. Cases III and IV are experimentally realizable. Case IV keeps the average molecular weights more comparable across the phase diagram. We will not detail the calculations of these diagrams here but refer the reader to the original theory papers referenced above.

Only two phase diagrams of a living polymer in a poor solvent have been published. One is the phase diagram of living poly(α-methylstyrene) initiated by n-butyllithium (see Fig. 4) in methylcyclohexane [76]. The solution also contained a small amount of tetrahydrofuran in order to prevent the agglomeration of the n-butyllithium molecules [3]. The phase diagram (shown in Fig. 8) was determined by making 14 samples with different values of x_m^0 but with the same r value of 0.008, then placing the samples in a temperature-controlled bath and observing (a) the appearance of a new phase as the temperature was lowered in order to map the phase separation lines and (b) the fall time of a glass ball at each temperature to detect the increase of viscosity at the onset of polymerization and thus map the polymerization line [77].

Figure 8 shows the resulting phase separation curves and polymerization line for living poly(α-methylstyrene) in methylcyclohexane. The viscosity data used to determine the polymerization line for three of the samples are plotted in Fig. 9. The polymerization line points are also included in Fig. 2. There is more scatter in the data than we would like, but these are difficult measurements. The upper critical solution point was determined from the relative volumes of coexisting phases to be at $T_c = 274 \pm 1$ K and at $x_m^0 = 0.18 \pm 0.02$. As can be seen in Fig. 8, at this x_m^0, the polymerization temperature T_p is 285 K, well above the critical temperature. The polymerization line meets the coexistence curve at about $x_m^0 = 0.12$ and does not intersect the liquid–liquid critical point to make a tricritical point.

Figure 8 shows also the predictions of the mean-field theory based on

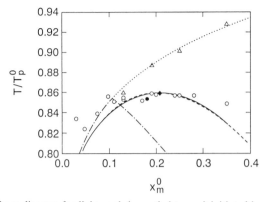

Figure 8. Phase diagram for living poly(α-methylstyrene) initiated by n-butyllithium in methylcyclohexane [76]. Parameter T/T_p^0 is temperature divided by polymerization temperature for pure α-methylstyrene (321 K), and x_m^0 is mole fraction of initial monomer. Open triangles are experimental polymerization temperatures. Dotted line is polymerization line calculated from mean-field theory for $r = 0$, where r is ratio of mole fraction initiator to mole fraction initial monomer. Open circles are experimental phase separation temperatures. Dot-dashed line is mean-field theory prediction for coexistence curve with $r = 0$. Dashed and solid lines are mean-field theory predictions for coexistence curves with $r = 0.008$: points on left solid curve have, as their corresponding equilibrium phases, points connected by horizontal tie-lines on right dashed curve; points on right solid curve have, as their corresponding equilibrium phases, points connected by horizontal tie-lines on left dashed curve. Solid circle is experimentally determined critical point. Solid diamond is theoretically calculated critical point.

case IV above. The free parameters in the calculation were $T_1 = 91.7$ K (note that this is not the value quoted in Section IV.B.3 because the solvent is not the same), and $T_p^0 = 321$ K (compared to 329 ± 6 K in Section IV.B.3). The theory is not very sensitive to ΔH_p^0, and ΔS_p^0 cancels out of the calculation. The mean-field theory for case IV can be seen to give a good qualitative model of the phase diagram of living poly(α-methylstyrene) in methylcyclohexane, initiated by n-butyllithium plus tetrahydrofuran.

The other published phase diagram for a living polymer in a poor solvent is that shown in Fig. 10 for living poly(styrene) in cyclohexane, initiated by n-butyllithium, with tetrahydrofuran added to prevent agglomeration of the initiator [78, 79]. The reaction mechanism is the same as that in Fig. 4, but for styrene monomer rather than α-methylstyrene monomer. The experimental procedure was exactly the same as than for the phase diagram for living poly(α-methylstyrene) in methylcyclohexane shown in Fig. 8. The conditions were those of case IV, with $r = 0.005$ for all samples. The viscosity measurements used to develop the phase

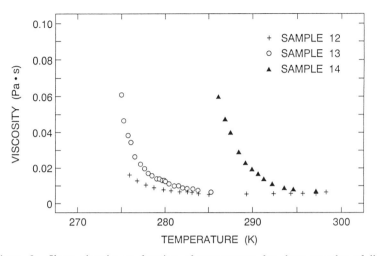

Figure 9. Shear viscosity as function of temperature for three samples of living poly(α-methylstyrene) in methylcyclohexane, with n-butyllithium as initiator [76]. Viscosity data are calculated with accuracy of 10% from fall times of glass balls in samples [14]. Samples 12, 13, and 14 have x_m^0 values of 0.19, 0.25, and 0.35, respectively, and T_p values of 285, 289, and 299 K, respectively. Mole fraction of initiator relative to that of monomer is 0.008 for all samples.

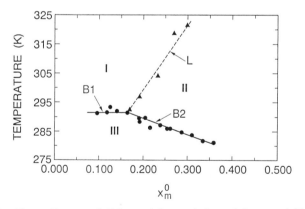

Figure 10. Phase diagram of living poly(styrene) in cyclohexane, initiated by n-butyllithium [78, 79]. Parameter x_m^0 is mole fraction of initial monomer in solvent. Ratio of moles of initiator to moles of initial monomer $r = 0.005$ for all samples. B1 and B2 are two branches of liquid–liquid phase separation curve. L is line marking points at which background viscosities of polymer solutions deviate from Arrhenius behavior.

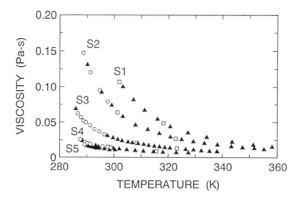

Figure 11. Shear viscosity as function of temperature for five samples of living poly(styrene) in cyclohexane, with n-butyllithium as initiator [59]. Viscosity data are calculated with accuracy of 10% from fall times of glass balls in samples. Samples S1, S2, S3, S4, and S5 have x_m^0 values of 0.30, 0.27, 0.23, 0.19, and 0.17, respectively. Mole fraction of initiator relative to that of monomer is 0.005 for all samples. Symbols indicate initial cooling run (○), subsequent heating run (▲), and final cooling curve (□).

diagram are shown in Fig. 11 [59]. However, it turns out that the "line" taken to be the polymerization line in this phase diagram (L in Fig. 10) is most probably *not* the polymerization line. We note the following about this system:

(a) While no experimental polymerization line like that given in Fig. 2 has ever been reported for living poly(styrene), the polymerization line can be estimated using Eq. (6) and the parameters given in Table I. This estimated line lies at very much higher temperatures. For example, at $x_m^0 = 0.25$, $T_p \approx 628$ K.

(b) Later experimental analysis by Ruiz-Garcia and Greer [79] indicated that polymeric styrene was present in the samples a temperatures above the line L in Fig. 10.

(c) Five years after the measurements in Fig. 10 were made, the same samples were examined again [80]. The samples had retained the red color characteristic of active living polymer. The viscosity measurements were repeated for samples S1 and S5; the viscosity data had shifted down by 20% from the original measurements but retained the same qualitative behavior as a function of T. Then the samples were deliberately terminated and the viscosity measurements repeated: The behavior as a function of T did not change! This result indicates that the strong dependence of viscosity on temperature seen in Fig. 11 is not related to the living nature of the polymers in solution.

We conclude that the measurement of the phase separation lines shown in Fig. 10 for living poly(styrene) in cyclohexane is accurate and is one of the two such measurements of a phase diagram for a living polymer solution, the other being Fig. 8 for living poly(α-methylstyrene) in methylcyclohexane. The polymerization line for poly(styrene) has not been measured but is expected to lie well above the phase separation lines. The theoretical calculation of the phase diagram for living poly(styrene) in cyclohexane has not been done. The behavior of the viscosity of this solution near the phase separation is interesting and not understood.

V. STRUCTURE: CORRELATION LENGTH

In the previous section, we concentrated on the macroscopic, thermodynamic properties of living polymer solutions. There is also much to consider with regard to the microscopic, structural nature of these solutions. The solutions above T_p are relatively simple solution of small molecules, some which are ionic, initiated monomers with their counterions. When the temperature is lowered below T_p, the initiated monomers begin to form small polymers. These small polymers will at first form a "dilute" polymer solution of independent, noninteracting polymer coils [36]. As the temperature is lowered further, the living polymer molecules grow larger, increasing in average molecular weight and in volume fraction. At some temperature, the polymer molecules will become large enough that they will begin to entangle with one another. After this "overlap" point, the polymer molecules are no longer independent of one another and constitute instead a "semidilute" solution. At equilibrium at any temperature, the living polymers will have a distribution of molecular weights.

One probe of the microscopic structure is the characteristic length or correlation length as determined by the scattering of radiation from the solution. We will consider here the small-angle scattering of neutrons: The same experiment used in Section IV.B.3 to determine a thermodynamic quantity (the concentration susceptibility) yields at the same time a structural property, the correlation length ξ, from fits of the scattering intensity to the Ornstein–Zernike expression:

$$I(q, T) = \frac{I(0, T)}{1 + q^2 \xi^2(T)} . \qquad (22)$$

For a *dilute* solution of polymers, ξ, as obtained from the scattering of radiation in the regime $q\xi \ll 1$, is a measure of the radius of gyration R_G of the individual, independent polymer coils [64]. For the *semidilute* regime, there is a macroscopically uniform distribution of flexible polymer chains in the solvent, and the system can be viewed as a mesh of polymer chains [36]. The correlation length ξ as obtained from the scattering of radiation such that $q\xi \approx 1$ will be the average size of that mesh [81].

For the living polymer solutions we study here, we expect $\xi(T)$ to *increase* as the temperature is lowered through the polymerization temperature and polymers start to grow. When the largest polymers begin to overlap, $\xi(T)$ will begin to *decrease* as the mesh is formed and becomes "finer". Thus the *maximum* in $\xi(T)$ for a living polymer solution will indicate the point of overlap and the transition from dilute to semidilute regimes. We will indicate the degree of polymerization at overlap as DP* and the volume fraction of polymer at overlap as ϕ_v^*. We see that the situation will be complicated by the polydispersity of the living polymer solution: The first chains to interpenetrate will be the largest ones.

The mean-field theory and the non–mean-field dilute $n \to 0$ magnet theory discussed in Section IV can also be applied to the prediction of $\xi(T)$ [54]. The mean-field equation is

$$\xi^2(T) = S(0, T)(\tfrac{1}{6}a^2)\left(\frac{4T_1}{T} + A[mK_p(T)][mK_p(T) + h]^2\right)\Bigg/$$
$$x_m^0\left[mK_p(T) + \frac{h}{2}\right]\{[mK_p(T) + h]^2 + h\}\Bigg) \tag{23}$$

where

$$A = 1 + \left(mK_p(T)\{1 - \tfrac{1}{2}[mK_p(T) + h]^2\}\Big/\left[mK_p(T) + \frac{h}{2}\right]\right.$$
$$\times \{[mK_p(T) + h]^2 + h\}\bigg) \tag{24}$$

where all the symbols have been defined in Section IV, but we note that $h = 2x_p/m = x_i/m$ and $m^2 = 2x_m(x_m^0 - x_m)$ and a is a new parameter related to the size of the monomer molecule. The calculation of $\xi(T)$ from the non–mean-field dilute $n \to 0$ magnet theory is

$$\xi = \xi_0(\theta^2)r_p^{-v} \tag{25}$$

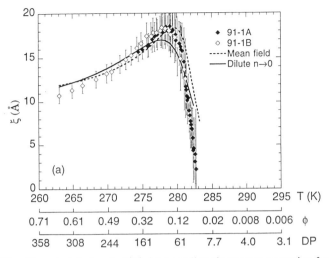

Figure 12. Characteristic length $\xi(T)$ from small-angle neutron scattering for samples of living poly-α-methylstyrene in tetrahydrofuran-d_8 [63]. Symbols are experimental values of $\xi(T)$ as obtained by fits to Eq. (22) for sample 91-1, where A and B refer to initial and second cooling runs. Sample 91-1 characterized by $x_m^0 = 0.12 \pm 0.01$, $r = (3.9 \pm 0.5) \times 10^{-3}$, and $T_p = 282 \pm 2$ K. Theoretical predictions for mean-field theory [Eq. (23)] using parameters discussed in text are shown as dashed lines. Theoretical predictions for non–mean-field, dilute $n \rightarrow 0$ magnet theory [Eq. (25)] using parameters discussed in test shown as solid lines. Axes also indicate extent of the polymerization ϕ and number-average degree of polymerization DP as calculated from mean-field theory.

where $\xi_0(\theta^2)$ is taken to be a constant, ξ_0; r_p is again the parametric variable; and the critical exponent $v = 0.589$ for $n \rightarrow 0$ [82].

Figure 12 compares the experimental measurements of $\xi(T)$ for the same sample as Fig. 7 with the theoretical predictions from Eqs. (23) and (25). The parameters are the same as for Fig. 7, with the addition of $a = 8 \pm 3$ Å and $\xi_0 = 1.8 \pm 0.6$ Å. We note first the expected increase, peak, and decrease in the measured values of $\xi(T)$ as the temperature decreases, reflecting quite dramatically the dilute, overlap, and semidilute regimes of the living polymer solution. We note further that both theories give a good qualitative description of the data, although neither is quantiatively accurate.

We include also in Fig. 12 the conversion of the temperature axes to the extent of polymerization, $\phi(T)$, and to the number-average degree of polymerization, DP(T), where ϕ and DP have been calculated from the mean-field theory, as discussed in Section IV.B.1. We note that $\phi(T)$ and DP(T) are not linear in T and are related to one another as DP $= 2\phi/r$. The maximum ξ^* of about 20 Å occurs at $\phi^* \approx 0.2$ and DP$^* \approx 100$; the corresponding overlap concentration c^* is about $0.03\,\mathrm{g\,cm^{-3}}$. An equivalent solution of monodisperse "dead" polymers with a degree of polymerization of 80 should have a much larger overlap concentration c^* and a much smaller correlation length ξ^* [83]. In a very polydisperse system such as a living polymer system, the overlap is determined by the longest polymers. In addition, the ionic nature of the living polymers should cause them to be more extended and thus to overlap at a lower average molecular weight [36].

VI. CONCLUSIONS

A. Where Are We Now?

The last 15 years have brought considerable understanding of the physicochemical and microscopic structural properties of living polymer solutions. Let us review where we are now.

1. Macroscopic Properties

In trying to understand the macroscopic thermodynamic properties of living polymer systems, we have taken the view suggested by Kennedy and Wheeler [51] that equilibrium polymerization can be modeled as a second-order phase transition in the $n \rightarrow 0$ universality class. We have compared the predictions of mean-field and non–mean-field forms of the theory to experimental measurements on one living polymer system of the extent of polymerization, the mass density, the concentration susceptibility, and phase diagrams in poor solvents. The general result is that the agreement between theory and experiment is quite satisfying. The non–mean-field form of the theory tends to be somewhat better, especially for the concentration susceptibility, but the mean-field theory is simple and useful. In fact, it is surprising that the theories work as well as they do, since they include assumptions of ideal solutions, thermodynamic parameters that are independent of molecular weight, and asymptotic parametric forms of the equation of state. It makes one wonder whether somehow the choice of values for various variables used in the predictions is somehow compensating for the assumptions of the theory. An experimentalist is therefore motivated to try to better determine those values by direct measurements. We also expect more differentiation between theories if we can make measurements on samples with smaller concentrations of initiator.

2. Microscopic Properties

The only microscopic property of a living polymer system that has been studied to date is the correlation length. The measurements of the correlation length from SANS show the increase in size of the living polymer molecules as the temperature is lowered below the polymerization temperature, the striking maximum in the correlation length when the polymers become large enough to overlap and thus pass from a dilute regime to a semidilute regime, and the subsequent decrease in the correlation length as the mesh of entangled polymers becomes finer and finer. Mean-field and non–mean-field theories describe the behavior qualitatively.

B. What Are Some Interesting Questions?

Almost all the experiments discussed above involved the same living polymer system: living poly(α-methylstyrene) in the good, polar solvent tetrahydrofuran. The exceptions were living poly(α-methylstyrene) in the poor, nonpolar solvent methylcyclohexane and living poly(styrene) in cyclohexane, for both of which phase diagrams were studied. While it is valuable to have a variety of measurements on the same system, it is also of interest to investigate other systems, such as the other monomers listed in Table I. There are also cases for which a change of initiator may prove informative, as will be discussed below.

1. Macroscopic Properties

(a) What is the behavior of the heat capacity near the polymerization transition? The measurement of the heat capacity C_p as a function of temperature near the polymerization transition will be most telling, because the heat capacity is a good indicator of the nature of a phase transition [84]. No such measurements have ever been made. Kennedy and Wheeler [51] predict a lambdalike anomaly in C_p at T_p, similar to that common to the other second-order phase transitions. The anomaly is expected to become sharper as the initiator concentration decreases.

(b) How is the development of the living polymers reflected in the shear viscosity? Figure 9 shows an increase in the viscosity as the living polymers grow. We note that as the temperature is lowered, both the average molecular weight of the polymer molecules and their volume fraction increase. Moreover, the system is a dilute solution near T_p, but becomes semidilute only a few degrees below T_p (see Fig. 12). The solution is complicated further by its broad distribution of molecular weights and by its ionic nature. Careful rheometry will be needed to guide the development of a theoretical analysis of the viscosity of living polymer solutions.

2. Microscopic Properties

(a) How does a living polymer solution approach equilibrium? In 1965, Miyake and Stockmayer [23] solved numerically the kinetic equations for equilibrium polymerization without transfer or termination. They found a three-stage process, starting with a Poisson molecular weight distribution, developing relatively quickly to the equilibrium extent of polymerization, and evolving into a Flory–Schulz molecular weight distribution over a much longer time. No experimental test of this prediction has ever been published. Das et al. [58, 85] have in progress a study of the time development of the molecular weight distribution and of the residual monomer concentration by gel permeation chromatography and gas chromatography, respectively, for the system poly(α-methyl-styrene), initiated by sodium naphthalide in tetrahydrofuran.

There remains much experimental work to explore the time development of a living polymer system and to reconcile that development with literature measurements of the chemical rate constants. The time development will depend on such factors as the ion pair equilibria (and thus solvent dielectric constant). The kinetic equations governing the time development can now be reconsidered in the hope that they can be solved with fewer approximations than in 1965 [23].

(b) What is the equilibrium distribution of molecular weights in a living polymer solution? On what factors does the distribution depend? An understanding of the equilibrium distribution of molecular weights (MWD) is critical to an understanding of living polymer solutions. Indeed, such properties as the shear viscosity or the diffusion constants are not decipherable without the MWD. As mentioned in the paragraph above, the MWD is expected to be the Flory–Schulz "most probable" distribution, based on both statistical [86] and kinetic [23] arguments. The Flory–Schulz distribution has the form [87]

$$N(n) = p^{n-1}(1-p) \qquad (26)$$

where $N(n)$ is the number of polymer molecules containing n monomers, $p = 1/(1 - M_n)$, and M_n is the number-average molecular weight. Renormalization group calculations by Schäfer [88] predict a MWD that differs from the Flory–Schulz one and depends upon the extent of chain overlap. Experiments (mentioned above) are in progress to study the actual distribution.

(c) What determines the tacticity of a living polymer molecule? How does the tacticity affect other properties? Brownstein et al. [24] have shown by proton resonance spectroscopy that the anionic polymerization of poly(α-methylstyrene) produces polymer molecules of mixed tacticity: For example, sodium naphthalide initiator at 195 K yields 13% isotactic, 48% heterotactic, and 39% syndiotactic polymers. The factors determining tacticity (e.g., steric hindrance, initiator, solvent) deserve further attention. The effects of tacticity on other properties (e.g., viscosity, approach to equilibrium, MWD) have not been considered.

(d) *When does Coulombic association occur in a living polymer solution, and how is it reflected in the macroscopic properties?* The presence of ionic active ends on the living polymer molecules and the presence of the associated counterions means that there can be Coulombic associations in the living polymer solutions. The development of such higher level aggregations will depend on the dielectric constant of the solvent and the nature of the counterion [3].

One kind of Coulombic association is the formation of ring polymers by the intramolecular "trapping" of a counterion between the two active ends of a living polymer molecule. There has been evidence of such ring formation from conductivity measurements [89]. The formation of such ring polymers is of particular interest because we would then expect a change of universality class from $n \to 0$ to $n = 1$ [52] or, for the presence of polymer rings and polymer chains, the possibility of bicritical phenomena [8, 53].

(e) *How do living polymers in solution move?* Dynamic light-scattering measurements [90] of the diffusion coefficient of a living polymer solution should reflect the changing microscopic structure of the solution as a function of temperature, as is also reflected in the correlation length (Fig. 12). Near T_p, the measured diffusion coefficient should be that of isolated polymer coils. Further below T_p, the measured diffusion coefficient should be the collective diffusion coefficient of the polymer mesh. The overlap point should manifest itself as an obvious transition from one behavior to the other. The experimental difficulty in dynamic light scattering (and a difficulty that probably precludes static light scattering) is the problem of filtering out dust, scattering from which can interfere with the scattering signal from the polymer molecules.

(f) *What if the living polymers are not flexible?* The only considerations of equilibrium polymerization of rigid living polymers are the two-dimensional models of Jaric and Bennemann [91, 92], who predict phase separations in such cases. The possibility of living liquid crystal polymers would be very interesting.

(g) *Can computer simulation help to answer some of the above questions?* Livne [93], has addressed connections between computer simulations and the $n \to 0$ magnet model. Some work on two-dimensional equilibrium polymerization has been done by Jaric and Tuthill [94, 95]. It would be of particular interest to address via simulation some of the experimental issues discussed above, for example, the factors that determine the nature of the equilibrium molecular weight distribution.

Milchev et al. [96, 97] have begun such simulations, albeit so far for systems which allow scission and recombination along the polymer chain.

ACKNOWLEDGMENT

It is a pleasure to acknowledge the support of the Chemistry Division of the National Science Foundation.

REFERENCES

1. M. Szwarc, *Nature* **178**, 1168 (1956).
2. F. S. Dainton and K. J. Ivin, *Nature* **162**, 705 (1948).
3. M. Szwarc, *Carbanions, Living Polymers, and Electron Transfer Processes*, Wiley, New York, 1968.
4. J. C. Wheeler, S. J. Kennedy, and P. Pfeuty, *Phys. Rev. Lett.* **45**, 1748 (1980).
5. J. C. Wheeler and P. Pfeuty, *Phys. Rev. A* **24**, 1050 (1981).
6. J. C. Wheeler and P. Pfeuty, *Phys. Rev. Lett.* **46**, 1409 (1981).
7. J. C. Wheeler and P. Pfeuty, *J. Chem. Phys.* **74**, 6415 (1981).
8. J. C. Wheeler, R. G. Petchek, and P. Pfeuty, *Phys. Rev. Lett.* **50**, 1633 (1983).
9. S. J. Kennedy and J. C. Wheeler, *J. Chem. Phys.* **78**, 1523 (1983).
10. J. C. Wheeler, *Phys. Rev. Lett.* **53**, 174 (1984).
11. J. C. Wheeler, *J. Chem. Phys.* **81**, 3635 (1984).
12. F. H. Stillinger, T. A. Weber, and R. A. LaViolette, *J. Chem. Phys.* **85**, 6460 (1986).
13. E. M. Anderson and S. C. Greer, *J. Chem. Phys.* **88**, 2666 (1988).
14. J. Ruiz-Garcia, E. M. Anderson, and S. C. Greer, *J. Phys. Chem.* **93**, 6980 (1989).
15. K. M. Zheng and S. C. Greer, *J. Chem. Phys.* **96**, 2175 (1992).
16. S. Bywater and D. J. Worsfold, *J. Poly. Sci.* **58**, 571 (1962).
17. K. M. Zheng and S. C. Greer, *Macromolecules* **25**, 6128 (1992).
18. D. E. Roberts and R. S. Jessup, *J. Res. Natl. But. Stnds.* **46**, 11 (1951). See also R. M. Joshi and B. J. Zwolinski, *Macromolecules* **1**, 25 (1968).
19. K. P. Andrews, Ph. D. Thesis, University of Maryland at College Park, 1994.
20. K. J. Ivin, *Trans. Far. Soc.* **51**, 1273 (1955).
21. P. Dreyfuss and M. P. Dreyfuss, *Adv. Polymer Sci.* **4**, 528 (1967).
22. M. Szwarc and M. Van Beylen, *Ionic Polymerization and Living Polymers*, Chapman and Hall, New York, 1993.
23. A. Miyake and W. H. Sockmayer, *Makrom. Chem.* **88**, 90 (1965).
24. S. Brownstein, S. Bywater, and D. J. Worsfold, *Makromol. Chem.* **48**, 127 (1961).
25. M. E. Cates, *Macromolecules* **20**, 2289 (1987).
26. F. Oosawa and S. Asakura, *Thermodynamics of the Polymerization of Protein*, Academic Press, New York, 1975.
27. B. Meyer, *Chem. Rev.* **76**, 376 (1976).
28. H. Morawetz, *Polymers: The Origins and Growth of a Science*, Wiley, New York, 1985.
29. K. Ziegler and K. Baehr, *Chem. Ber.* **61**, 253 (1928).

30. A. Abkin and S. Medvedev, *Trans. Far. Soc.* **32**, 286 (1936).

31. P. J. Flory, *J. Am. Chem. Soc.* **62**, 1561 (1940).

32. R. G. Beaman, *J. Am. Chem. Soc.* **70**, 3115 (1948).

33. M. Szwarc, M. Levy, and R. Milkovich, *J. Am. Chem. Soc.* **78**, 2656 (1956).

34. O. W. Webster, *Science* **251**, 887 (1991).

35. J. Bronowski, *Science and Human Values*, Harper & Row, New York, 1965.

36. P. G. de Gennes, *Scaling Concepts in Polymer Physics*, Cornell University Press, Ithaca, NY, 1979.

37. M. Fontanille, in *Chain Polymerization, Part I, Vol. 3*, G. C. Eastman, A. Ledwith, S. Russo, and P. Sigwalt, eds., Pergamon Press: Oxford, 1989.

38. A. Vrancken, J. Smid, and M. Szwarc, *Trans. Far. Soc.* **58**, 2036 (1962).

39. Z. Salajka and M. Kucera, *Coll. Czech. Chem. Comm.* **48**, 3041 (1983).

40. K. F. O'Driscoll and A. V. Tobolsky, *J. Poly. Sci.* **35**, 259 (1959).

41. S. Bywater and D. J. Worsfold, *Can. J. Chem.* **40**, 1564 (1962).

42. F. S. Dainton and K. J. Ivin, *Trans. Far. Sco.* **46**, 331 (1950).

43. A. V. Tobolsky, *J. Polym. Sci.* **25**, 220 (1957).

44. A. V. Tobolsky and A. Eisenberg, *J. Am. Chem. Soc.* **81**, 2302 (1959).

45. A. V. Tobolsky, A. Rembaum, and A. Eisenberg, *J. Poly. Sci.* **45**, 345 (1960).

46. A. V. Tobolsky and A. Eisenberg, *J. Am. Chem. Soc.* **82**, 289 (1960).

47. A. V. Tobolsky and A. Eisenberg, *J. Coll. Sci.* **17**, 49 (1962).

48. F. Oosawa, K. Asakura, K. Hotta, N. Imai, and T. Ooi, *J. Poly. Sci.* **37**, 323 (1959).

49. P. G. de Gennes, *Phys. Lett.* **38A**, 339 (1972).

50. J. des Cloizeaux, *J. Phys.* **36**, 281 (1975).

51. S. J. Kennedy and J. C. Wheeler, *J. Chem. Phys.* **78**, 953 (1983).

52. R. Cordery, *Phys. Rev. Lett.* **47**, 457 (1981).

53. R. G. Petschek, P. Pfeuty, and J. C. Wheeler, *Phys. Rev. A* **34**, 2391 (1986).

54. J. C. Wheeler and P. M. Pfeuty, *Phys. Rev. Lett.* **71**, 1653 (1993).

55. L. J. Fetters, *J. Res. Natl. Bur. Stnds.* **70A**, 421 (1966).

56. S. Ndoni, C. M. Papadakis, F. S. Bates, and K. Almdal, *Rev. Sci. Instrum.* **66**, 1090 (1995).

57. A. P. Andrews, Ph. D. Thesis, University of Maryland at College Park, 1993.

58. S. S. Das, Ph. D. Thesis, University of Maryland at College Park, 1994.

59. J. Ruiz-Garcia, Ph. D. Thesis, University of Maryland at College Park, 1989.

60. S. S. Das, A. P. Andrews, and S. C. Greer, *J. Chem. Phys.* **102**, 2951 (1995).

61. M. P. Dreyfuss and P. Dreyfuss, *J. Poly. Sci.* **4**, 2179 (1966).

62. D. Sims, *J. Chem. Soc.* 864 (1964).

63. A. P. Andrews, K. P. Andrews, S. C. Greer, F. Boué, and P. Pfeuty, *Macromolecules* **27**, 3902 (1994).

64. J. S. Higgins and H. C. Benoit, *Polymers and Neutron Scattering*, Clarendon Press, Oxford, 1994.

65. F. Boué, J. P. Ambroise, R. Bellissent, and P. Pfeuty, *J. Phys. I France* **2**, 969 (1992).

66. P. Pfeuty, F. Boué, J. P. Ambroise, R. Bellissent, K. M. Zheng, and S. C. Greer, *Macromolecules* **25**, 5539 (1992).

67. R. L. Scott, *J. Phys. Chem.* **69**, 261 (1965).

68. A. V. Tobolsky and A. Eisenberg, *J. Am. Chem. Soc.* **81**, 780 (1959).

69. P. J. Flory, *J. Phys. Chem.* **10**, 51 (1942).

70. P. J. Flory, *J. Chem. Phys.* **12**, 425 (1944).

71. M. L. Huggins, *Ann. N. Y. Acad. Sci.* **43**, 1 (1942).

72. C. M. Knobler and R. L. Scott, in *Phase Transitions and Critical Phenomena*, Vol. 9, C. Domb and J. L. Lebowitz, Eds., Academic Press, New York, 1984.

73. I. D. Lawrie and S. Sarbach, in *Phase Transitions and Critical Phenomena*, Vol. 9, C. Domb and J. L. Lebowitz, eds., Academic Press, New York, 1984.

74. R. B. Griffiths and J. C. Wheeler, *Phys. Rev. A* **2**, 1047 (1970).

75. L. R. Corrales and J. C. Wheeler, *J. Phys. Chem.* **96**, 9479 (1992).

76. K. M. Zheng, S. C. Greer, L. R. Corrales, and J. Ruiz-Garcia, *J. Chem. Phys.* **98**, 9873 (1993).

77. J. A. Larkin, J. Katz, and R. L. Scott, *J. Phys. Chem.* **71**, 352 (1967).

78. J. Ruiz-Garcia and S. C. Greer, *Phys. Rev. Lett.* **64**, 1983 (1990).

79. J. Ruiz-Garcia and S. C. Greer, *Phys. Rev. Lett.* **64**, 3204 (1990).

80. A. Prados and S. C. Greer, unpublished.

81. M. Daoud, J. P. Cotton, B. Farnoux, G. Jannink, G. Sarma, H. Benoit, R. Deplessix, C. Picot, and P. G. de Gennes, *Macromolecules* **8**, 804 (1975).

82. J. C. Le Guillou and J. Zinn-Justin, *J. Phys. Lett.* **46**, L-137 (1985).

83. J. des Cloizeaux and G. Jannink, *Polymers in Solution: Their Modelling and Structure*, Clarendon Press, Oxford, 1990.

84. A. B. Pippard, *Elements of Classical Thermodynamics*, Cambridge University Press, Cambridge, 1966.

85. S. S. Das, A. P. Andrews, S. C. Greer, C. Guttman, and W. Blair, unpublished.

86. P. J. Flory, *Principles of Polymer Chemistry*, Cornell University Press, Ithaca, NY, 1953.

87. L. H. Peebles, Jr., *Molecular Weight Distibutions in Polymers*, Wiley, New York, 1971.

88. L. Schäfer, *Phys. Rev. B* **46**, 6061 (1992).

89. D. N. Bhattacharyya, J. Smid, and M. Szwarc, *J. Am. Chem. Soc.* **86**, 5024 (1964).

90. K. S. Schmitz, *An Introduction to Dynamic Light Scattering by Macromolecules*, Academic Press, New York, 1990.

91. M. V. Jaric and K. H. Bennemann, *Phys. Rev. A* **27**, 1228 (1983).

92. M. V. Jaric and K. H. Bennemann, *Phys. Lett.* **95A**, 127 (1983).

93. S. Livne, *Macromolecules* **27**, 5318 (1994).

94. M. F. Jaric and G. F. Tuthill, *Phys. Rev. Lett.* **55**, 2891 (1985).

95. G. F. Tuthill and M. V. Jaric, *Phys. Rev. B* **31**, 2981 (1985).

96. A. Milchev and Y. Rouault, *J. Phys. II France* **5**, 343 (1995).

97. Y. Rouault and A. Milchev, *Phys. Rev. E* **51**, 5905 (1995).

TRANSPORT AND KINETICS IN ELECTROACTIVE POLYMERS

MICHAEL E. G. LYONS

Electroactive Polymer Research Group Physical Chemistry Laboratory University of Dublin, Trinity College Dublin, Ireland

CONTENTS

Advances in Chemical Physics, Volume XCIV, Edited by I. Prigogine and Stuart A. Rice.
ISBN 0-471-14324-3 © 1996 John Wiley & Sons, Inc.

I. INTRODUCTION

A. Electroactive Polymer Materials

Electroactive polymer films have attracted considerable attention in the electrochemical community in recent years, due largely to the wide range of possible applications of these materials in the areas of electrocatalysis, molecular electronics, chemical and biosensor technology, and energy conversion and storage and as media for controlled drug release.

Electroactive polymers may be classified into *three* major types: redox polymers, electronically conducting polymers (plastic metals), and loaded ionomers. For the purposes of the review we shall consider thin films of electroactive polymers deposited onto the surface of a support electrode. The combination of deposited polymer film and support constitutes a *chemically modified electrode*. The latter involves the deliberate immobilization of a chemical microstructure on a host electrode surface to perform a specific task. Hence one is dealing with "tailor-made electrochemistry" using surface-deposited multilayer redox active microstructures.

As implied in the title of this chapter, we shall discuss the mechanism of charge percolation through surface-deposited polymer films. This topic is of central importance, in that the rate of charge percolation through the polymer matrix will generally dictate the operational characteristics of the deposited microstructure when used in a practical application. In this review the fundamentals of charge percolation in thin electroactive polymer layers will be discussed along with a description of a number of electrochemical techniques that can be used to quantify the rate of charge percolation in these materials.

The allied topic of heterogeneous mediated electrocatalysis using surface-deposited electroactive polymer films will also be discussed in some detail in this chapter. Electorcatalysis of electrode reactions at macromolecular layers involves the direct participation of the polymer material. Instead of a direct electron transfer between the Fermi level of the metallic support electrode and the redox active species present in the solution phase,* the electron transfer is mediated by species contained within the surface-immobilized film. Furthermore, the overpotential at which a given substrate reaction occurs at an appreciable rate may be appreciably lowered at a polymer-modified electrode, compared to that obtained at an uncoated electrode. We shall therefore be interested in

---

* This, of course, is the classical mechanism of interfacial electron transfer discussed in monographs on electrode kinetics.

examining the kinetics of mediated electron transfer reactions at chemically modified electrode surfaces.

A number of useful reviews dealing with either charge percolation/conductivity processes of polymer-modified electrodes or mediated electrocatalysis at polymer-modified electrodes have previously been published. We briefly mention some representative literature. The review by Murray [1] provides a good summary of early work in the area. The reviews by Albery and Hillman [2] and Hillman [3] are also useful. More recent reviews have been written by Abruna [4], Evans [5], Smyrl and Lien [6], Kaner [7], Pethrick [8], and on a more general level Lyons [9]. A monograph on the electrochemistry of electroactive polymers has recently been published under the editorship of Murray [10]. In this volume a useful chapter by Majda [11] dealing with the dynamics of electron transport in polymeric assemblies of redox centers can be found. In the same volume a chapter coauthored by Oyama and Ohsaka [12] deals with a survey of experimental methods used to quantitate transport in electroactive polymer films. There is much useful experimental data tabulated in this work. In the same volume Murray [13] produces a comprehensive overview of the electrochemistry of electroactive polymers, whereas the contribution by Andrieux and Saveant [14] considers electrocatalytic applications that will form the subject of Chapter 3 of this book. A recent volume of *Faraday Discussions* [15] dealing with charge transfer in polymeric systems is also of interest and contains a good overview of progress in the area up to 1989. A comprehensive monograph on the electrochemistry of electroactive polymer systems will shortly appear [16]. Lyons [17] has recently summarized current progress in mediated electrocatalysis using electroactive polymer electrodes.

We shall also briefly consider the important developing field of polymer ionics and polymer electrolytes. This class of materials consists of polar macromolecular solids in which one or more of a wide range of salts has been dissolved. A classic example that has been much studied is the combination of poly(ethylene oxide) (PEO) containing LiX salt as solute. The reader is referred to a recent monograph edited by Scrosati [18] and to review articles by Vincent [19], Linford and Scrosati [20], and Owen [21] and to a volume edited by MacCallum and Vincent [22] for further information on this rapidly expanding area of polymer science. The major focus of this chapter, however, is on electroactive polymers used as *electrode materials*. Polymeric *electrolytes*, although important both in a technological and fundamental sense, present different problems to those encountered with surface-deposited electroactive thin films, and so, because of space limitations, we largely restrict discussion to electroactive-polymer-based chemically modified electrodes.

Early work on chemically modified electrodes concentrated on mono-layer derivatized surfaces. In this case reaction was confined to a two-dimensional region near the electrode surface. From a catalytic view-point, this is not a very effective strategy. Homogeneous chemical reactions occur in a three-dimensional zone. Consequently the reaction flux is much greater. However, if one uses a multilayered polymer film, one has a three-dimensional dispersion of active redox sites throughout the material. Consequently a high local concentration of redox active sites is achieved even though the total amount of active material is small ($\sim 10^{-7}$ mol cm^{-2}). For a layer typically 1 μ thick, the redox site concentration can be quite high. The following calculation is instructive. The site concentration c (units: moles per cubic centimeters) is given by $c = \Gamma/L$, where Γ is the surface coverage (units: moles per square centimeters) and L is the layer thickness. Hence for $\Gamma = 10^{-7}$ mol cm^{-2} and $L = 10^{-4}$ cm we obtain $c = 10^{-3}$ mol cm$^{-3} = 1.0 M$. Hence the redox site concentration is quite high. In contrast, the surface coverage for a monolayer derivatized system is a factor of 1000 less, with $\Gamma = 10^{-10}$ mol cm^{-2}. Hillman [3] has noted that in order to pass a current of 1 A cm^{-2} from a typical film containing 10^{-7} mol cm^{-2} of redox sites acting as a catalytic mediator, one requires a turnover rate of 100 s^{-1}. This number implies many turnovers for a useful layer lifetime.

A number of examples of typical electroactive polymer materials are illustrated in Fig. 1. Let us examine some general characteristics of each type of material. Redox polymers are localized state conductors con-taining redox active groups covalently bound to an electrochemically inactive polymeric backbone. In these materials electron transfer occurs via a process of sequential electron self-exchange between neighboring redox groups. This process is termed electron hopping. In contrast with electronically conducting polymers, the polymer backbone is extensively conjugated, which results in considerable charge delocalization. Charge transport (via polarons and bipolarons) along the polymer chain is rapid, and interchain charge transfer is rate limiting. Redox polymers, such as poly(vinyl ferrocene), exhibit the interesting effect that they remain conductive only over a limited range of potential. Maximum conductivity is observed when the concentrations of oxidized and reduced sites in the polymer are equal. This will occur at the standard potential of the redox centers in the polymer. This is a characteristic of redox conduction. In contrast, electronically conducting polymers such as poly(pyrrole) display quasi-metallic conductivity and remain conductive over an extended potential range. The window of conductivity will be governed to a large extent by the chemical nature of the polymer and may therefore be synthetically controlled. Redox polymers are usually preformed and

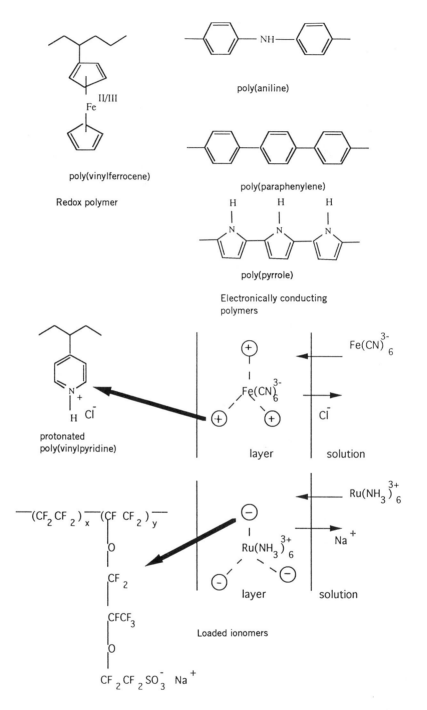

Figure 1. Selection of typical electroactive polymers.

302

subsequently deposited onto the support electrode surface via dip or spin coating. In contrast, electronically conducting polymers are usually generated via in situ electrodeposition. In this case one has electropolymerization of a redox active monomer. However, one can also form redox polymers via electropolymerization, and electronically conductive polymers may be made via chemical synthetic routes. The method of synthesis adopted will depend largely on the material and on the application envisaged. In loaded ionomers one has redox active species electrostatically incorporated in an ionomer (ion exchange polymer) matrix. In this case the redox active component is a counterion to a polyionic (anionic or cationic) polymer material. In this case the conductivity is due either to local electron hopping between fixed redox sites that remain immobile, as in redox polymers, or to physical diffusion of the incorporated redox moeities followed by electron transfer. Loaded ionomers can be prepared by placing an electrode modified with the polymeric ion exchange material in a solution containing the redox active ion. The ion exchange polymer can then extract the ion from the solution and electrostatically incorporate it into the film. Thus, we note that the conductivity mechanism can differ depending on the type of material being examined.

In all cases the process of *redox switching*, that is, the transition from an insulating to a conducting form, is accomplished via an electrochemically induced change in the oxidation state of the layer. This change in oxidation state has an immediate consequence. Electroneutrality within the film must be maintained. Hence the oxidation state change is accompanied by the ingress or egress of charge-compensating counterions. This occurs in a macroscopic sense and may be conveniently measured by various methods such as probe beam deflection spectroscopy (PBDS) or using the electrochemical quartz crystal microbalance (EQCM). These techniques will be described later in the review. The important point to note at this stage is that electroactive polymers are *mixed conductors*. They exhibit both electronic and ionic conductivity. Consequently, in order to fully characterize the material, one has to be able to determine both of these quantities.

B. Outline of Present Review

The present review contains a number of sections. Our theme is *transport* and *kinetics* in electroactive polymers. The former is discussed in the first parts of the review, whereas the latter is examined in the final sections.

In Section II we discuss some basic notions and concepts associated with charge percolation in electroactive polymers. In particular, some simple physical models describing the latter process will be presented and evaluated. We shall concentrate on redox polymers and electronically conducting polymers. In Section III some complicating factors that influence the transport of charge in electroactive polymer materials (such as redox site interaction effects and solvent transport) will be examined. In Section IV the nucleation and growth of electronically conducting polymer films on inert support electrode surfaces is discussed. The topic of electrochemical phase formation is classical, but it has only been recently realized that the concepts developed in this area could be applied to the nucleation and growth of electronically conducting polymer films. In Section V, the important technique of complex impedance spectroscopy is presented, and the equivalent circuit representation of electroactive polymer films is discussed in a physically meaningful way. In particular, the theoretical description of the impedance response observed for redox and electronically conducting polymer materials is presented in some detail. The major focus of this part of the chapter will be on the presentation of a consistent theoretical picture of conduction (both electronic and ionic) in these interesting and important materials.

We then move on in succeeding sections to a discussion of the use of electroactive polymer films in the important area of electrocatalysis. In Section VI, the discussion is focused on the consideration of three-dimensional catalytic microstructures using polymeric films. In this section the well-known Albery–Hillman model is discussed in detail. In Sections VII and VIII, we consider more complex catalytic systems utilizing electroactive polymer films. In Section VII, we consider an example of complex catalytic kinetics (a Michaelis–Menten mechanism), whereas in Section VIII, multicomponent microheterogeneous catalytic systems are reviewed. Our man aim will be in the presentation of simple analytical models that may be used to quantitatively describe the processes of substrate transport and catalytic kinetics in these three-dimensional catalytic systems.

II. CHARGE PERCOLATION IN ELECTROACTIVE POLYMERS: BASIC CONCEPTS

A. Redox Polymers and Loaded Ionomers

1. Introduction

We begin with a discussion of redox conduction. This process is reasonably well understood. A good general discussion of this topic has been provided by Murray and co-workers [13, 23]. It is best to consider a specific example. Consider the metallo-polymer illustrated in Fig. 2a. This material has the general representation $[M(bpy)_2(Pol)_n Cl]Cl$, where M =

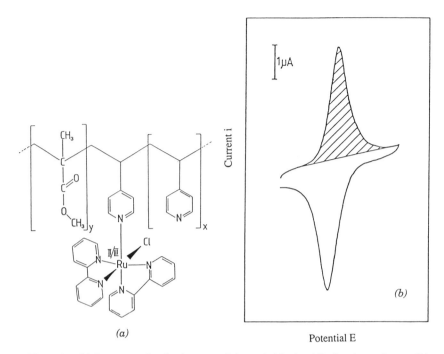

(a)

Potential E

Figure 2. (*a*) Structure of ruthenium containing poly(vinylpyridine) redox polymer. (*b*) Typical cyclic voltammetric response exhibited by redox active metallopolymer illustrated in (*a*). Potential limits are 0.4 and 1.1 V (vs. Ag/AgCl). Supporting electrolyte, 0.1 M HClO$_4$. Surface coverage (obtained via integration of shaded peak area) $\Gamma = 6.8 \times 10^{-8}$ mol cm^{-2}.

Ru or Os and bpy = 2,2′-bipyridyl. Furthermore Pol = poly-4-vi-nylpyrdine (PVP), poly-N-vinylimidazole (PVI), or a copolymer of 4-vinylpyridine and styrene or methyl methacrylate. This material has been extensively investigated by Vos and co-workers [24–26]. The electrochemical properties of the material depend largely on the nature of the polymer backbone, the metal center and the metal–polymer ratio (*n*). This type of redox polymer material is prepared via spin coating the preformed polymer onto a support electrode surface. This approach confers the advantage that the polymer may be well characterized using a variety of spectroscopic techniques [27] before deposition onto the support surface.

The cyclic voltammetric response of $[Ru(bpy)_2(PVP)_5Cl]Cl$ in 0.1 M HClO$_4$ is outlined in Fig. 2*b*.* The potential limits are 400–1100 mV

---

* At this stage we merely introduce cyclic voltammetry as a form of electrochemical spectroscopy.

versus Ag–AgCl. The voltammetric peaks are well defined and correspond to the Ru(II)/Ru(III) redox transformation. Note that E^0 [Ru(II/III)] = 712 mV. Hence redox conduction occurs via electron hopping between neighboring Ru sites. A schematic representation of the electron-hopping process is outlined in Fig. 3. This mechanism was proposed originally by Levich [28] and Ruff [29] some time ago for electron transfer reactions in solution. Kaufman and co-workers [30] adopted this approach and proposed a model for the transport of charge through redox polymer films involving electron self-exchange reactions between oxidized and reduced neighbors within the film.

As outlined in Fig. 3, two processes are involved in the transport of charge across the layer. The first is charge injection at the polymer–support electrode interface. The second is charge transport via electron self-exchange, or "hopping," through the layer. The former process occurs between redox sites located close (~1 nm) to the electrode surface. This charge injection process is potential driven and follows Butler–Volmer kinetics. The latter process of charge transport, is to a first approximation, concentration gradient driven and may be quantified in terms of a quasi-diffusional process. Hence in the most simple terms, the rate of charge transport is quantified in terms of a charge transport diffusion coefficient D_{CT}.

3. Electron Hopping: Quasi-diffusional Model

The transport of electrons through macromolecular structures containing redox active sites occurs via two mechanisms: physical displacement of the redox molecules and electron hopping from one reduced molecule to an adjacent oxidized species (termed donor–acceptor electron self-exchange). A good recent summary of this topic has been provided by Majda [11]. Now for a redox reaction in solution, the process of physical displacement will be most important and the contribution to the observed

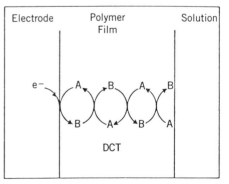

Figure 3. Schematic representation of charge injection/extraction at electrode–polymer interface and quasi-diffusional charge percolation process (electron hopping) in surface-immobilized polymer film.

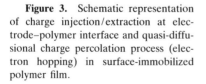

diffusion coefficient due to electron self-exchange processes will be insignificant. However, for a macromolecular film, both may be significant. We now show that the process of electron self-exchange may, to a first approximation, be modeled in terms of a simple diffusion equation.

In order to model the process of charge percolation it is convenient to assume that the redox centers A and B, where A is an electron acceptor (oxidized species) and B is an electron donor (reduced species), are distributed uniformly throughout the layer. Let the concentration of oxidized species be denoted by a and the reduced species by b, respectively. We also assume that, to a first approximation, we can neglect local electric field effects between neighboring sites. One then assumes that the layer can be divided into a series of slices, $1, 2, 3, \ldots, j - 1, j$. Each slice contains a redox site. The overall layer thickness is given by L. The process of charge percolation through the layer may then be represented in terms of the following reaction involving the self-exchange of electrons between oxidized and reduced sites in neighboring layers:

$$
\begin{array}{ccc}
A_{j-1} & A_j & A_{j+1} \\
B_{j-1} & B_j & B_{j+1}
\end{array}
\qquad (2.1)
$$

$$
\leftarrow \delta \rightarrow
$$

where δ represents the distance between neighboring sites. One notes therefore that bimolecular kinetics are used to describe the process of electron exchange between sites in adjacent layers. The dynamics of the electron exchange is quantified in terms of a second-order bimolecular rate constant k (units: $cm^3 mol^{-1} s^{-1}$). In this simple analysis we will neglect any potential difference between sites.

Let us examine a typical exchange reaction involving sites in slice $j - 1$ and j:

$$
B_{j-1} + A_j \xrightarrow{k} A_{j-1} + B_j \qquad (2.2)
$$

The net electron flux (units: $mol\, cm^{-2}\, s^{-1}$) j_Σ is given by

$$
j_\Sigma = k\delta(b_{j-1}a_j - a_{j-1}b) \qquad (2.3)
$$

where a_j and b_j represent the concentrations of the redox sites in slice j. These quantities may now be written as

$$
b_j = b_{j-1} + \delta \frac{db_{j-1}}{dx} \qquad (2.4)
$$

and

$$a_j = a_{j-1} + \delta \frac{da_{j-1}}{dx} \qquad (2.5)$$

One therefore obtains the following expression for the flux:

$$j_\Sigma = k\delta^2 \left(b_{j-1} \frac{da_{j-1}}{dx} - a_{j-1} \frac{db_{j-1}}{dx} \right) \qquad (2.6)$$

The latter expression is valid for any slice in the film, and so in general we write

$$j_\Sigma = k\delta^2 \left(b \frac{da}{dx} - a \frac{db}{dx} \right) \qquad (2.7)$$

The total redox site concentration in the film, c_Σ, is related to the sum of the concentrations of the reduced and oxidized sites, and so we write that $c_\Sigma = a + b$; hence Eq. (2.7) may be written as

$$j_\Sigma = k\delta^2 c_\Sigma \left(\frac{b}{c_\Sigma} \frac{da}{dx} - \frac{a}{c_\Sigma} \frac{db}{dx} \right) \qquad (2.8)$$

We can now define the electron-hopping diffusion coefficient D_E (units: centimeters squared per second) as

$$D_E = k\delta^2 c_\Sigma \qquad (2.9)$$

Hence from Eqs. (2.8) and (2.9) we obtain

$$j_\Sigma = D_E \left(\frac{b}{c_\Sigma} \frac{da}{dx} - \frac{a}{c_\Sigma} \frac{db}{dx} \right) = D_E \left(\frac{a+b}{c_\Sigma} \frac{da}{dx} \right)$$

$$= D_E \frac{da}{dx} = -D_E \frac{db}{dx} \qquad (2.10)$$

where we used the relationships: $db/dx = -da/dx$ and $a + b/c_\Sigma = 1$. The expression presented in Eq. (2.10) is just the steady-state Fick diffusion equation. Hence the redox conductivity may be modeled in terms of a *quasi-diffusional* process.

One now utilizes the continuity equation, which in one dimension has the form

$$\frac{\partial a}{\partial t} = \frac{\partial j_\Sigma}{\partial x} = \frac{\partial}{\partial x} \left(D_E \frac{\partial a}{\partial x} \right) \qquad (2.11)$$

Simplification of Eq. (2.11) results in the time-dependent one-dimensional Fick diffusion equation

$$\frac{\partial a}{\partial t} = D_E \frac{\partial^2 a}{\partial x^2} \tag{2.12}$$

This analysis was originally described by Andrieux and Saveant [31] and Laviron [32].

The derivation just described has concentrated on one-dimensional nearest-neighbor hopping. However, as noted by Murray and co-workers [23], one can also have hopping in two or three dimensions. In the general situation one may write

$$D_E = \theta k \delta^2 c_\Sigma \tag{2.13}$$

where θ represents a geometric factor whose numerical value depends on the dimensionality of the hop. For one-dimensional hopping, $\theta = 1$. For two-dimensional hopping $\theta = \frac{1}{4}$, whereas for three-dimensional hopping, $\theta = \frac{1}{6}$ [17, 33, 34].

It is difficult to measure the distance over which the electron hops, δ. One can of course assume that δ is given by the average intersite distance. If this approach is adopted, then one may determine δ from a knowledge of the total redox site concentration c_Σ. To do this, one must make an assumption about the nature of the redox site packing. The most simple situation assumes a cubic packing (Fig. 4). In such a situation one may write that $N_A c_\Sigma \delta^3 = 1$, and so δ is given by

$$\delta = \left(\frac{1}{N_A c_\Sigma}\right)^{1/3} \tag{2.14}$$

where N_A is the Avogadro constant. If the redox site concentration is quite high, then this approximation is reasonable.

This problem of electron self-exchange and the more general phenomemon of species transfer reactions have been attacked in general terms in two recent papers by Ruff and co-workers [33, 34]. In this analysis the effect of an exchange reaction of the type $AX + A \rightleftharpoons A + AX$ on the transport of species X was discussed for the general three-dimensional situation where the transport process is driven by the gradient of chemical potential $\nabla \mu_{AX}$ of species AX. The special cases of isothermal diffusion and that of electrical conductivity were also discussed. Two approaches were adopted: an essentially macroscopic thermodynamic treatment and a microscopic analysis based on a three-dimensional random walk on a regular lattice. The reader is referred to

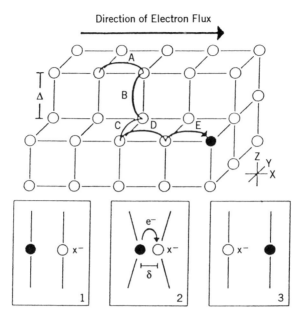

Figure 4. Intersite electron hopping in three-dimensional cubic lattice. Also illustrated is microscale counterion displacement associated with individual electron-hopping event between nearest-neighbor redox sites. Note that Δ represents cubic lattice parameter, whereas δ denotes electron-hopping distance. To a good approximation, we may set $\Delta = \delta$.

the original papers for full details of the analysis. However, in essence, both approaches yielded the θ factor correction previously outlined in Eq. (2.13). Ruff has noted that the θ factor arises quite readily from the random-walk formulation of the theory, whereas if a macroscopic thermodynamic approach is adopted, then the factor arises from the solution of rather complicated integrals involving the gradient of the chemical potential and the temperature (the latter only occurring if nonisothermal diffusion is considered).

The motivation behind the concept of electron self-exchange [30] arose initially from early work conducted by Dahms [35] and Ruff and Friedrich, [36] who examined charge transport processes arising from electron hopping between freely diffusing redox centers. Hence both physical displacement and electron hopping contributed to the overall diffusion coefficient. Furthermore, the electron-hopping contribution to the diffusion coefficient depended in a linear manner on the concentration of redox species. This result has been utilized in the polymer-modified electrode area in recent years, and a common perception now exists that the overall measured charge transfer diffusion coefficient D_{CT}

quantifying the transport of charge throughout an electroactive polymer film consists of two components:

$$D_{CT} = D_P + D_E = D_P + \theta k \delta^2 c_\Sigma \qquad (2.15)$$

where D_P denotes the diffusion coefficient due to the physical displacement of redox sites. Equation (2.15) is termed the Dahms–Ruff equation.* The latter predicts that the charge transport diffusion coefficient should vary in a linear manner with redox site loading provided that electron self-exchange processes are important in determining the overall rate of charge percolation. Now in an aqueous solution $D_P \approx 10^{-5}$–10^{-6} cm^2 s^{-1}, whereas D_E is typically 10^{-12}–10^{-7} cm^2 s^{-1}. Hence, it is clear that physical displacement will predominate in such media. However, if the redox sites are chemically or electrostatically anchored to a polymeric backbone, D_P decreases in magnitude, the situation is less clear-cut, and the D_E contribution will play an increasingly more important role in determining the overall rate of charge percolation.

The expression outlined in Eq. (2.15) is deceptively simple. In many experiments the predictions of the Dahms–Ruff equation have not been observed. The variation of D_{CT} with redox center concentration c_Σ may not be linear. For instance, Sharp et al. [37] have recently examined the process of charge propagation through Nafion films loaded with Os(bpy)$_3^{2+/3+}$ centers using potential step chronocoulometry and complex impedance spectroscopy. For low loading values, D_{CT} was found to increase linearly with increasing redox center concentration c_Σ. However, for concentration values greater than 0.5 M, D_{CT} was found to increase very rapidly with redox center concentration. This observation is not in accord with the simple Dahms–Ruff prediction. In fact, as pointed out recently by Blauch and Saveant [38], the interplay between the D_P and the D_E factors can be subtle and complex. We shall examine this analysis in the next section, where we examine intersite electron hopping and physical displacement in more detail.

3. Intersite Electron Hopping: Percolation Models and Long-Range Electron Transfer Considerations

The process of intersite electron hopping has been discussed in Section II.A.2 in terms of a simple model involving quasi-diffusional process. We now take a more detailed view of the intersite electron transfer reaction in a fixed-site redox polymer. In particular we examine recent work

---

* Note that in the early literature, the Dahms Ruff equation was written $D_{CT} = D_P + \frac{1}{4}\pi k \delta^2 c_\Sigma$. This expression is now known to be incorrect.

reported by Blauch and Saveant [38], Fritsch-Faules and Faulkner, [39, 40] and He and Chen [41].

The Dahms–Ruff expression presented in Eq. (2.15) strictly pertains to the situation of charge transport arising from electron hopping between freely diffusing redox centers. The situation of charge transport within an electroactive polymer film as modeled via the Andrieux–Saveant–Laviron approach [31, 32] is quite different in that charge transport arises from electron–hopping events between redox centers that are irreversibly attached to a supermolecular structure. In fact, we are looking at two extremes of behavior: totally mobile and totally immobile redox sites. It is reasonable to suppose that many electroactive polymer films will exhibit behavior intermediate between these two extreme descriptions. The question is, how may such systems be modeled?

Blauch and Saveant [38] have proposed that the *percolation* aspect of the charge transport problem be addressed. Percolation theory has proved to be extremely useful in understanding transport processes in disordered media. Both bond and site percolation models are usually used to discuss transport in a rigid but disordered material. Blauch and Saveant [38] point out that charge transport between strictly immobile centers is a static percolation problem [42–45]. In such a formalism, hops between sites are either forbidden or allowed with specific fixed probabilities. In short, a random distribution of redox centers produces an ensemble of clusters in which each molecule in a cluster is accessible by hops from molecules occupying adjacent sites. Consequently, an electron can readily travel throughout a specific cluster but can never escape from the confines of that cluster. This means that charge transport will only take place across a region determined by the size of the largest cluster. Hence, below a critical concentration called the *percolation threshold*, all clusters are of a finite microscopic size, and therefore charge transfer across macroscopic distances is forbidden. However, above the percolation threshold, a dominant cluster termed the percolation cluster spans the entire system regardless of the system's dimensions, and so charge transport across macroscopic distances becomes possible.

The situation becomes more complex when the redox centers are able to freely diffuse. Under such conditions, the clusters within the system are constantly changing and reorganizing. In other words, the local polymeric environment changes with time: One has percolative electron hopping within a dynamically disordered structure. One therefore has a *dynamic percolation* situation. In dynamic percolation the time-varying structural changes in the host structure in which the hoppers are located is specifically included in the mathematical analysis. The formalism describing such a situation has been recently developed by a number of workers in recent years, most notably by Bug and co-workers [46, 47] and Nitzan, Ratner, and co-workers [48–53]. The former group were especially interested in ionic transport in water–oil emulsions, whereas the latter examined ionic transport in polymeric electrolytes.

Blauch and Saveant [38] point out that the Dahms–Ruff analysis is particularly appropriate under dynamic percolation conditions. When the rate of physical motion of the redox sites in an electroactive polymer increases relative to that of intersite electron hopping, a stage will ultimately be reached where the reorganization of redox sites will be sufficiently rapid and complete, such that any correlation between the cluster geometry existing between successive electron-hopping events will be eliminated. This scenario corresponds to a *mean-field approximation*. We now follow Blauch and Saveant [38] and derive an expression for the charge transport diffusion coefficient that is valid under mean-field conditions.

We assume that the redox system can be modeled as a simple lattice (either square or cubic depending on whether hopping occurs in two or three dimensions), with δ denoting the distance between adjacent lattice sites. Electron exchange is assumed to occur only between molecules occupying adjacent lattice sites. One can define a fractional site loading X given by $X = N_m/N_\Sigma$, where N_m denotes the total number of redox molecules and N_Σ is the total number of lattice sites. As before, the total concentration of lattice sites c_Σ is given by $c_\Sigma = 1/N_A\delta^\nu$, where ν denotes the dimensionality of the hopping. Electron hopping between adjacent lattice sites is modeled as a Poisson process with a characteristic relaxation time τ_e (this is the average time between hops).

We begin our discussion using the general random-walk formula for D_{CT} [54]:

$$D_{CT} = \frac{1}{2\nu} f \langle \bar{r}.\bar{r} \rangle \qquad (2.16)$$

where f denotes the frequency of the random jumps and $\langle \bar{r}.\bar{r} \rangle$ represents the mean-square displacement of each jump. Now we note that the random-jump frequency is equal to the frequency of electron hops. The rate at which a given electron hops between molecules is simply the product of the frequency of attempted electron hops $(1/\tau_e)$ and the probability that the destination site is occupied by a molecule (X). Hence we can write

$$f = \frac{X}{\tau_e} \qquad (2.17)$$

Note that the rate of electron hopping may also be quantified in terms of the electron-hopping diffusion coefficient D_E as follows:

$$D_E = \frac{\delta^2}{2\nu\tau_e} = \frac{k_{ET}c_\Sigma\delta^2}{2\nu} \tag{2.18}$$

where k_{ET} denotes the bimolecular activation-controlled electron transfer rate constant with $k_{ET} = 1/c_\Sigma\tau_e$.

Furthermore, the displacement vector $\bar{r}$ of an electron between two successive hops may be regarded as the sum of the displacement vector of the electron resulting from the hop and the displacement vector of the host molecule between electron hops. If it is assumed that electron hops and physical hops of the host molecule are uncorrelated, we obtain

$$\langle \bar{r}.\bar{r} \rangle = \langle \bar{r}_e.\bar{r}_e \rangle + \langle \bar{r}_p.\bar{r}_p \rangle = \delta^2 + \langle \bar{r}_p.\bar{r}_p \rangle \tag{2.19}$$

since electron hops always occur over the distance δ. The question now is to evaluate the mean-square displacement of a freely diffusing molecule $\langle \bar{r}_p.\bar{r}_p \rangle$. We first note that the rate of physical displacement is characterized by a relaxation time τ_p that represents the average time between attempted molecular hops. Also, the rate of physical motion of the redox molecules is quantified in terms of a diffusion coefficient D_P given by

$$D_P = \frac{\delta^2}{2\nu\tau_p} \tag{2.20}$$

Furthermore, we can relate τ_p to the conventional bimolecular diffusion-limited rate constant k_D via

$$k_D = 4\nu N_A \delta^{\nu-2} D_P = \frac{2}{\tau_p c_\Sigma} \tag{2.21}$$

According to Blauch and Saveant [38], the mean-squared displacement of a freely diffusing molecule during a time period t is

$$\langle \bar{r}_p.\bar{r}_p \rangle = (1-X)F_c\delta^2\frac{t}{\tau_p} = 2\nu D_p(1-X)F_c t \tag{2.22}$$

Note that a "blocking" factor $1 - X$ has been introduced that accounts for the reduction in the frequency of molecular displacements due to obstruction by other molecules. An attempted molecular displacement will only be successful if the destination site is unoccupied. Hence the effective displacement frequency is $(1 - X)/\tau_p$. Clearly when $X = 1$, no physical motion can occur because all lattice sites are occupied. A further complication arises. The obstruction of molecular motion by other molecules gives rise to correlation effects that are quantified in terms of a correlation factor F_c. The issue of correlation effects in the absence of electron hopping has been addressed in some detail in the literature [55–59]. One important manifestation of the correlation effect is an increased tendency for a molecule to backtrack and return to the site that it has most recently vacated. Hence the overall distance traveled by the molecule will be reduced, and this will mean that the diffusion coefficient will be less than that expected.

We now calculate the distribution of times t between successive electron hops, which is governed by Poisson statistics. Hence the probability that an electron waits a time t between successive hops is $(X/\tau_e)\exp(-Xt/\tau_e)$, and so the weighted mean-square physical displacement between electron-hopping events is

$$
\begin{aligned}
\langle \bar{r}_p \cdot \bar{r}_p \rangle &= \int_0^\infty \frac{X}{\tau_e} \exp\left(-\frac{Xt}{\tau_e}\right) \langle \bar{r}_p \cdot \bar{r}_p \rangle_t \, dt \\
&= \int_0^\infty \frac{X}{\tau_e \tau_p} (1-X) F_c \delta^2 t \exp\left(-\frac{Xt}{\tau_e}\right) dt
\end{aligned}
\tag{2.23}
$$

We now recall the standard identity

$$
\int_0^\infty t^n \exp(-\alpha t) \, dt = n! \alpha^{-(n+1)}
\tag{2.24}
$$

where $\mathrm{Re}(\alpha) > 0$. Using the latter identity, we note that Eq. (2.23) reduces to

$$
\langle \bar{r}_p \cdot \bar{r}_p \rangle = \frac{1-X}{X} F_c \delta^2 \frac{\tau_e}{\tau_p}
\tag{2.25}
$$

Hence from Eq. (2.16) we obtain:

$$
\begin{aligned}
D_{CT} &= \frac{X}{\tau_e} \frac{\langle \bar{r} \cdot \bar{r} \rangle}{2\nu} = \frac{X}{2\nu\tau_e} \left(\delta^2 + \frac{\tau_e}{\tau_p} \frac{1-X}{X} F_c \delta^2 \right) \\
&= \frac{X\delta^2}{2\nu\tau_e} + (1-X) \frac{\delta^2}{2\nu\tau_p} F_c \\
&= D_E X + D_P (1-X) F_c
\end{aligned}
\tag{2.26}
$$

Hence we note that Dahms–Ruff type behavior arises as a natural consequence of a mean-field approximation. We note from Eq. (2.26) that when $X \to 0$, $D_{CT} \to D_P$ and when $X \to 1$, $D_{CT} \to D_E$. Futhermore it has been shown by Nakazato and Kitahara [57] that when $\tau_e/\tau_p \to \infty$ or $D_E \to 0$, the correlation factor F_c is given by

$$F_c = \frac{(2 - X)F_{c,X=1}}{2F_{c,X=1} + (1 - 2F_{c,X=1})X} \tag{2.27}$$

where the correlation factor at full fractional loading $F_{c,X=1}$ has values 0.466942 and 0.653109 for square and simple cubic lattices, respectively.

The results of a simulation study on free diffusion of redox molecules coupled with electron hopping performed by Blauch and Saveant [38] are presented in Fig. 5, where the variation of the apparent charge transport diffusion coefficient D_{CT} with fractional loading X for increasing values of the ratio $\tau_e/\tau_p = D_P/D_E$ is examined. The important point to note from these curves is that as the rate of physical motion of the redox molecules increases relative to that of electron hopping, the critical behavior characteristic of static percolation is lost. The static percolation response ($D_{CT} = 0$ below the percolation threshold followed by the sudden onset of redox conduction at the critical fractional loading with a subsequent linear increase in D_{CT} with increasing loading X) is illustrated by the dashed lines in Fig. 5. The solid lines represent the mean-field response predicted by the theory just presented with $F_c = 1$, whereas the dotted lines represent the mean-field behavior with F_c given by Eq. (2.27). We note that the simulations show that correlation effects may be reasonably neglected when the rate of electron hopping approaches that of physical displacement. Hence, when $\tau_e \approx \tau_p$, we can set $F_c = 1$. The important point to note from this simulation is that mean-field Dahms–Ruff type behavior is only attained when D_P is of an equal magnitude to D_E. Conversely, whenever physical diffusion is slower than electron hopping, percolation effects are observed. The transition from static percolation to mean-field conditions defines the region of dynamic percolation. The bottom line is that for many electroactive polymers where the donor–acceptor sites are rigidly bound to a polymeric backbone (thereby making D_P small), *static percolation behavior rather than classical Dahms–Ruff behavior is applicable*. This is a very significant result.

Blauch and Saveant [38] have suggested that the concept of electron hopping coupled with bounded diffusion of donor–acceptor species is a more appropriate basis on which to model charge transport in electroactive polymer systems. In this approach, each redox molecule is attached to its lattice site via an imaginary spring (a harmonic approximation is

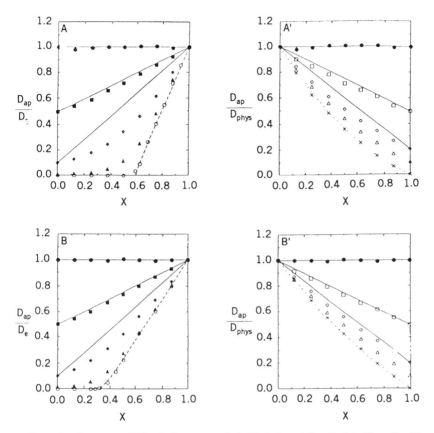

Figure 5. Summary of Blauch–Saveant analysis. Variation of $D_{CT}D_E$ (A, B) or D_{CT}/D_P (A', B') with X for square (A, A') and simple cubic (B, B') lattices for free diffusion. Discrete points are obtained from simulation results for $\tau_e/\tau_p = 0$ ($\bigcirc$), 0.01, ($\blacktriangle$), 0.1 ($\blacklozenge$), 0.5 ($\blacksquare$), 1.0 ($\bullet$), 2.0 ($\square$), 5 ($\diamondsuit$), 10 ($\triangle$), and ∞ ($\times$). Dashed lines represent static percolation behavior. Dotted lines represent mean-field behavior with F_c given by Eq. (2.27). Solid lines represent mean-field behavior with $F_c = 1$ (adapted from ref. 38).

used) with a force constant f_s. A parameter λ is introduced, where

$$\lambda = \sqrt{\frac{2k_B T}{f_s}} \tag{2.28}$$

The results of a simulation of bounded diffusion coupled with electron hopping are outlined in Fig. 6. We note from this figure that the variation of the apparent diffusion coefficient for charge transport depends on two parameters: τ_e/τ_p and λ/δ. The first parameter compares the rates of

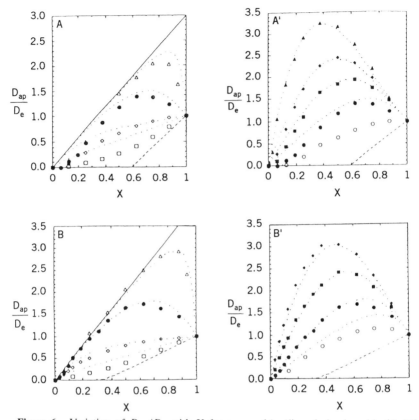

Figure 6. Variation of D_{CT}/D_E with X for square (A, A') and simple cubic (B, B') lattices for bounded diffusion according to Blauch–Saveant analysis. Points presented in A and B are simulated results obtained for $\lambda/\delta = 1.0$ and $\tau_e/\tau_p = 0.1$ ($\square$), 1.0 ($\lozenge$), 10 ($\bullet$), and 100 ($\triangle$). Points in A' and B' are simulation results for $\lambda/\delta = 10$ and $\tau_e/\tau_p = 0.7$ ($\bigcirc$), 1.0 ($\bullet$), 1.5 ($\blacksquare$), 2.0 ($\blacklozenge$), and 3.0 ($\blacktriangle$, A' only). Dashed line represents static percolation behavior. Dotted lines represent mean-field behavior [Eqs. (2.31)–(2.33)]. Solid lines represent limiting mean-field behavior for $\tau_e/\tau_p \to \infty$ [Eqs. (2.32) and (2.34)].

electron hopping and physical displacement, whereas the second compares the range of molecular motion permitted to the attached redox center by the polymeric backbone to the electron-hopping distance. Whenever one of these two parameters is small, percolation effects are observed, with static percolation effects being operative asymptotically as either parameter approaches zero. In contrast, mean-field behavior is predicted when both parameters become large. Under such conditions the rate and range of physical displacement are sufficiently large to cause a complete scrambling of the molecular clusters between electron-hopping events and percolative effects are eliminated.

The mean-field behavior under bounded diffusion conditions may be derived as follows. Again we begin with Eq. (2.16) and apply Eq. (2.19). One then notes

$$\langle \bar{r}_p \cdot \bar{r}_p \rangle_t = \nu \lambda^2 \left\{ 1 - \exp\left[-\frac{(1-X)F_c\delta^2 t}{\nu \lambda^2 \tau_p} \right] \right\} \tag{2.29}$$

where the classical one dimensional treatment of Chandrasekhar [60] has been adapted to the two or three dimensional situation taking blocking and correlation effects into account. Note that the limiting mean square displacement is given by $\langle \bar{r}_p \cdot \bar{r}_p \rangle_\infty = \nu \lambda^2$. We now insert Eq. (2.29) into Eq. (2.23) to obtain:

$$\langle \bar{r}_p \cdot \bar{r}_p \rangle = \int_0^\infty \frac{X}{\tau_e} \exp\left(-\frac{Xt}{\tau_e} \right) \nu \lambda^2 \left\{ 1 - \exp\left[-\frac{(1-X)F_c\delta^2 t}{\nu \lambda^2 \tau_p} \right] \right\} dt$$
$$= \frac{\nu \lambda^2 \rho}{1 + \rho} \tag{2.30}$$

where we note that

$$\rho = \frac{F_c\delta^2}{\nu \lambda^2} \frac{\tau_e}{\tau_p} \frac{1 - X}{X} \tag{2.31}$$

Hence from Eqs. (2.16) and (2.19) we obtain

$$D_{CT} = \frac{X}{2\nu\tau_e} \langle \bar{r} \cdot \bar{r} \rangle$$
$$= \frac{X}{2\nu\tau_e} (\delta^2 + \langle \bar{r}_p \cdot \bar{r}_p \rangle)$$
$$= \frac{X}{2\nu\tau_e} \left(\delta^2 + \frac{\nu \lambda^2 \rho}{1 + \rho} \right)$$
$$= D_E X \tag{2.32}$$

where the electron-hopping diffusion coefficient is now given by

$$D_E = \frac{1}{2\nu\tau_e} \left(\delta^2 + \nu \lambda^2 \frac{\rho}{1 + \rho} \right) \tag{2.33}$$

We see that, in general, D_E will vary with X in a rather complicated way due to the presence of the $\rho/(1 + \rho)$ term in Eq. (2.33). When physical displacements is much faster than electron hopping, $\rho \gg 1$ and Eq. (2.33) reduces to

$$D_E = \frac{\delta^2 + \nu\lambda^2}{2\nu\tau_e} = \frac{k_{ET}c_\Sigma}{2\nu}(\delta^2 + \nu\lambda^2) \tag{2.34}$$

The difference between the latter expression and Eq. (2.18) should be noted. The $\nu\lambda^2$ term represents a correction factor arising from the bounded physical displacement. The important point to note here is that there is a synergy between restricted physical displacement and electron exchange, in that physical motion continually rearranges the distribution of redox centers, thereby allowing redox donor and acceptor species to meet one another, while electron-hopping circumvents the diffusive restrictions imposed by the macromolecular backbone, allowing physical motion to contribute to charge transport. Note also that when $X \to 1$, $\rho \to 0$ and $D_{CT} \approx D_E$. A final point to be noted is that the mean-field expressions derived above will only be valid under conditions where $\tau_e/\tau_p \geq 1$. Also λ must be large enough to permit significant encounters between a molecule and its nearest neighbors. A *minimum* criterion in this respect is $\lambda = \delta$.

The overall self-exchange rate constant k may be expressed in terms of the Noyes equation:

$$\frac{1}{k} = \frac{1}{k_{ET}} + \frac{1}{k_D} \tag{2.35}$$

The main thrust of the Blauch–Saveant [38] analysis is that the inter-dependence between physical displacement and electron hopping in charge propagation in electroactive polymer systems leads to two limiting behaviors: percolation behavior ($k_D \ll k_{ET}$) and mean-field behavior ($k_D \gg k_{ET}$). The former behavior occurs when physical motion either is nonexistent or occurs at a much slower rate than electron transfer. In such a situation the microscopic distribution of redox centers will be fundamentally important with respect to the propagation of charge. The hopping electron has a memory of its previous environment, and adjacent site connectivity will be important. In contrast, mean-field behavior occurs when redox site motion is rapid. The molecular distribution is rearranged between successive electron hops and the electron does not carry a memory of its previous environment. The classical Dahms–Ruff approach used so often in the literature of electroactive polymers corresponds to these conditions and so cannot be used to describe charge transport in electroactive polymer systems where the redox sites are strongly bound to a polymeric backbone or supramolecular structure. Instead, the concept of bounded diffusion must be used. A mean-field analysis may be applied provided the rate of restricted diffusion of the bound redox groups is much greater than the rate of intersite electron hopping. Furthermore, the range of the redox group movement must be great enough to enable neighboring groups to encounter one another in an effective manner. Under such mean-field conditions the Laviron–Andrieux–Saveant [31, 32] expression presented in Eq. (2.13) will still be valid. The bimolecular rate constant k may be identified with the activation-controlled electron transfer rate constant k_{ET}. However, the mean-square distance is not equal to δ^2 but rather to $\delta^2 + \nu\lambda^2$, where λ characterizes the range of physical displacement permitted by the irreversible attachment of the donor–acceptor group to the supramolecular structure. Hence both δ and λ should be evaluated when determining k_{ET} values from experimentally derived D_E values.

In the discussion to date we have assumed that electron exchange only takes place across a distance δ, which represents the center-to-center distance of closest approach for the redox molecules. Numerous studies have shown [61] that long-distance electron transfer is possible. The variation in electron transfer rate constant with distance is given by

$$k_{ET}(r) = A \exp\left(-\frac{r - \delta}{\kappa}\right) \tag{2.36}$$

where the parameter κ characterizes the rate of drop-off in k_{ET} with distance r and the preexponential factor $A = k_{ET}(r = \delta)$. However, under mean-field conditions, it is probable that long-range electron transport will not be very important because $\kappa \ll (\delta^2 + \nu\lambda^2)^{1/2}$. If we wish to take a long-range electron transfer into account we can, to a first approximation, replace k_{ET} by $k_{ET}(1 - \kappa/\delta)^2$ and δ by $(\delta + \kappa)^2$ in our previous formulation. However, when $\lambda \ll \delta$, that is, when percolation effects become important, long-range electron transfer will play an important role by providing a mechanism by which an electron can escape from a cluster without significant rearrangement of the system. We now consider this scenario in greater detail.

The approach adopted here is due to Fritsch-Faules and Faulkner [39, 40]. These workers developed a microscopic model to describe the electron-hopping diffusion coefficient D_E in a rigid three-dimensional polymer network as a function of the redox site concentration c_Σ. The model takes excluded volume effects into consideration and is based upon a consideration of probability distributions and random-walk concepts. The microscopic approach was adopted by these workers in order to obtain parameters that could be readily understood in the context of the molecular architecture of the polymer. A previously published related approach was given by Feldberg [62].

In the Fritsch-Faules–Faulkner model [39], it is assumed that the redox centers are immobilized. This of course rules out electron transfer via physical diffusion of the center. The model is based on the notion of extended electron transfer. It is also assumed that the concentration, and therefore the spatial distribution of redox sites, will affect the diffusion coefficient D_E. A notable feature of the analysis is the explicit consideration of the finite volume of the redox center, which, for simplicity, is assumed to be a rigid sphere. Excluded volume effects will be important when the redox site concentration is high (recall that a typical value is in the range 0.1 to 1 M).

Since it is well known that in a microscopic sense diffusion may be modelled in terms of a random walk [54], in three dimensions, the electron-hopping diffusion coefficient may be expressed as

$$D_E = \tfrac{1}{6} f \langle l^2 \rangle \qquad (2.37)$$

where f represents the number of displacements per unit time, and so $f = 1/\tau$, where τ is the average residence time of an electron on any particular site. Hence one notes that τ is simply the inverse of the first-order rate constant k_{ET} for the extended electron transfer. It should be noted here that the latter rate constant is defined differently than that used in the simple macroscopic diffusional model. Furthermore, the quantity $\langle l^2 \rangle$ is the mean-square displacement distance.*

All redox centers are assumed to be identical, immobile, noninteractive hard spheres distributed randomly throughout a rigid three-dimensional homogeneous network. An individual electron-hopping event is pictured as follows. The central site is in the reduced form and may donate its electron only by extended electron transfer to one of a number of neighboring oxidized sites. In contrast, for hole hopping, an oxidized site may donate its hole to any one of several neighboring reduced sites.

One now assumes that each hop occurs over a distance equal to the average nearest-neighbor separation $\langle r_{nn} \rangle$. Hence one writes that $\langle l^2 \rangle =$

---

* We use $\langle l^2 \rangle$ here to denote the mean-square displacement distance rather than the vectorial notation used previously. This is in accordance with the original terminology used by the authors.

$\langle r_{nn}^2 \rangle$, and so Eq. (2.19) becomes

$$D_E = \tfrac{1}{6} k_{et} \langle r_{nn}^2 \rangle \tag{2.38}$$

Since the electron transfer is of an extended nature, the electron transfer rate constant will vary with distance. Miller and co-workers [61] and others [63] have established that k_{et} varies with distance in the following manner:

$$k_{et} = A \exp\left(\frac{\langle r_{nn} \rangle - r_0}{\kappa}\right) \tag{2.39}$$

where r_0 is the contact radius (Fig. 7). The quantity A is a preexponential factor and is related to the intrinsic kinetic facility for a particular system. The quantity κ represents a characteristic distance describing the spatial extent of the electronic coupling in the medium. Hence the overlap between wave functions will be the rate-limiting factor for intersite electron transfer. This topic has been discussed in a recent paper by Lewis [64].

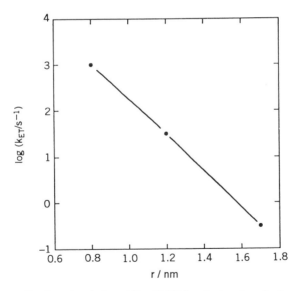

Figure 7. Verification of variation of Eq. (2.39) for electron transfer in metalloprotein systems (ref. 61). Individual data points correspond to different redox metalloproteins. Data analysis yields $A = 3 \times 10^6 \, \text{s}^{-1}$ and $\kappa = 0.105 \, \text{nm}$.

From Eqs. (2.38) and (2.39) we obtain

$$D_E = \tfrac{1}{6}\langle r_{nn}^2\rangle \exp\left(\frac{\langle r_{nn}\rangle - r_0}{\kappa}\right) \tag{2.40}$$

One now must determine an expression for $\langle r_{nn}\rangle$ using a statistical argument. The full details of the derivation have been provided in the original paper [39], and so only the final result will be outlined here. It has been shown [39] that

$$\langle r_{nn}\rangle = \int_0^\infty rf(r)\,dr \tag{2.41}$$

For a hard-sphere model the distribution function $f(r)$ is given by

$$f_{HS}(r) = 4\pi S(r)c_\Sigma r^2 \exp[-(\tfrac{4}{3})\pi c_\Sigma(r^3 - r_0^3)] \tag{2.42}$$

One now replaces $f_{HS}(r)$ into Eq. (2.41) and replaces the exponential term by a Taylor series expansion; evaluating the integrals yields the following expression for the nearest-neighbour separation $\langle r_{nn}\rangle$:

$$\langle r_{nn}^{HS}\rangle = \left(\frac{3}{4\pi c_\Sigma}\right)^{1/3} \exp\left[\gamma\right]\left[-\Gamma(\tfrac{4}{3}) - \sum_{n=0}^{\infty} \frac{(-1)^n \gamma^{(n+4/3)}}{n!(n+\tfrac{4}{3})}\right] \tag{2.43}$$

where $\Gamma(\tfrac{4}{3})$ is the gamma function and has the numerical value 0.893. The parameter γ is given by

$$\gamma = (\tfrac{4}{3})\pi r_0^3 c_\Sigma \tag{2.44}$$

This parameter is dimensionless and expresses the number of redox centers that, on average, occupy any volume equal to the excluded volume. The variation of $\langle r_{nn}^{HS}\rangle$ with concentration is illustrated in Fig. 8. Note that Eq (2.43) may be simplified to consider the case of point molecules by setting $r_0 = 0$ as follows:

$$\langle r_{nn}^{PT}\rangle = \left(\frac{3}{4\pi c_\Sigma}\right)^{1/2}\Gamma(\tfrac{4}{3}) = 0.554c^{-1/3} \tag{2.45}$$

This is a well-known result [60]. It should be noted that the model will no longer be totally valid in concentration regions approaching closest packing. In this circumstance the system becomes more ordered and volume exclusion by the second- and higher order nearest neighbors can no longer be ignored. Note also from Fig. 8 that at very low concentrations the contract radius r_0 becomes insignificant in comparison

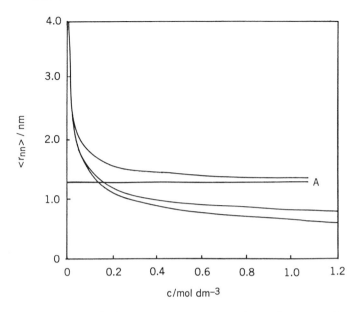

Figure 8. Variation of $\langle r_{nn} \rangle$ with redox site concentration c. From top to bottom r_0 is 1.3, 0.6, and 0 nm. Curves calculated using Eq. (2.43) (hard-sphere approximation) and Eq. (2.45) (point molecule approximation). Point A indicates $\langle r_{nn} \rangle$ and c_Σ for closest packing conditions.

with the nearest-neighbor distance $\langle r_{nn} \rangle$. Consequently the curves for hard spheres and point molecules merge at low concentrations.

The dependence of D_E on concentration c_Σ may now be determined from Eq. (2.40). Fritsch-Faules and Faulkner [39] have defined three types of diffusion coefficients as follows:

$$D_E^{HS} = (\tfrac{1}{6}A)(\langle r_{nn}^{hs} \rangle)^2 \exp\left(- \frac{\langle r_{nn}^{hs} \rangle - r_0}{\kappa} \right) \qquad (2.46)$$

$$D_E^{PT} = (\tfrac{1}{6}A)(\langle r_{nn}^{pt} \rangle)^2 \exp\left(- \frac{\langle r_{nn}^{pt} \rangle}{\kappa} \right) \qquad (2.47)$$

$$D_E^{HYB} = (\tfrac{1}{6}A)(\langle r_{nn}^{pt} \rangle)^2 \exp\left(- \frac{\langle r_{nn}^{pt} \rangle - r_0}{\kappa} \right) \qquad (2.48)$$

The manner in which the simpler expressions D_E^{PT} and D_E^{HYB} compare to the more rigorous expression D_E^{HS} is illustrated in Fig. 9. If the molecular diameter is very small, then $r_0 \to 0$ and both D_E^{PT} and D_E^{HYB} approach D_E^{HS}. A number of further observations may be made from analysis of

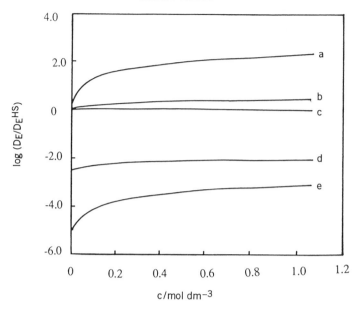

Figure 9. Comparison of normalized D_E^{HS}, D_E^{PT}, and D_E^{HYB} as functions of concentration: (a) log (D_E^{HYB}/D_E^{HS}), $r_0 = 1.3$ nm; (b) $\log(D_E^{HYB}/D_E^{HS})$, $r_0 = 0.6$ nm; (c) log (D_E^{HS}/D_E^{HS}); (d) $\log(D_E^{PT}/D_E^{HS})$, $r_0 = 0.6$ nm; (e) $\log(D_E^{PT}/D_E^{HS})$, $r_0 = 1.3$ nm.

Fig. 9. For any value of the diameter, one notes that $\langle r_{nn}^{pt} \rangle \rightarrow \langle r_{nn}^{hs} \rangle$ at infinite dilution. However, the ratio D_E^{HS}/D_E^{PT} does not approach unity at infinite dilution. Furthermore, as the concentration increases, the ratio $\langle r_{nn}^{pt} \rangle / \langle r_{nn}^{hs} \rangle$ decreases and is always less than unity. The ratio D_E^{PT}/D_E^{HS} is less than unity and increases with increasing concentration. Hence over the entire range of concentration D_E^{PT} never approximates D_E^{HS}. This has the implication that excluded volume effects can never be neglected.

Fritsch-Faules and Faulkner [39] have examined data published [65] on electron transfer processes in metalloprotein systems to test the predictions of their analysis. The original data were obtained from laser flash photolysis and pulse radiolysis measurements. Typical values of $A = 3.06 \times 10^6 \, \text{s}^{-1}$ and $\kappa = 0.105$ nm were used [65]. The value of r_0 chosen was 1.3 or 0.6 nm [the former corresponds to the approximate diameter of the Ru(bpy)$_3^{2+/3+}$ species, which may be electrostatically incorporated in Nafion or polystyrene sulfonate films and the latter to a ferrocene unit in polyvinylferrocene]. These are reasonable values for this quantity. If these values are substituted into Eq. (2.46), then the result illustrated in Fig. 10 is obtained for the variation of D_E^{HS} with concentration over the range 0–1.07 M. The magnitude of D_E and the curve shape depend on

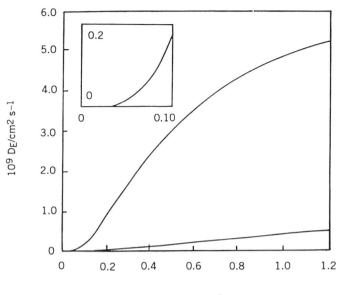

Figure 10. Variation of hard-sphere electron-hopping diffusion coefficient D_E with redox site concentration. Curves were calculated from Eq. (2.46) in text with A and κ values used in analysis the same as those obtained from Figure 7. For upper curve and inset $r_0 = 1.3$ nm, whereas for lower curve $r_0 = 0.6$ nm.

the magnitude of r_0. For $r_0 = 1.3$ nm, D_E^{HS} varies from 1.9×10^{-10} cm^2 s^{-1} at $0.1\,M$ to 5.0×10^9 cm^2 s^{-1} at $1.07\,M$. For the lower value of r_0 (0.6 nm) the magnitudes are considerably lower; in this case D_E^{HS} varies from 2.8×10^{-12} cm^2 s^{-1} at $0.1\,M$ to 4.6×10^{-10} cm^2 s^{-1} at $1.07\,M$. A maximum value of $D_E^{HS} = 1.5 \times 10^{-9}$ cm^2 s^{-1} is obtained at a concentration of $10.9\,M$, which corresponds to the situation of closest packing. These values are in good agreement with D_E measurements obtained for chemically modified electrodes [66, 67]. Note that at low concentrations ($<0.1\,M$) the curve profiles exhibit approximately exponential behavior. Furthermore, at any particular concentration, larger redox sites have higher D_E values. This may be readily rationalized as follows. One recalls that the expression for the diffusion coefficient includes an exponential term in separation. This is the dominant term. The larger the redox center, the more volume it occupies and the smaller the interstitial volume that results. This implies that the edge-to-edge separation will be smaller, and thus one will have a larger electron transfer rate.

In summary, the Fritsch-Faules–Faulkner model [39] predicts that the electron-hopping diffusion coefficient should initially rise exponentially with redox site loading and then "roll over" when the redox site concentration is large. The exponential rise occurs because the increasing site concentration provides smaller nearest-neighbor distances, which promote intersite electron transfer with an exponentially increasing probability. The rollover reflects the finite size of the redox centers. At sufficiently high concentration of the latter, each redox center has a nearest neighbor practically in contact. Further increase in loading will then not appreciably change the mean nearest-neighbour separation, and consequently one will have little further rise in the D_E value.

The range of possible D_E values is very large. Values may range from $\sim 10^{-11}$ to 10^{-2} cm^2 s^{-1}. These workers note [39] that if the intersite coupling is allowed to become very strong (the latter being achieved by chemical linkages), then the polymer network must cease to be considered as a set of weakly interacting electron traps. One then has an electronically conducting polymer, and a very high electronic conductivity will be observed.

The predictions obtained using this model differ significantly from the more traditional results based on bimolecular kinetics discussed previously. In the latter approach one notes from Eq. (2.13) that D_E should vary in a *linear* manner with c_Σ. Indeed, this is the basis of the so-called Dahms–Ruff expression.

Sharp et al. [37] have recently examined the process of charge propagation through Nafion films loaded with Os(bpy)$_3^{2+/3+}$ centers using potential step chronocoulometry and complex impedance spectroscopy. For low values of mediator concentration, D_{CT} was found to vary linearly with c_Σ. For concentration values greater than $0.5\,M$, D_{CT} increased very rapidly with concentration. However, the rate of increase was not as marked as that predicted by the Fritsch-Faules–Faulkner [39] model. The magnitudes of the diffusion coefficients obtained for the loaded ionomer were similar to those generally predicted using the Fritsch-Faules–Faulkner model [39].

Alternative approaches to quantify the relationships between the diffusion coefficient and redox site concentration have been proposed. For instance, the work of He and Chen [41] is of interest. These workers based their analysis on Nafion films loaded with either Ru(bpy)$_3^{2+/3+}$ or Os(bpy)$_3^{2+/3+}$ redox centers. The second-order bimolecular rate constant for electron exchange is rather high for these redox couples, typically $10^9\,M^{-1}\,s^{-1}$ [66]. In order that the electron may hop, the redox sites may have to physically diffuse toward each other. Typically, the physical diffusion coefficient for the M(bpy)$_3^{2+/3+}$ species is $\sim 10^{-11}$ cm^2 s^{-1} [68]. Hence if the diffusion process is slower than the self-exchange reaction, then the observed bimolecular rate constant k is given by [see Eq. (2.35)]

$$k = \frac{k_{ET}k_D}{k_{ET} + k_D} \tag{2.49}$$

where k_{ET} and k_D denote the exchange and diffusional rate constants, respectively. The latter may be estimated via the Smoluchowski equation [69]:

$$k_D = \frac{4\pi N_A R D}{10^3} \tag{2.50}$$

where R is the collision radius, which has a value of $\sim 1.36 \times 10^{-7}$ cm for the $Ru(bpy)_3^{2+/3+}$ center. Hence, typically, $k_D = 10^4\, M^{-1}\, s^{-1}$. Hence, at least for the specific system considered by He and Chen [41], the observed net bimolecular rate constant would be governed by local physical diffusion, rather than by simple exchange.

He and Chen [41] now consider the situation as the redox center concentration increases. The sites crowd in on each other and there may well be a number of redox sites in close proximity to each other. Hence one can visualize a hemisphere of radius equal to the collision radius centered about the primary acceptor site. The volume contained within this hemisphere may contain a number of further redox sites. Hence the electron may be readily shuttled along these adjacent sites due to the fact that k_{ET} is very large. Hence for each diffusional encounter between sites one may have several electron hops. The Poisson distribution is now invoked. The probability $P(n)$ of finding a volume V with n adjacent sites is

$$P(n) = \frac{\xi^n}{n!} \exp(-\xi) \tag{2.51}$$

where ξ represents the mean number of redox centers contained in the volume V. Note that ξ is given by

$$\xi = \frac{c_\Sigma V N_A}{10^3} \tag{2.52}$$

The probability that no other sites are nearby is

$$P(0) = \exp(-\xi) \tag{2.53}$$

In their model He and Chen [41] propose an enhancement factor F

given by the expression

$$F = P(0) + \{[1 - P(0)] - [1 - P(0)]^2\}(1 + \tfrac{1}{2}) + \{[1 - P(0)]^2 - [1 - P(0)]^3\}$$

$$\times [1 + (\tfrac{1}{2}) \cdot 2] + \cdots + \{[1 - P(0)]^m - [1 - P(0)]^{m+1}\}[1 + (\tfrac{1}{2})m] + \cdots$$

$$(2.54)$$

This enhancement factor F contains the quantity $P(0)$, which from Eq. (2.53) depends on the concentration c_Σ. On the right-hand side (rhs) of Eq. (2.54) one has a series of terms reflecting the multihop electron transfer arising from each diffusion-controlled local encounter between neighboring donor and acceptor sites. Note that the quantity $[1 - P(0)]^m$ represents the fraction of molecules that contribute to charge transfer at state m. At low concentrations the F factor will be very close to unity. It is only at higher concentrations that the correction factor will become appreciable. The electron diffusion coefficient may therefore be written as*

$$D_E = \theta k(\delta F)^2 c_\Sigma \qquad (2.55)$$

A plot of Eq. (2.55) is illustrated in Fig. 11a. It is interesting to note that this model predicts a different response from that obtained experimentally [37]. In Fig. 11b the experimental results obtained by Sharp and co-workers [37] via complex impedance spectroscopy and chronocoulometry for loaded Nafion films are superimposed on the theoretical D_E/c_Σ profile arising from the He–Chen model [41]. This result leads to the conclusion that the He–Chen model [41] is incomplete as presently formulated.

We have noted that the observed charge percolation diffusion coefficient D_{CT} consists of contributions arising from electron exchange, D_E and that from physical diffusion, D_p. One or the other may be rate determining, especially for ionomers loaded with redox active complex ions. It is possible to propose [68] a rather simple criterion that will enable one to distinguish between these two possible rate-determining processes. If the observed diffusion coefficient for the oxidized and reduced components of the redox couple are equal, then the physical process that is rate determining is almost certainly electron hopping.

---

* In the He–Chen paper the incorrect form of the Dahm–Ruff equation was quoted. In our presentation here we have therefore modified the analysis and also introduced the geometric factor θ, which takes the dimensionality of the hopping process into account [refer to Eq. (2.13)].

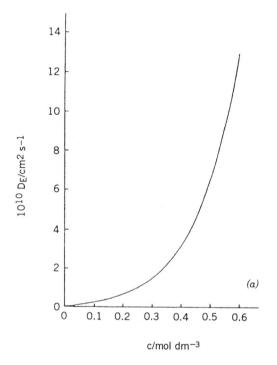

Figure 11. (*a*) Variation of D_E with redox site concentration according to He–Chen model [Eq. (2.55)]. (*b*) Comparison between He–Chen prediction and experimental data for electron-hopping diffusion coefficients for $Os(bpy)_3^{2+}$ loaded in Nafion films obtain via complex impedance spectroscopy by Sharp and co-workers. [37].

However, the observation of larger diffusion coefficients for the less highly charged component of the redox couple signals that molecular diffusion dominates. A good discussion of recent experimental data on D_{phys} and D_E has been provided by Majda [11].

The rate constant k for the electron exchange process may be modeled in terms of Marcus theory [70–73]. In this approach the electron transfer is modeled in terms of precursor and successor complexes. This can be done by writing

$$k = K_P k_{ET} \qquad (2.56)$$

where K_P is the precursor equilibrium constant and k_{ET} is the unimolecular rate constant for electron transfer. This modification of the original Marcus theory has been developed in recent years by Weaver and co-workers [74] and Sutin et al. [75, 76]. The electron transfer rate

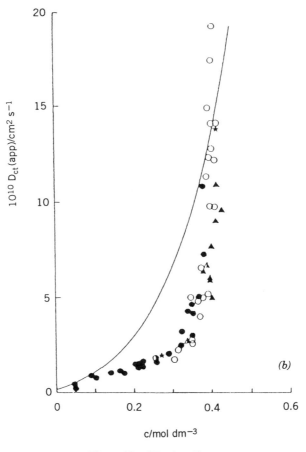

Figure 11. (*Continued*).

constant is given by

$$k_{ET} = \kappa_E \nu_n \Gamma_n \exp\left(-\frac{\Delta G^*}{RT}\right) \quad (2.57)$$

where ΔG^* denotes the activation free energy and κ_E and ν_n (unit: reciprocal second) denote the electronic transmission coefficient and nuclear frequency factor, respectively. The quantity Γ_n denotes the nuclear tunneling factor. The latter quantity typically approaches unity at room temperature, although it may differ significantly from this value at low temperature. The activation energy for the elementary electron transfer process is given by the Marcus relation:

$$\Delta G^* = \frac{(\Delta G_\lambda + \Delta G_E^0)^2}{4\,\Delta G_\lambda} \qquad (2.58)$$

where ΔG_λ denotes the reorganization energy and ΔG_E^0 denotes the Gibbs energy of electron transfer. The latter is an *internal quantity* [77] representing the thermodynamic driving force for the precursor–successor reaction. As outlined by Albery [77], this quantity may be readily related to the overall *external* thermodynamic free-energy change ΔG_{TD}^0 via the expression

$$\Delta G_E^0 = \Delta G_{TD}^0 + W_S - W_P \qquad (2.59)$$

where W_P and W_S represent the work done to bring the reactants from the bulk to form the precursor and the work required to form a successor complex from the products.

The preequilibrium constant can be expressed as a product of a term describing the work expended to overcome electrostatic repulsion in the formation of the preequilibrium precursor complex and the statistical probability of forming such a complex. Thus, if we assume that the reactants admit a spherical geometry, then it has been shown that [74–76]

$$K_P = \frac{4\pi}{1000}\,N_A R^2\,\Delta R\,\exp\!\left(-\frac{W_P}{RT}\right) \qquad (2.60)$$

In this expression N_A is the Avogadro constant, R is the average separation between the reacting centers upon electron transfer, and ΔR is the effective reaction zone thickness. This latter factor has been discussed extensively in the papers written by Weaver and co-workers [74]. It is immediately obvious that R is the same as the intersite jump distance δ discussed previously. The reaction layer thickness ΔR expresses a range of nuclear separations within which the electron transfer reaction proceeds with a rate constant k_{ET}. Majda [11] has noted that the magnitude of ΔR is strongly linked to the electronic transmission factor κ_E. Its value is larger for strongly adiabatic reactions where $\kappa_E = 1$ than for nonadiabatic processes where $\kappa_E \ll 1$. Weaver et al. [74] have noted that a reasonable choice for ΔR is 1×10^{-8} cm. In general, if the electron transfer is nonadiabatic, then κ_E will vary with reactant separation r according to

$$\kappa_E(r) = \kappa_0\,\exp[-\alpha(r - \sigma)] \qquad (2.61)$$

where κ_0 is the value of $\kappa_E(r)$ at the plane of closest approach and σ is the separation distance at this point. The coefficient α is approximately $(1.8-1.4) \times 10^8 \, \text{cm}^{-1}$ for reactions between metal ions. If typically we take the mean value of $1.6 \times 10^8 \, \text{cm}^{-1}$ for α, then from Eq. (2.61) we see that $\kappa_E(r)$ drops to only 20% of κ_0 when $r - \sigma = \Delta R = 1 \times 10^{-8} \, \text{cm}$. Hence we expect intersite electron hopping to be strongly nonadiabatic. The nuclear frequency factor ν_n is $\sim 1 \times 10^{13} \, \text{s}^{-1}$. This value is obtained from the expression

$$\nu_n^2 = \frac{\nu_{os}^2 \, \Delta G_{os}^* + \nu_{is}^2 \, \Delta G_{is}^*}{\Delta G_{os}^* + \Delta G_{is}^*} \tag{2.62}$$

where ν_{os} and ΔG_{os}^* are the characteristic frequency and free energy of activation associated with outer shell solvent reorganization and ν_{is} and ΔG_{is}^* denote the characteristic frequency and activation energy for inner sphere bond vibrations. For an aqueous medium $\nu_{os} = 10^{11} \, \text{s}^{-1}$ and $\nu_{is} = 10^{13} \, \text{s}^{-1}$ for a typical ligand–metal stretching frequency. Typically, $\Delta G_{is}^* > 0.25 \, \Delta G_{os}^*$. Hence we note from Eq. (2.62) that $\nu_n \approx 10^{13} \, \text{s}^{-1}$. The nuclear frequency factor is an important quantity. In simple terms it describes the dynamics of all nuclear motions involved in the formation of the transition state. As we see from Eq. (2.62), in cases where the inner shell contribution arising from bond distortion of the reactants is small compared with the other shell solvent reorganization component, ν_n describes the solvent dynamics. Hence one could expect that electron transfer involving redox centers imbedded in a polymer matrix may proceed with a rate constant substantially different from that expected for a redox process in a well-characterized solvent medium. Typical nuclear frequency factors have not been estimated to any great extent for electron transfer processes in polymeric matrices. However, but any numerical results obtained will have limited predictive power.

4. Local Field Corrections: Diffusion–Migration Effects in Electron Hopping

To this point we have considered electron-hopping processes in isolation to other charge transport phenomena occurring in electroactive polymer films. This, of course, is not valid. One must also consider the transport of charge-compensating counterions within the polymer matrix. These species move to ensure that electroneutrality within the film is maintained. Electroactive polymers are therefore *mixed conductors*: They exhibit both electronic and ionic conductivity. Since the propagation of electrons is accompanied by a movement of electroinactive counterions so as to maintain electroneutrality, it is possible that overall control of charge percolation through the material could be governed by the transport of counterions. Hence electron hopping and counterion displacement are *coupled* in some manner. The quantitative analysis of this effect is difficult and has been tackled by Saveant in a number of papers [78–83]. The problem has also been addressed by Buck [84–87], Baldy and co-workers [88], and Albery et al. [89]

A fundamental feature of all of these analyses is that electron hopping is not only driven by a concentration gradient, it is also field assisted. One therefore has a diffusional term and a migrational term. Hence the potential difference between sites is taken into account. Thus the electron-hopping flux j_Σ must contain a migrational contribution. In the following, we adopt the analysis presented by Albery and co-workers [89].

We again assume that the polymer consists of neutral sites A that can be oxidized to form positive charge carriers B^+. At the same time a charge-compensating counterion X^- enters the polymer layer from the solution as follows:

$$—A—A—A—+X^-(aq) \longrightarrow —A—B^+—A—+e^-$$
$$X^-$$

We consider three adjacent sites in the polymer located at $x - \delta$, x, and $x + \delta$. As before bimolecular electron exchange occurs between these sites and is quantified by a second-order rate constant k. The transition states are located at $x - (1 - \alpha)\delta$ and $x + \alpha\delta$, where α is a symmetry factor. Hence one adopts the following scheme:

$$
\begin{array}{cccccc}
A & \ddagger & A & \ddagger & A \\
B^+ & & B^+ & & B^+ \\
x - \delta & x - \delta + \alpha\delta & x & x + \alpha\delta & x + \delta
\end{array}
$$

We now can write the expression

$$\frac{db_x}{dt} = -\vec{k}_x a_{x-\delta} b_x - \vec{k}_x a_{x+\delta} b_x + \vec{k}_{x-\delta} a_x b_{x-\delta} + \vec{k}_{x+\delta} a_x b_{x+\delta} \quad (2.63)$$

The rate constants depend on the potential as follows:

$$k_x = k \exp[\theta_x - \theta_\ddagger] \approx k f_x \quad (2.64)$$

where we have written* $\theta = FE/RT$ and where the kinetic factor f_X is given by

$$f_X = 1 + \theta_X - \theta_{\ddagger} \tag{2.65}$$

Note that in Eq. (2.64) we have expanded the exponential because δ is small. We can now expand the concentrations, the potential terms, and the kinetic factors in terms of a Taylor expansion as follows:

$$a_{x \pm \delta} = a_x \pm \delta \frac{\partial a}{\partial x} + \frac{1}{2} \delta^2 \frac{\partial^2 a}{\partial x^2} \tag{2.66}$$

$$b_{x \pm \delta} = b_x \pm \delta \frac{\partial b}{\partial x} + \frac{1}{2} \delta^2 \frac{\partial^2 b}{\partial x^2} \tag{2.67}$$

$$\theta_{x \pm \delta} = \theta_x \pm \delta \frac{\partial \theta}{\partial x} + \frac{1}{2} \delta^2 \frac{\partial^2 \theta}{\partial x^2} \tag{2.68}$$

$$\theta_{x - \delta + \alpha \delta} = \theta_x - (1 - \alpha)\delta \frac{\partial \theta}{\partial x} + \frac{1}{2}(1 - \alpha)^2 \delta^2 \frac{\partial^2 \theta}{\partial x^2} \tag{2.69}$$

$$\theta_{x + \alpha \delta} = \theta_x + \alpha\delta \frac{\partial \theta}{\partial x} + \frac{1}{2}\alpha^2\delta^2 \frac{\partial^2 \theta}{\partial x^2} \tag{2.70}$$

$$\vec{f}_{x - \delta} = 1 - \alpha\delta \frac{\partial \theta}{\partial x} + \alpha\left(1 - \frac{1}{2}\alpha\right)\delta^2 \frac{\partial^2 \theta}{\partial x^2} \tag{2.71}$$

$$\tilde{f}_x = 1 + (1 - \alpha)\delta \frac{\partial \theta}{\partial x} - \frac{1}{2}(1 - \alpha)^2\delta^2 \frac{\partial^2 \theta}{\partial x^2} \tag{2.72}$$

$$\vec{f}_x = 1 - \alpha\delta \frac{\partial \theta}{\partial x} - \frac{1}{2}\alpha^2\delta^2 \frac{\partial^2 \theta}{\partial x^2} \tag{2.73}$$

$$\tilde{f}_{x + \delta} = 1 + (1 - \alpha)\delta \frac{\partial \theta}{\partial x} + \frac{1}{2}(1 - \alpha)^2\delta^2 \frac{\partial^2 \theta}{\partial x^2} \tag{2.74}$$

If we substitute these results into Eq. (2.63) and simplify, and if we

---

* In this section we use the symbol θ to denote a normalized potential. We assume a simple one-dimensional hopping problem and so do not have to introduce a numerical factor in the expression for D_E as was done in Eq. (2.13).

ignore terms in δ^3 and δ^4, which will be small, then one obtains

$$\frac{\partial b_x}{\partial t} = D_E\left(\frac{\partial^2 b_x}{\partial x^2} + \left(\frac{a_x - b_x}{c_\Sigma}\right)\frac{\partial b_x}{\partial x}\right)\frac{\partial \theta}{\partial x} + \frac{a_x b_x}{c_\Sigma}\frac{\partial^2 \theta}{\partial x^2} \qquad (2.75)$$

where, as before,

$$D_E = kc_\Sigma \delta^2 \qquad (2.76)$$

and where one also notes that

$$c_\Sigma = a_x + b_x \qquad (2.77)$$

This implies that the total site concentration is constant. This fact allows us to write

$$\frac{\partial a_x}{\partial x} = -\frac{\partial b_x}{\partial x} \qquad \frac{\partial^2 a_x}{\partial x^2} = -\frac{\partial^2 b_x}{\partial x^2} \qquad (2.78)$$

Note that the expression outlined in Eq. (2.75) is similar to that previously derived by Saveant [78]. Using Eq. (2.78) in Eq. (2.75), one obtains

$$\frac{\partial b}{\partial t} = D_E\frac{\partial[\partial b/\partial x + (ab/c_\Sigma)(\partial\theta/\partial x)]}{\partial x} \qquad (2.79)$$

where we have dropped the subscript x. Under steady-state conditions $\partial b/\partial t = 0$ and so the steady-state flux j_Σ is given by

$$j_\Sigma = D_E\left(\frac{\partial b}{\partial x} + \frac{ab}{c_\Sigma}\frac{\partial \theta}{\partial x}\right) \qquad (2.80)$$

This is not the standard form of the Nernst–Planck equation for a diffusion–migration process. This point was first realized by Saveant [78]. The first term in Eq. (2.80) is simply that due to diffusion. The second term carries the effect of the electric field and also reflects the bimolecular nature of the intersite electron exchange. One should also note that because of the coupled reaction between A and B the term in ab/c_Σ will pass through a maximum when $E = E^0(A/B)$, the standard potential of the immobilized redox couple A/B. At this potential $a = b = \frac{1}{2}c_\Sigma$. One will have maximum redox conductivity in this region. The term ab/c_Σ will be very small for either a fully oxidized or a fully reduced layer.

If the polymer layer is at equilibrium, then the flux j_Σ is zero. In this case the Nernst equation is valid and one may write

$$\ln\left(\frac{b}{a}\right) = -(\theta - \theta_0) \tag{2.81}$$

where θ_0 is the normalized potential when $a = b = \frac{1}{2}c_\Sigma$. Differentiation of Eq. (2.81) and using the result obtained in Eq. (2.78), one obtains

$$\frac{ab}{c_\Sigma}\frac{\partial\theta}{\partial x} = -\frac{\partial b}{\partial x} \tag{2.82}$$

If we substitute this expression into the equation for the total flux given in Eq. (2.80), we note that the net flux is zero, as indeed it should be. One can now use the result obtained in Eq. (2.82) to express Eq. (2.80) in the form

$$j_\Sigma = D_E \frac{ab}{c_\Sigma}\frac{\partial[\theta + \ln(b/a)]}{\partial x} \tag{2.83}$$

As previously noted by Albery [89], this form of the flux expression places an emphasis on the coupled nature of the bimolecular exchange process. It also illustrates the important fact that the electric field described by the θ factor is modified by the Nernst concentration term, $\ln(b/a)$, which provides the correct driving force for the flux through the polymer.

Hence the diffusion–migration flux may be written as

$$
\begin{aligned}
j_\Sigma &= -D_E\left[\frac{\partial a}{\partial x} + \frac{nF}{RT}a\left(1 - \frac{a}{c_\Sigma}\right)\frac{\partial\Phi}{\partial x}\right] \\
&= D_E\left(\frac{\partial b}{\partial x} - \frac{nF}{RT}b\left(1 - \frac{b}{c_\Sigma}\right)\frac{\partial\Phi}{\partial x}\right]
\end{aligned}
\tag{2.84}
$$

The process of counter ion transport is described by the classical Nernst-Planck equation. This may be written as follows using the equation of continuity:

$$\frac{\partial c_{X^-}}{\partial t} = -D_I\frac{\partial}{\partial x}\left(\frac{\partial c_{X^-}}{\partial x} + \frac{zF}{RT}c_{X^-}\frac{\partial\Phi}{\partial x}\right) \tag{2.85}$$

Hence the counter ion flux is given by:

$$j_X = -D_I\left(\frac{\partial c_{X^-}}{\partial x} + \frac{zF}{RT}c_{X^-}\frac{\partial\Phi}{\partial x}\right) \tag{2.86}$$

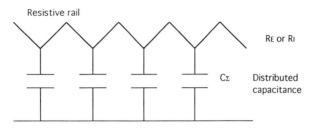

Figure 12.   Classical single-rail transmission line model.

In the latter expressions D_I denotes the counterion diffusion coefficient and z is the valence of the ion. The ion flux and the electron-hopping flux are coupled via the electroneutrality condition.

This leads to a description of the mixed-conduction characteristics of an electroactive polymer in terms of a simple equivalent circuit. A particularly appealing mode of representation utilizes the concept of a *transmission line*. This type of approach has been advocated by a number of workers such as Rubinstein [90], Buck [91], and Albery et al. [92–95] and Fletcher [96, 97]. The classical transmission line is illustrated in Fig. 12. Note that the distributed capacitance C_Σ connects a resistive line to a wire of zero resistance. This circuit element is appropriate when one has two types of charge carrier, one of which is much more mobile than the other. In this model, the more mobile carrier is modeled by the wire of zero resistance, whereas the resistive line describes the less mobile carrier. However, from a more general viewpoint it is preferable to propose a dual-rail transmission line, as illustrated in Fig. 13. In this case

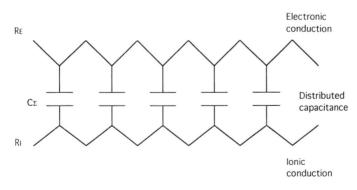

Figure 13.   Albery–Mount dual-rail transmission line model.

each line has its own resistance, R_E and R_I, which can be used to model the mobilities of the two carriers (electrons and counterions) without any restriction on their relative mobilities. This essentially has been the approach adopted in the work of Albery and co-workers [92–95]. A somewhat more general approach has been recently published by Fletcher [96, 97], who based his analysis on general ladder network theory [98]. The transmission line adopted by Fletcher is illustrated in Fig. 14 and is especially applicable for the analysis of charge percolation in electronically conducting polymer systems. This type of transmission line modeling forms the basis of the description of the complex impedance response of electroactive polymers. We discuss the impedance response in Section III of this review.

The electronic resistance R_E is given by the expression

$$R_E = \frac{RT}{n^2 F^2} \frac{c_\Sigma}{ab} \frac{L}{AD_E} \quad (2.87)$$

where A is the area of the electrode and L is the layer thickness. Furthermore, the ionic resistance R_I is given by

$$R_I = \frac{RT}{n^2 F^2} \frac{L}{AD_I b} \quad (2.88)$$

The distributed capacitance is given by the expression

$$\frac{1}{C_\Sigma} = \frac{1}{C} + \frac{RT}{n^2 F^2} \frac{c_\Sigma}{ab} \frac{1}{AL} + \frac{RT}{n^2 F^2} \frac{1}{b} \frac{1}{AL} \quad (2.89)$$

In this expression the first term on the rhs describes the distributed capacitance between the polymer strands and the aqueous pores. If C is very large, then this term will be insignificant and there will be no

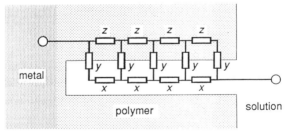

Figure 14. More sophisticated dual-rail transmission line representation due to Fletcher for charge percolation in electroactive polymer film.

potential drop between the polymer and the pore electrolyte. The second term arises from the Nernst equation and simply describes the response of the A/B couple to the change in potential. This term will pass through a minimum at the half-oxidation state. The final term arises from the Donnan potential needed to bring the counterions into the film [99].

Note that R_E and R_I vary with the extent of film oxidation in different ways. As the layer becomes oxidized R_E will pass through a minimum at the half-oxidation state, since the number of AB exchange reactions is at a minimum at $E = E^0$. On the other hand, if we adopt a rather simple view, R_I should decrease in a monotonic manner as the layer becomes more oxidized, due to the fact that greater quantities of counterions are present in the film as the degree of oxidation increases.

An important point should be noted from the analysis just presented. The modified form of the Nernst–Planck equation derived for intersite electron hopping gives rise to the following. Detailed analysis conducted by Saveant and co-workers [78–84] has indicated that under both steady-state and transient conditions, the presence of a "migration" term will always lead to an *enhancement* in the rate of intersite electron hopping. Furthermore, it is to be expected that the enhancement will increase as the mobility of the electroinactive counterion decreases. Thus it appears that earlier conclusions in the literature that charge transport rates in electroactive polymers might be controlled by the *slower* of the two coupled processes of electron hopping and electroinactive counterion displacement are somewhat inappropriate. One has instead to contend with the notion that the slower the movement of electroinactive counterions, the faster the electron hopping and the larger the current density observed. This important conclusion has only recently been taken fully on board by workers in this area.

Saveant [81] has noted that the chronoamperometric response to a step in potential (a method usually used to experimentally determine D_{CT} values) will exhibit a Cottrell response; namely, i versus $t^{-1/2}$ will be linear at short times but the value for the diffusion coefficient obtained from the Cottrell slope will contain an enhancement factor brought about by any mismatch between the rates of electroinactive counterion displacement and electron hopping. These diffusion coefficients will increase with redox center concentration more steeply than that predicted by the simple Dahms–Ruff equation.

The major question to be answered therefore is the exact mechanism by which a coupling is achieved between the local intersite electron hop and the counterion displacement.

A further important point should be noted at this stage. The models just described explicitly ignore activity effects in the polymer. The models deal with the dynamics of electron hopping in concentration and potential gradients. What is assumed, however, is that interactions between each ion and its neighbors remain invariant as the composition, that is, the redox state, of the film is changed. This assumption is very likely to be invalid due to the fact that polymer films will contain quite a large ion concentration. Hence interaction and activity effects cannot be simply ignored.

Saveant [84] has considered the effect of ion pairing between the immobile redox groups covalently attached to the polymer backbone and the mobile electroinactive counterions on the rate of overall charge percolation through the film. This aspect has not been considered by other workers. This is surprising given the fact that the redox site concentration in electroactive polymers is generally high and that reasonably large concentrations of counterions may also be contained within the polymer matrix [99]. Many regions of the polymer will consist of regions of low dielectric constant and be weakly polar. Hence one would expect that ions contained in redox polymers or loaded ionomers should undergo extensive ion pairing or indeed form aggregates of higher complexity. Hence it is reasonable to assume that activity effects may be due to the operation of ion association processes.

The model proposed by Saveant is illustrated in Fig. 15. The ion-paired redox centers (D) are assumed to be immobile and therefore do not participate in intersite electron hopping. The free fixed ions (A) can participate in electron hopping. The electroinactive counterions (C) are mobile. The situation where one has fixed immobile electroinactive ions (F) that can pair with the mobile electroinactive counterions was also considered. The transport equations for this system were derived. We shall not present the details of the analysis here but simply quote the final results as follows:

$$\frac{\partial a}{\partial t} = \frac{D_E}{c_\Sigma} \frac{\partial}{\partial x} \left(b \frac{\partial a}{\partial x} - a \frac{\partial b}{\partial x} + \frac{abF}{RT} \frac{\partial \phi}{\partial x} \right) - k_A ac + k_D d \qquad (2.90)$$

$$\frac{\partial b}{\partial t} = \frac{D_E}{c_\Sigma} \frac{\partial}{\partial x} \left(a \frac{\partial b}{\partial x} - b \frac{\partial a}{\partial x} - \frac{abF}{RT} \frac{\partial \phi}{\partial x} \right) \qquad (2.91)$$

$$\frac{\partial c}{\partial t} = D_I \frac{\partial}{\partial x} \left(\frac{\partial c}{\partial x} - \frac{cF}{RT} \frac{\partial \phi}{\partial x} \right) - k_A ac + k_D d \qquad (2.92)$$

$$\frac{\partial d}{\partial t} = k_A ac - k_D d \qquad (2.93)$$

In these expressions a, b, c and d represent the concentrations of the

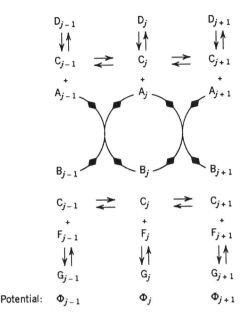

Figure 15. Detailed view of electron hopping in electroactive polymer films. Effect of ion-pairing equilibria also outlined.

various species referred to in Fig. 15. Note also that ϕ denotes the potential difference between adjacent sites. Furthermore D_E is the electron-hopping diffusion coefficient given as before by $D_E = kc_\Sigma \delta^2$, and D_I is the mobile electroinactive ion diffusion coefficient given by $D_I = k_I \delta^2$. As before c_Σ represents the total redox site concentration.

In the steady state of fluxes are given by

$$j_A = D_E \left(\frac{\partial a}{\partial x} + \frac{ab}{c_\Sigma} \frac{F}{RT} \frac{\partial \phi}{\partial x} \right) \qquad (2.94)$$

$$j_B = D_E \left(\frac{\partial b}{\partial x} - \frac{ab}{c_\Sigma} \frac{F}{RT} \frac{\partial \phi}{\partial x} \right) \qquad (2.95)$$

$$j_C = -D_I \left(\frac{\partial c}{\partial x} - \frac{cF}{RT} \frac{\partial \phi}{\partial x} \right) \qquad (2.96)$$

and

$$j_D = 0 \qquad (2.97)$$

In these expressions we assume that the mobile counterion is uninegative and the electron-hopping process involves the transfer of a single electron. It is also assumed that the covalently attached redox center is unipositive in its oxidized form and neutral in its reduced form. Hence the redox center experiences ion pairing in the oxidized state. The usual form of the Nernst–Planck equation applies to the mobile counterion C, which has a valence of -1. Of course extension to the case of a fixed mononegative reduced form of the redox center paired with a monopositive mobile electroinactive counterion and a fixed neutral oxidized form is immediate. The steady-state responses were obtained by noting the electroneutrality and mass conservation conditions:

$$a = c \qquad a + b + d = c_\Sigma \qquad (2.98)$$

Furthermore under steady-state conditions, ion pairing remains at equilibrium and this results in the additional relation

$$d = Kac \qquad (2.99)$$

where K is the association equilibrium constant given by $K = k_A / k_D$.

The foregoing expressions can be used to determine the transient and steady-state responses of an electroactive polymer when ion pairing is significant.

If a fixed electroinactive ionic species F is also present, then it will play a role similar to that of a supporting electrolyte. In this slightly more complicated situation the expressions just outlined are still valid, but one must modify the equation for the transport of the mobile electroinactive counterion as follows:

$$\frac{\partial c}{\partial t} = D_I \frac{\partial}{\partial x} \left[\frac{\partial c}{\partial x} - \frac{F}{RT} \left(c \frac{\partial \phi}{\partial x} \right) \right] - k_A ac + k_D d - k_A' fc + k_D' g \qquad (2.100)$$

where k_A' and k_D' represent the association and dissociation rate constants for the ion pairing reaction $F + C = G$. In addition one notes that

$$\frac{\partial f}{\partial t} = -k_A' af + k_D' g \qquad (2.101)$$

and

$$\frac{\partial g}{\partial t} = k_A' af - k_D' g \qquad (2.102)$$

Conservation of matter leads to an additional expression as follows:

$$f + g = f_\Sigma \qquad (2.103)$$

where f_Σ denotes the total concentration of supporting ions including ion pairs. Under steady-state conditions, the set of expressions previously outlined for the flux remain valid. The second ion pairing reaction $F + C = G$ remains in equilibrium, and so one may write

$$g = K'fc \qquad (2.104)$$

where the equilibrium constant K' is given by the ratio k'_A / k'_D.

Saveant [84] examined quantitatively the steady-state situation in which one had a redox polymer membrane sandwiched between two parallel planar electrodes. This particular experimental arrangement is used to determine electron-hopping diffusion coefficients and will be outlined later in this review. In general, the steady-state current–voltage response for such a thin-film membrane is sigmoidal, and the diffusion coefficient may be obtained from the limiting current plateau. The reader is referred to the original paper [84] for a full description of the mathematical analysis. We shall only present a summary of the results obtained. The characteristics of the steady-state voltammogram was computed as a function of the redox composition of the film, the extent of ion pairing of the electroactive redox centers by the mobile electroinactive counterions, the addition of an attached electroinactive ion (the "supporting ion"), and the extent of pairing of the latter with the mobile electroinactive counterions.

One major conclusion was that the charge transport rate fell quite rapidly as the extent of ion pairing between the redox centers and the mobile counterions increased. A parameter κ was introduced that quantitatively described the extent of ion association. This was given by $\kappa = Kc_\Sigma$. Hence the counterion affects the charge percolation rate, not because of its mobility, but via the affinity that the counterion may have for ion association. In short, the stronger the ion pair formed, the slower the charge transport through the layer. This leads to another important conclusion. As ion association becomes more marked the charge transport rate becomes very small. In such a situation one has to invoke a mechanism in which the ion pairs participate directly in the electron-hopping mechanism.

Let us consider the mechanism illustrated in Fig. 16. This is a scheme of squares in which the subscripts 1 and 2 denote adjacent sites in a redox polymer film. Electron hopping occurs vertically and electroinactive counterion displacement occurs horizontally. Note that the species A_1^+, BC_2^-, BC_1^-, and A_2^- are produced by strongly uphill reactions, and therefore these may be considered to fulfil the steady-state condition. It should also be noted that the $A_1^+ + BC_2^-$ and $BC_1^- + A_2^+$ states of the

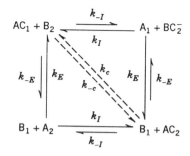

Figure 16. Scheme of squares outlining ion association processes in electroactive polymer films.

system have the same free energy. Thus $k_I/k_{-I} = k_E/k_{-E} = K$. Under these conditions one may write

$$\frac{d[AC_1]}{dt} = \frac{d[B_2]}{dt} = -\frac{d[AC_2]}{dt} = -\frac{d[B_1]}{dt} = k_c([AC_2][B_1] - [AC_1][B_2])$$

(2.105)

where the rate constant for the concerted process k_c is given in the context of the steady-state approximation by

$$\frac{1}{k_c} = \frac{1}{k_{-1}} + \frac{1}{k_{-E}}$$

(2.106)

We note therefore that charge transport is formally equivalent to a diffusion of the immobile redox centers, AC and B, each having the same diffusion coefficient:

$$D_c = k_c c_\Sigma \delta^2 = \frac{k_{-E}k_{-I}}{k_{-E} + k_{-I}} c_\Sigma \delta^2$$

(2.107)

The relationship illustrated in Eq. (2.107) is typical of a situation of charge transport via mixed control. Electron hopping is characterized by the rate constant k_{-E}, whereas counterion displacement is characterized by the rate constant k_{-I}. If $k_{-E} \gg k_{-I}$, then charge transport will be controlled via counterion displacement; if $k_{-I} \gg k_{-E}$, then electron hopping will be rate determining.

This model is formulated by assuming that the two fundamental steps in the charge percolation process occur in a sequential manner. However, this mechanism implies that the system will pass through two rather energetically unfavorable states (bottom left and top right in Fig. 16). Hence it appears that an energetically more favorable situation would involve a concerted pathway (the dotted path in Fig. 16). Saveant has modelled this situation in terms of Marcus theory. The expressions outlined in Eqs. (2.94) and (2.95) are still valid. However, in this case the expression for the diffusion coefficient D_E must be replaced by D_c defined on Eq. (2.107).

Ion association effects may be summarized as follows. The work of Saveant has indicated that the activation energy barrier for electron hopping between fully associated ions will be large due mainly to the fact that the electron transfer between the oxidized and reduced components will require dissociation of an ion pair linkage and the formation of a new one. Consequently, predissociation mechanisms in which the fully ion paired oxidized half of the redox couple dissociate prior to the electron transfer step, which will then be a simple outer sphere process, will be energetically more favorable. What about the D_{CT}/c_Σ relationship? Since the fraction of dissociated ion pairs increases with the concentration of redox sites, the apparent diffusion coefficient will exhibit a steep increase with increasing c_Σ value.

B. Electronically Conducting Polymers

1. Introduction

Electronically conducting polymers are conjugated organic materials and differ from redox polymers in that the polymer backbone is itself electronically conducting. This conductivity is imparted due to the addition of dopants, in relatively large quantities, into the polymer matrix and may be accomplished electrochemically or via purely chemical pathways. Hence, in principle, the magnitude of the observed electronic conductivity should be considerably greater than that observed for redox polymer materials. We are in effect dealing with molecular wires or plastic metals. A useful overview of the fundamental properties of electronically conducting polymers has been recently provided by Evans [5].

Both theoretical calculations and experimental studies indicate that the precise nature of charge carriers in conjugated polymer materials depends on the type of polymer. However, some useful generalizations can be made at this stage. The charge carriers (solitons, polarons, or bipolarons) are defects that are delocalized over a number of repeat units on the polymer chain. Roth and co-workers [100] have recently noted that macroscopic charge transport in a conducting polymer matrix represents a superposition of local transport mechanisms. For instance, one may have carrier transport within a conjugated strand, from strand to strand, and if the polymer morphology is fibrillar, from fiber to fiber. These transport processes are illustrated in Fig. 17.

Roth and co-workers [100] define an intrinsic conductivity that refers

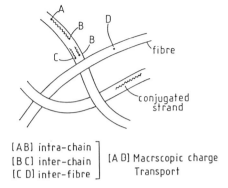

Figure 17. Charge transport processes (microscopic and macroscopic) is electronically conducting polymers.

$$\left.\begin{array}{l} \text{[A B] intra-chain} \\ \text{[B C] inter-chain} \\ \text{[C D] inter-fibre} \end{array}\right] \begin{array}{l} \text{[A D] Macrscopic charge} \\ \text{Transport} \end{array}$$

to conduction processes within a strand. The latter process is governed by the physics of conjugated double bonds and that of one-dimensional metals and includes such topics as metal–nonmetal transitions, solitons, polarons, bipolarons, and electron–phonon coupling [101]. This intrinsic conductivity will be very high. The intrinsic conductivity can be improved by ensuring that the polymer chains are well aligned and contain the minimum amount of defects. The molecular weight of the polymer must also be high. These are synthetic problems. If improved synthetic methodologies can be developed, then there is good reason to expect that high-performance polymers may be fabricated with electronic conductivities significantly greater than those of even the best conventional metals.

Various hopping and tunneling models have been proposed to describe the nonintrinsic (interstrand and interfibre) conduction process [102–104]. For instance if charge carriers move by quantum-mechanical tunneling between individual localized states, a variable range hopping (VRH) model is appropriate [105]. Hence the macroscopic conductivity is described in terms of a model of random hopping of charge carriers between localized states of adjacent chain segments of different strands. Mechanistic analysis is usually aided by determining the temperature dependence of the DC and AC conductivity over a large temperature range [106]. According to hopping theory, two different temperature regimes may be distinguished. At low temperatures the charge carriers tend to hop not only to nearest-neighbor but also to more remote sites, in order to minimize the energy required for the hop. At higher temperatures nearest-neighbor hopping almost exclusively occurs.

The correlation between the voltammetric response exhibited by electronically conducting polymers and the redox chemistry of the charge–discharge reactions involving polarons and bipolarons is also a topic of considerable interest. The vexing question of distinguishing between capacitive and non-capacitive current in the voltammetric response in potential regions where the polymer is conducting is also a topic of current controversy, and definitive answers have not yet been obtained. Furthermore, ion transport in the conducting polymer matrix can in many cases be rate determining. The fundamental features of the chemistry underpinning the process of redox switching in a conductive polymer have not yet been fully elucidated. The transition between insulating and conducting behavior (redox switching) is in many respects analogous to a phase transition. One begins with a totally insulating material, a perturbation (such as a step in potential) is applied, and the material switches to the conducting state. Hence the polymer matrix is not a homogeneous medium but will consist of conducting and insulating regions. Hence it is reasonable to suppose that redox switching could be described in terms of percolation theory. This topic has been considered in two recent reviews [107, 108]. In simple terms, the percolation of charge through a two-phase system, only one phase of which is conducting, is dependent on the concentration of the latter. At low concentrations of the conducting phase, the oxidized regions remain isolated in a largely insulating matrix. A current may flow only if a connected pathway exists through the matrix. Hence the sample will be essentially electronically insulating. The current response in the voltammogram will be low. As the concentration of the conducting phase increases, a connected path will be formed and the conductivity of the sample will show a large increase. The critical concentration value corresponding to the so-called percolation threshold will depend on the detailed structure of the two-phase system.

All of the aforementioned topics will be discussed in the succeeding sections of the present review.

2. Doping Processes in Electronically Conducting Polymers

It is well established that the electrical conduction properties of elemental semiconductors such as Si may be rigorously controlled by the addition of very small quantities of foreign atoms into the host semiconductor lattice. The host semiconductor may be made n type or p type depending on the nature of the added dopant atoms, that is, whether the latter has an excess or deficit of electrons. New dopant energy levels are introduced into the band gap and conduction is facilitated. The conductivity level attained depends strongly on the concentration of donor or acceptor species incorporated. Excess conduction electrons or holes are generated in the material as a result of the doping and so the conductivity is enhanced. A similar terminology has been applied to conjugated polymers. These materials may also be "doped", and indeed the conductivity level obtained will depend on the doping level. However, the doping mechanism differs considerably from that observed for elemental semi-conductors. First, the doping levels attained in conjugated polymers are significant (doping levels can be as large as 10 mol %. Second, one has charge transfer between the incorporated dopant atom and the polymer chain. Hence the latter is partially oxidized or reduced.

The partial oxidation of the polymer chain is termed p doping. The basic process involves the removal of electrons to form a positively charged repeat unit:

$$-(P)_x-+xy\, A^- \rightarrow -[(P^{y+})A_y^-]x-+xy\, e^-$$

where P represents the basic monomeric repeat unit in the polymer. Furthermore, if the polymer chain is partially reduced, one has n-type doping as follows:

$$-(P)_x-+xy\, e^- + xy\, M^+ \rightarrow -[M_y^+(P^{y-})]_x-$$

It is clear that partial oxidation of the backbone may be achieved either electrochemically via application of a potential or chemically via use of a gas-phase oxidizing or reducing again. The former method is to be preferred due to the regions control one can have over the degree of doping. The requirement that electroneutrality be maintained requires that ionic transport also be considered.

The process of doping may be quantitatively described in terms of nonlinear diffusion–reaction equations. In this type of approach the polymer is assumed to be a thin homogeneous film. One has a coupled diffusion and reversible binding of dopants to immobile sites within the layer. This problem has been examined quite recently in papers by Kim and co-workers [109, 110] and Prock and Giering [111, 112]. The most comprehensive treatment to date has been provided by Bartlett and Gardner [113].

In the latter work the general problem of a species. A diffusing into a planar homogeneous film of constant thickness L that contained a uniform concentration of immobile reactive sites S was considered. The local chemical reaction (the dopant–site interaction) between the diffusing species and the sites was described via the following expression:

$$A + S \underset{k_r}{\overset{k_f}{\rightleftharpoons}} AS \qquad (2.108)$$

where k_f is a second-order rate constant (units: $cm^3\, mol^{-1}\, s^{-1}$) describing the binding of the species A at an unoccupied site S and k_r is a first-order rate constant (units: s^{-1}) describing desorption of the species A from an occupied site AS. One may also define an equilibrium constant $K = k_f/k_r$ that is a measure of the affinity of the sites for the species A. When K is large, the sites become saturated at low concentrations of A, whereas when K is small, the sites interact weakly with A and high concentrations of the dopant are required to saturate the sites in the film.

The mobile dopant obeys the following diffusion–reaction equation:

$$D \frac{\partial^2 a}{\partial x^2} - k_f a(1 - \theta)N + k_r \theta N = \frac{\partial a}{\partial t} \qquad (2.109)$$

where $a(x, t)$ denotes the concentration of dopant A in the layer, $q(x, t)$ is the fraction of occupied sites, and N is the concentration of binding sites within the film. If we consider the kinetics of reaction at the sites, one can write

$$N \frac{\partial \theta}{\partial t} = k_f a(1 - \theta)N - k_r \theta N \qquad (2.110)$$

Combining Eqs. (2.109) and (2.110), we obtain

$$D \frac{\partial^2 a}{\partial x^2} - \frac{\partial a}{\partial t} = N \frac{\partial \theta}{\partial t} \qquad (2.111)$$

The pertinent initial and boundary conditions are as follows: At time $t = 0$ and for $0 < x < L$, $q = 0$; furthermore for $t > 0$ at $x = 0$, $da/dx = 0$, and $a = a^\infty$ at $x = L$, where we have assumed that species A is present in the external phase at a bulk concentration a^∞ and the partition coefficient is unity. We further assume that there is no kinetic barrier at the film–solution interface to the passage of dopant A into the film.

The mathematical analysis is facilitated by introduction of the following dimensionless parameters:

$$\chi = \frac{x}{L} \qquad \tau = \frac{Dt}{L^2} \qquad u = \frac{a}{a^\infty} \qquad \lambda = Ka^\infty \qquad \kappa = \frac{k_f NL^2}{D} \qquad \eta = KN$$

$$(2.112)$$

Note that the parameter κ describes the balance between the adsorption $(k_f N)$ and diffusional (D/L^2) kinetics in the film. For example, when $\kappa \gg 1$, the adsorption process is much faster than that of diffusion. When $\kappa \ll 1$, the opposite situation pertains. The parameter λ describes the position of equilibrium. For $\lambda \ll 1$, most of the sites will be unoccupied at equilibrium, whereas for $\lambda \gg 1$, all the sites will be occupied at equilibrium. Finally, η is the product of the adsorption equilibrium constant and the concentration of binding sites and will therefor be a property of the material.

Using these dimensionless parameters, Eq. (2.111) may be recast into the form

$$\frac{\partial^2 u}{\partial \chi^2} - \frac{\partial u}{\partial \tau} = \frac{\eta}{\lambda} \frac{\partial \theta}{\partial \tau} \tag{2.113}$$

whereas Eq. (2.110) may be written as

$$\eta \frac{\partial \theta}{\partial \tau} = \kappa \lambda u (1 - \theta) - \kappa \theta \tag{2.114}$$

The initial and boundary conditions are now given by

$$\text{At} \quad \tau = 0: \quad \theta = 0 \quad u = 0 \quad \text{for} \quad 0 \leq \chi \leq 1$$
$$\forall \tau > 0: \quad \left[\frac{\partial u}{\partial \chi} \right]_{\chi = 0} = 0 \quad \text{and} \quad u = 1 \quad \text{at} \quad \chi = 1 \tag{2.115}$$

The expressions outlined in Eqs. (2.113) and (2.114) are coupled nonlinear partial differential equations and do not admit an exact analytical solution. However, Gardner and Bartlett [113] have identified a number of limiting cases for which analytical expressions for the dopant concentration profiles and site occupancy profiles may be derived. Six cases have been identified. The detailed mathematical treatment is complex and therefore will not be described here. The reader is referred to the original paper for full details [113]. Approximate analytical expressions for the dopant concentration profiles u and site occupancy functions θ are outlined in Table I.

The six cases are identified according to the magnitudes of the parameters κ, λ, and η. We now consider each of these cases in turn. In case I ($\lambda < 1$, $\eta > 1$, and $\kappa > 1$) the film is unsaturated at equilibrium and the kinetics of the reaction are fast when compared to diffusion. Consequently, the site occupancy function follows the diffusion of the species A into the film. The amount of dopant A bound on sites is small compared to the amount of unbound A in the film, and therefore the diffusion of dopants is not affected by the binding kinetics. Case II ($\lambda < 1$, $\eta > 1$, $\kappa > 1$) considers the situation where the film is unsaturated at equilibrium and the reaction kinetics are rapid compared to the rate of diffusional transport of dopant in the layer. In this case a large proportion of A is bound to sites in the film. This leads to the prediction of a reduced

TABLE I
Approximate Analytical Expressions for Dopant Concentration and Site Occupancy

$u(\chi, \tau)$	$\theta(\chi, \tau)$
Case I: $(\lambda < 1: \eta < 1; \kappa > \eta)$	
$1 - \dfrac{2}{\pi} \displaystyle\sum_{n=0}^{\infty} \dfrac{\cos[(n + \frac{1}{2})\pi\chi]\exp[-(n + \frac{1}{2})^2\pi^2\tau]}{(-1)^n(n + \frac{1}{2})}$	$\lambda\gamma$
Case II: $(\lambda < 1: \eta > 1; \kappa > 1)$	
$1 - \dfrac{2}{\pi} \displaystyle\sum_{n=0}^{\infty} \dfrac{\cos[(n + \frac{1}{2})\pi\chi]\exp[-(n + \frac{1}{2}^2\pi^2\tau/\eta]}{(-1)^n(n + \frac{1}{2})}$	$\lambda\gamma$
Case III: $(\lambda < 1: \eta < 1; \kappa > \eta)$	
$1 - \dfrac{2}{\pi} \displaystyle\sum_{n=0}^{\infty} \dfrac{\cos[(n + \frac{1}{2})\pi\chi]\exp[-(n + \frac{1}{2})^2\pi^2\tau]}{(-1)^n(n + \frac{1}{2})}$	$\lambda[1 - \exp(-\kappa\tau/\eta)]$
Case IV: $(\kappa < 1: \lambda > 1)$	
$1 - \dfrac{2}{\pi} \displaystyle\sum_{n=0}^{\infty} \dfrac{\cos[(n + \frac{1}{2})\pi\chi]\exp[-(n + \frac{1}{2})^2\pi^2\tau]}{(-1)^n(n + \frac{1}{2})}$	$1 - \exp(-\lambda\kappa\tau/\eta)$
Case V: $(\kappa > 1: 1 < \lambda^2 < \eta)$	
$1 - \dfrac{2}{\pi} \displaystyle\sum_{n=0}^{\infty} \dfrac{\cos[(n + \frac{1}{2})\pi\chi]\exp[-(n + \frac{1}{2})^2\pi^2\tau]}{(-1)^n(n + \frac{1}{2})}$	$\dfrac{\lambda\gamma}{1 - \lambda\gamma}\left[1 - \exp\left(-\dfrac{(1 + \lambda\gamma)\kappa\tau}{\eta}\right)\right]$
Case VI: $(\kappa > 1: 1 < \lambda^2 < \eta)$ Unsaturated region of film, $\chi \leq \chi^*$:	
$u_1 = \dfrac{\gamma^*}{\mathrm{erfc}(\zeta\eta^{1/2})}\,\mathrm{erfc}\left(\dfrac{\eta^{1/2}(1 - \chi)}{2\tau^{1/2}}\right)$	
Saturated region of film, $\chi > \chi^*$:	
$u_2 = 1 - \dfrac{1 - \gamma^*}{\mathrm{erf}\,\zeta}\,\mathrm{erf}\left(\dfrac{1 - \chi}{2\tau^{1/2}}\right)$	$\lambda\gamma/(1 + \lambda\gamma)$

TABLE I *(Continued)*

$u(\chi, \tau)$	$\theta(\chi, \tau)$

Position of moving boundary is $1 - \chi^* = 2\zeta\sqrt{\tau}$, where ζ is the root of

$$\frac{\exp(-\zeta^2)}{\operatorname{erf}\zeta} - \frac{\gamma^* \exp(-\zeta^2\eta)}{\eta^{1/2}(1-\gamma^*)\operatorname{erfc}(\zeta\eta^{1/2})} = \frac{\eta\pi^{1/2}\zeta}{\lambda(1-\gamma^*)}$$

When the boundary hits the substrate (at $\tau = \tau^*$), we have diffusion:

$$u(\chi, \tau) = 1 + \frac{2}{\pi}\sum_1^\infty \frac{\cos(n\pi)-1}{n}\sin[\tfrac{1}{2}n\pi(\chi+1)]\exp[-\tfrac{1}{4}n^2\pi^2(\tau-\tau^*)]$$

$$+ 2\sum_1^\infty \sin[\tfrac{1}{2}n\pi(\chi+1)]\exp[-\tfrac{1}{4}n^2\pi^2(\tau-\tau^*)]\int_0^1 \gamma_2(\chi', \tau^*)\sin[\tfrac{1}{2}n\pi(\chi+1)]\,d\chi'$$

diffusion coefficient, and the characteristic time for A to diffuse into the film is greater by a factor of h than that predicted in case I. In case III ($\lambda < 1$, $\kappa \ll 1$, all h values) the layer is unsaturated at equilibrium, but in this case the reaction kinetics are slow when compared to diffusion. Therefore dopant species A diffuse into the layer and subsequently attain a constant concentration before reaction occurs. The reaction then proceeds on a slower time scale in a homogeneous manner throughout the film. Cases I–III correspond to linear diffusion/reaction. In case IV ($\kappa < 1$, $\lambda > 1$) the film is saturated at equilibrium but the reaction kinetics are still slow when compared to diffusional transport. Hence, again A diffuses into the film and attains a constant concentration before the sites are filled. The reaction then proceeds homogeneously throughout the film to completely fill the sites. In case V ($\kappa > 1$, $\lambda > 1$, $\lambda^2 > \eta$) the sites are saturated at equilibrium, and the kinetics are fast so that the occupancy of the sites follows the diffusion of dopant species into the film. The reaction does not significantly perturb the diffusion because only a small fraction of A ends up in the bound form. Consequently the region of occupied sites spreads into the film from the outside at a rate governed by the diffusion of A. Finally the situation described by case VI is somewhat complex but is very interesting. In this case one has $\kappa > 1$ and $1 < \lambda^2 < \eta$. Once again the sites are fully occupied at equilibrium and the reaction kinetics are fast. However, in this case the fraction of dopant species bound to sites is significant. Thus the reaction slows down diffusion of A into the film. The region of occupied sites spreads through the film as a moving boundary from the outside in. In this case the differential equations to be solved are

$$\frac{\partial \theta}{\partial \tau} = \frac{\lambda}{(1 + \lambda u)^2} \frac{\partial u}{\partial \tau} \qquad (2.116)$$

and

$$\frac{\partial^2 u}{\partial \chi^2} = \frac{(1 + \lambda u)^2 + \eta}{(1 + \lambda u)^2} \frac{\partial u}{\partial \tau} \qquad (2.117)$$

This case has also been considered by Kim and co-workers, 109 who derived a nonlinear partial differential equation of a form similar to that outlined in Eq. (2.117). The problem was also considered by Hermans [114] many years ago. The method of solution proposed by Bartlett and Gardner [113] is perhaps the most comprehensive suggested to date and is based on the Neumann analysis of the phase transformation problem in a semi-infinite diffusion space described in the classic text written by Carlsaw and Jaeger [115]. The reader is referred to the original paper for full details of the analysis. One notes therefore that the process of dopant transport and reaction in a polymer matrix can be complex.

The case diagrams for all six cases are illustrated in Fig. 18a, whereas the computed concentration profiles for substrate u and site occupancy θ for each of the six cases are outlined in Fig. 18b. Bartlett and Gardner [113] have utilized this basic analysis to examine the operation of a chemiresistor for the detection of gases and vapors such as NH_3, alcohols and NO_2. Th chemiresistor detects the vapor or gas via monitoring the conductance change brought about by the diffusion and adsorption of the analyte in the polymer film that serves as the active sensing element.

3. Charge Carriers and Conductivity in Electronically Conducting Polymers

We now consider the various types of charge carriers that may be found in electronically conducting polymers. As previously noted, both experimental and theoretical evidence suggest that the precise nature of the charge carriers present in conjugated polymer systems depends to a very large extent on the type of polymer. We shall discuss two representative polymer materials, polyacetylene and polypyrrole. These systems have been the subject of considerable study.

A brief summary of band theory is useful at this point [116, 117]. Electrical conductivity depends on a number of fundamental parameters such as the number density of mobile charge carriers n, the carrier charge

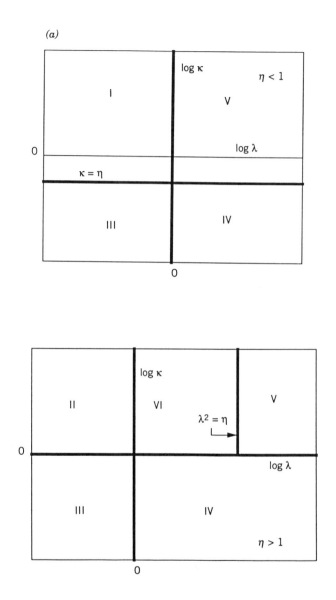

Figure 18. (*a*) Kinetic case diagrams for dopant transport and reaction in electronically conducting polymer films according to Bartlett–Gardner model. Thick lines separate different approximate solutions to transport/kinetic problem. Six distinct cases are noted. (*b*) Computed concentration profiles $u(\chi)$ and site occupancy functions θ for six cases presented in part (*a*). (*c*) Schematic representation of moving boundary problem (case VI).

CASE I: Normal diffusion/fast kinetics

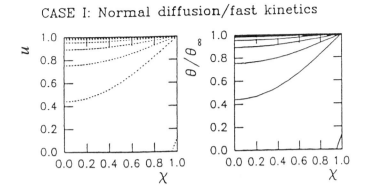

CASE II: Modified diffusion/fast kinetics

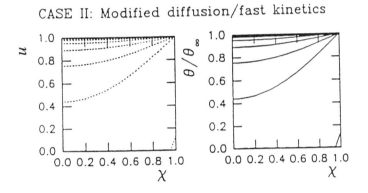

CASE III: Fast diffusion/linear kinetics

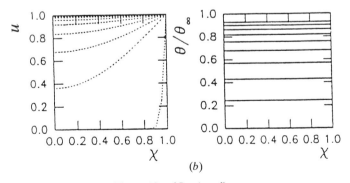

(b)

Figure 18. (*Continued*).

CASE IV: Fast diffusion/saturated kinetics

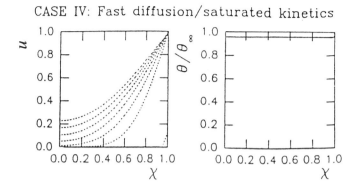

CASE V: Fast diffusion/nonlinear kinetics

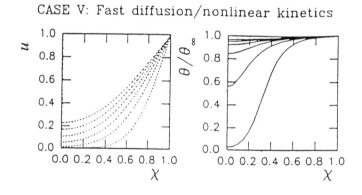

CASE VI: Modified diffusion/modified kinetics

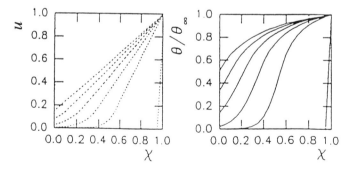

Figure 18. (*Continued*).

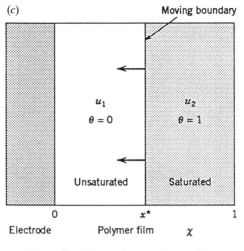

Schematic of the moving boundary problem

Figure 18. (*Continued*).

q, and the carrier mobility μ. The relationship between conductivity σ and the latter quantities is expressed via the general relationship

$$\sigma = nq\mu \qquad (2.118)$$

Conduction in solids is usually expressed in terms of the well-established band theory. In this model it is assumed that a solid consists of an N-atom system where the atoms are tightly packed together. Each individual electronic energy state splits into N levels. Typically there are about 10^{22} atoms cm^{-3} in a crystalline solid, and so the energy levels are spaced very close together. Hence one may ignore the discreteness of the levels and consider a continuous energy band. Due to the periodic nature of a crystalline solid, one will have energy gaps between the various energy bands. The highest occupied energy band is termed the valence band, whereas the lowest unoccupied energy band is the conduction band. Only charge carries with energies near the top of the valence band (near the so-called Fermi energy) can contribute to electronic conduction by being thermally promoted to the empty conduction band where they are free to move under the influence of an applied electric field. Metallic conductors are characterized either by a partially filled valence band or by the presence of a marked degree of overlap between the valence and conduction bands. On the other hand, semiconductors and insulators are characterized by the presence of an appreciable band gap between the top of the valence band and the bottom of the conduction band. Hence the ease of thermal promotion of electrons across the band gap to generate mobile conducting carriers in the conduction band will depend on the magnitude of the band gap. In simple terms semiconductors have reasonably low band gaps, whereas the gap for insulators is rather large. The situation is outlined schematically in Fig. 19.

We now relate these basic concepts to conjugated organic polymers such as polyacetylene and polyparaphenylene.

Polyacetylene is perhaps the simplest type of electronically conducting polymer, at least from the viewpoint of structure. A simple picture of the bonding in polyacetylene is as follows. Two of the three p orbitals of the carbon atoms in polyacetylene are in the form of sp^2 hybrid orbitals, two of which give rise to the s bonding framework of the polymer, with the third entering into a covalent bond with the hydrogen atom s orbitals. The third p orbital (labeled p_z) forms an extended p system along the carbon chain. The latter can, in principle, produce a quasi-metallic material with a half-filled conduction band. This would be the case if all of the carbon–carbon bonds in the polymer were identical. Hence, if all bond lengths along the polymer backbone were equal, with each bond having a partial double-bond character, then the polymer would behave as a quasi-one-dimensional metal having good conductive properties. This is not the case, however. Analysis of the physics of one-dimensional

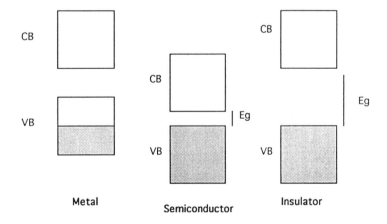

Figure 19. Schematic representation of band structure of metal, semiconductor, and insulator. VB and CB represent valence and conduction bands, respectively. Band gap energy E_g is small for semiconductor and large for insulator and concept does not apply for metallic conductor. Shaded regions denote filled electronic states, whereas unshaded regions represent unoccupied electronic states.

metals has led to the conclusion that this type of configuration is unstable, and so the one-dimensional system will undergo lattice distortion by alternating compression and extension of the linear chain. The Peierls theorem states that a one-dimensional metal will be unstable and an energy gap will form at the Fermi level due to the occurrence of the lattice distortion so that the material becomes either a semiconductor or an insulator. Elastic energy is used during lattice distortion, which is compensated for by a lowering in the electronic energy of the occupied states and the generation of a band gap. This is illustrated in Fig. 20. The application of this idea to polyacetylene is immediate. Hence in trans polyacetylene, there will be a periodic alternation of the carbon–carbon bond length along the polymer chain resulting in a stable structure of low energy. In simple terms the carbon–carbon atom spacing in the polymer backbone is altered to produce a system of alternating long and short bonds. In an approximate way we can visualize this effect as consisting of

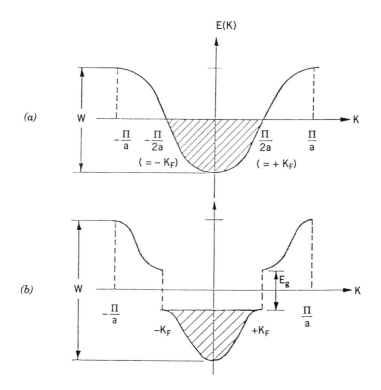

Figure 20. (a) Half-filled band of metallic polyacetylene. (b) Filled Peierls distorted band of semiconducting polyacetylene.

a sequence of alternating double and single bonds. However, one must keep in mind that the p_z electrons are not completely localized.

One can therefore have a number of possible structures for polyacetylene taking this bond alternation idea on board. This is done in Fig. 21. We note from this figure that one has two trans structures that turn out to be energetically *degenerate*. Both are energetically equivalent and both are thermodynamically stable. One can also have two cis structures that are not energetically equivalent. It has been shown that the trans–cis structure is of higher energy than the cis–trans structure. As a consequence only the latter is thermodynamically stable and so cis poly-

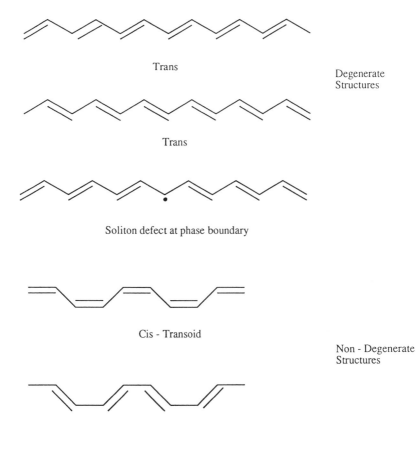

Trans

Degenerate
Structures

Trans

Soliton defect at phase boundary

Cis - Transoid

Non - Degenerate
Structures

Trans - Cisoid

Figure 21.   Various polyacetylene structures. Degenerate and non degenerate structures shown.

acetylene has a *nondegenerate* ground state. The observation that the trans form of polyacetylene has a degenerate ground state is important. This gives rise to the implication that one can have the presence of *structural defects*, or *kinks*, in the polymer chain located in regions where there is a *change in the sense of bond alternation*. This scenario is also illustrated in Fig. 20. This phenomenon is termed the *Peierls distortion effect*. This defect gives rise to a single unpaired electron (located in a nonbonding orbital) at the phase boundary between the two degenerate trans phases of polyacetylene where the bond alternation has been reversed. The overall charge remains zero, however. Translating this chemical picture into the language of band theory, we note that the presence of the defect results in the generation of a new energy state in the energy gap. This energy level is located midway up the gap. The neutral defect is termed a *soliton*. The energy level is singly occupied, and therefore the defect state has an associated spin value of $\frac{1}{2}$. Bredas and co-workers [118] have noted from theoretical calculations that the defect is delocalized over some 15 carbon atoms. It is the presence of these neutral solitons that gives trans polyacetylene the electrical characteristics of a semiconductor with an intrinsic conductivity of about 10^{-7}–$10^{-8} \, \text{S cm}^{-1}$. The neutral soliton is paramagnetic.

We saw previously that the conductivity of the polymer may be considerably enhanced by doping (via either a chemical or an electrochemical route). The soliton energy level can accommodate zero, one, or two electrons, and so the soliton may also be positively or negatively charged. This gives rise to the interesting observation that charged solitons have no spin, whereas neutral solitons have spin but no charge. The three classes of solitons are outlined in Fig. 22. When the electron in the localized state is removed, for example by acceptor doping or by electrochemical oxidation, the soliton is positively charged with spin zero and is nonmagnetic. The positive soliton is equivalent to a stabilized (via delocalization) carbonium ion on the polyacetylene chain. In a similar way, double occupancy induced by donor doping or electrochemical reduction will lead to a spin-zero negatively charged state. The negative soliton is therefore equivalent to a stabilized carbanion. Theoretical calculations have indicated that the formation of charged solitons on doping is more energetically favorable than the formation of electron–hole pairs. Furthermore, when charge is added to or removed from the polymer chain by doping, it will be located in the midgap states as the latter provide the highest occupied molecular orbital (HOMO) for charge removal and the lowest unoccupied molecular orbital (LUMO) for charge injection. It should also be noted that the defect is mobile and can move quite readily along the chain. Hence, one can see that the intrinsic

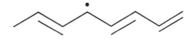

Neutral soliton

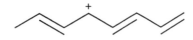

Positive soliton

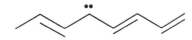

Negative soliton

(a)

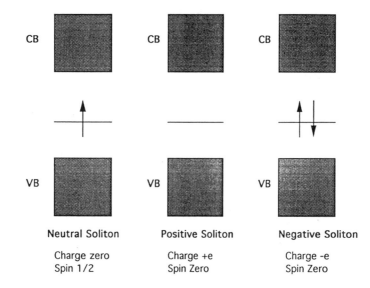

Figure 22. (*a*) Schematic representation of soliton structures in polyacetylene. (*b*) Schematic band structure for neutral, positive, and negative solitons.

conductivity of the polymer can be modeled in terms of the movement of the defects along the chain.

Bredas and co-workers [118] have noted that two neutral solitons located on the same chain will tend to recombine, leaving no deformation. However, two charged solitons will tend to repel each other and therefore give rise to two isolated charged defects. Furthermore, a neutral soliton and a charged one can achieve a minimum-energy configuration by pairing [118, 119]. This will occur when they are both located on the same chain. This pairing gives rise to a *polaron*, which is a radical cation [120]. The generation of a polaron gives rise to two energy states in the band gap that are symmetrically placed about the midgap energy. If the number of charges on the chain is increased due to continued doping, a stage will eventually be reached when the polaron states begin to interact. At a high enough doping level one would expect that the polaron states would suffer recombination to form two charged solitons that would subsequently separate [118].

We see therefore that soliton transport can be used in general terms to rationalize electrical conductivity in polyacetylene. If the carrier concentration is low, then it has been established that the rate-limiting step is interchain transfer of carriers. However, this raises a difficulty. Solitons are *topologically restricted* in that they cannot move from one chain to another. Two models have been developed to eliminate this difficulty. The first of these is due to Kivelson [121]. In this approach it is assumed that a substantial concentration of neutral solitons is present. Interchain chare transfer occurs via the transfer of charge from a charged soliton on one chain to a neutral soliton on another adjacent chain. The process is termed *intersoliton hopping* and is specific to polyacetylene. The second model is one in which the charged solitons are present in pairs on a single chain and are constrained to remain close to one another. This doubly charged excitation may then be treated as a *bipolaron*, which in chemical terms is a dication. The process of interchain transport is then controlled by the rate at which these bipolarons can hop or possibly tunnel between chains [122]. The bipolarn model is now thought to have the most general validity and can be used to rationalize the electrical conductivity of conjugated polymers whether or not they exhibit a degenerate ground state. Chance et al. [123] have noted that like charged soliton pairs are analogous to bipolarons and they are not topologically restrained from interchain hopping. Friend and Burroughs [124] have noted that the interchain transport will involve the intermediate step in which one of the two charges transfers to the adjacent chain and the instantaneous description is that of two polarons located on adjacent chains. If the second charge then follows the first, then the bipolaron has moved from one chain to another. The energy barrier to be surmounted here is that equivalent to the stabilization energy of the bipolaron or the soliton–antisoliton pair. This barrier is typically 0.31 eV [122].

It should be noted that trans polyacetylene is unique in the degeneracy of its ground state. All other conjugated polymers have nondegenerate ground states. One cannot in these cases utilize the concept of soliton transport, since if two regions separated by a topological defect are not energetically degenerate, then formation of single solitons is energetically unfavorable [118]. The energetically preferred configuration is that of pairing [119]. This can be illustrated by considering polyparaphenylene (Fig. 23), which can be represented either as a benzenoid or quininoid (this is of higher energy) structure. We see from Fig. 23 that unpaired electrons are generated where the benzenoid and quininoid structures meet. It is well established in solid-state physics that if a charge carrier is localized and trapped, it tends to polarize the local environment, which then relaxes into a new equilibrium position. This local deformed section of the polymer chain and the charge carrier then are termed a polaron [120]. Unlike the soliton, the polaron must overcome an energy barrier before it can move, and so it undergoes a hopping process along the chain or between chains. In polyparaphenylene we see that the solitons are trapped due to the changes in the polymer structure (since the latter are energetically nonequivalent), and so a polaron is created that is an isolated charge carrier. The polaronic defect may be delocalized over approximately five ring units in the chain. If both defects are charged, then they may pair up to form a *bipolaron* [120, 125], which is a doubly charged defect that again is delocalized over about five rings. At high doping levels the bipolarons (the energy states of which are located in the band gap) interact to form bipolaron bands within the energy gap [120, 125–127]. This process is also illustrated in Fig. 23.

Hence a general picture for polymers with a nondegenerate ground state is as follows. The neutral polymer has full valence and empty conduction bands separated by a band gap. Electrochemical doping removes one electron and results in the generation of a polaron level located at midgap. Further oxidation results in the removal of a second electron to generate a bipolaron. Still further oxidation results in the generation of bipolaron energy bands in the band gap. Electronic conductivity is rationalized in terms of bipolaron hopping. The identity of the various classes of charge carriers has also been confirmed from spectroscopic studies [128]. The reader is directed to the review by Bredas and Street [125] for further details on this topic.

The analysis can be taken one step further. It is well established in the field of polymer science that dramatic improvements in the physical

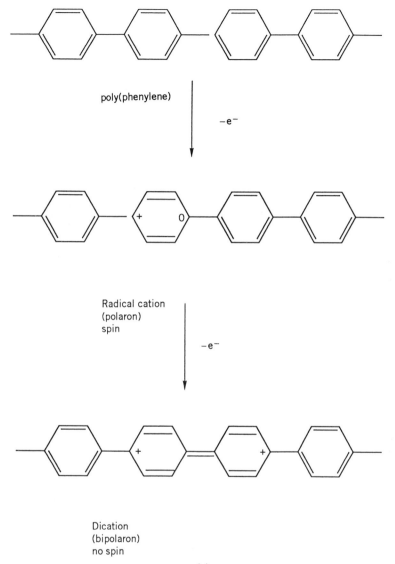

poly(phenylene)

$-e^-$

Radical cation
(polaron)
spin

$-e^-$

Dication
(bipolaron)
no spin

(a)

Figure 23. (a) Generation of polaron and bipolaron defects in conjugated organic polymer such as poly(phenylene). Removal of one electron from poly(phenylene) forms cation–radial pair (termed polaron); removal of second electron forms spinless dication (termed bipolaron). Spatial extent of bipolaron will be determined by competition between electrostatic repulsion of like charges and energetic disadvantage of quininoid resonance structure. (b) Band structure of conjugated polymer as function of doping level illustrating polaronic and bipolaronic states in band gap. Electronic conduction involves both polarons and bipolarons.

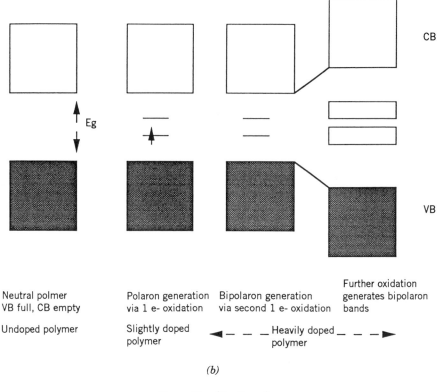

Neutral polmer	Polaron generation	Bipolaron generation	Further oxidation generates bipolaron
VB full, CB empty	via 1 e- oxidation	via second 1 e- oxidation	bands
Undoped polymer	Slightly doped polymer	◄ _ _ _ Heavily doped _ _ _ ► polymer	

(b)

Figure 23. *(Continued)*.

properties of polymers may be brought about via chain extension and alignment. It has been shown that a dramatic improvement in DC electrical conductivity may be achieved if the chains are stretched and aligned. The physics of conductivity optimization in conjugated polymer systems has been analyzed by Heeger [129]. The essential aim is to synthesize conducting polymers with a sufficient quality such that the mean free path is limited by intrinsic scattering events from thermal vibrations of the lattice (phonons). He noted that a principal problem is that of localization. Quasi-one-dimensional electronic systems are prone to localization of electronic states due to disorder. Consider an array of polymer chains, each with a few defects of the sp^3 type. We assume that the typical distance δ between defects is many lattice sites, but it is still considerably less than the total length L of the polymer chain. A charge carrier on a typical chain will then move with the Fermi velocity v_F until it meets a defect site, at which point it back scatters and moves in the opposite direction until it meets another defect site on the same chain and back scatters again. This multiple-resonant scattering localizes the electronic wave function, and the resultant conjugation length is much less than the chain length L. The result is electronic localization, with carrier transport limited by phonon-assisted hopping, according to the Mott model.

In the Mott model of variable range hopping the conductivity is given by

$$\sigma = \sigma_0 \exp\left[-\left(\frac{T}{T_0}\right)^{-\gamma} \right] \qquad (2.119)$$

where σ_0 and T_0 are constants and γ is a number, where $\frac{1}{4} < \gamma < \frac{1}{2}$, related to the dimensionality d of the hopping process via

$$\gamma = \frac{1}{1+d} \qquad (2.120)$$

Note that σ_0 depends on the electron–phonon coupling constant and $T_0 = 16/k_B \xi^2 N(E_F)$, where $N(E_F)$ denotes the localized density of states near the Fermi energy F_F and ξ denotes the decay length of the wave function. In such a situation the conductivity will be rather low and would decrease to zero as the temperature $T \to 0$, in contrast to the behavior exhibited by a metal. This type of conductivity behavior has been verified for many conjugated polymer systems. A particularly good summary of transport properties in conjugated polymers such as polyacetylene has been provided in a comprehensive review written by Heeger et al. [130].

The problem of localization is overcome due to the fact that interchain charge transfer occurs. Heeger [129] has noted that if the molar mass of the polymer is fairly high and if only a few sp^3-type defects are present, then even if one has only relatively weak interchain coupling, one dimensional localization can be avoided, and the transport problem becomes three dimensional. The mean free path will become quite large and will no longer be limited by static defects and imperfections. Under such conditions a marked increase in conductivity will be observed. The electronic mean free path will be determined by phonon-scattering events, as is the case for true metals. Under such conditions the conductivity σ will be given by [131].

$$\sigma = \frac{nq^2 b^2}{4\pi\hbar} \frac{M\omega_0 t_0^2}{\beta^2 \hbar} \exp\left(\frac{\hbar\omega_0}{k_B T}\right) \qquad (2.121)$$

In the latter expression b denotes the lattice constant, M is the molar mass of the polymer repeat unit, ω_0 is the frequency of the lattice phonon, β is the electron–phonon coupling constant, and t_0 denotes the π electron transfer matrix element ($4t_0$ is the π bandwidth). For polyacetylene $\beta = 4.1$ eV Å^{-1} and $\hbar\omega_0 = 0.12$ ev. For a typical carrier density $n = 10^{21}$ cm^{-3} one can estimate a room temperature value for the intrinsic conductivity for trans polyacetylene of close to 2×10^6 S cm^{-1}, which is about four times greater than that of copper. This gives one an idea of the potential conductivities that may be achieved for a doped conjugated polymer. However, for the best polymers prepared to date, room temperature conductivities are in the range 0.2–1.0×10^5 S cm^{-1} and do not exhibit such a strong temperature dependence on temperature as that predicted by Eq. (2.121). Hence this means that polymer fabrication techniques have still not been developed sufficiently that will produce materials with near intrinsic conductivities.

4. *Polymer Molar Mass and Chain Orientation Effects on Conductivity Behavior of Electronically Conducting Polymers: Pearson Model*

We have previously noted that both intrachain and interchain transport of carriers must be considered when examining the electronic conductivity of conjugated polymer materials. We now describe a theoretical model recently developed by Pearson and co-workers [132] that can be used to examine the effect of polymer molar mass and polymer orientation on the conductivity behavior of conjugated polymer materials.

If interchain transport occurs less readily than motion along the polymer chain, then one would expect that the macroscopic conductivity would be limited by the overall size of the polymer and so would depend to a large extent on the molar mass of the material. On the other hand, if interchain transport occurs more readily than intrachain transport, then the observed conductivity would be independent of the polymer molar mass. The competition between inter- and intrachain transport has been investigated by de Gennes [133]. Conductivity is quantified in terms of two characteristic times: τ_c, the mean lifetime of the charge carrier on the polymer chain, and τ_i, the time taken for the charge carrier to completely traverse the length of the polymer chain. When $\tau_c \gg \tau_i$, the conductivity is limited by interchain hopping. Conversely, when $\tau_i \gg \tau_c$, intrachain conductivity is rate determining.

The charge transport problem may be approached by solving the Fick diffusion equation in one dimension for the flow of the probability density ρ of the diffusing charge along a polymer modeled as a Gaussian coil, where one has a unit instantaneous plane source at $t = 0$ and reflecting boundary conditions at $s = 0$ and $s = L$, where L is the total length of the polymer chain and s denotes the displacement [134]. If we let $\rho(s, s_0, t)$

represent the probability density that the diffusing charge is located between s and $s + ds$ at a time t given that it was at location s_0 at time $t = 0$, then we have

$$\rho(s, t; s_0) = \frac{1}{L}\left[1 + 2\sum_{j=1}^{\infty}\cos\left(\frac{j\pi s}{L}\right)\cos\left(\frac{j\pi s_0}{L}\right)\exp\left(-\frac{j^2 t}{\tau_i}\right)\right] \quad (2.122)$$

Here we note that $\tau_i = L^2/\pi^2 D_i$, where D_i denotes the diffusion coefficient of the carrier along the chain. If the charge carrier hops onto the chain at a random position, then we may describe the average displacement $\langle s(t)\rangle = |s - s_0|$ as

$$\langle s(t)\rangle = \frac{1}{L}\int_0^L ds_0 \int_0^L \rho(s, t; s_0)|s - s_0|\, ds$$

$$= \frac{L}{3}\left[1 - \frac{6}{\pi^2}\sum_{j=1}^{\infty}\frac{1}{j^2}\exp\left(-\frac{j^2 t}{\tau_i}\right)\right] \quad (2.123)$$

where we have averaged over all possible starting points on the chain. One can also evaluate the mean-square displacement $\langle s(t)^2\rangle$ as follows:

$$\langle s^2(t)\rangle = \frac{1}{L}\int_0^L ds_0 \int_0^L (s - s_0)^2\rho(s, t; s_0)\, ds$$

$$= \frac{L^2}{6}\left[1 - \frac{90}{\pi^4}\sum_{j=1}^{\infty}\frac{1}{j^4}\exp\left(-\frac{j^2 t}{\tau_i}\right)\right] \quad (2.124)$$

We now assume that $m(t)$ denotes the probability density that the charge carrier leaves the polymer chain at time t. One assumes, for instance, that $m(t)$ is given by

$$m(t) = \frac{1}{\tau_c}\exp\left(-\frac{t}{\tau_c}\right) \quad (2.125)$$

where the time τ_c is given by

$$\tau_c = \int_0^{\infty} tm(t)\, dt \quad (2.126)$$

Now the average distance that the charge carrier diffuses along the chain between hopping on and off is equal to $\langle L(\tau_c) \rangle$ and is given by

$$\langle L(\tau_c) \rangle = \int_0^\infty m(t) \langle s(t) \rangle \, dt$$

$$= \frac{2L}{\pi^2} \sum_{j=1}^\infty \frac{\tau_c}{\tau_i + j^2 \tau_c}$$

$$= L \left(\frac{1}{\alpha} \coth \alpha - \frac{1}{\alpha^2} \right) \tag{2.127}$$

where $\alpha = \pi \sqrt{\rho_i / \tau_c}$. Note also that in deriving Eq. (2.127) we have averaged over the function $m(t)$. One can also show that $\langle L^2(\tau_c) \rangle$ is given by

$$\langle L^2(\tau_c) \rangle = \frac{5L^2}{2} \left(\frac{1}{\alpha^2} - \frac{3}{\alpha^2} \coth \alpha + \frac{3}{\alpha^4} \right) \tag{2.128}$$

We now derive an expression for the macroscopic DC conductivity σ. We recall from Eq. (2.118) that the conductivity is given by

$$\sigma = nq\mu = nq^2 \frac{D_c}{k_B T} \tag{2.129}$$

where D_c denotes the interchain diffusion coefficient. Using the random-walk model, this diffusion coefficient may be estimated from the expression

$$D_c = \frac{\langle X^2(\tau_c) \rangle}{2\tau_c} \tag{2.130}$$

where $\langle X^2(\tau_c) \rangle$ is the mean-square step length along the direction of conduction. For an isotropic polymer coil one may show that

$$\langle X^2(\tau_c) \rangle = \tfrac{1}{3} a \langle L(\tau_c) \rangle \tag{2.131}$$

where the parameter a denotes the persistance length of the polymer chain.* Hence from Eqs. (3.129)–(2.131) we obtain

$$\sigma = \frac{nq^2 a}{6\tau_c k_B T} \langle L(\tau_c) \rangle$$

$$= \frac{nq^2 aL}{6\tau_c k_B T} \left(\frac{1}{\alpha} \coth \alpha - \frac{1}{\alpha^2} \right) \tag{2.132}$$

We now recall that $\tau_i = L^2/\pi^2 D_i$ and $\alpha = \pi\sqrt{\tau_i/\tau_c} = L/\sqrt{D_i \tau_c}$; hence Eq. (2.132) may be written in the form

$$\sigma = \frac{nq^2 aL}{6k_B T \tau_c \alpha} \left(\coth \alpha - \frac{1}{\alpha} \right)$$

$$= \frac{nq^2 a}{6k_B T} \sqrt{\frac{D_i}{\tau_c}} \left(\coth \alpha - \frac{1}{\alpha} \right)$$

$$= \sigma^\infty \left(\coth \alpha - \frac{1}{\alpha} \right) = \sigma^\infty L_G(\alpha) \tag{2.133}$$

where we note that $L_G(\alpha)$ is the Langevin function, given by

$$L_G(\alpha) = \coth \alpha - \frac{1}{\alpha} \tag{2.134}$$

and σ^∞, which corresponds to the conductivity of an infinitely long chain, is given by

$$\sigma^\infty = \frac{nq^2 a}{6k_B T} \sqrt{\frac{D_i}{\tau_c}} \tag{2.135}$$

A plot of Eq. (2.133) for $\sigma(\alpha)$ as a function of α is presented in Fig. 24. Note that the parameter α has a rather simple physical meaning. It is equal to the ratio of the total length of the polymer chain L to the distance traversed by a charge carrier in a time τ_c on a chain of infinite length.

We now examine some limiting forms of Eq. (2.133) for the conductivity. Now, when the time between interchain hops is much greater than the time taken to explore the entire polymer chain (i.e., $\tau_c \gg \tau_i$), the

---

* The persistence length a of a polymer chain requires brief explanation. Let the angle between adjacent links of a polymer chain be given by θ. Assume that this angle is close to zero. Let the length of each link be x. The persistence length is then defined as $a = x/(1 - \cos \theta)$.

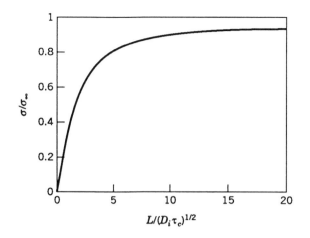

Figure 24. Conductivity of Gaussian coil as function of polymer length. Conductivity has been made dimensionless with reference to conductivity of infinitely long chain. Curve is calculated from Eq. (2.132) in text. Length is expressed in terms of dimensionless parameter $\alpha = L/(D_i\tau_c)^{1/2}$. Parameter $(D_i\tau_c)^{1/2}$ is distance that charge-carrying defect would travel along infinitely long chain in characteristic time τ_c (adapted from ref. 132).

parameter α will be small. Under such conditions we note that

$$\coth \alpha \cong \frac{1}{\alpha}\left(1 + \frac{\alpha^2}{3}\right)$$

Hence

$$L_G(\alpha) = \coth \alpha - \frac{1}{\alpha} \cong \frac{1}{\alpha} + \frac{\alpha}{3} - \frac{1}{\alpha} = \frac{\alpha}{3}$$

and so the conductivity reduces to

$$\sigma(\alpha) \cong \sigma^\infty \frac{\alpha}{3} = \frac{nq^2 aL}{18 k_B T \tau_c} \tag{2.136}$$

Hence in this low regime we see that the DC conductivity depends in a linear manner on L and hence on the molar mass M of the polymer. Alternatively, when α is large, the charge carrier hops from chain to chain much more rapidly than it explores a single-polymer strand and so $\coth \alpha \approx 1$. Hence, $L_G(\alpha) \approx 1$ since $1/\alpha \to 0$ when α is very large. In this case the DC conductivity is given by

$$\sigma \cong \frac{nq^2 a}{6k_B T} \sqrt{\frac{D_i}{\tau_c}} = \sigma^{\infty} \qquad (2.137)$$

and we note that the DC conductivity is independent of the molar mass of the polymer.

The Pearson model [132] may be extended to describe the way in which the conductivity is affected by mechanical stretching, which increases chain orientation. If the polymer is stretched so that the distance between the ends of each chain is increased by a factor of λ along the conduction direction, then the factor $\langle X^2(\tau_c) \rangle$ is given by

$$\langle X^2(\tau_c) \rangle = \frac{a}{3} \langle L(\tau_c) \rangle + \frac{a}{2L} \langle L^2(\tau_c) \rangle (\lambda^2 - 1) \qquad (2.138)$$

For $\tau_c \gg \tau_i$, using the definitions outlined in Eqs. (2.127) and (2.128) and expanding the coth α function for small α, simplifying, and substituting the result into the conductivity expression

$$\sigma = nq^2 \frac{\langle X^2(\tau_c) \rangle}{2k_B T \tau_c} \qquad (2.139)$$

one obtains the following result for conductivity:

$$\sigma = \frac{nq^2}{k_B T} \frac{aL}{18\tau_c} \left[1 + \frac{3}{4} (\lambda^2 - 1) \right] \qquad (2.140)$$

If the stretching ratio λ is large, then the increase in conductivity can be substantial. Pearson [132] has calculated that, for $\lambda = 3.6$, one obtains an order-of-magnitude increase in conductivity. The maximum stretch ratio for a Gaussian coil is $\lambda_m \approx \sqrt{L/a}$. Hence near this point we note that σ varies approximately with L^2.

Alternatively, if $\tau_i \gg \tau_c$, then we have the large-α limit and the conductivity takes the limiting form

$$\sigma = \frac{nq^2 a}{6k_B T} \sqrt{\frac{D_i}{\tau_c}} \left[1 + \frac{15}{4} \frac{\sqrt{D_i \tau_c}}{L} (\lambda^2 - 1) \right] \qquad (2.141)$$

Again we see that in the large-α limit the conductivity can be considerably enhanced by stretching the polymer chains. If we look at the intermediate situation when the time taken to explore the polymer chain is approximately equal to the lifetime of a charge carrier, that is, $\tau_c \approx r_i$, we note that stretching resulting in chain orientation will have the most marked effect on DC conductivity. Hence, one will obtain a maximum in σ as a function of α for certain values of λ. This effect is illustrated schematically in Fig. 25a. Furthermore a plot of σ as a function of λ for various values of α is given in Fig. 25b. Note that the overall shape of the σ-versus-λ curves is in good accord with experimental data reported by a number of workers [135, 136].

It is also possible that stretching may affect the conjugation length of the polymer chain. This will also affect the conductivity. If pulling the

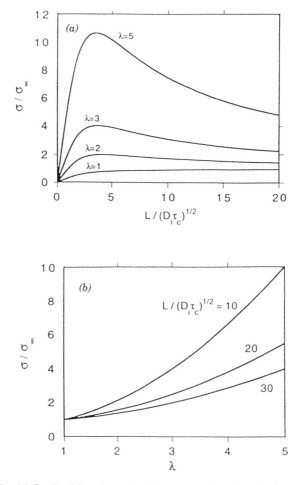

Figure 25. (a) Conductivity of stretched Gaussian coils vs. length of polymer chain for various values stretch ratio λ. (b) Conductivity of stretched Gaussian coil as function of stretch ratio λ for various values of normalized chain length α. (adapted from ref. 132.)

polymer aligns the π-conjugated orbitals in a way that enhances the charge transfer integrals [137], there will be further delocalization of electrons and an increase in the persistence length a. Stretching the polymer and the subsequent alignment of neighboring bonds along the chain should also increase the rate at which the charge carrier diffuses. Under such conditions the diffusion coefficient D_i is given by

$$D_i = \frac{l^2}{\tau_0} \exp(\Delta t) \qquad (2.142)$$

where l denotes the length of a dimer of conjugated bonds, τ_0 denotes the hopping time when the bonds are fully aligned, and Δt represents the difference in transfer integrals between the fully aligned and the actual positions of the bonds. Hence reducing Δt is equivalent to the removal of defects that slow the diffusion of charge along the polymer chain. The greatest increase in conductivity should occur in the region where the conductivity is independent of the molecular mass of the polymer, in which case σ is proportional to $D_i^{1/2}$.

Pearson and co-workers [132] also considered the situation where one has defects present in the polymer chain that actually block the propagation of charge. Such charge blocks may well be missing double bonds that were not formed during synthesis or that were lost in subsequent reactions during exposure to the environment. The block might also be due to a local steric arrangement of a dimer that effectively makes the charge transfer integrals so large that charge diffusion is stopped. This type of situation could arise if one has large side-chain substituents on neighboring repeat units. The analysis in this case is rather complex, and the reader is referred to the original paper for further details.

Much of the analysis described above refers to soluble conjugated polymers. These materials [such as the poly(3-alkylthiophenes)] have only been developed in recent years [138] and are of much interest from both a practical polymer processing and theoretical standpoint. Of course the behavior of nonconjugated polymers in solution has been studied for many years [139], but the polymer dynamics of conjugated polymers in solution is a relatively unexplored area. Heeger and co-workers [136] have noted that for a conjugated polymer in solution, one has an interesting balance between the conformational entropy which gives rise to chain flexibility, and the electronic energy of delocalised π electrons which tends to straighten the chain. Detailed models describing the statistical mechanics of conjugated polymer chain conformation in solvents are only now being developed, and some basic theoretical approaches have been described by Heeger and co-workers [136], Viallat

and Pincus [137c] and Pincus et al. [135]. An important concept introduced is that termed a *conformon*. An electron is assumed to be
associated with a local rigid region of the chain (the conformon)
[135, 137a,b]. This idea is similar to the polaron concept used in solid-
state physics. The reader is referred to the original literature for further
details.

5. Redox Switching in Conjugated Polymer Thin Films

In this section we examine the mechanism of potential induced redox
switching in electronically conducting polymer films. The term *redox
switching* refers to the transition from an electronically insulating to an
electronically conductive state.

It is now established that the process of redox switching in conjugated
organic polymer materials involves both a change in oxidation state of
sites on the polymer backbone as well as transport of charge-compensating counterions through the polymer matrix. Both anions and cations
have been identified as charge-compensating counterions in electronically
conducting polymers. Indeed, as we will discuss later, the mixed conducting nature of these materials is best represented in terms of a dual-rail
transmission line model. The voltammetric response of electronically
conducting polymer materials has been shown to depend on the nature of
the charge-compensating counterion. For instance, redox switching in
PPy/Cl^- materials (Fig. 26a) is accompanied by anion transport in the
polymer film and can be described in terms of the reaction

$$PPy + X^-(aq) \rightarrow PPy^+ X^-(pol) + e^-$$

Hence polymer chain oxidation results in the generation of positive
charges (polarons) on the backbone that are delocalized over about four
monomer units. In order that electroneutrality is preserved, ingress of
anions from the solution into the polymer matrix occurs. The reverse
occurs on reduction: Anions are ejected from the polymer matrix into the
solution. In contrast, for redox switching in PPy/DBS^- (where DBS^-
represents the dodecylbenzenesulfonate ion) films (Fig. 26b), polymer
oxidation is accompanied by cation ejection and reduction by cation
injection as follows:

$$PPy - C^+RSO_3^-(pol) \rightarrow PPy^+RSO_3^- + C^+ + e^-$$

The DBS^- ion is quite large and exhibits a low diffusive mobility through
the polymer matrix. Hence redox switching is accompanied by cation
transport in this case. It is clear that the shape of the voltammetric profile

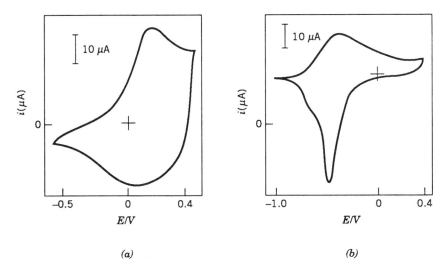

Figure 26. (*a*) Cyclic voltammetric response of 1.5-μm-thick PPy/Cl$^-$ coated glassy carbon electode in 0.1 *M* NaCl. Sweep rate, 20 mV s^{-1}. (*b*) Cyclic voltammetric response of 1.5-μm-thick PPy/DBS$^-$ coated glassy carbon electrode in 0.1 *M* NaCl. Sweep rate, 20 mV s^{-1}.

depends on whether anions or cations are involved in the redox switching process.

The process of redox switching clearly involves ionic transport. However, an even more important aspect of the switching process involves the dynamics of charge percolation along the polymer backbone. We now discuss a model originally developed by Aoki and co-workers [140–143] that examines the latter aspect of the redox switching process from a fundamental physical point of view.

The dynamics of redox switching is in general quite a complex problem to analyze, and it has only recently been examined in a series of papers by Aoki et al. [140–143]. The problem has also been examined by Kalaji and co-workers [144]. A number of specific observations may be made at this stage. The composition of the layer will be nonuniform during elec-trolysis: Certain regions of the film will be conducting and others will be insulating. Aoki et al. [140–143] have presented a model based on the propagation of a conductive zone under charge transfer control. A number of distinct steps must be considered. One initially has the convention of a resistive domain very close to the support electrode surface into a conductive zone due to interfacial electron transfer at the electrode–film interface. This is a triggering process. This conductive region then subsequently functions an a *quasi-electrode*. One has a charge transfer reaction at the interface between the conductive and the nonconductive regions. A new conductive region is generated, the quasi-electrode grows, and one has the development of a conductive front through the film. This type of conductive front propagation model differs significantly from the intergroup electron-hopping process used to quantify redox switching in redox polymer films. The latter is a quasi-diffusional process. The distribution of oxidized and reduced sites in the film arising from the electron-hopping process is illustrated in Fig. 27a. We see from this diagram that the conductive regions (designated as dots) are at a high concentration in regions near the electrode surface, but the distribution of conductive locations becomes more dilute the further out from the support electrode one goes. Furthermore the outermost conductive regions are disconnected electrically from the substrate electrode and are therefore effectively "dead" from the point of view of further growth. The essential point to note is that in order to develop a conductive zone, all the conductive domains must be *electrically connected* to the substrate electrode.

This idea of connectedness immediately leads to ideas of site and bond percolation [145] as being a suitable framework to describe the redox

Oxidised site concentration   Random fibrils   One dimensional fibrils
distribution

A   B   C

Figure 27. (*a*) Schematic representation of oxidized site distribution in electronically conducting polymer film according to Aoki phase propagation model. (*b*) Conductive front propagation modeled as random distribution of oxidized chains, each chain starting from random location on support electrode surface. (*c*) Assembly of conductive one-dimensional pillar fibrils of various lengths.

switching process. Hence a simple distribution model for the conductive front propagation is that of an assembly of electronically conductive, randomly oriented fibrils, each fibril starting at a random location on the support electrode and growing at a well-defined rate governed by the Butler–Volmer equation. This idea, proposed by Aoki, is illustrated in Fig. 27b. An even more simple model is an assembly of one-dimensional pillar fibrils of various lengths. This is illustrated in Fig. 27c. Fibril growth in the normal direction stops when the tip of the fibril reaches a distance $x = L$, the layer thickness. This simple fibrillar growth model is analogous to two-dimensional nucleation or electrochemical phase formation. The conducting strands initially propagate preferentially normal to the electrode surface as a result of a high local electric field present at the strand tip. Once strands have propagated across the film, the transformation can then proceed via expansion of the conductive cylinders as illustrated in Fig. 28. Hence the analogy to two-dimensional nucleation and growth is obvious.

We now examine the conductive phase propagation model in more detail. For the sake of simplicity we examine a continuous model and neglect the fine detail of the strand structure. Let us assume a very simple example, that of a potential step chronoamperometric experiment. Let the layer thickness be L, and we assume that initially the film is totally reduced. If one now applies a potential step perturbation to the film, one will have charge extraction at the metal–polymer interface characterized

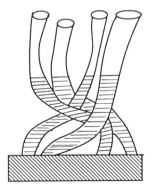

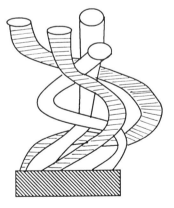

Intramolecular interface   Intermolecular interface

Figure 28. Schematic representation of conductive cylinders. Shaded regions are conductive, unshaded regions insulating. Both intramolecular and intermolecular interfaces are illustrated.

by a first-order rate constant k_E (units: reciprocal seconds). This generates a conductive zone in the region of the inner interface, which, as was stated, acts as a metallike electrode. One then has charge transfer across the conductive–nonconductive phase boundary, which is quantified by a propagation rate constant k_p (units: centimeters per second). The overall current time response to the potential step will depend on the balance between k_E and k_p. One makes the further assumption that the oxidized form is sufficiently conductive such that there is no potential drop in the conductive phase. Furthermore, we make the simplifying assumption that the film morphology has no effect on the rate of oxidation.

Let a and b denote the concentrations of reduced and oxidized species present in the film, and at any time t we have that the total site concentration $c_\Sigma = a + b$, whereas at time $t = 0$, $b = 0$ and $c_\Sigma = a$. Now charge transfer across the conducting–nonconducting boundary is given by the Butler–Volmer expression

$$j_p = v_p c_\Sigma = k_p \{a \exp(\alpha \vartheta) - b \exp[-(1 - \alpha)\vartheta]\} \qquad (2.143)$$

where j_p denotes the propagation flux, v_p is the velocity of propagation, and α is the transfer coefficient. The function θ is given by

$$\vartheta = \frac{nF}{RT}(E - E^0_{A/B}) \qquad (2.144)$$

We now consider a time period Δt during the potential step. The increment in surface coverage of the oxidized species at the conducting–nonconducting interface $\Delta \Gamma_B$ is

$$\Delta \Gamma_B = c_\Sigma [v_p(x + \Delta x) - v_p(x)] \Delta t \qquad (2.145)$$

Hence the change in oxidized species concentration Δb is given by

$$\Delta b = \frac{\Delta \Gamma_B}{\Delta x} = c_\Sigma \frac{v_p(x + \Delta x) - v_p(x)}{\Delta x} \Delta t \qquad (2.146)$$

In the limit as $\Delta x \to 0$ we obtain

$$\Delta b = \frac{\partial v_p}{\partial x} c_\Sigma \Delta t \qquad (2.147)$$

We now evaluate the quantity $\partial v_p / \partial x$. We note that

$$v_p = \frac{k_p}{c_\Sigma} \{ a \exp(\alpha\vartheta) - b \exp[-(1-\alpha)\vartheta] \}$$

$$= \frac{k_p}{c_\Sigma} \{ (c_\Sigma - b) \exp(\alpha\vartheta) - b \exp[-(1-\alpha)\vartheta] \} \qquad (2.148)$$

Hence we obtain

$$\frac{\partial v_p}{\partial x} = -\frac{k_p}{c_\Sigma} \{ \exp(\alpha\vartheta) + \exp[-(1-\alpha)\vartheta] \} \frac{\partial b}{\partial x}$$

$$= -\frac{\lambda}{c_\Sigma} \frac{\partial b}{\partial x} \qquad (2.149)$$

where λ is given by

$$\lambda = k_p \{ \exp(\alpha\vartheta) + \exp[-(1-\alpha)\vartheta] \}$$

$$= k_p \exp(\alpha\vartheta)[1 + \exp(-\vartheta)] \qquad (2.150)$$

Substituting Eq. (2.149) into Eq. (2.147), we obtain

$$\Delta b = -\lambda \frac{\partial b}{\partial x} \Delta t \qquad (2.151)$$

which in the limit as $\Delta t \to 0$ reduces to

$$\frac{\partial b}{\partial t} = -\lambda \frac{\partial b}{\partial x} \qquad (2.152)$$

Hence we finally obtain

$$\frac{\partial b}{\partial t} + \lambda \frac{\partial b}{\partial x} = 0 \qquad (2.153)$$

Hence we see that the differential equation quantifying the propagation of conductive sites through the film is first order with respect to distance and time. It is in effect the equation of continuity, and it is considerably different in form from the Fick diffusion equation, which is second order in the space coordinate and first order in time. This expression is valid for $0 < x < L$.

This expression must now be solved. Since it is a first-order linear partial differential equation, one may use the method of characteristics

[146] as follows. The partial differential equation is equivalent to

$$\frac{dt}{1} = \frac{dx}{\lambda} = \frac{db}{0} \qquad (2.154)$$

Equating the first and the last terms, we obtain $db = 0$ and integrating we obtain $b/c_\Sigma = \beta_1$, where β_1 is a constant independent of the time. Equating the first and the middle expressions in Eq. (2.130), we obtain $dt = dx/\lambda$ and so $dx = \lambda \, dt$, which on integration yields $x = \lambda t + \beta_2$, where β_2 is another integration constant. When $x = 0$, we note that $b = b_0$ and so $\beta_1 = b_0/c_\Sigma$. Also when $x = 0$, $\beta_2 = -\lambda t$. Consequently, to proceed further, we must evaluate b_0, the concentration of oxidized species at $x = 0$.

We now examine the situation at $x = 0$. In this case the charge transfer is again described by the Butler–Volmer expression as follows:

$$
\begin{aligned}
\frac{db_0}{dt} &= k_E\{a_0 \exp(\alpha\vartheta) - b_0 \exp[-(1-\alpha)\vartheta]\} \\
&= k_E\{(c_\Sigma - b_0)\exp(\alpha\vartheta) - b_0\exp[-(1-\alpha)\vartheta]\} \qquad (2.155) \\
&= k_E(c_\Sigma \exp(\alpha\vartheta) - b_0\{\exp(\alpha\vartheta) + \exp[-(1-\alpha)\vartheta]\})
\end{aligned}
$$

We now recall that $\lambda/k_p = \exp(\alpha\vartheta) + \exp[-(1-\alpha)\vartheta]$; hence Eq. (2.155) simplifies to

$$\frac{db_0}{dt} = k_E\left[c_\Sigma \exp(\alpha\vartheta) - \frac{b_0\lambda}{k_p} \right] \qquad (2.156)$$

This ordinary differential equation may be readily integrated to obtain

$$\int \frac{db_0}{c_\Sigma \exp(\alpha\vartheta) - \lambda b_0/k_p} = k_E t + \beta \qquad (2.157)$$

We now use the following fundamental identity:

$$\int \frac{dx}{\gamma x + \delta} = \frac{1}{\gamma} \ln(\gamma x + \delta) \qquad (2.158)$$

Setting $\gamma = -\lambda/k_p$, $\delta = c_\Sigma \exp(\alpha\vartheta)$, we obtain

$$-\frac{k_p}{\lambda} \ln\left[c_\Sigma \exp(\alpha\vartheta) - \frac{\lambda b_0}{k_p} \right] = k_E t + \beta \qquad (2.159)$$

When $t = 0$, $b_0 = 0$, and so $\beta = -(k_p/\lambda)\ln[c_\Sigma \exp(\alpha\vartheta)]$, and Eq. (2.159) simplifies to

$$\frac{c_\Sigma \exp(\alpha\vartheta)}{c_\Sigma \exp(\alpha\vartheta) - \lambda b_0/k_p} = \exp\left(\frac{\lambda k_E}{k_p}\right) = \exp(\lambda\kappa t) \qquad (2.160)$$

where $\kappa = k_E/k_p$. We can finally solve for b_0 to obtain

$$b_0 = \frac{k_p c_\Sigma}{\lambda} \exp(\alpha\vartheta)[1 - \exp(-\kappa\lambda t)] \qquad (2.161)$$

Recalling the definition of λ from Eq. (2.150), we obtain

$$\begin{aligned} b_0 &= \frac{c_\Sigma \exp(\alpha\vartheta)}{\exp(\alpha\vartheta) + \exp[-(1-\alpha)\vartheta]} [1 - \exp(-\kappa\lambda t)] \\ &= \frac{c_\Sigma}{1 + \exp(-\vartheta)} [1 - \exp(-\kappa\lambda t)] \end{aligned} \qquad (2.162)$$

We have therefore finally arrived at our expression for b_0.

We now return to the problem of determining the analytical form of the concentration profile $b(x, t)$ for the oxidized species throughout the layer.

We recall that

$$\frac{b(x, t)}{c_\Sigma} = \beta_1 \qquad (2.163)$$

where $\beta_1 = b_0/c_\Sigma$ and $\beta_2 = -\lambda t$ when $x = 0$. Hence from Eq. (2.163) we obtain

$$\beta_1[1 + \exp(-\vartheta)] = 1 - \exp(\kappa\beta_2) \qquad (2.164)$$

Hence the expression for $b(x, t)$ reduces to

$$b(x, t) = c_\Sigma \beta_1 = \frac{c_\Sigma[1 - \exp(\kappa\beta_2)]}{1 + \exp(-\vartheta)} = \frac{c_\Sigma[1 - \exp(\kappa x)\exp(-\kappa\lambda t)]}{1 + \exp(-\vartheta)} \qquad (2.165)$$

where we have used the fact that for any value of the space coordinate x, $\beta_2 = x - \lambda t$. Hence we wrote that Eq. (2.165) provides an expression describing the concentration profile of oxidized species in the layer as a function of distance and time.

We now transform Eq. (2.165) into a nondimensional form as follows.

We set

$$u = \frac{b}{c_\Sigma} \qquad \chi = \frac{x}{L} \qquad \tau = \kappa \lambda t \tag{2.166}$$

Hence Eq. (2.165) reduces to

$$u(\chi, \tau) = \frac{1}{1 + \exp(-\vartheta)} [1 - \exp(\kappa L\chi) \exp(-\tau)]$$

$$= F(\vartheta)[1 - \exp(\mu\chi) \exp(-\tau)] \tag{2.167}$$

where we have set $F(\vartheta) = 1/[1 + \exp(-\vartheta)]$ and $\mu = \kappa L$. Clearly the value of $F(\theta)$ depends on the magnitude of the potential step applied. If the step amplitude is large, then $\theta \gg 1$ and so $\exp(-\theta) \to 0$ and $F(\theta) \to 1$. The concentration profiles $u(\chi, \tau)$ are illustrated in Fig. 29 for various values of τ. We note from these profiles that for $\tau < \mu$, the conductive zone has not yet reached the outside of the film. For $\tau > \mu$ the conductive zone has reached the outside of the film, and the concentration profiles become rather uniform.

The total quantity of oxidized species in the layer is given by the integrated charge Q, which is

$$\frac{Q}{nFA} = \Gamma_B = \int_0^L b(x, t) \, dx \tag{2.168}$$

and the current is given by

$$i(t) = \frac{dQ}{dt} = nFA \frac{d\Gamma_B}{dt} = nFA \frac{d}{dt} \int_0^L b(x, t) \, dx \tag{2.169}$$

Hence in order to obtain an expression for the amperometric current response $i(t)$, we must integrate $b(x, t)$ with respect to the space coordinate x and subsequently differentiate with respect to time.

To make the algebra simple, we work in terms of nondimensional quantities. We define a dimensionless surface coverage γ as $\gamma = \Gamma_B/c_\Sigma L$. Furthermore, we note that

$$\frac{d\Gamma_B}{dt} = c_\Sigma \lambda \kappa L \frac{d\gamma}{d\tau} = c_\Sigma \lambda \mu \frac{d\gamma}{d\tau}$$

Furthermore, the normalized current response $y(\tau)$ is given by $y(\tau) =$

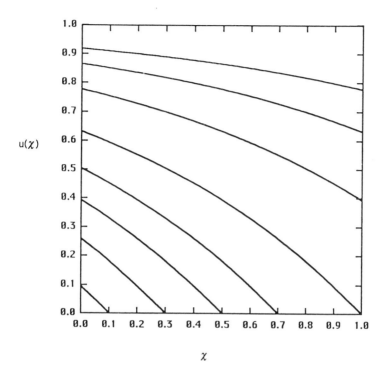

Figure 29. Normalized concentration profile $u(\chi)$ for oxidized sites as function of normalized distance χ in layer. Profiles were calculated form Eq. (2.167) in text, setting $\mu = 1$, and are presented for various values of normalized time τ. From bottom to top $\tau = 0.1, 0.3, 0.5, 0.7, 1.0, 1.5, 2.0, 2.5$.

$i/nFA\mu\lambda c_\Sigma = d\gamma/d\tau$. One further complication must be considered. When $\tau \leq \mu$, corresponding to a time $t \leq L/\lambda$, the oxidation front has not reached the outside of the film, and so the effective upper limit in the integral in Eq. (2.169) is $x = \lambda t$. In dimensionless terms this limit corresponds to $\chi = \tau/\mu$. Hence for $\tau \leq \mu$ the expression for the non-dimensional surface coverage of oxidized species γ is given by

$$\gamma = F(\vartheta) \int_0^{\tau/\mu} [1 - \exp(-\tau) \exp(\mu\chi)] \, d\chi$$

$$= F(\vartheta) \left\{ \frac{\tau}{\mu} - \frac{1}{\mu} [1 - \exp(-\tau)] \right\}$$

(2.170)

Transforming back to dimensional quantities, we obtain

$$\Gamma_B = c_\Sigma F(\vartheta)\{\lambda t - \kappa^{-1}[1 - \exp(-\kappa\lambda t)]\}$$

$$= \frac{c_\Sigma}{1 + \exp(-\vartheta)}\{\lambda t - \kappa^{-1}[1 - \exp(-\kappa\lambda t)]\} \tag{2.171}$$

On the other hand, when $\tau > \kappa L = \mu$, corresponding to a time $t > L/\lambda$, the upper limit of the integral defining the surface coverage is $x = L$ or $\chi = 1$. In this case we have

$$\gamma = F(\vartheta)\int_0^1 [1 - \exp(-\tau)\exp(\mu\chi)]\,d\chi$$

$$= F(\vartheta)\left\{1 - \frac{\exp(-\tau)}{\mu}[\exp(\mu) - 1]\right\} \tag{2.172}$$

Hence the critical time t_c is given by the quantity L/λ. Transforming Eq. (2.172) back into dimensioned variables, we obtain

$$\Gamma_B = \frac{c_\Sigma}{1 + \exp(-\vartheta)}\{L - \kappa^{-1}\exp(-\kappa\lambda t)[\exp(\kappa L) - 1]\} \tag{2.173}$$

Finally the current response is obtained by differentiating Eqs. (2.171) and (2.173) with respect to time. Working in terms of nondimensional quantities, we see that for $\tau < \mu$ we have

$$y = \frac{d\gamma}{d\tau} = \frac{1}{\mu}F(\vartheta)[1 - \exp(-\tau)] \tag{2.174}$$

or

$$i = \frac{nFAc_\Sigma\lambda}{1 + \exp(-\vartheta)}[1 - \exp(-\kappa\lambda t)] \tag{2.175}$$

Since we note that $\lambda = k_p \exp(\alpha\vartheta)[1 + \exp(-\vartheta)]$, Eq. (2.175) reduces to

$$i = nFAc_\Sigma k_p \exp(\alpha\vartheta)[1 - \exp(-k\lambda t)]$$

$$= nFAc_\Sigma k_p \exp(\alpha\vartheta)(1 - \exp\{-k_E \exp(\alpha\vartheta)[1 + \exp(-\vartheta)]t\})$$

$$(2.176)$$

If θ is large, then Eq. (2.176) reduces to

$$i = nFAc_\Sigma k_p \exp(\alpha\vartheta)\{1 - \exp[-k_E \exp(\alpha\vartheta)t]\} \qquad (2.177)$$

This expression will be valid for times less than the critical time t_c. On the other hand, when the conductive zone has reached the outside of the layer, corresponding to $\tau > \mu$ or $t > t_c$, the normalized current is given by

$$y = \frac{d\gamma}{d\tau} = \frac{\exp(-\tau)}{\mu}[\exp(\mu) - 1] \qquad (2.178)$$

Hence in terms of the actual current we obtain

$$i = \frac{nFAc_\Sigma \lambda \exp(-\kappa\lambda t)}{1 + \exp(-\vartheta)}[\exp(\kappa L) - 1]$$

$$= nFAc_\Sigma k_p \exp(\alpha\vartheta)\exp\{-k_E \exp(\alpha\vartheta)[1 + \exp(-\vartheta)]t\}$$

$$\times \left[\exp\left(\frac{k_E L}{k_p}\right) - 1\right] \qquad (2.179)$$

When $\theta \gg 1$, the latter simplifies to

$$i = nFAc_\Sigma k_p \exp(\alpha\vartheta)\exp[-k_E \exp(\alpha\vartheta)t]\left[\exp\left(\frac{k_E L}{k_p}\right) - 1\right] \qquad (2.180)$$

We see therefore that the expression for the chronoamperometric response is rather complicated: The actual expression used depends on whether the conductive zone has reached the outside of the film.

We can make one further simplifying assumption at this stage. We assume that the rate of electron transfer between the conductive region and nonconductive region is quite rapid. In this case the propagation rate constant $k_p \to \infty$. This means that $k_E \ll k_p$ and so the ratio $\kappa \ll 1$, and consequently $\mu \ll 1$. Let us also assume that $\theta \gg 1$ and so $F(\theta) = 1$. The total normalized current y is given by the product of $y(\tau < \mu)$ and $y(\tau > \mu)$, and so from Eqs. (2.174) and (2.178) we obtain

$$y(\tau < \mu) = \frac{1}{\mu}[1 - \exp(-\tau)] \approx \frac{1}{\mu}[1 - (1 - \tau)] = \frac{\tau}{\mu} \qquad (2.181)$$

In deriving the latter expression, we note that $\tau \ll 1$ as well since, by definition, $\tau < \mu$ and $\mu \ll 1$. When $\tau > \mu$, we have

$$y(\tau > \mu) \approx \frac{\exp(-\tau)}{\mu}[\exp(\mu) - 1] \approx \frac{\exp(-\tau)}{\mu}(1 + \mu - 1)$$

$$= \exp(-\tau) \qquad (2.182)$$

Hence the total normalized current is given by

$$y(\mu \ll 1) = y(\tau \leq \mu) \cdot y(\tau > \mu) = \frac{\tau}{\mu}\exp(-\tau) \qquad (2.183)$$

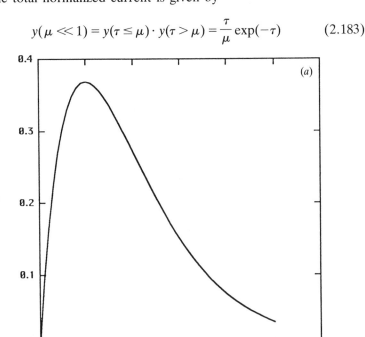

Figure 30. (a) Plot of normalized current y vs. normalized time τ. Curve was calculated using Eq. (2.183) in text setting $\mu = 1$. (b) The y vs. τ curves for various μ values. From top to bottom $\mu = 0.1, 1.0, 10$.

This is a very simple expression and is plotted in Fig. 30. It is immediately evident from this figure that a very characteristic current response is observed. At short times the current rises rapidly, reaches a maximum, and then subsequently exhibits a gradual decay. This maximum occurs at $\tau = \mu$. The rate-determining step is the charge transfer reaction at the support electrode–polymer interface at $x = 0$. The rapid increase in current observed at short times is due to the fact that the area of the quasi-metallic electrode (the conductive region) increases. Once the oxidation front reaches the outer boundary of the film, the area no longer increases. Since the amount of reactant is limited in the film, the current will decrease after the peak. The critical time corresponding to the current maximum is given by $t = L/\lambda$. Hence this time should vary with the upper limit of the potential step due to the potential dependence of λ. If should also vary in a linear manner with the layer thickness L. For $\theta \gg 1$ we note that $t_c = (L/k_p) \exp(-\alpha\vartheta)$, and so taking logarithms, we obtain

$$\ln t_c = \ln\left(\frac{L}{k_p}\right) - \alpha\vartheta = \ln\left(\frac{L}{k_p}\right) - \frac{\alpha n F}{RT}(E - E^0) \qquad (2.184)$$

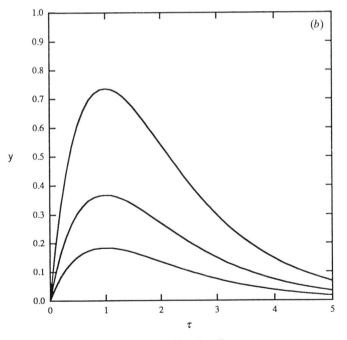

Figure 30. (*Continued*).

Hence a plot of $\ln t_c$ versus E is linear, the slope yields αn, and the intercept yields the propagation rate constant k_p provided that the layer thickness L may be determined. Hence we note that the variation of the peak line with layer thickness and applied potential is diagnostic.

Plots similar in shape to those illustrated in Fig. 30 have been obtained experimentally for a number of conducting polymers such as poly-(pyrrole) and poly(aniline). The currents passed during redox switching are usually quite large, and therefore the experiments have been conducted on polymer-coated ultra-microelectrodes. Some typical experimental transients recorded for polyaniline-coated microelectrodes are illustrated in Fig. 31. In each case the layer was initially reduced at $-100\,\text{mV}$ for 10 min before the oxidizing potential pulse was applied. In all cases a bell-shaped current transient was observed, as predicted by the phase propagation model. The peak maximum was found to shift to shorter times with increasing amplitude of the potential step applied. This observation is also in accord with the predictions of the phase propagation model. The most effective way of analyzing the experimental data is to utilize the reduced variable representation where the experimental transient is normalized with respect to peak current i_m and peak time t_m. One can compare theory and experiment by plotting i/i_m versus t/t_m. One notes that

$$\frac{i}{i_m} = \frac{t}{t_m} \exp[-\kappa\lambda(t - t_m)] \tag{2.185}$$

Furthermore we also recall that $\tau_m = \kappa\lambda t_m$ and so $\kappa\lambda = \tau_m/t_m$. We also note that $\tau = \tau_m$ when we have $dy/d\tau = 0$, and so from Eq. (2.183) we have $(1/\mu)\exp(-\tau)(1 - \tau_m) = 0$ and hence $1 - \tau_m = 0$ and $\tau_m = 1$. This means that $\kappa\lambda = 1/t_m$, and so Eq. (2.185) reduces to

$$\frac{i}{i_m} = \frac{t}{t_m} \exp\left(-\frac{t - t_m}{t_m}\right) \tag{2.186}$$

The experimental and computed normalized current transients for redox switching at polyaniline-coated microelectrodes are compared in Fig. 32. The full line corresponds to the theory [Eq. (2.186)] whereas the discrete points are experimental data. The fit is quite good over an extended range of t/t_m. In contrast, Kalaji and co-workers [144] have suggested that oxidative redox switching may be modeled in terms of a nucleation and growth model; the model chosen is instantaneous two-dimensional nucleation and growth. In terms of a reduced-variable representation one can show that instantaneous nucleation and two-dimensional growth is

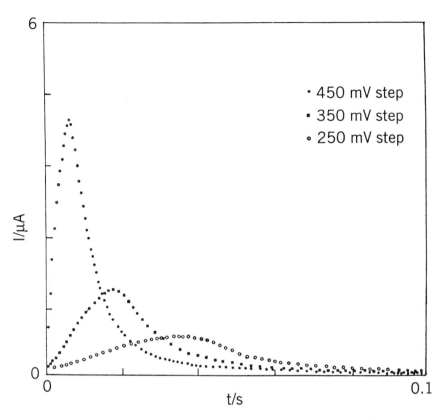

Figure 31. Typical chronoamperometric current transients for redox switching obtained for poly(aniline)-coated ultramicro Pt electrode (10 μm diameter) in 1.0 M HCl. In each case film was initially reduced at -300 mV for 10 min before application of potential pulse. Step amplitude is indicated in diagram.

described via

$$\frac{i}{i_m} = \frac{t}{t_m} \exp\left(-\frac{t^2 - t_m^2}{2t_m^2} \right)$$

(2.187)

The experimentally determined normalized transients are directly compared with the theoretical transient defined via Eq. (2.187) in Fig. 32. Note that the fit, although good at short times, is only approximate, especially for t/t_m values greater than unity. A better fit is obtained for all values of the normalized time if one uses the Aoki phase propagation

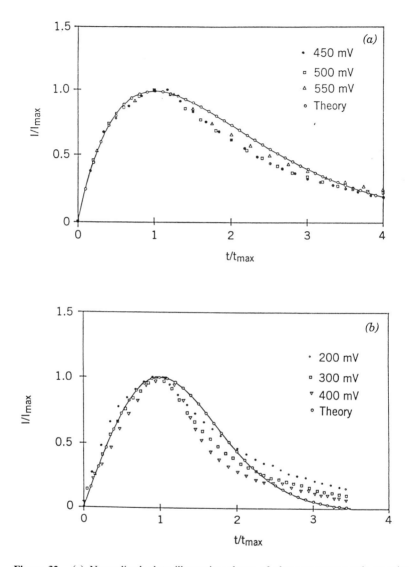

Figure 32. (*a*) Normalized plots illustrating shape of chronoamperometric transient response for redox switching at poly(aniline) microelectrodes. Discrete data points are experimental measurements. Full line corresponds to normalized transient computed in context of Aoki phase propagation model [Eq. (2.186)]. Note good correspondence between theory and experiment. (*b*) Normalized transient computed using Peter two-dimensional instantaneous nucleation model [Eq. (2.187)] compared with experimental data recorded via chronoamperometry at poly(aniline) microelectrodes. Note poor agreement between theory and experiment for t/t_m values greater than unity.

model. These results seem to indicate that some elements of nucleation theory may be used to describe redox switching in conducting polymer films, but the approach does not tell the entire story. One important drawback of the nucleation picture is the specific dependence of the shape of the current transient on the geometry of the expanding growth center. This of course will limit the applicability of the model. We conclude therefore that the Aoki linear propagation model may perhaps better describe the oxidative redox switching process in surface-deposited electronically conducting polymer films.

In a recent communication Aoki et al. [141] extended the theoretical analysis of the phase propagation model to consider the analytical form of the current response under conditions of a linear potential sweep. In this case the dimensionless potential θ is given by

$$\vartheta = \frac{nF}{RT}(E_i - E^0) + \frac{nFvt}{RT} \qquad (2.188)$$

and the charge transfer reaction at $x = 0$ is described by the Nernst equation

$$\frac{b_0}{c_\Sigma - b_0} = \exp \vartheta \qquad (2.189)$$

Again the equation of continuity, given by

$$\frac{\partial b}{\partial t} + k_p \exp(\alpha\vartheta)\frac{\partial b}{\partial x} = 0 \qquad (2.190)$$

is solved in a manner analogous to that described previously to obtain the following expression for the concentration profile of oxidized species:

$$b(x, t) = \frac{c_\Sigma}{1 + \Psi[\exp(\alpha\vartheta) - \alpha x/L\zeta]^{-1/\alpha}} \qquad (2.191)$$

where the function $\Psi(u)$ for any general argument u is given by

$$\Psi(u) = \begin{cases} u & \text{for } u \geq 0 \\ 0 & \text{for } u < 0 \end{cases} \qquad (2.192)$$

Furthermore the kinetic parameter ζ is given by

$$\zeta = \frac{k_p RT}{nFvL} \qquad (2.193)$$

and is directly related to the propagation rate constant k_p. Again the total current is given by

$$i = nFA \frac{\partial}{\partial t} \int_0^L b(x, t) \, dx \qquad (2.194)$$

Substituting Eq. (2.191) into Eq. (2.194), integrating with respect to the space coordinate x, and then differentiating with respect to time, one obtains

$$\begin{aligned}
i &= \frac{n^2 F^2 A c_\Sigma L \upsilon}{RT} \\
&= \zeta \exp(\alpha \vartheta) \left\{ \frac{1}{1 + \Psi[\exp(\alpha \vartheta) - \alpha/\zeta]^{1/\alpha}} - \frac{1}{1 + \exp(\vartheta)} \right\} \qquad (2.195)
\end{aligned}$$

Figure 33 is a plot of the dimensionless current $y = iRT/n^2F^2Ac_\Sigma L\upsilon$ as a function of the dimensionless potential $\alpha\theta = (\alpha nF/RT)[E(t) - E^0]$ for several different values of ζ and α. A number of factors should be noted from this diagram. First, the shape of the voltammograms are independent of the transfer coefficient α. Second, the shape greatly depends upon the numerical value chosen for the kinetic parameter ζ. For $\zeta > 1$, corresponding to $k_p > (nFL/RT)\upsilon$, the voltammograms exhibit a characteristic bell shape and the peak current is located at $\theta = 0$ or $E = E^0$. Under such conditions the current response reduces to

$$i = \frac{n^2 F^2 A c_\Sigma L \upsilon}{RT} \frac{1}{4 \cosh^2(\vartheta/2)} \qquad (2.196)$$

This is the characteristic response for the linear potential sweep voltammogram for an electroactive polymer exhibiting rapid charge percolation. As the value of the kinetic parameter ζ decreases (in effect one proceeds from the reversible to the quasi-reversible and ultimately irreversible regimes), the shape of the voltammetric response changes. The peak symmetry is lost and the peak becomes very sharp. For $\alpha\zeta < 0.1$ a very characteristic response is obtained: One has an exponential increase in current with increasing potential leading to the peak, followed by a very sharp drop in current after the peak. Furthermore, for $\zeta < 0.1$, the voltammetric response shifts in a positive direction along the potential axis in a linear manner with a decrease in log ζ. The shape of the voltammetric profile remains invariant under these conditions. Aoki et al. [141] have claimed that if the pseudocapacitive contribution to the voltammetric response is subtracted from the total observed current response, then this type of needle current response is observed in

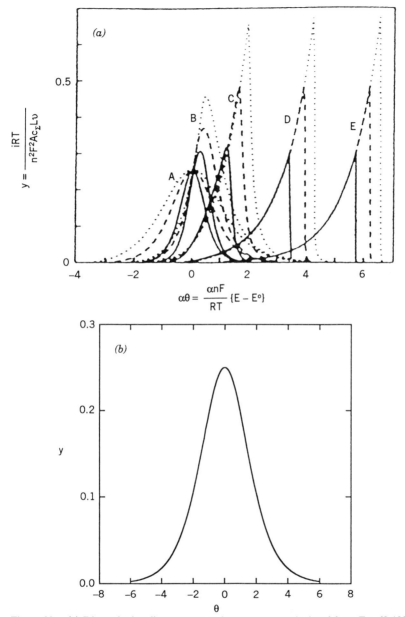

Figure 33. (a) Dimensionless linear sweep voltammograms calculated from Eq. (2.195) for parameter ζ values of 10 (A), 0.5 (B), 0.1 (C), 0.01 (D), and 0.001 (E). Curves for various α values are also illustrated: $\alpha = 0.3$ (- - - -), 0.5 (———), and 0.7 (.....). (Adapted from Ref. 141.) (b) Plot of normalized voltammetric response for situation where $\zeta > 1$. Curve is calculated from Eq. (2.196).

experimentally obtained voltammograms. They do not present any such deconvoluted voltammograms in their paper to support this assertion, however. The reason for this very sharp decrease in current may be explained with reference to the concentration profiles illustrated in Fig. 34. This diagram is drawn to correspond to the situation when half of the redox couple in the layer is oxidized. The transfer coefficient α is assigned the value of 0.5. For large ζ values (10 and 1, say, curves A and B) the concentration profiles are fairly uniform. However, this situation changes when $\zeta \ll 1$. Consider curve D, corresponding to $\zeta = 0.01$. The concentration profile for the oxidized species, B, exhibits a rather drastic change from a value c_Σ to zero at $\chi = x/L = 0.5$. One therefore has a well-defined interface in this region of the film between the conductive and resistive parts of the film. The rate of propagation j_p is given by $j_p = k_p b \exp(\alpha\vartheta)$. Hence prior to the peak the current will rise exponentially with potential due to the Tafel dependence. When the oxidation front reaches the top of the film, corresponding to the situation postpeak, there is no further reduced species present and the current drops very rapidly to zero.

Aoki [142] has also considered the stochastic aspects of the phase

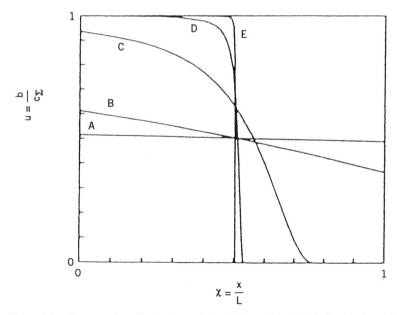

Figure 34. Concentration distribution calculated from Eq. (2.191) for ζ values of 10 (A), 1 (B), 0.1 (C), 0.01 (D), and 0.001 (E) assuming that $\alpha = 0.5$. (Adapted from ref. 141.)

propagation mechanism and has related his analysis to the theory of percolation and the fractal dimension of the system. In this approach the Nernst equation for charge transfer at the substrate–film interface was used to compute the probability of the presence of a conductive seed or nucleus. When the potential is incremented, this seed could then grow in a one-dimensional manner governed by the propagation rate constant k_p or the kinetic parameter ζ to form a conductive pillar of a definite length. Also, new nucleii can form at the support electrode–film interface during the potential increment. The location of the seed nucleus is assumed to be governed by a stochastic process. The concentration of oxidized species was replaced by the average length $\langle x \rangle$ of a one-dimensional pillar. The phase propagation process has been simulated using the Monte Carlo technique. An assembly of one-dimensional conductive pillars was examined. We shall not present the details of the analysis here. The reader is referred to the original publication [143] for full details. Figure 35 is a schematic of the results of his computation. When $\zeta = 3$ most pillars have reached the top of the film corresponding to $x = L$ because of generation of almost all the seed nucleii at the substrate–film interface results in the full growth of the pillars. Hence we see from Fig. 35a that the conductive zone is dispersed throughout the entire film, and visual inspection of the

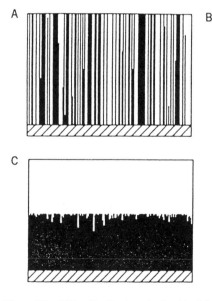
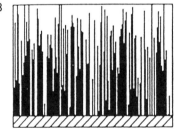

Figure 35. Pillar distribution obtained by Monte Carlo simulation of phase propagation problem in context of lps voltammetry for ζ values of 0.01 (A), 0.279 (B), and 3.0 (C) assuming that $\alpha = 0.5$. (Adapted from ref. 142.)

latter at a macroscopic level may confer the impression of uniform oxidation. If this one-dimensional picture is transformed to three dimensions, then one has a network of conductive fibrils within which there exists resistive microdomains. Furthermore one would expect that conductivity between the substrate electrode and the top of the layer would be good at any potential due to the extensive net of connected pathways through the film. This corresponds to the bell-shaped current response observed at ζ values greater than unity and reversible propagation kinetics. Conversely, when $\zeta \ll 1$, the conductive pillars are localized in a fairly uniform region near the substrate electrode. A well-defined front exists between the conductive and insulating regions in the film. This situation corresponds to Fig. 35c, for which $\zeta = 0.01$. For intermediate values of the kinetic parameter ξ we see from Fig. 35b (corresponding to $\zeta = 0.279$, which is the kinetic condition defining the percolation threshold) that the conductive pillars exhibit a range of lengths varying from zero to L. The analysis has been extended to examine the effect of layer morphology on the charge percolation process [142]. In a further paper [143] the potential–composition relationship (where the latter is described in terms of the mole fraction of oxidized sites) is derived via stochastic models and Monte Carlo simulation. A Nernst-type relationship is obtained. The analysis also indicated that when the mole fraction was below 0.31, the conductive sites were localized near the support electrode, whereas for mole fractions greater than 0.32 the conductive sites were percolated over the entire layer. Hence the mole fraction value of 0.31 corresponds to the film composition where there is a second-order phase transition from localized conduction to global conduction in the layer. This is the percolation threshold.

To this point we have discussed the transformation of a film from an initially insulating (I) state of a conductive (C) state. What about the reverse, $I \rightarrow C$, transformation?

In two recent papers [146, 147] Aoki and co-workers have examined the cause of the hysteresis effect observed in the cyclic voltammetric profiles recorded for conducting polymer electrodes in the potential region associated with redox switching. This hysteresis is usually associated with an asymmetry in peak shape when the oxidation ($I \rightarrow C$ transformation) and reduction ($C \rightarrow I$ transformation) voltammetric peaks are compared. Furthermore, it is often observed that the voltammetric profile recorded during the first sweep differs significantly from that observed on the second and subsequent sweeps. We illustrate this effect by presenting, in Fig. 36, a typical voltammmetric profile recorded for poly(aniline)-modified electrodes. This asymmetry is not only manifested in the cyclic voltammetric response. It also appears in ultraviolet–visible

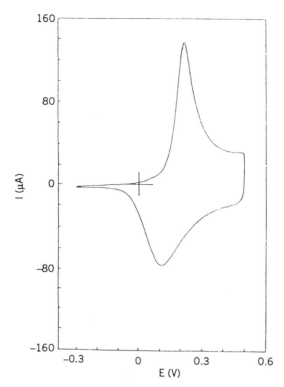

Figure 36. Voltammetric response of poly(aniline)-coated glassy carbon electrode in 1.0 M HCl at 298 K. Sweep rate, 20 mV s^{-1}. Hysteresis observed in current response between forward oxidizing and reverse reducing sweeps is apparent.

(UV/VIS) spectroelectrochemical measurements [148], probe beam deflection studies [149], conductivity measurements [150], and spin density measurements obtained via electrochemical electron spin resonance (ESR) [151] techniques. It is important to understand the nature of this hysteresis effect because electronically conductive polymer materials show promise as active components in advanced battery systems.

Aoki and co-workers [146, 147] have established that the C→I transformation occurs throughout the entire bulk of the polymer film. This is in contrast to the I→C transformation discussed previously where a well-defined conduction front between the C and I regions propagates through the layer. They have shown that electrochemical reduction is incomplete, with as much as 25% of the film consisting of conducting regions after electrolysis. To explain this observation, they introduced the concept of a C-cluster, which is defined as an assembly of conducting regions in electrical contact with each other. These conducting regions will of course be in contact with insulating (I) regions, and some C clusters will also be in electrical communication with the underlying support electrode. The C→I conversion can occur at any point at the C–I region interface. When the transformation in a region where the C cluster is very narrow, the cluster can split and parts of it can become electrically disconnected from the support electrode. Hence the number of C clusters increase, but their average size decreases. As the number of electrically disconnected clusters increases, the net C→I conversion rate decreases, and the conversion will finally halt when the percolation threshold is attained. Hence electrochemical reduction will be necessarily incomplete. Further C→I conversion then occurs over quite a long time scale and is assumed to involve a slow relaxation process involving rearrangement, via thermal fluctuations, of the electrically disconnected clusters to form new clusters that are electrically connected to the support electrode. These species may then be subsequently reduced until complete reduction of the film occurs.

III. CONDUCTION IN ELECTROACTIVE POLYMERS: COMPLICATING FACTORS

A.   Nonideality in Electroactive Polymer Films

1. Introduction

In this section we discuss nonideality in electroactive polymer films. This effect is manifested in the observation of deviations from ideal Nernst-type responses in plots of potential versus redox composition and in peak broadening or peak narrowing in linear potential sweep or cyclic voltammetry experiments. These deviations from ideal behavior may be ascribed to the operation of interactive forces between the charged redox sites in the polymer film. In the following section we shall assume that charge percolation through the film is rather rapid and the potential–composition relationship is well described by the Nernst equation. Thus we shall only investigate the way that redox site interactions affect the redox potential of the layer. We shall discuss two theoretical approaches that have been developed to quantify this phenomenon. These models have been developed by Brown and Anson [152] and Albery et al. [153]. Further work on this topic has been presented by Laviron [154] and Tokuda and co-workers [155, 156]. In the work presented by the latter researchers a more general formulation was established in which the more complex situations of quasi-reversible and totally irreversible charge percolation were considered. The reader is referred to the original publications for full details of this more comprehensive approach. We shall only consider the simple situation in this section.

3. Brown–Anson Model

Let us assume that the redox activity in a surface-deposited electroactive polymer film is quantified via the Nernst equation suitably modified to take redox site interaction effects into account [152]. We can use the Nernst equation if the charge percolation through the layer is rapid. We consider the simple redox process $A \rightarrow B + ne^-$. If interactions are absent, then we can write the Nernst equation in the form

$$\frac{b}{a} = \exp\left[\frac{nF}{RT}(E - E^0)\right]$$ (3.1)

This expression may be written in a more simple form as follows. We set $y = \ln(b/a)$ and $\vartheta = (nF/RT)(E - E^0)$ to obtain

$$y = \vartheta$$ (3.2)

If interactions are present within the electroactive polymer, the Nernst equation is modified to read

$$y = \xi\vartheta$$ (3.3)

where we have introduced the parameter ξ as a measure of the departure from ideality. Clearly, in the ideal system $\xi = 1$. We now wish to obtain an expression for this parameter ξ in the context of a simple model.

In the Brown–Anson approach [152] it is assumed that the degree of redox center interaction depends on the extent of oxidation or reduction of the film. We can write the Nernst equation in the form

$$\frac{a_B}{a_A} = \frac{\gamma_B b}{\gamma_A a} = \exp(\vartheta)$$ (3.4)

where a_k and γ_k represent the activity and activity coefficient of species k. One now utilizes the Frumkin assumption that the activity coefficients depend on the connection of reduced and oxidized site concentrations as follows:

$$\gamma_A = \exp[-(r_{AA}a + r_{AB}b)]$$
$$\gamma_B = \exp[-(r_{BB}b + r_{BA}a)]$$ (3.5)

In Eq. (3.5) the *interaction parameters* r_{AA}, r_{BB}, r_{AB}, and r_{BA} have been introduced, which quantify the interaction between redox sites. One should note that r values may be negative or positive. If *repulsive* interactions operate, then r values are *negative*, whereas if *attractive* interactions pertain, then r values are *positive*. Some authors use the opposite convention. We note that Eq. (3.4) may be written as

$$y = \ln\left(\frac{\gamma_A}{\gamma_B}\right) + \vartheta \tag{3.6}$$

Introducing the expressions for the activity coefficients from Eq. (3.5) into Eq. (3.6), we obtain

$$\begin{aligned} y &= r_{AA}a - r_{AB}b + r_{BB}b + r_{BA}a + \vartheta \\ &= (r_{BA} - r_{AA})a + (r_{BB} - r_{AB})b + \vartheta \end{aligned} \tag{3.7}$$

We now make some simplifying assumptions. We define

$$r_A = r_{AA} - r_{AB} \qquad r_B = r_{BB} - r_{BA} \tag{3.8}$$

Now $r_{AB} = r_{BA}$ and so $r_{BA} - r_{AA} = -(r_{AA} - r_{AB}) = -r_A$ and $r_{BB} - r_{AB} = r_{BB} - r_{BA} = r_B$; hence Eq. (3.7) reduces to

$$y = r_B b - r_A a + \vartheta \tag{3.9}$$

We now introduce the mole fractions X_A and X_B as follows:

$$X_A = \frac{a}{c_\Sigma} \quad X_B = \frac{b}{c_\Sigma} \tag{3.10}$$

where $c_\Sigma = a + b$ denotes the total redox site concentration. From Eqs. (3.10) and (3.9) we obtain

$$\begin{aligned} y &= r_B c_\Sigma X_B - r_A c_\Sigma X_A + \vartheta \\ &= \vartheta + \sigma_B X_B - \sigma_A X_A \\ &= \vartheta + \sigma_B X_B - \sigma_A(1 - X_B) \end{aligned} \tag{3.11}$$

where we have defined a dimensionless interaction parameter $\sigma_k = r_k c_\Sigma$. This parameter is related to the Gibbs energy of interaction G via the relation

$$\sigma_k = \frac{G_k}{RT} \tag{3.12}$$

If we assume that $\sigma_A = \sigma_B = \sigma$ or $G_A = G_B = G$, then Eq. (3.11) reduces to

$$y = \vartheta + \sigma X_B - \sigma(1 - X_B)$$
$$= \vartheta + \sigma(2X_B - 1) \tag{3.13}$$

We now need to evaluate X_B, the mole fraction of oxidized sites. Since $b/a = b/(c_\Sigma - b) = X_B/(1 - X_B)$, $y = \ln(b/a) = \ln[X_B/(1 - X_B)]$, and so $X_B/(1 - X_B) = \exp(y)$. Hence we solve for X_B to obtain

$$X_B = \frac{\exp(y)}{1 + \exp(y)} \tag{3.14}$$

Hence we note that

$$2X_B - 1 + \frac{2\exp(y)}{1 + \exp(y)} - 1 = \frac{\exp(y) - 1}{\exp(y) + 1}.$$

We now recall the fundamental identity

$$\tanh x = \frac{1 - \exp(-2x)}{1 + \exp(-2x)} = \frac{\exp(2x) - 1}{\exp(2x) + 1}.$$

Hence we note that

$$2X_B - 1 = \frac{\exp(y) - 1}{\exp(y) + 1} = \tanh\left(\frac{y}{2}\right) \tag{3.15}$$

and so Eq. (3.13) reduces to:

$$y = \vartheta + \sigma \tanh\left(\frac{y}{2}\right) \tag{3.16}$$

Hence

$$\vartheta = \gamma - \sigma \tanh\left(\frac{y}{2}\right)$$
$$= y\left(1 - \frac{\sigma}{2}\left(\frac{\tanh(y/2)}{y/2}\right)\right) \tag{3.17}$$

We recall that the modified Nernst equation may be written in the form

$$\vartheta = \frac{1}{\xi} y \tag{3.18}$$

Hence we immediately note that the parameter ξ is given by

$$\xi = \left(1 - \frac{\sigma}{2}\frac{\tanh(y/2)}{y/2}\right)^{-1} \tag{3.19}$$

This is the fundamental expression in the Brown–Anson model [152]. Equation (3.17) is plotted in graphical form in Fig. 37. In this plot we have set $\sigma = 4.67$. Note that a linear relationship is observed between θ and y for small values of y corresponding to the potential range $-50\,\text{mV} < E - E^0 < 50\,\text{mV}$. However, for larger values of y the curve deviates from linearity. When y is small, $\tanh(y/2) \approx (y/2)$ and so

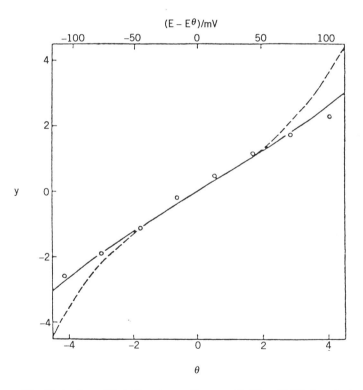

Figure 37. Plot of modified Nernst equation $y = \xi\theta$ [Eq. (3.3)] for electroactive polymer film when redox site interactions taken into account. Data points recorded for poly(thionine) film. Dashed line computed from Eq. (3.17) (Brown–Anson model) with ξ given by Eq. (3.19), whereas full line obtained from Albery–Colby model, with ξ given by Eq. (3.38). Good agreement between experiment and theory obtained using Albery–Colby model for $\sigma = 4.67$.

$\tanh(y/2)/(y/2)$. Hence Eq. (3.19) reduces to

$$\xi \approx \frac{1}{1 - \sigma/2} \tag{3.20}$$

for small y. This is a useful way of evaluating the interaction coefficient σ from the experimentally obtained ξ value. In the small-y limit Eq. (3.18) reduces to

$$\vartheta = (1 - \tfrac{1}{2}\sigma)y \tag{3.21}$$

and the linear relationship between θ and y is confirmed. For larger values of y, typically $|y| > 2$, linearity is not observed. In the situation of very large y we have that $\tanh(y/2)/(y/2) \rightarrow 0$, and so $\xi \approx 1$. Hence we see from Fig. 37 that the Brown–Anson model predicts a strong curvature in the θ-versus-y plot for large θ values.

These Nernst-types plots may be obtained experimentally if one can estimate y in an accurate manner. This can be done in two ways. If the components of the redox couple exhibit well-defined UV/VIS absorption spectra, then the absorbance at a given wavelength may be used as an estimate of y. These absorbance readings are then obtained as a function of potential. Alternatively one may use coulometry to determine y as a function of potential. Both procedures yield similar results.

3.   Albery–Colby Model

This approach [153] proposes that interactions between redox sites in the layer are responsible for the nonideal Nernst behavior of $\xi < 1$, but it is assumed that these interactions do not depend on the degree of oxidation or reduction of the film. Instead it is assumed that each redox site has its own standard potential value E_j^0 and that these values are distributed about the observed value E^0. Hence the *heterogeneity* of the layer is emphasized in this approach. The spread of standard potentials is assumed to be *Gaussian*.

We consider a set of redox couples A/B with a single formal potential E_j^0 obeying the Nernst equation such that the mole fraction of couples X_j in this set is given by

$$X_j = X_{B,j} + X_{A,j} \tag{3.22}$$

For each set we write

$$E = E_j^0 + \frac{RT}{nF} \ln\!\left(\frac{X_{B,j}}{X_{A,j}}\right) \tag{3.23}$$

We also can write Eq. (3.23) in the form

$$\vartheta = \vartheta_j + \ln\left(\frac{X_{B,j}}{X_{A,j}}\right) \tag{3.24}$$

where

$$\vartheta_j = \frac{nF}{RT}(E_j^0 - E^0) \tag{3.25}$$

Hence we note that

$$\exp(\vartheta - \vartheta_j) = \frac{X_{B,j}}{X_{A,j}} = \frac{X_{B,j}}{X_j - X_{B,j}} \tag{3.26}$$

Solving for $X_{B,j}$ we obtain

$$X_{B,j} = \frac{X_j}{1 + \exp(\vartheta_j - \vartheta)} \tag{3.27}$$

Summing over all possible sets of nonequivalent oxidized sites gives the total mole fraction of the oxidized form in the layer:

$$X_B = \sum_j X_{B,j} = \sum_j \frac{X_j}{1 + \exp(\vartheta_j - \vartheta)} \tag{3.28}$$

Macrosopically, the net observed behavior is given by the modified Nernst equation:

$$\ln\left(\frac{X_B}{X_A}\right) = \ln\left(\frac{X_B}{1 - X_B}\right) = \xi\vartheta \tag{3.29}$$

Hence solving for X_B, we obtain

$$X_B = \frac{1}{1 + \exp(-\xi\vartheta)} \tag{3.30}$$

Equating the expression obtained for X_B in Eqs. (3.28) and (3.30) we immediately obtain

$$\sum_j \frac{X_j}{1 + \exp(\vartheta_j - \vartheta)} = \frac{1}{1 + \exp(-\xi\vartheta)} \tag{3.31}$$

We now assume that X_j is Gaussian distributed and write

$$X_j = \frac{\exp(-\zeta^2 \vartheta_j^2) \, \Delta \vartheta_j}{\Sigma_j \exp(-\zeta^2 \vartheta_j^2) \, \Delta \vartheta_j} \tag{3.32}$$

where the parameter ζ represents the spread of the Gaussian distribution. Note also that $\Delta \vartheta_j = (nF/RT)(E^0 - \bar{E}_j^0)$.

Hence substitution of Eq. (3.32) into Eq. (3.31) yields

$$\frac{1}{1 + \exp(-\xi \vartheta)} = \frac{\displaystyle\sum_j \frac{\exp(-\zeta^2 \vartheta_j^2) \, \Delta \vartheta_j}{1 + \exp(\vartheta_j - \vartheta)}}{\displaystyle\sum_j \exp(-\zeta^2 \vartheta_j^2) \, \Delta \vartheta_j} \tag{3.33}$$

Replacing the summation over j with an integral over all values of θ_j gives

$$\frac{1}{1 + \exp(-\xi \vartheta)} = \frac{\displaystyle\int_{-\infty}^{\infty} \frac{\exp(-\zeta^2 \vartheta_j^2)}{1 + \exp(\vartheta_j - \vartheta)} \, d\vartheta_j}{\displaystyle\int_{-\infty}^{\infty} \exp(-\zeta^2 \vartheta_j^2) \, d\vartheta_j} \tag{3.34}$$

We now let $\eta = \zeta_j$. Then $\vartheta_j = \eta/\zeta$ and $d\vartheta_j = d\eta/\zeta$. Hence Eq. (3.34) reduces to

$$\frac{1}{1 + \exp(-\xi \vartheta)} = \frac{\displaystyle\int_{-\infty}^{\infty} \frac{\exp(-\eta^2)}{1 + \exp(\eta/\zeta - \vartheta)} \, d\eta}{\displaystyle\int_{-\infty}^{\infty} \exp(-\eta^2) \, d\eta} \tag{3.35}$$

We now note the standard integral

$$\int_{-\infty}^{\infty} \exp(-\eta_2) \, d\eta = \sqrt{\pi} \tag{3.36}$$

Hence Eq. (3.35) reduces to

$$\frac{1}{1 + \exp(-\xi \vartheta)} = \frac{1}{\sqrt{\pi}} \int_{-\infty}^{\infty} \frac{\exp(-\eta^2)}{1 + \exp(\eta/\zeta - \vartheta)} \, d\eta \tag{3.37}$$

We now solve this expression for ξ to obtain

$$\xi = -\frac{1}{\vartheta} \ln \left[\frac{\sqrt{\pi}}{\displaystyle\int_{-\infty}^{\infty} \frac{\exp(-\eta^2)}{1 + \exp(\eta/\zeta - \vartheta)} \, d\eta} - 1 \right] \qquad (3.38)$$

The integral in Eq. (3.38) may be solved numerically to obtain an expression for ξ as a function of ζ and θ. Albery and Colby have shown [153] that, for $\xi < 0.7$ and $\theta = 0$, Eq. (3.38) reduces to

$$\xi = \text{erf}[2\zeta] \qquad (3.39)$$

This approximate expression may be used to obtain an estimate of the spread parameter ζ from a knowledge of the interaction parameter ξ.

It should be noted that the Albery–Colby approach [153] has not received the same degree of attention as the Brown–Anson model [152] in the literature. A characteristic feature of the former model is that the θ-versus-y plot does not exhibit the same degree of curvature for large θ values. This is illustrated in Fig. 36. In this figure we also include experimental potential-versus-composition data for a polythionine film in contact with an aqueous solution of low pH. One immediately notes that for this electroactive polymer at least, the Gaussian spread model is the most appropriate one to use. In this case the y parameter is given by $y = \ln(([Th]/[L])$, where Th is thionine and L is leucothionine. A good fit between the experimental data points and the Gaussian model was obtained for the spread parameter $\zeta = 0.14$.

Information on interaction parameters is most conveniently obtained using linear potential sweep or cyclic voltammetry.

4. Chidsey–Murray Model for Redox Conduction

We now return to the topic of redox conduction in electroactive polymer materials. We shall see how redox site interaction phenomena may be incorporated into a formal model of redox conduction. The approach we will describe has been developed by Chidsey and Murray [157]. This approach attempts to quantify a number of characteristic features observed for electroactive polymer films. Now, these materials have the ability to store charge to a greater or lesser extent. Chidsey and Murray quantify this charge storage capability in terms of a concept termed *redox capacity*. We have previously noted that electroactive polymer materials are *mixed conductors*, in that they exhibit both DC electronic and ionic conductivity. This mixed-conduction behavior is a necessary requisite for bulk charge storage. A quantitative macroscopic treatment of DC conduction in electroactive materials is rather complex due to the fact that these materials exhibit a compositional gradient of either ions or electrons under steady-state conditions under the influence of an applied potential bias. In the work of Chidsey and Murray, the macroscopic concepts of redox capacity ρ and DC electron conductivity κ_e are developed in terms of a model based on statistical mechanics. These concepts are then related to the quasi-diffusion of electrons through the polymer, which is quantified by the electron diffusion coefficient D_E.

In the following analysis we will be concerned with a description of the following type of experiment. The electroactive polymer film is sandwiched between two inert metallic electrodes. A potential bias is then applied across the film, and the DC conductivity corresponding to lateral electron flow across the layer is measured. The physics is quite simple. After the potential bias is applied, an electric field exists in the film and electrons begin to flow in response to the field. However, since the polymer is a mixed conductor, ions also respond to the field. This results in polarization at the film–electrode interfaces that has the knock-on effect of decreasing the electric field felt in the bulk of the polymer film. The sample now begins to exhibit a compositional gradient: The electroactive material becomes more reduced at the negative electrode and more oxidized at the positive electrode. This leads to the situation where the electrons are driven by the concentration gradient of reduced material (which corresponds to an electron concentration gradient) as well as by the residual field in the bulk of the material.

We now discuss the charge storage capability of an electroactive polymer. The redox capacity ρ (units: $F\,cm^{-3}$) is defined as

$$\rho(E_{eq}) = \frac{dq}{dE_{eq}} = e^2 \frac{dn_e}{d\mu_{ei}} \tag{3.40}$$

where q represents the charge stored per unit volume, E_{eq} denotes the equilibrium cell potential, e is the electronic charge, and n_e defines the number density of electrons in the electroactive material. The quantity μ_{ei} represents the electrochemical potential of the electron and the counter-ions required for the preservation of net electroneutrality. Hence we have

$$\mu_{ei} = \mu_e + (1/z_i)\mu_i \tag{3.41}$$

where we note that μ_e, μ_i, and z_i denote the electrochemical potentials of the electron, counterion, and counterion valence, respectively. The redox capacity is obtained readily from cyclic voltammetric measurements of peak current i versus sweep rate v: $i = \rho Vv$, where V denotes the volume of the deposited polymer film. If there is a large excess of supporting electrolyte present in the film, then the electrochemical potential of the counterions is constant and does not vary with changes in redox state of the material, and the definition of ρ reduces to

$$\rho = e^2 \frac{dn_e}{d\mu_e} \tag{3.42}$$

If all of the sites in the material that accept electrons are equivalent and if one neglects interelectron interactions, then the equilibrium cell potential is given by the Nernst equation:

$$E_{\mathrm{eq}} = E^0 - \frac{k_B T}{e} \ln\left(\frac{X_e}{1 - X_e}\right) \tag{3.43}$$

where X_e denotes the fraction of sites occupied by electrons. This equation may be readily solved for X_e to obtain

$$X_e = \frac{\exp(\vartheta)}{1 + \exp(\vartheta)} \tag{3.44}$$

where, as before, we define $\vartheta = (e/k_B T)(E_{\mathrm{eq}} - E^0)$, where E^0 is the standard potential of the sites. Furthermore the redox capacity ρ is given by

$$\rho = \frac{n_s e^2}{k_B T} X_e(1 - X_e)$$

$$= \frac{n_s e^2 \exp(\vartheta)}{k_B T[1 + \exp(\vartheta)]^2} \tag{3.45}$$

This function is plotted in a schematic way in Fig. 38. Clearly we see that the maximum redox capacity ρ_m occurs when $\theta = 0$ and, from Eq. (3.45), this is given by $\rho_m = n_s e^2/4k_b T$. A characteristic feature of the redox capacitance response is the peak width at half-height δ. This quantity is given by $\delta = 2\theta_{1/2}$, where we note that $\theta = \theta_{1/2}$ when $\rho = \frac{1}{2}\rho_m$. Under such conditions we note from Eq. (3.45) that

$$[\exp(\theta_{1/2})]^2 - 6\exp(\vartheta_{1/2}) + 1 = 0 \tag{3.46}$$

The quadratic in $\exp(\theta_{1/2})$ may be solved to yield

$$\exp(\vartheta_{1/2}) = 3 + 2\sqrt{2} \tag{3.47}$$

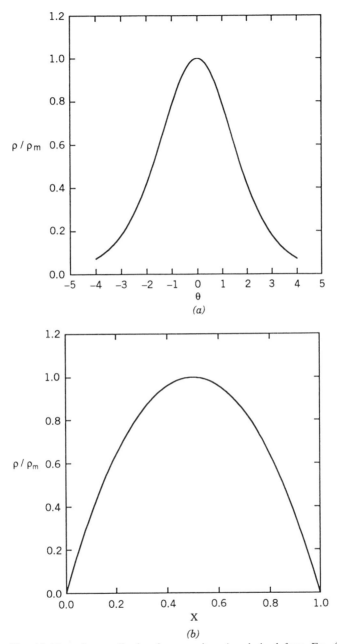

Figure 38. (*a*) Plot of normalized redox capacity ρ/ρ_m derived from Eq. (3.45) as function of normalized potential θ. (*b*) Plot of normalized redox capacity ρ/ρ_m derived from Eq. (3.45) as function of mole fraction of oxidized sites X.

Hence the width of the capacitance peak at half-height is given by

$$\delta = \frac{k_B T}{e} 2 \ln(3 + 2\sqrt{2}) \approx 90 \text{ mV} \qquad (T = 298 \text{ K}) \qquad (3.48)$$

This is a characteristic feature of the redox capacitance in the absence of interaction effects. We now examine the more realistic situation where interaction effects are taken into account.

Chidsey and Murray [157] picture the electroactive polymer film as a simple cubic lattice of oxidized sites each with a charge $z_s e$, charge-compensating counterions of charge $z_i e$, and electrons of charge $-e$. The number density of sites is designated n_s and the number density of electrons is n_e, where $n_e = n_s X_e$. The number density of counterions in the film is given by $n_i = n_x X_i = n_s (X_e - z_s)/z_i$, where X_i denotes the number of counterions per site. Interactions between occupied sites are quantified in terms of potential u between neighboring lattice sites that are both occupied. The number of nearest neighbors to a given site is given by γ. It is assumed that only electrons interact with each other. Interaction effects between counterions are neglected. Chidsey and Murray [157] have shown, using a statistical mechanical mean-field approximation [158, 159], that the chemical potential of the interacting electrons is given by

$$\mu_e = \varepsilon X_e - e\phi + k_B T \ln\left(\frac{X_e}{1 - X_e}\right) \qquad (3.49)$$

The first term in Eq. (3.49) quantifies the contribution of occupied site–occupied site interactions to the chemical potential. The occupied site interaction energy $\varepsilon = \gamma u$ has been introduced. For a simple cubic lattice $\gamma = 6$. This is similar to the interaction parameter σ previously introduced, and one has $\varepsilon = -\sigma$. The second and third terms in Eq. (3.49) represent the electrostatic (where ϕ is the local electrostatic potential between sites) and entropic contributios to the chemical potential of the hopping electrons. When $\varepsilon = 0$ we have the ideal situation. This will be valid for $\varepsilon \ll kBT$. If ε is large and positive, then one has repulsive interactions, the occupied sites will tend to avoid each other, and the alternating site occupancy can ultimately give rise to the situation of *compound formation*. If ε is large and negative, attractive interactions pertain, and ultimately the sitution of *phase separation* may be encountered. The chemical potential of the noninteracting counterions is given by

$$\mu_i = k_B T \ln X_i + z_i e\phi = k_B T \ln\left(\frac{X_e - z_s}{z_i}\right) + z_i e\phi \qquad (3.50)$$

Hence the equilibrium potential is given by

$$\begin{aligned}
E_{\text{eq}} &= E^0 - \frac{1}{e}(\mu_e + z_i^{-1}\mu_i) \\
&= E^0 - \frac{\varepsilon X_e}{e} - \frac{k_B T}{e}\left[\ln\left(\frac{X_e}{1 - X_e}\right) + z_i^{-1}\ln\left(\frac{X_e - z_s}{z_i}\right)\right]
\end{aligned} \qquad (3.51)$$

This expression may, in principle, be used to define a composition–potential relationship. The first term on the rhs is the standard electrode potential, the second quantifies occupied site interactive effects. The last logarithmic tems describe the usual Nernst factor and also take into account that the activity of the counterion may vary with extent of redox conversion.

One may show [157], by differentiating Eq. (3.51) with respect to X_e, that the redox capacity is in general given by

$$\rho = \frac{n_s e^2}{k_B T}\left(\frac{\varepsilon}{k_B T} + \frac{1}{X_e} + \frac{1}{1 - X_e} + \frac{1}{z_i(X_e - z_s)}\right)^{-1} \qquad (3.52)$$

This general expression takes both interaction effects and the chemical potential of the charge-compensating counterions into account. We outline in Fig. 39 a plot of ρ versus normalized potential θ for several values of z_s. We also present a plot of ρ versus normalized potential θ for various values of the interaction parameter ε. Note that the curve for $z_s = \infty$ corresponds to the case of having a large excess of supporting electrolyte present in the film. We show the situation obtained for $\varepsilon = 0$ and for negative and positive values of ε. The important point to note here is that nearest-neighbor occupied site interactions lead to *symmetric* broadening or narrowing of the redox capacitance curves.

In the context of the Chidsey–Murray model it is also possible to show that the steady-state electron flux j_e is given by

$$j_e = -n_s e\delta\langle \vec{p} - \overleftarrow{p}\rangle \qquad (3.53)$$

where δ denotes the intersite hopping distance and $\vec{p}$, $\overleftarrow{p}$ denote the probability of an electron hop in the forward and reverse directions,

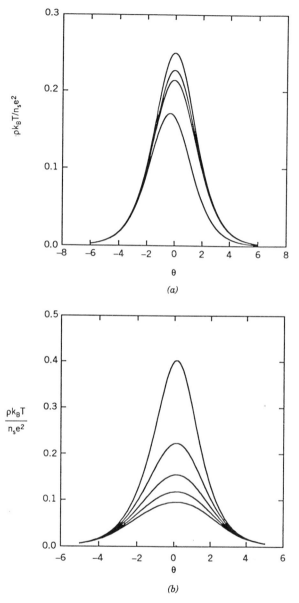

(a)

(b)

Figure 39. (a) Plot of redox capacity ρ calculated from Eq. (3.52) for various values of z_s, valence of oxidized redox center. We assume that $\varepsilon = 0$. From top to bottom $z_s = \infty$ (corresponding to large excess of supporting electrolyte present in film) or $z_s = 3, 2, 1$. (b) Plot of redox capacity ρ calculated from Eq. (3.52) for various values of interaction parameter ε. We assume that $z_s = 3$ and $z_i = -1$. From top to bottom $\varepsilon/k_B T = -4, -2, 0, 2, 4$.

respectively. We note that

$$\vec{p} = \vec{k} X_e(x)[1 - X_e(x + \delta)]$$
$$\vec{p} = \vec{k} X_e(x + \delta)[1 - X_e(x)]$$

(3.54)

where the forward and reverse hopping rate constants are related by

$$\frac{\vec{k}}{\vec{k}} = \exp\left(-\frac{U(x + \delta) - U(x)}{k_B T}\right)$$

(3.55)

Substituting this relationship into Eq. (3.54), expanding the exponentials to first order, and performing a Taylor expansion to the first order of the quantities $X_e(x + \delta)$ and $U(x + \delta)$ about x, one finally obtains, after some algebra, that the electron-hopping flux is

$$j_e = k\delta^2 n_s e\left(\frac{dX_e}{dx} + \frac{X_e(1 - X_e)}{k_B T}\left\langle\frac{dU}{dx}\right\rangle\right)$$

(3.56)

This equation has been seen previously. It bears a marked similarity to Eq. (2.84), which described the steady-state diffusion–migration flux. Now the average enegy change per unit displacement of the electron $\langle dU/dx\rangle$ will depend on the local gradient in electrostatic potential as well as on the probability of having neighboring sites occupied, and so we may state that

$$\left\langle\frac{dU}{dx}\right\rangle = \varepsilon\frac{dX_e}{dx} - e\frac{d\phi}{dx}$$

(3.57)

Hence the steady-state electron flux becomes

$$j_e = n_s e k\delta^2\left[\frac{dX_e}{dx} + \frac{X_e(1 - X_e)}{k_B T}\left(\varepsilon\frac{dX_e}{dx} - e\frac{d\phi}{dx}\right)\right]$$

(3.58)

Equation (3.58) is an extension of Eq. (2.80) to take nearest-neighbor interactions into account. It may also be shown after some algebra that the steady-state electron flux may be written as

$$j_e = -\sigma_e\frac{dE_{eq}}{dx}$$

(3.59)

where the redox conductivity σ_e is given by

$$\sigma_e = k\delta^2 \frac{n_s e^2}{k_B T} X_e(1 - X_e) \tag{3.60}$$

Hence we see that the electron-hopping conductivity depends on the hopping distance δ, the field-independent hopping frequency k, and the degree of oxidation or reduction of the film (via the X_e terms). The redox conductivity is related to the redox capacity ρ via

$$\sigma_e = D_E \rho \tag{3.61}$$

where D_E denotes the electron-hopping diffusion coefficient. We note therefore that in the context of the Chidsey–Murray model we can make the following identification:

$$D_E = k\delta^2 \left[1 + \left(\frac{1}{z_i(X_e - z_s)} + \frac{\varepsilon}{k_B T} \right) X_e(1 - X_e) \right] \tag{3.62}$$

Note that in the limit of zero nearest-neighbor interactions corresponding to $\varepsilon = 0$ and setting $z_S = \infty$ corresponding to having a vast excess of supporting electrolyte in the layer, the diffusion coefficient reduces to $D_E = k\delta^2$, which is a true constant. We note, therefore, from Eq. (3.62) that the redox diffusion coefficient varies with the composition of the layer. In Fig. 40 we show the variation of D_E with potential for various values of ε and z_S. Note that D_E does not vary very much with potential, but generally one expects to see a maximum in the D_E value at the standard redox potential corresponding to $\theta = 0$.

The Chidsey–Murray model of redox conduction provides a reasonable first-order approach to the quantitation of conductivity in electroactive polymer films, in that it considers intersite interactions as well as the variation of the electrochemical potential of the charge-compensating counterions with redox composition. The useful feature of the approach is that it establishes a theoretical framework from which the experimental determination of ρ and σ_e values as a function of potential may be interpreted.

B. Solvent, Salt, and Ion Transport in Electroactive Polymer Films

1. Introduction

It is well established that the process of redox switching of electroactive polymer films is complex. For example, in addition to electron transfer, one must also consider processes such as ion, salt, solvent, and other neutral molecule transfers as well as polymer configurational changes.

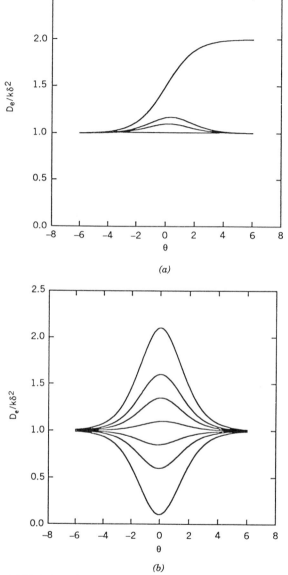

Figure 40. (*a*) Variation of electron-hopping diffusion coefficient D_E with normalized potential θ calculated from Eqs. (3.62) and (3.44) for various values of z_s. From top to bottom $z_s = 1$, 2, 3, ∞. Assume that $\varepsilon = 0$. (*b*) Variation of electron-hopping diffusion coefficient D_E with normalized potential θ calculated from Eqs. (3.62) and (3.44) for various values of ε/k_BT. From top to bottom $\varepsilon/k_BT = 4$, 2, 1, 0, -1, -2, -4. Set $z_s = 3$ and $z_i = 1$.

Any of these processes may be rate determining. Electrochemical techniques only provide information on the transport of charged species. The current flowing across the electrode–polymer interface provides a convenient measure of the rate of electron transfer. However, the exclusive reliance on electrochemical techniques implies that the specific role of electroinactive species such as counterions, ion pairs, and solvent may not be directly deduced. One can, of course apply a range of in situ spectroscopic techniques [160] such as Fourier transform infrared (FTIR) spectroscopy or ellipsometry to the electroactive polymer during redox switching. However, spectroscopic techniques such as FTIR can suffer from the fact that it is difficult to distinguish between species present in the polymer film and the large excess of the same species present in the adjacent solution phase. Ellipsometry is surface specific and very sensitive but expensive. A less expensive option is to directly measure changes in film mass during redox switching. This is accomplished using the Electrochemical Quartz Crystal Microbalance (EQCM) methods.

A further point should be noted at ths stage. The time scales for electron transfer, and counterion transport, which is coupled to the electron transfer process via the operation of the electroneutrality condition, will, in general, be quite different from the time scale involved in the transfer of heavy species such as solvent, neutral molecule, or salt. Hence the question of whether an electroactive film is at complete thermodynamic equilibrium where the activities of all mobile species present in the film must be the same as those of their counterparts in solution can be difficult to ascertain. A diagnostic method that enables one to answer this question is therefore required. One also wishes to be able to identify the way in which solvation can affect the dynamics of redox switching. All of these questions can be adequately addressed by a combination of standard electrochemical and EQCM methods. Consequently, in this section we shall describe the EQCM technique and present a discussion of redox switching in the context of recent studies conducted using the EQCM method.

The technique of Probe Beam Deflection Spectroscopy (PBDS) has also been recently developed to examine the ingress or egress of charge-compensating counterions during redox switching in electroactive polymer films. This method will also be discussed in this section of the chapter. We shall illustrate the application of these methods by discussing some recent studies on electroactive polymer films in which useful information on the mechanism of redox switching has been obtained. This discussion will not be exhaustive but will concentrate on a limited number of examples. We only wish to illustrate the usefulness of the methodology developed.

2. EQCM and PBDS Methods: Principles and Case Studies

We see from the considerations just outlined that a complete description of redox switching in electroactive polymers requires, at the very least, consideration of concomitant changes in ion and solvent populations. This can be achieved using the EQCM method [161–169]. The quartz crystal microbalance is a *piezoelectric* device capable of monolayer mass sensitivity. The method involves sandwiching a quartz crystal between two electrodes, one of which is used as the working electrode of an electrochemical cell. These electrodes are used to impose a radio frequency (RF) electric field across the crystal at a resonant frequency determined by the dimensions and the total mass loading of the crystal. A change in the mass of the working electrode causes a change in the resonant frequency of the device, which can then be used to determine the quantity of added mass. This effect is quantified via the Sauerbrey equation [170], which relates the change in resonant frequency Δf from its value f at the start of the experiment and the mass change ΔM per unit area of the electrode as follows:

$$\Delta f = -\frac{2}{\rho v} \Delta M \, f^2 \qquad (3.63)$$

where ρ denotes the density of the crystal and v is the velocity of the acoustic shear wave within the polymer. The negative sign indicates that Δf decreses as ΔM increases. The Sauerbrey equation will be valid when the deposited polyme layer behaves as a rigid layer, that is, when the film is perfectly elastic (zero viscosity) and thin enough so that the frequency change is small compared with the resonant frequency. If the layer is very thick, then this linearity between frequency change and mass may not be observed due to viscous damping of the acoustic shear wave within the polymer film. The great advantage of the EQCM method is that a simultaneous in situ determination of working electrode mass and electrochemical parameters may be obtained. Thus one may electrochemically induce mass changes in thin electroactive polymer layers (such as might be caused by solvent or ion transport during redox switching) and monitor these mass changes concurrently with the conventional electrochemical measurements such as cyclic voltammetry or chronoamperometry.

The EQCM technique has been used recently to study both overall thermodynamic changes in ion and solvent populations within electroactive polymer films [164, 166, 169, 171–173] and the dynamics of ion and solvent motion within these films [174–176]. This had led to the development of a qualitative picture of the mechanism of mobile species transport and polymer relaxation during redox switching in electroactive polymers [177].

We now briefly consider the technique of *probe beam deflection spectroscopy*. The fundamental theory has been described by Royce and co-workers [178], Pawliszyn et al. [179–181], and Russo and co-workers [182]. This technique yields information about concentration and thermal gradients adjacent to an electrode, by examining the deflection of a laser beam aligned parallel and very close to a planar electrode. The beam will undergo deflection (refraction) due to the existence of refractive index gradients in the interface region. These latter gradients are set up due to the generation or consumption of species in the electrolyte phase.

The deflection Ψ of a laser beam traveling along a concentration profile adjacent to a planar electrode surface is given by the expression

$$\Psi = \frac{w}{n} \frac{dn}{dc} \frac{dc}{dx} \tag{3.64}$$

where c represents the concentration, n is the refractive index of the electrolyte solution, and w is the path length over which the laser beam interacts with the concentration profile. Under most conditions, w, dn/dc, and n are constant, and so the deflection Ψ is mainly proportional to the concentration gradient dc/dx. A detailed calculation of the path of a laser beam probing an interfacial concentration gradient has been provided by Mandelis and Royce [183]. It should be noted that a positive deflection of the probe beam results from a positive concentration gradient at the electrode–solution interface, whereas a negative deflection is obtained if the interfacial concentration gradient is negative. Consequently, the *sign* of the beam deflection tells one whether an interfacial electrochemical reaction is accompanied by a net ion flux away from the electrode surface or towards the electrode surface. In combination with the sign of the simultaneously measured current obtained from a cyclic voltammetry experiment, it is possible to determine whether anions or cations are exchanged at the interface. Hence the PBDS method can be applied to polymer-modified electrodes where ion transport plays a major role in redox switching.

The PBDS technique probes concentration gradients near the electrode surface. However, it is not possible to distinguish between a concentration gradient produced by ion transfer or by solvent transfer in the opposite direction. In practice, additional experiments with differing electrolyte concentrations or techniques such as the EQCM or radiotracer measurements are necessary to evaluate sovlent transfer contributions to the total beam deflection.

The PBDS and EQCM techniques have been used to resolve a long-standing controversy regarding redox switching in surface-deposited poly(aniline) films. The redox switching process in poly(aniline) films has been studied extensively in recent years. Ions are known to participate in redox switching from the insulating leucoemeraldine to the conducting emeraldine forms. This redox transition occurs in the potential range −200–400 mV (vs. saturated calomel electrode (SCE)). The degree of protonation of the polymer and the applied electrode potential define the degree of conductivity exhibited by polyaniline [184]. The electroneutrality condition also dictates that charge-compensating counterions move through the polymer matrix to compensate for changes in oxidation state of the polymer backbone. Hence proton as well as anion transport in the film will be important in governing the redox switching rate. Hence the fundamental question one may ask is are protons expelled or are anions inserted as the polymer is switched from an insulating to a conducting state?

Simple electrochemical measurements have only provided a partial answer to this question. Kitiani and co-workers [185] proposed that only anions are inserted during redox switching, assuming that the reduced form of poly(aniline) is totally deprotonated. This assertion was supported by investigations involving the use of radiotracer techniques [186] and cyclic voltammetric peak potential measurements as a function of solution pH [187]. Kobayashi and co-workers [188] proposed that the reduced polyaniline is protonated at solution pH values below 3–4, and protons are ejected during oxidation at low positive potentials [below 0.45 V (vs. reversible hydrogen electrode (RHE))]. Anion insertion was observed at higher potentials. At alternative model was proposed by Huang and co-workers [189]. In this approach anion ingress was postulated at pH > 1 whereas only proton expulsion occurred for pH values less than −0.2. Finally Shimazu et al. [190] suggested that protons together with anions are released from the film at low pH whereas anions are injected at higher pH values.

An attempt toward resolution of this issue has been provided in recent EQCM and PBDS studies. In the EQCM study of poly(aniline)-modified electrodes reported by Orata and Buttry [166] a mass increase was observed during the oxidation of leucoemeraldine to emeraldine. This could support the assertion that oxidation is accompanied by anion incorporation. This assertion was confirmed by examining frequency and therefore mass changes during oxidation when the poly(aniline) film was in contact with solutions of various acids such as HCl, HBr, H_2SO_4, $HClO_4$, and CF_3CO_2H. In all cases examined the frequency changes scaled with the mass of the anions. This is convincing evidence for an anion insertion mechanism. Furthermore, the mass change that accompanies the oxidation was found to track the charge consumption over a wide range of sweep rates. Also it was observed that the relative mass change corresponding to Cl^- insertion indicated that the incoming anion was hydrated to a certain extent. This observation is in accord with the notion that Cl^- is reasonably strongly hydrated due to its high charge-to-radius ratio. However, significantly, it was observed that a potential region existed (located prior to the rapid increase in current leading to the oxidation peak in the voltammogram; see Fig. 41) where redox charge was developed, but the latter was not accompanied by any increase in mass. It should be noted that proton transport (via a Grotthus mechanism) cannot be detected using the EQCM technique because of the low mass of H^+. Hence it was concluded that the oxidation of the polymer chains in this low-potential region was probably associated with proton exulsion as dictated by the electroneutrality condition. Very little

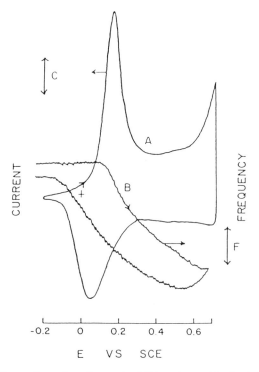

Figure 41. Composite cyclic voltammetry (A)/quartz crystal microbalance (B) data for electrodeposited poly(aniline) films in $1.0\,M$ H_2SO_4 supporting electrolyte. Sweep rate, $100\,\text{mV s}^{-1}$; current sensitivity $C = 100\,\mu A$, $F = 20\,\text{Hz}$. Curve B is offset 40 mV to left with respect to curve A. [166].

solvent is transported into the layer over the potentail range where the leucoemeraldine/emeraldine redox transition occurs (-200–400 mV SCE).

The PBDS method detects ion transport across the polymer–solution interface and so can detect cations (including protons) and anions. As noted by Barbero and co-workers [191], expulsion of cations from the film during redox switching will give rise to an increase in concentration at the polymer–solution interface. The concentration gradient will be negative, and so the path of the laser beam will be deflected in a negative manner (i.e., toward the electrode). The corresponding voltammetric current will be positive (oxidation). A positive beam deflection away from the electrode (reflecting a positive interfacial concentration gradient) in combination with a positive voltammetric current reflects anion insertion into the film. Furthermore, a negative beam deflection in combination with a negative (reduction) current indicates anion expulsion from the layer, and a positive deflection in combination with a negative current, cation insertion. The experimental data reported by Barbero and co-workers are illustrated in Fig. 42. We note from this figure that the PBDS signal in the cyclic deflectogram [192] exhibits both negative and positive deflections during the anodic sweep. This observtion provides clear evidence that proton expulsion and anion insertion occur during the oxidation of the polymer. Hence one has proton release preceding anion uptake during the anodic scan, and subsequently there is consecutive anion ejection and proton insertion during the reverse cathodic sweep.

The effect of solution pH on the redox switching process in poly-(aniline) films may also be rationalized from PBDs measurements. Cyclic deflectograms reacorded during polymer oxidation for a number of different acid concentrations are also outlined in Fig. 42. The deflection becomes increasingly more negative as the acid concentration is increased. It should also be noted that positive deflections are observed at more anodic potentials even in concentrated acid solutions. Hence proton expulsion is the dominant mode for preservaton of electroneutrality in concentrated acid media. However, anion insertion also occurs.

The redox chemistry is summarized in Fig. 43. Reduced poly(aniline) is partially protonated below a certain pH level and will contain a significant quantity of charge compensating anions. Upon oxidation x protons are expelled and $y - x$ anions A^- are inserted consecutively to maintain electroneutrality. The relative contribution of protons and anions in maintaining electronuetrality during oxidation depends on the degree of protonation of the polymer backbone, which is higher in solutions of low pH. In the case of fully protonated reduced poly(aniline), where $x = y$, only protons are expelled during oxidation, whereas for the fully deproto-

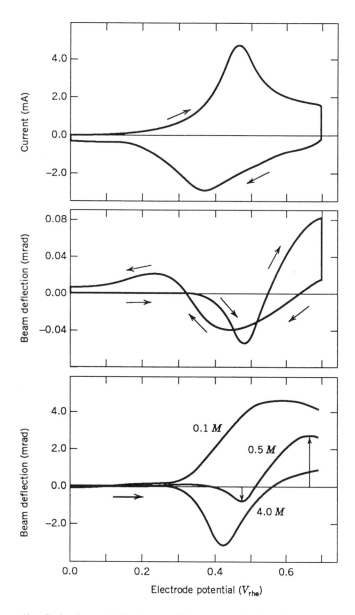

Figure 42. Probe beam deflection applied to examination of redox chemistry of poly(aniline) modified electrods. (i) Cyclic voltammogram (*a*) and corresponding cyclic deflectogram (*b*) recorded for poly(aniline) film electrodeposited on Au support electrode in $1.0\,M$ HCl. Sweep rate, $50\,\text{mV s}^{-1}$; surface coverage, $1.4 \times 10^{-8}\,\text{mol cm}^{-2}$. (ii) Forward scan of cyclic deflectogram obtained for poly(aniline) films in contract with various HCl solutions. Concentrations of supporting electrolyte indicated. Sweep rate, $50\,\text{mV s}^{-1}$. Upward and downward pointing arrows indicate positive and negative deflections of laser beam respectively. (Adapted from ref. 192).

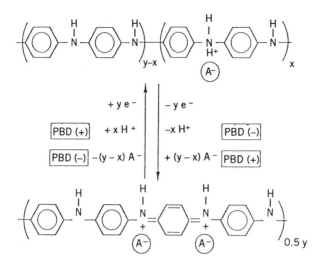

Figure 43. Redox chemistry of poly(aniline) and correlation with PBD data illustrated in Fig. 42.

nated reduced poly(aniline), where $x = 0$, only anions are inserted. Hence the PBDs data confirm the mechanism proposed by Huang and co-workers [189] and correlated well with the EQCM data reported by Orata and Buttry [166].

An important point must be noted at this stage. The observation that electronic or redox equilibrium is established over a given experimental time scale is *a necessary but insufficient* condition that all mobile species have achieved their equilibrium populations. This assertion will now be examined in the context of work carried out by Hillman and co-workers on the redox polymer material poly(thionine) [168].

In Fig. 44 we outline the results of simultaneous EQCM–cyclic voltammetry measurements carried out at a slow scan rate on poly-(thionine) films in contact with 0.1 M acetic acid (pH 2.90) and 0.05 M $HClO_4$. We note that the mass changes observed for the layer during redox switching when it is in contact with the different acid solutions proceed in different directions. Layer reduction results in a *decrease* in mass in acetic acid media (Fig. 44a) whereas reduction is accompanied by an *increase* in mass in perchlorate solution (Fig. 44b).

Let us first consider the situation in acetic acid media. The redox stoichiometry in acetic acid media is given by

$$[(TH^+A^-).H_2O.X.(HA)]_p + 2e^- + 2H_3O_s^+ \rightarrow [TH_4^{2+}(A^-)_2.X]_p + 3H_2O_s$$

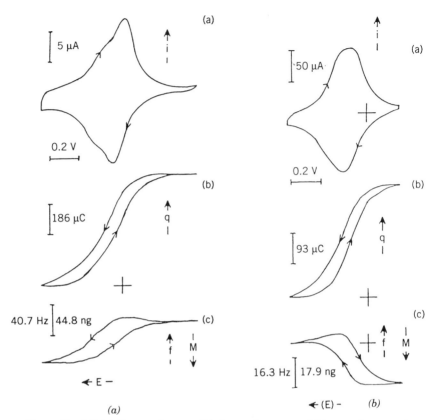

Figure 44. (*A*) (*a*) Current change–, (*b*) charge change–, and (*c*) mass change–voltage curves recorded under cyclic voltammetric conditions for poly(thionine)-modified electrode. Supporting electrolyte is $0.1\,M$ CH_3CO_2H, pH 2.9. Sweep rate, $5\,mV\,s^{-1}$. (*B*) (*a*) Current change–, (*b*) charge change–and (*c*) mass change–voltage curves recorded under cyclic voltammetric conditions for poly(thionine)-modified electrode. Supporting electrolyte is $0.05\,M$ $HClO_4$. Sweep rate, $100\,mV\,s^{-1}$. [168].

where X may be acetic acid or water and the subscripts p and s denote the polymer and solution phases, respectively. This reaction differs from the simple $2e^- - 3H^+$ redox stoichiometry and one would expect for the thionine–leucothionine reaction in low-pH media. In the latter case one would expect that the film mass should *increase* upon reduction by an amount $(3M_{H^+} + M_{A^-})\,g\,mol^{-1}$. Furthermore, in the reaction outlined above, T represents a monomeric unit of the thionine neutral base and TH^+, TH_4^{2+} represent the predominant protonation state of the oxidized and reduced polymer for solution pH <5. Hence we see that there is a net expulsion of one water molecule and ingress of two protons per redox site. Furthermore, the *counterion source* is undissociated acetic acid within the film. The latter is an important observation. Gas-phase acetic acid absorption experiments have indicated that acetic acid is incorporated and stronly bound within poly(thionine) films to the extent of about two molecules per redox site. Hence anions do not have to enter into the film during reduction. They are already present in the layer as HA. The latter species also provides one of the protons required to balance the stoichiometry. Hence during layer reduction, the mass decreases monotonically and the mass–charge relationship is linear. The situation for layer oxidation is different. First, we note that there is an initial mass excursion (a decrease) away from the final state and, second, the mass response during the remainder of the conversion lags the charge response.

These observations have led to the following general conclusion: *In the absence of film structural changes, ejection of a given species from the film is more facile than ingress of that species.* This is termed the difficult-ingress, rapid-egress (DIRE) mechanism. Hence during reduction of TH^+ in acetic acid media, water expulsion and proton ingress are rapid on the experimental time scale. The protons may move rapidly within the film via a Grotthus-type process. Hence the mass and charge changes track each other. In contrast, during TH_4^{2+} oxidation, the rates of water ingress and proton expulsion are very different. Proton expulsion is very rapid because it involves a Grotthus-type transport process involving water and/or amine sites on the poly(thionine). This initial proton ejection produces the observed initial mass decrease. Subsequent water ingress into a compact layer is much slower and is unaided by the field, and so charge and mass changes will not track each other.

Let us now consider the situation in perchloric acid. We note from Fig. 44b that layer reduction is now accompanied by a net mass increase, although the net change is much smaller than that anticipated for a redox stoichiometry involving the transfer of $3H^+ + ClO_4^-$. In this situation Hillman and co-workers [168] postulated that there was again an anion source within the film that was proposed to be the ion pair $H_3O^+ClO_4^-$. This hydronium perchlorate within the film serves as a source of counterions and protons during reduction. The breakup of the ion pair results in the liberation of water within the film. Hillman and co-workers [168] have shown that the extent of water expulsion and solution-phase hydronium and perchlorate participation depend largely on the concentration of acid used in the aqueous phase. The overall reaction stoichiometry is given by

$$[TH^+(ClO_4^-).(H_2O)_a.(H_3O^+ClO_4^-)_b]_p + 2e^- + cH_3O_s^+ + dClO_{4s}^-$$
$$\rightarrow [TH_4^{2+}(ClO_4^-)_2.(H_2O)_x.(H_3O^+ClO_4^-)_y]_p + zH_2O_s$$

We note from Fig. 44*b* that during the reduction cycle the mass response is nonmonotonic and decreases initially. The mass increases during the remainder of the reduction scan. The charge and mass changes do not remain in step. The mass lags behind the charge. In contrast, during oxidation, the mass and charge responses are more closely related. This again is in accord with the DIRE hypothesis. The redix transformation involving net mass loss (oxidation) exhibits a relatively linear mass–charge correlation. The reverse reduction transformation requiring a net mass increase exhibits a nonmonotonic and subsequently lagging mass response not simply correlated with the charge.

The main point to note from this work is that film sources of counterions may be significant. The extent of participation of these species (undissociated acid or ion pairs) will depend on the nature and concentration of the bathing electrolyte. In general terms the high concentration and close proximity of film sources of counterions will lead to the expectation that they will take temporal precedence over solution-phase sources. Of course over a longer time scale solution-phase sources will make a significant contribution and the imbalance between the different sources of supply will be redressed. This aspect of redox switching has not been fully realized in much of the literature published on the topic. The demands of electroneutrality will generally make the field-assisted transport of charged species more rapid than the transport of neutral species. Consequently solvation equilibria may only be slowly established. Finally for any species, its ingress into the film is more difficult than its egress. The special transport properties of protons in the layer should also be noted. In a hydrated electroactive polymer, protons may alone assume responsibility for maintaining electroneutrality upon initial injection of charge. The bottom line is that equilibria associated with electronic, ionic, and solvation processes may be established on quite different time scales. However, given a long enough time scale, thermodynamics will prevail and global equilibrium will be established.

From the examples presented we see that the EQCM technique can yield quite an amount of information on the redox switching process in an electroactive polymer film. We now discuss a diagnostic scheme proposed by Hillman and Bruckenstein [175, 176] that uses combined electro-chemical–EQCM measurements to establish whether global equilibrium exists during redox switching or, if it does not exist, to identify the nature of the rate-determinining step, which can be associated with either electron-hopping, counterion, or neutral species transport.

We recall that EQCM frequency data contain inforamtion about the transfer from a bathing solution and the polymer film of all mobile species both charged and neutral, whereas data obtained using conventional electrochemical measurements pertain only to charged species. Of course all of these transport processes are driven by electrochemical potential gradients of the various species: Net neutral species transport rates are determined by their activity gradient alone, whereas the mobility of charged species also depends on the potential gradient within the film.

The net instantaneous mass flux at the polymer–solution interface is given by

$$\frac{dM}{dt} = \dot{M} = \sum_k m_k \dot{\Gamma}_k = \sum_n m_n \dot{\Gamma}_n + \sum_i m_i \dot{\Gamma}_i \qquad (3.65)$$

where we have set $\dot{\Gamma} = d\Gamma/dt$. The individual mass fluxes $m_k \dot{\Gamma}_k$ may be decomposed into fluxes due to net netural species (subscript n) and ions (subscript i). The net instantaneous charge flux, the electron flux at the electrode–polymer interface, is given by

$$i = \frac{dQ}{dt} = \dot{Q} = -\sum_i z_i F \dot{\Gamma}_i \qquad (3.66)$$

where i denotes the current density and $z_i F$ denotes the charge carried per mole of ion i. A positive charge flux corresponds to oxidation. We may now combine Eqs. (3.65) and (3.66) to define a new function $\dot{\Phi}_j$ for some selected ion j as follows:

$$\dot{\Phi}_j = \dot{M} + i\frac{m_j}{z_j F} = \sum_n m_n \dot{\Gamma}_n + \sum_i \left(m_i - \frac{m_j z_i}{z_j}\right)\dot{\Gamma}_i \qquad (3.67)$$

where m_j denotes the molar mass of some selected ion j (usually that with the largest transport number t_j) of charge z_j. This expression quantifies the weighted difference of the differential mass and current responses (units: $g\,cm^{-2}s^{-1}$).

One can integrate Eqs. (3.65) and (3.66) to obtain

$$\Delta M = \sum_k m_k \Delta\Gamma_k = \sum_n m_n \Delta\Gamma_n + \sum_i m_i \Delta\Gamma_i \qquad (3.68)$$

and

$$Q = -\sum_i z_i F \Delta\Gamma_i \qquad (3.69)$$

The latter eaxpressions may be combined to define a function Φ_j for a specific ion j as follows:

$$\Phi_j = \Delta M + Q\,\frac{m_j}{z_j F} = \sum_n m_n\,\Delta\Gamma_n + \sum_i \left(m_i - \frac{m_j z_i}{z_j}\right)\Delta\Gamma_i \qquad (3.70)$$

We note from the expressions outlined above that neutral species contribute to the mass flux but not to the charge flux. One also notes that one can make a free choice of ion j from any of the ions present in the system. The contribution of this ion to the functions Φ_j and $\dot{\Phi}_j$ is by definition zero. This property enables one to isolate the contribution of each ion j to overall mass transfer. Hence if ion species j is the only mobile species present, then the quantities Φ_j and $\dot{\Phi}_j$ will be zero throughout the course of the redox transformation. This is a necessary but insufficient condition for permselectivity without solvent transfer. Since ion j contributes nothing to the functions Φ_j and $\dot{\Phi}_j$, the larger the contribution of ion j to maintenance of electroneutrality, the smaller the values of Φ_j and $\dot{\Phi}_j$ that one will obtain. In short, the value of Φ_j/m_j will be smallest for the ion with the largest transport number. Let us now consider the situation where one has neutral species (such as solvent) transfer during redox switching. A permselective film immersed in a solution containing a single electrolyte will yield on analysis Φ_j and $\dot{\Phi}_j$ factors that only depend on the transport of neutral species. In such a situation the latter functions enable one to directly monitor solvent transfer during redox switching.

The reader is referred to the original publications [175, 176] for further detailed analysis of the Hillman–Bruckenstein functions $\dot{\Phi}$ and Φ.

The general approach illustrated above using the Hillman–Bruckenstein functions (which are linear combinations of charge and mass change or current and mass flux) is useful since the functions have the special property that the contribution of the selected charged species to the functions can be eliminated. We will now show that simple diagnostic plots involving Q, ΔM, and Φ (or the time derivates of these quantities) may be used to test in an unequivocal manner for the existence of global equilibrium. If global equilibrium is not present then one can identify from the plots the nature of the rate-determining step during redox switching. The diagnostic scheme developed may be applied to cyclic voltammetric and chronoamperometric measurements.

The Hillman–Bruckenstein diagnostic scheme is illustrated in Fig. 45. The designations Y and N in the scheme refer to the observation of *tracking* between specific experimental quantities. These workers define the term tracking to signify that a plot of one experimental quantity versus another is independent of the scan rate and scan direction. They

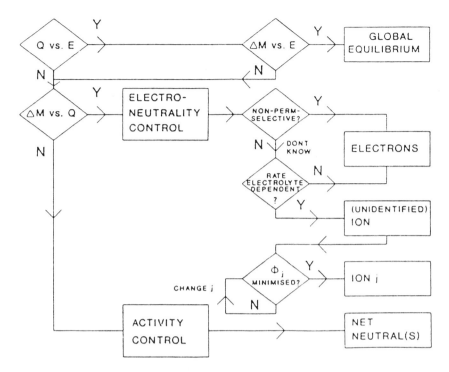

Figure 45. Hillman–Bruckenstein diagnostic flow chart (in terms of integral quantities Q, ΔM, and Φ) for evaluation of rate-determining process in redox switching in electroactive polymer films.

also note that the existence of tracking merely signifies that a pseudo-equilibrium exists on the selected time scale of the experiment. Slower polymer relaxation processes may not be detected electrochemically.

Now, a plot of Q versus E provides a test of global (electron and mobile species) equilibrium. However, tracking of ΔM-versus-E data demonstrates that all species, electrons, neutrals, and ions, are in global equilibrium. Let us now assume that there is tracking of ΔM-versus-Q data but not of ΔM-versus-E data. Tracking of ΔM-versus-Q data implies that mobile species transfer is in equilibrium with the instantaneous oxidation state of the film. Hence under these circumstances, failure of ΔM-versus-E data to track signifies that transport of electrons or coupled ionic charge is rate limiting. Hence we have *elctroneutrality control*. To isolate the rate-determining step (electron hopping or counterions), one must know whether the film is permselective or not. One can readily evaluate whether permselectivity is operative by examining whether or not the overall mass transfer rate with respect to time depends on the natrue of the electrolyte ions present. Conversely, if the polymer is nonpermselective, then electron hopping will be rate determining. If unidentified ion motion is indicated as being rate limiting, then one must introduce the Hillman–Bruckenstein functions. Minimization of these functions will identify the ion. If the ΔM-versus-Q plots do not track, then the situation corresponds to *activity control*, and net neutral species transfer will be rate determining.

Hence we see that this scheme provides a very convenient method for isolating the rate-determining step for the redox switching process. The analysis has been applied to redox switching in poly(bithiophene) films [175] and poly(vinylferrocene) layers [176]. The reader is referred to these papers for further details.

3. Effect of Observational Time Scale and Film History on Switching Kinetics

We finish this section with a discussion of the interrelationship between mobile species transport and polymer relaxation processes in permselective electroactive polymer films when the latter undergo redox switching. This discussion will be based to a large extent on a paper recently presented by Hillman and Bruckenstein [177].

A number of general observations may be made initially. We have previously noted that redox switching may involve a number of fundamental processes such as electron transfort, counterion, coion, and neutral species (such as solvent) transport and film structural changes. When the time scale is long, thermodynamics prevails and one attains a state of global equilibrium. On the other hand, when the experimental time scale is very short, only the most rapid processes have enough time to occur and these will be rate determining in the context of the particular time scale. We note therefore that the polymer system may pass from the initial state via a series of intermediate metastable states to the final state corresponding either to that of global thermodynamic equilibrium or pseudoequilibrium, depending on the time scale. Electron injection/ percolation and counterion transport processes can be quite rapid, whereas solvent and other neutral species transport occurs on a much longer time scale. Redox switching involves a change in the oxidation state of the film, which in turn can have a marked effect on polymer layer structure. Polymer relaxation processes [such as those involving extended segmental motion and chain (dis)entanglement] will only be manifest on the longest time scales. Local segmental motion occurs much more rapidly and may be a prerequisite for ion and neutral species transport. We conclude that redox switching is therefore quite a complex process.

We now consider the situation of redox switching involving an electroactive polymer film that is permselective. Let us also assume that electron transfer occurs only between oxidized O and reduced R species having the same nuclear geometry. Let us first examine the two processes, coupled electron–ion transfer and solvent transport resulting in solvation and desolvation of the redix active centres. Hillman and Bruckenstein [177] represented this simple situation in terms of a two-dimensional 1×1 square scheme (Fig. 46). The X coordinate in the square represents electron–ion tranfer and the Y coordinate the chemical solvent transport process. One notes from this figure that there are four distinct states: R, O, R^S, and O^S. Let us assume that the initial reduced state R is lyophobic and that the final state O^S is lyophilic. This is the situation encountered, for example, with poly(pyrrole)-modified electrodes. Hence the stability of the R state will increase as the extent of desolvation increases, whereas the stability of the O state will increase as the degree of solvaton increases. One can have two possible paths on proceeding from the initial R state to the final O^S state. One has, for the oxidation half-cycle,

$$EC: R \rightarrow O \rightarrow O^S$$

$$C'E': R \rightarrow R^S \rightarrow O^S$$

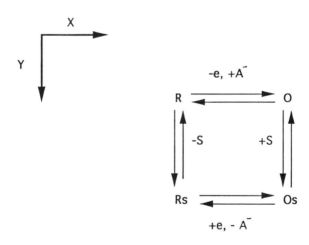

X = electron / ion transfer
Y = solvent transfer

Figure 46. Two-dimensional (1×1) square scheme representation of electron–ion and solvent transfer processes that operate during redox switching in electroactive polymer film.

Alternatively, if the reduced initial state is lyophilic and the final oxidized state is lyophobic, then we have

$$E'C: R^S \rightarrow O^S \rightarrow O$$

$$C'E: R^S \rightarrow R \rightarrow O$$

Hence since we have two possible initial states R and R^S, one has a total of four possible pathways by which the redox transformation may be accomplished in a half-cycle. Correspondingly for the reduction half-cycle we can write

$$D'C': O^S \rightarrow R^S \rightarrow R$$

$$CE: S^S \rightarrow O \rightarrow R$$

$$EC': O \rightarrow R \rightarrow R^S$$

$$CE': O \rightarrow O^S \rightarrow R^S$$

Hence for any complete cycle we must consider a total of eight possible mechanistic pathways for the two-dimensional 1×1 square scheme. This is the simplest possible situation.

The analysis may be extended to describe the more complicated process in which the switching dynamics must be described in terms of three coordinates: X corresponding to coupled electron–ion transfer, Y corresponding to solvation–desolvation, and Z defining polymer configurational changes. Hence a three-dimensional $1 \times 1 \times 1$ square scheme of the type illustrated in Fig. 47 must be considered. In this diagram we consider two possible polymer configurations α and β. One notes that the edges of the cube represent the reaction coordinate between the states located at the corners. Consider the following reaction sequence that takes us from an initial lyophobic R_α state to a final reconfigured lyophilic state O_β^S, involving oxidation, solvation, and polymer reconfiguration: $R_\alpha \rightarrow O_\alpha \rightarrow O_\alpha^S \rightarrow O_\beta^S$. This sequence assumes that electron transfer occurs first. A sequence of free-energy profiles for this path is illustrated in Fig. 48. The height of these energy barriers will determine the magnitudes of the rate constants for each of the three elementary steps: electron–ion transfer, solvation, and polymer reconfiguration. We note from Fig. 48 that the barrier to electron transfer is typically the lowest.

One can of course suggest other pathways for the R_α–O_β^S interconversion. This will depend on whether the electron transfer occurs first, second, or third in the reaction sequence. The possibilities are outlined in

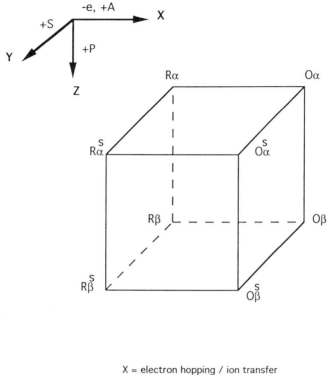

X = electron hopping / ion transfer
Y = solvent transfer
Z = polymer reconfiguration

Figure 47.   Three-dimensional $(1 \times 1 \times 1)$ cube scheme representation of electron–ion transport, solvent transfer, and polymer strand relaxation/reconfiguration processes that occur during redox switching in electroactive polymer film.

Table II. For the oxidation half-cycle there are six possible pathways. There are another six pathways for the reverse $O_\beta^S \rightarrow R_\alpha$ reduction. This makes a total of 12 paths for a complete cycle for these particular initial and final states. The complexity does not end here, however. There are three other possible R-to-O state interconversions that one may consider. These are (1) lyophilic R_α^S and lyophobic reconfigured O_β, (2) lyophobic reconfigured R_β and lyophilic O_α^S, and (3) lyophilic reconfigured R_β^S and lyophobic O_α. The description of these three interconversions (in both directions) is analogous to that presented in Table II for the $R_\alpha - O_\beta^S$ transformation. Hence we see that the situation can be very complex,

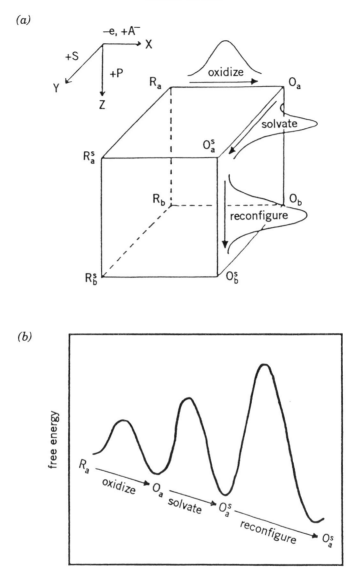

reaction coordinate

Figure 48. (*a*) Three-dimensional cube representation with schematic free-energy profiles for conversion of species R_α to O_β^s via path $R_\alpha \to O_\alpha \to O_\alpha^s \to O_\beta^s$, corresponding to electron–ion transfer followed by solvent transfer and finally polymer reconfiguration. (*b*) Schematic representation of free energy versus reaction coordinate for process presented in (*a*).

TABLE II
Mechanism for R_a/O_b^S Interaction

Position of Electron–Ion Transfer Step	Possible Mechanisms	Redox state(s) of Intermediates	Face of Competing Chemical Steps
	Oxidation		
First	$R_a \to O_a \to O_a^S \to O_b^S$ $R_a \to O_a \to O_b \to O_b^S$	O	right
Second	$R_a \to R_a^S \to O_a^S \to O_b^S$ $R_a \to R_b \to R_b^S \to O_b^S$	R and O	a
Third	$R_a \to R_b \to R_b^S \to O_b^S$ $R_a \to R_a^S \to R_b^S \to O_b^S$	R	left
	Reduction		
First	$O_b^S \to R_b^S \to R_b \to R_a$ $O_b^S \to R_b^S \to R_a^S \to R_a$	R	left
Second	$O_b^S \to O_b \to R_b \to R_a$ $O_b^S \to O_a^S \to R_a^S \to R_a$	O and R	a
Third	$O_b^S \to O_b \to O_a \to R_a$ $O_b^S \to O_a^S \to O_a \to R_a$	O	right

^{a} The two chemical steps are separated by electron–ion transfer, so they occur on opposite faces of the cube.

each conversion (half-cycle) may proceed via 3! pathways in each direction. Since there are four possible initial states, this yields a total of 144 possible mechanisms for a complete redox cycle.

The Hillman–Bruckenstein model [177] may be readily extended to situations in which there are more than two states along a given coordinate. The situation may be described qualitatively in terms of a cluster of fused $1 \times 1 \times 1$ cubes. If the number of redox, solvation, and configuration states are u, v, and w, respectively, when the number of required cubes is $(u-1)(v-1)(w-1)$. Depending on the experimentally accessed states, the cubes may be joined by corners and/or edges and/or faces. The situation in general can get quite complex!

Hillman and Bruckenstein [177] have utilized this type of analysis to discuss "break-in" effects, charge and mass trapping, structural evolution with redix cycling, kinetic decoupling of ion and solvent transfer, and variations in apparent charge transport rate and formal potential with experimental time scale. The reader is referred to the original paper for further details.

IV. ELECTROPOLYMERIZATION MECHANISMS: NUCLEATION AND GROWTH OF ELECTRONICALLY CONDUCTING POLYMER FILMS

A. Introduction

The fundamental properties of electrodes modified with electroactive polymer films have been examined in some detail in previous sections of this chapter. It is now generally agreed, however, that the properties exhibited by the electrodeposited layer will greatly depend on the morphology of the deposit, which will in turn depend on the conditions of electropolymerization employed. It is therefore necessary and important to be in a position to fully define the electrodeposition process. If this can be done, an accurate correlation between the layer morphology and the electrochemical–physicochemical properties may be elucidated.

The fact that electronically conducting polymer films may be formed electrochemically (they may also be formed via chemical routes) has the useful advantage that the process of electrodeposition and film formation may be readily followed in real time using relatively simple electrochemical techniques. Furthermore, the experimental data may be quantiatively analyzed to obtain information on the kinetics and mechanism of the deposition process. In short, a major advantage of the electrochemical approach toward polymer synthesis is that the electrode can act simulataneously as an initiator, monitoring agent, and deposition substrate.

In this section we concentate on electropolymerization as a synthetic strategy. We shall not describe the alternative chemical synthetic pathways that have been developed since these are considered to be outside the scope of the chapter. This is not to say that nonelectrochemical synthetic routes are not important. They are. One should realize that while many conductive polymers may be prepared via a variety of methods (both chemical and electrochemical), many, and most notably poly(acetylene), are still only accessible via direct chemical synthesis. In reacent years chemical synthetic strategies have involved the generation of polymeric precursor species that may be readily purified and cast onto substrates. Subsequent conversion to the desired product is then achieved via heat treatment. The elegant work of Edwards and Feast [193] on poly(acetylene) synthesis (the so-called Durham route) provides a good example of the latter strategy. The precursor route is also attractive in that the material may be readily processed during synthesis. One particularly interesting aspect here is the phenomenon of stress orientation: The material is stressed while being converted into the desired form. This manipulation results in an alignment of the polymer chains producing a material that exhibits a high degree of order and therefore greater conductivity.

Electronically conducting polymers are traditionally deposited as thin films via anodic oxidation of the relevant redox active monomer (the latter being dissolved in a conductive electrolyte solution) onto a support electrode surface. The basic requirements are as follows. First, the monomer must exhibit an oxidation potential that is accessible via a suitable solvent system. Second, the radical cation formed as a product of the initial oxidation step must react more quickly with other monomers to form oligomeric products than with other nucleophiles present in the deposition solution. Third, the oxidation potential of the polymer must be lower than that of the monomer. The choice of deposition substrates is not critical. Typical materials include glassy carbon, platinum, gold, or Indium tin oxide glass. The electrodeposition is generally accomplished via potentiostatic (constant-potential) or galvanostatic (constant-current) techniques. Alternatively, the deposition may be accomplished via potentiodynamic techniques such as cyclic voltammetry. In the latter method a repetitive triangular potential waveform is applied to the support electrode surface and the resultant current/potential response is monitored. The former methods are easier to describe quantitatively and have therefore been much utilized to examine the mechanism of layer nucleation and macrogrowth. The latter potentiodynamic method has been used mainly to obtain general qualitative information about the redox chemistry involved in the initial stages of the polymerization reaction as well as serving as a tool for examining the electrochemical response of the polymer layer once it is formed.

Although it is now over 12 years since it was established that heterocyclic conductive polymers could be synthesized electrochemically, the full details of the mechanism of electropolymerization are still subjects of active discussion in the literature. In this section of the review we address two outstanding problems. First the overall features of the electropolymerization mechanism are discussed from a molecular point of view. Second, the question of the mechanism of macrogrowth is examined in the context of electrochemical phase formation models. These general considerations will be amplified by examining two respresentative examples [poly(pyrrole) and poly(aniline)] drawn from work conducted in the author's laboratory. Even though the polymer systems chosen are specific, the conclusions presented in the chapter will be quite general and will apply to a wide range of polymer systems. Our intention is to give the reader an appreciation of the type of information that may be obtained on the kinetics and mechanism of electropolymerization and film formation of conductive polymers using simple electrochemical methods.

B.   General Features of Electropolymerization Mechanism of Electronically Conductive Heterocyclic Polymers

It is generally agreed that the mechanism of polymerization of heterocyclic polymers is a complex multistep $E(CE)_n$ process [194–198]. The overall reacton pathway is illustrated in Scheme 1. The first step involves the irreversible electrochemical oxidation of the neutral monomer species R to form a radical cation $R^{+\cdot}$. The latter species are unstable and rather reactive. It is obvious that a high concentration of radical cations may be continuously maintained at the electrode–solution interface region via the steady-state diffusion of R from the bulk of the solution. In an early fundamental study on the electropolymerization of heterocyclic monomers, Waltman and co-workers [199, 200] noted that there is an optimum range of monomer oxidation potential that leads to polymer formation [typically 1.2–2.1 V (vs. SCE) in 0.1 M tetrabutylammonium perchlorate–acetonitrile media (TBAP)]. This behavior was rationalized in terms of the stability or reactivity of the radical cation intermediate in the interfacial region near the electrode surface. A competitive kinetic analysis was applied. Under steady-state conditions the following possibilities arise as outlined in Scheme 2. The $R^{+\cdot}$ intermediate may either react with the electrode surface to form oligomer and subsequently polymer (quantified in terms of a rate constant k_p) or, if stable enough, diffuse away from the electrode (rate constant k_D), or as a final alternative, enter into side reactions with the solvent S or with anions X^-, which are located in the interface region (quantified by a rate constant k_S). These workers showed that the fraction f_p of cation radicals that take part in electropolymerization is given by

$$f_p = \frac{k_p}{k_p + k_D + k_S(s + x)} \qquad (4.1)$$

where s and x denote the solvent and anion concentrations, respectively. The useful window in oxidation potential can now be rationalized. The lower cutoff limit is due to enhanced stability of the radical cation. When $k_D > k_p + k_S(s + x)$, diffusion of $R^{+\cdot}$ from the electrode results in the production of soluble products. The upper cutoff in potential occurs when $k_S(s + x) \gg k_p + k_D$. Under such conditions the radical cation intermediate becomes unstable and reacts with solvent or anions. The useful potential window corresponds to the condition where $k_p \gg k_D + k_S(s +$

1. Monomer oxidation

[A]   [B]

2.Resonance forms

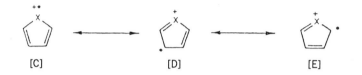

[C]   [D]   [E]

3.Radical - radical coupling

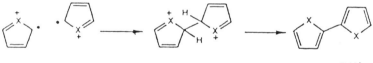

+ 2 H+

4.Chain propagation

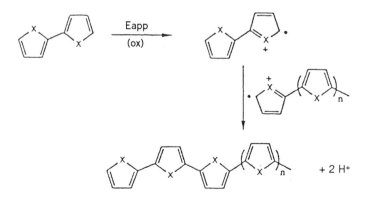

+ 2 H+

Scheme 1.   Electropolymerization of heterocyclic monomers (such as pyrrole) at inert electrodes in aqueous solution.

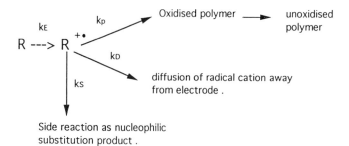

Scheme 2. Fate of radical cation species produced via electrooxidation at inert electrode surface.

x). Under such conditions $f_p \approx 1$. Waltman and co-workers [199] noted that substituent effects can be important in determining the ease of monomer oxidation. A general conclusion was that substituents, even when removed from the reactive position on the monomer such that steric effects are unimportant, nevertheless can suppress electropolymerization via electronic effects.

The delocalized radical cation species may be expressed via a number of resonance forms, as outlined in Scheme 1. It has been shown by Waltman and Bargon [197] that the declocalized radical cation has the highest spin density at the α position, and so, of the three possible resonance forms I–III illustrated in Scheme 1, form III is the most stable. The next step in the polymerization sequence involves either radical/radical coupling or radical/monomer coupling. Radical–radical coupling at the α position is now thought to be most favoured. Radical–radical coupling is then followed by expulsion of two H^+ ions to produce the neutral dimer. The driving force for this step is the return to aromaticity.

In recent years, considerable experimental evidence has built up supporting the radical–radical coupling mechanism. According to Genies and co-workers [194], if after initiating polymerization at a potential where pyrrole is readily oxidized, the potential is lowered to a value intermediate between the oxidation potential of the monomer and the polymer, no further polymerization occurs. If radical–monomer coupling was viable, further polymerization would have been observed. Another indicator for the operation of a radical–radical coupling mechanism has been provided in the work reported by Baker and Reynolds [195, 198]. These workers have shown that the copolymerization of pyrrole with substituted pyrroles is potential dependent. If radical–neutral monomer coupling pertained, then a copolymer would have been generated at the potential where only one the monomer types was oxidized. Finally, in a recent study, Andrieux and co-workers [201] examined the early stages of the electrochemical polymerization of three substituted pyrroles using ultrafast double-potential step chronoamperometry. A number of features are striking in this work. Now, the substituted pyrroles used in the study (see Scheme 3) were chosen such that their radical catons exhibited a relatively long lifetime. Now cation radicals are usually short lived and are therefore difficult to study using conventional electrochemical techniques. However, in the work reported by Andrieux and co-workers [201], ultramicrodisc electrodes were used, and a recently developed very fast scan rate cyclic voltammetry procedure was employed to examine the redox chemistry of radical formation. The fact that very small electrodes are used enables one to extract useful data at very high sweep rates with a minimum of complications due to Ohmic drop considerations. The operation of the latter has often invalidated much cyclic voltammetry work conducted at high scan rates at conventional macrosize electrodes. The cyclic voltammetry results obtained these workers for TISPP in acetonitrile–NEt_4ClO_4 as a function of sweep rate in the range 20–100 V s^{-1} are presented in Fig. 49a. Note that as the sweep rate is increased, the system passes from almost complete irreversibility (the condition usually encountered) to complete reversibility. The electrochemistry therefore becomes more well behaved at very fast sweep rates (corresponding to very short times). A systematic analysis of the initial polymerization kinetics using double-potential step chronoamperometry indicated that for all three substituted pyrrole species examined, the cation radicals, rather that the neutral radicals, which would result from deprotonation of the latter, are involved in the initial carbon–carbon bond formation. Furthermore, it was established, by comparing normalized experimental chronamperometric data with theoretically obtained normalized working curves computed for radical–radical and radical–monomer coupling processes (Fig. 49b), that the best correlation between

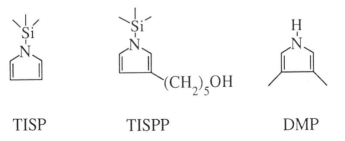

TISP   TISPP   DMP

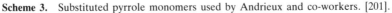

Scheme 3. Substituted pyrrole monomers used by Andrieux and co-workers. [201].

446 M.E.G. LYONS

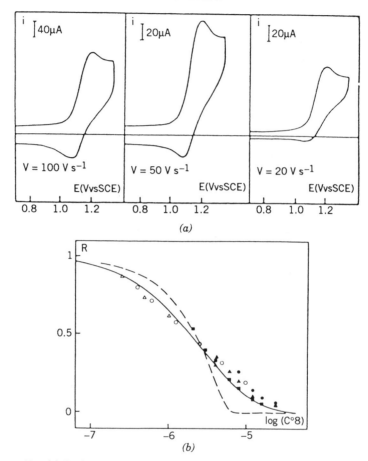

Figure 49. (*a*) Typical cyclic voltammetric response of TISPP (2 m*M*) in acetonitrile–NEt$_4$ClO$_4$ (0.6 *M*) medium recorded at 1-mm-diameter Pt electrode at 293 K at three different scan rates. (*b*) Double-potential step chronoamperometric results obtained for TISPP system. Same medium as in (*a*). Potential step sequence: $0.70 \rightarrow 1.45$ V and 1.45 V $\rightarrow 0.70$ V. Full curve shows variation of current ratio R with parameter log $(c^0\theta)$, where θ is characteristic reversal time and c^0 denotes bulk concentration of substituted pyrrole monomer in solution. After initial forward potential step, potential is subsequently stepped back again to its initial value at time θ. Anodic current $i_a(\theta)$ is determined at time θ and cathodic current $i_c(2\theta)$ is then determined at time 2θ. Current ratio $i_c(2\theta)/i_a(\theta)$ is then normalized with respect to ratio expected for simple diffusion-controlled reaction in which there are no following chemical steps. Characteristic ratio $R = i_c(2\theta)/i_a(\theta)/\{[i_c(2\theta)/i_a(\theta)]_D\}$. We outline variation of R with log($c^0\theta$) recorded at 1-mm-diameter Pt disc electrode (filled squares, circles, and triangles) and at smaller 20-μm-diameter Pt disc electrode (open circles and triangles) for various values of monomer concentration c^0 in range 1–10 m*M*. Full line represents theoretical working curve derived by assuming radical–radical coupling mechanism, whereas dashed line in working curve computed for radical–monomer coupling process.

theory and experiment was obtained for the radical–radical coupling case. These results provide the strongest confirmation to date of the validity of the radical–radical coupling mechanism.

Returning to Scheme 1, we note that chain propagation subsequently proceeds via oxidation of the neutral dimer to form the dimer radical, which can then combine chemically with other monomeric dimeric or oligomeric radicals to extend the chain. The chemical coupling–electrochemical oxidation process repeats itself until chain growth is terminated. The overall process is described in terms of an $E(CE)_n$ mechanism, where E represents the initial monomer oxidation step and the CE designation corresponds to the chemical coupling–electrochemical oxidation sequence. The latter is repeated n times until termination occurs. The overall sequence may be represented as follows:

$$2A \xrightarrow{-2e^-} A^{\cdot+} + A^{\cdot+} \to A^+ - A^+ \to A - A + 2H^+$$

$$A^{+\cdot} + [A - (A)_n - A]^{+\cdot} \to [A - (A)_n - A - A]^{2+} \to (A)_{n+3} + 2H^+$$

In the parlance of polymer science the mechanism may be classified as an *oxidative polycondensation* process.

The following question arises at this point: Where does the polymerizaton occur? Is the neutral monomer initially adsorbed and then undergoes oxidative polycondensation, or does the polycondensation process occur in the interfacial region? The answer to this problem is not trivial. Recent studies using rotating ring–disc electrodes (RRDEs) and in situ differential ellipsometry have lent support to the hypothesis that the first stages of polymer deposition involved polymerization in solution followed by precipitation of olgomeric chains onto the electrode surface [202, 203]. Very recent FTIR and ellipsometry studies reported by Hamnett and co-workers [204] have lent support to this proposal. These workers claim that poly(pyrrole) grows on Pt support electrodes in aqueous perchlorate solutions via the deposition of oligomers of an average length of 34 monomer units. It should also be noted that the assertion of soluble intermediate formation is supported from coulombic efficiency measurements conducted during electrodeposition. It has been established that electrodeposition is less than 100% efficient. The latter can be attributed to transport of oligomeric species away from the electrode surface.

To this point we have examined some global features of the chemistry of the polycondensation process. Ultimately, a new phase is formed on the surface of the support electrode. This occurs via a process of nucleation and layer growth. The latter process is already well established as a valid mechanism to describe metal electrodeposition. In the next section we discuss nucleation/growth or electrochemical phase formation processes at a fundamental level and indicate how the kinetics of nucleation may be examined using simple electrochemical techniques such as potential step chronoamperometry.

C. Fundamentals of Nucleation and Layer Growth

1. Introduction

Electrochemical phase formation is now an established area of physical electrochemistry and has been extensively investigated over the last 30 years. Classical reviews covering the early literature and the theoretical fundamentals have been written by Fleischmann and Thirsk [205] and Harrison and Thirsk [206]. More reacently the topic has been reviewed by de Levie [207]. Additional information may be found in a *Faraday Symposium* on electrocrystallization [208] and in a special issue of *Electrochimica Acta* [209]. Pletcher has written an excellent account at a basic level [210]. In this discussion we concentrate on metal deposition processes since this area is well established, and the results obtained may be directly applied to the electrodeposition of electronically conducting polymers.

The phenomenon of electrochemical phase formation [such as, e.g., metal deposition via reduction of a metal ion in solution onto an inert support surface (Cu deposition on carbon)] is an extremely complex multistep process and may be decomposed into a number of fundamental steps, any of which may be rate determining. We must differentiate between the formation of a stable nucleus and the formation of a macroscopic three-dimensional layer. Both are complex multistep processes. With respect to the former, the following sequence of events is envisaged (Fig. 50a). The solvated cations diffuse to the electrode surface and then undergo electron transfer allied with partial or full desolvation to form an ad-atom. One then has diffusion of ad-atoms along the support surface. The latter species may then cluster together to form a nucleus of sufficient size (the critical nucleus) to exhibit stability. The nucleus grows via incorporation of ad-atoms at favorable sites in the lattice structure of the metal. With respect to macrogrowth (Fig. 50b) we begin with nucleus formation at sites on the support surface and the subsequent expansion of isolated nucleii. At some stage these isolated nucleii overlap, and one has the formation of a continuous layer over the entire support electrode surface. The latter then thickens to form a three-dimensional film. It is clear that ideas developed to describe the growth, overlap, and subsequent layer growth of metallic films may be

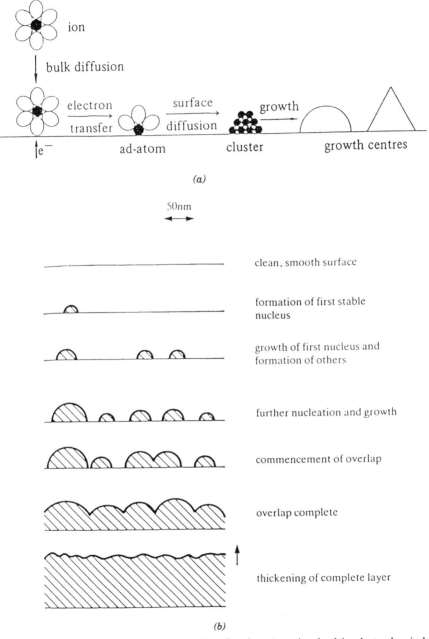

Figure 50. (*a*) Schematic representation of various steps involved in electrochemical phase formation. (*b*) Schematic representation of macrolayer formation and growth on electrode surface.

extended to describe conductive polymer film formation. However, the mechanism involved in the formation of the critical nucleus of a metal will not be applicable to the conducting polymer case. Consequently, we shall concentrate more on macrolayer growth in the following discussion. Our presentation here is influenced by Pletcher [210].

2. General Comments on Fundamental Thermodynamics and Kinetics of Nucleation

We begin our presentation of fundamental principles with some general comments on the thermodynamcis of nucleus formation. It should be obvious that the formation of nucleii of a new phase will be an improbable event that must be forced to occur: This is because these small centers of the new phase are *thermodynamically unstable*. The total Gibbs energy of formation of the new phase ΔG may be decomposed into two components: A surface component ΔG_S and a bulk component ΔG_B. Both components are additive, and we may write

$$\Delta G = \Delta G_S + \Delta G_B \qquad (4.2)$$

The first term on the rhs of Eq. (4.2) is positive and will be proportional to the surface area, whereas the second term is negative and will be proportional to the volume of the new phase. Small nucleii are therefore unstable because of the following simple reason: They exhibit a large ratio of surface area to volume. In short, ΔG is positive. Hence in order for the nucleus to attain stability, it must reach a critical size. In the case of metal electrodeposition, this may only be achieved if the ad-atoms collect together very rapidly. For the case of electrodeposition of metals, a high surface ad-atom concentration is favored by two factors: a high concentration of metal ions in the solution and the magnitude of the applied overpotential η. Indeed, the overpotential is the electrochemical equivalent of supersaturation, the latter being an essential condition for crystal growth in nonelectrochemical situations. We recall from the previous section that initiation of electropolymerization, and indeed the rate of electropolymerization of conductive polymers, will also depend on the magnitude of the applied overpotential.

These simple thermodynamic ideas may be quantified as follows. We assume a spherical nucleus geometry. Consequently, the surface component of the Gibbs energy is given by $\Delta G_S = 4\pi r^2 \gamma$, where γ denotes the molar surface free energy (the surface tension) and r is the radius of the spherical nucleus. Furthermore, the bulk component of the Gibbs energy is given by $\Delta G_B = \frac{4}{3}\pi r^3 \Delta G_V = -\frac{4}{3}\pi r^3 [zF\eta/V_m]$, where V_m denotes the molar volume of the deposit (note that $V_m = M/\rho$, where M is the molar mass and ρ is the density of the deposit) and z denotes the valence of the metal ion. Note also that F is the Faraday constant. Note that ΔG_V denotes the Gibbs energy per unit volume associated with the formation of the bulk phase. It is obvious that the strategy of using values of V_m and γ for the bulk phase of the electrodeposit as suitable quantities for very small nucleii is open to question. However, to a first approximation this strategy is reasonable.

For a given value of overpotential, the surface and volume contributions combine to produce a maximum in ΔG as a function of the size of the nucleus, as illustrated in Fig. 51. From this, we can define two important parameters. The first is the critical radius r_c and the second is the critical Gibbs energy ΔG_c. In short, for a nucleus to evolve into a stable entity, its size must exceed r_c. To do this, a critical energy ΔG_c must be overcome, and so the latter quantity is a rough measure of the activation energy for nucleation.

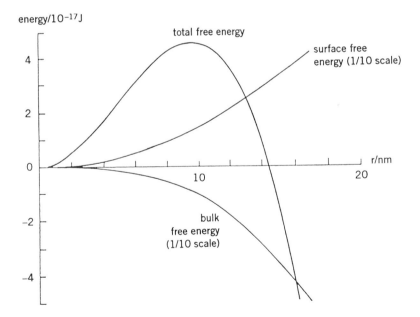

Figure 51.   Free-energy curves for homogeneous nucleation and growth of spherical Hg droplet from vapor phase at room temperature for supersaturation ratio of 3.2. Contribution to free energy from surface and volume processes presented on reduced scale. (Adapted from ref. 212.)

The net free energy ΔG is given by

$$\Delta G = 4\pi r^2 \gamma - \frac{4\pi \rho r^3 z F \eta}{3M} \qquad (4.3)$$

Differentiating Eq. (4.3) with respect to (wrt) r, we obtain

$$\frac{\partial \Delta G}{\partial r} = 8\pi r \gamma - \frac{4\pi \rho r^2 z F \eta}{M} \qquad (4.4)$$

Now $r = r_c$ when $\partial \Delta G / \partial r = 0$. Hence from Eq. (4.4) we obtain

$$r_c = \frac{2M\gamma}{zF\rho\eta} \qquad (4.5)$$

Hence the critical Gibs energy for stable nucleus formation is obtained by substituting Eq. (4.5) into Eq. (4.3) as follows:

$$\Delta G_c = \frac{16\pi \gamma^3 M^2}{3z^2 F^2 \rho^2 \eta^2} \qquad (4.6)$$

An important result obtained from this simple analysis is that the critical free energy for stable nucleus formation depends inversly on the square of the applied overpotential.

We can extend this idea and consider the situation where a nucleus is formed on a preexisting surface of the same material. We consider the free-energy changes associated with the formation of a circular patch of radius r and height h. In this case we write

$$\Delta G = 2\pi r h \gamma - \frac{\pi r^2 h \rho z F \eta}{M} \qquad (4.7)$$

Utilizing the same procedure as before, we arrive at the following expressions for the critical radius and Gibbs energy:

$$r_c = \frac{M\gamma}{zF\rho\eta} \qquad \Delta G_c = \frac{\pi h M \gamma^2}{zF\rho\eta} \qquad (4.8)$$

We recall that ΔG_c behaves pretty much as an activation energy. Hence in a very simple way we may write that the nucleation rate constant k_n is given by the following type Arrhenius expression:

$$k_n = A \exp\left(-\frac{\Delta G_c}{k_B T}\right) \qquad (4.9)$$

where k_B is the Boltzmann constant, T is the temperature, and A is a preexponential factor.

A somewhat more rigorous discussion of the kientics of critical nucleus formation has been presented in the calssical work of Zeldovich [211]. In this approach the formation of clusters of atoms via the successive addition of ad-atoms or monomers was examined. In the steady state, the rates of addition and loss of monomers are in balance and the distribution of clusters is determined by the following type of coupled equilibria:

$$\longleftrightarrow n-1 \underset{\beta_n}{\overset{\alpha_{n-1}}{\longleftrightarrow}} n \underset{\beta_{n+1}}{\overset{\alpha_n}{\longleftrightarrow}} n+1 \longleftrightarrow$$

where the α and β factors are individual rate factors. Now as the cluster size increases toward a critical value, the point is approached when further addition of monomer units results in the formation of a stable growth center. Once the critical size is reached, the stable center will no longer be involved in the steady-state distribution. This fact is taken on board by assuming that "supercritical" clusters are removed and replaced by the equivalent amount of monomer. Using this idea, we can show that the steady-state nucleation rate is given by

$$j = Z\alpha_n \cdot c_n \cdot \qquad (4.10)$$

where α_{n^*} represents the rate at which monomer species are added to the critical cluster containing n^* atoms and c_{n^*} denotes the equilibrium concentration of critical-size clusters. The parameter Z is called the Zeldovich factor. It is dimensionless and takes account of the fact that clusters are removed from the steady-state distribution when they reach critical size. The equilibrium concentration of critical clusters is related to the concentration of monomers c_1 via the Boltzmann distriution:

$$c_n^* = c_1 \exp\left(-\frac{\Delta G_{n^*}}{k_B T}\right) \qquad (4.11)$$

Hence from Eqs. (4.10) and (4.11) we note that

$$j = Z\alpha_{n^*} c_1 \exp\left(-\frac{\Delta G_{n^*}}{k_B T}\right) \qquad (4.12)$$

Again we see that the nucleation rate is described via an Arrhenius-like expression, and we can identify ΔG_{n^*}, the activtion energy for the formation of a n^* cluster with ΔG_c the activation energy for critical nucleus formation derived in the corresponding thermodynamic treatment.

The discussion just presented is quite qualitative but conveys the essential features of the thermodynamics and kinetics of homogeneous nucleation processes. In this discussion we have assumed that the surface is perfect. In reality, this is not the case. Real surfaces may contain a large number of defect sites that may act as preferential nucleation sites. Indeed, the nucleation free energy ΔG_{n*} may exhibit a range of values. It will also depend on overpotential.

3. Quantitiative Analysis of Nucleation/Growth Processes

The following approach is often used in classifying nucleation processes. If N_0 denotes the number of active sites on an electrode surface, then if we assume that the rate of appearance of stable growth centers follows simple first-order kinetics, we can propose the following expression for the time variation of the number density of nuclei:

$$N(t) = N_0[1 - \exp(-k_n t)] \qquad (4.13)$$

Two limiting cases may be distinguished. Now, when $k_n t \ll 1$, we can expand the exponential term in a Taylor series and, keeping only the first two terms, write $\exp(-k_n t) \approx 1 - k_n t$. Under such conditions, Eq. (4.13) reduces to

$$N(t) \approx N_0 k_n t \qquad (4.14)$$

We see that the number density of nuclei increase in a linear manner with time. This limiting situation is termed *progressive* nucleation. On the other hand, when $k_n t \gg 1$, we note that $\exp(-k_n t) \to 0$, and so Eq. (4.13) reduces to

$$N(t) \approx N_0 \qquad (4.15)$$

Here the number density of nuclei is independent of time. This limiting situation is called *instantaneous* nucleation. In instantaneous nucleation all the nuclei are formed at the same time, whereas in progressive nucleation, the nuclei are formed gradually. One immediate knock-on effect arising from these two limiting types of nucleation is the *size distribution* of nuclei. After a given period of time one expects a monodispersion in nucleus size if instantaneous nucleation prevails. Alternatively, if progressive nucleation is operative, then a spread or dispersion in growth center size is expected.

The discussion to date has concentrated on the formation of stable nuclei. Our immediate interest in this section is in macroscopic phase formation. Consequently, we now examine the growth of nuclei on an electrode surface. Before we attempt a quantitative discussion, a somewhat more qualitative description is in order.

Nucleation/growth processes are usually investigated using potential step chronoamperometry [212]. In this technique, the potential is stepped from a value where nucleation does not occur to a value where nucleation and subsequent growth occur. The shape of the resultant current–time transient is examined to determine the nature of the nucleation/growth process. Besides being a very simple experimental technique, the form of the chronoamperometric current response is also quite readily quantified in terms of simple mathematical models. It is also possible to examine nucleation/growth phenomena using more sophisticated electrochemical techniques such as linear sweep voltammetry [213] or complex impedance spectroscopy [214]. The latter methods are readily performed, but data analysis in the context of quantitative mathematical models can be complex. Consequently, we confine our discussion to an examination of the chronoamperometric current response in this chapter.

Once stable nuclei have been formed, they will expand and grow readily. Three types of growth process (one, two, or three dimensional) may be envisaged. One-dimensional needle-type growth is quite rare. Two- and three-dimensional growth is much more common. As outlined later in the chapter, examples of all three types of growth dimensionalty may be put forward when we consider the electrodeposition of electronically conducting polymer films. When growth is considered, the *geometry* assumed for the growth center is of crucial importance. For example, the formation of monolayer deposits usually occurs with a dimensionality of 2 and involves the formation and growth of disc-shaped nuclei that expand laterally and eventually coalesce. A three-dimensional film may also be formed via a two-dimensional growth mechanism, with the provision that the deposition proceeds in a layer-by-layer manner. In three-dimensional growth, the growth center geometry is usually assumed to be hemispherical or right circular conical. For the former geometry, the rate of growth is the same in the lateral and normal directions (a single-rate process), whereas for the latter, one has a different growth rate in the lateral and normal directions (i.e., a dual-rate process). For the case of metal electrodeposition, the rate of growth of nuclei will be determined by the rate of conversion of metal ions to ad-atoms at the expanding surface of the growth center. The rate of metal ion reduction can, of course, depend on either the heterogeneous interfacial electron transfer kinetics (activation control) or the diffusive transport of ions to the electrode surface (mass transport control). Similar rate limitations will apply for electroactive polymer electrodeposition. One expects to observe (under chronoamperometric conditions) an increase in current (i.e., deposition rate) during the initial stages of growth due to the fact that there is an increase in the area available for further deposition.

Once the stage has been reached where the electrode surface is covered with a large number of growing centers, a stage will be reached in which the individual growth centers will start to overlap and interact with one another. Under such conditions the current response will no longer increase at the same rate as observed during the initial stages of growth. Prior to the onset of overlap it is assumed, for the sake of simplicity, that each center grows independently and that the current observed contains contributions from each individual growth center. It is now realized [215] that this assumption is far from correct, but for our purposes it will suffice. The detailed calculation of the form of the current response taking nuclear overlap effects into account can be quite complex and will be discussed subsequently.

As a consequence, it is often the practice to examine the shape of the chronoamperometric current–time transient response profile at short times before significant growth center overlap occurs. This results in a simplified data analysis. The forms of the current transient at short times before appreciable overlap occurs, assuming charge transfer control, are as follows: $i \propto k_g^n t^{n-1}$ for instantaneous nucleation and n-dimensional growth and $i \propto k_g^n t^n$ for progressive nucleation followed by n-dimensional growth, where $n = 1, 2, 3$. In the latter expressions k_g denotes the growth rate parameter (units: $mol\,cm^{-2}\,s^{-1}$). This will depend on the magnitude of the overpotential via the Butler–Volmer equation.

Simplified analytical expressions for the current observed for nucleation/growth at short times before the onset of appreciable overlap are presented in Table III. A collage of typical amperometric current–time transient curves for various scenarios is outlined in Fig. 52. In Table III, the expressions for the current response are obtained by first evaluating the current flowing through a single center of well-defined geometry (needle, disc, hemisphere, or cone). This is accomplished as follows:

$$i = zFSk_g = \frac{zF\rho}{M}\frac{dV}{dr}\frac{dr}{dt} \tag{4.16}$$

where V and S represent the volume and area of the growth center, respectively. Both are well defined via the geometry assumed. Equation (4.16) may be integrated to obtain an analytical expression for the radius as a function of time $r(t)$, and the current response $i(t)$ may subsequently be determined from Eq. (4.16) since $S(r)$ is known. Subsequently, if the nucleation is instantaneous, the net current is obtained by multiplying the analytical expression for the current arising from a single growth center

TABLE III

Simplified Expressions for Transient Current Describing Nucleation and Growth at Short Times before Overlap Occurs between Growing Centers[a]

Nucleation Type	Growth Type One Dimensional (Needles)	Growth Type Two Dimensional (Discs)	Growth Type Three Dimensional	
			Charge Transfer Control[b]	Diffusion Control[c]
Instantaneous	nFN_0Sk	$2\pi nF(MI/p)hN_0k^2t$	$2\pi FN_0(MI/p)^2k^3t^2$	$nF\pi N_0(2Dc^\infty)^{3/2}M^{1/2}t^{1/2}/p^{1/2}$
Progressive	nFk_gN_0Skt	$nF\pi(MI/p)hk_gN_0k-2t^2$	$2\pi nFM^2k_gN_0k^3t^3/3p^2$	$4nF\pi kgN_0(Dc^\infty)^{3/2}M^{1/2}t^{3/2}/3p^{1/2}$

[a] Symbols: m = number of electrons transferred, F = Faraday constant, M = molar mass (g mol^{-1}), p = density (g cm^{-3}), C^∞ = bulk concentration of monomer (mol cm^{-3}), D = diffusion coefficient (cm^2 s^{-1}), k = growth rate constant, k_g = nucleation rate constant (s^{-1}), N_0 = number density of nucleation sites (cm^{-2}), k = growth rate parameter (mol cm^{-2}s^{-1}), h = height of growth center.

[b] Growth geometry assumed to be hemispheric. Single-rate model.

[c] Spheric diffusion pertains.

457

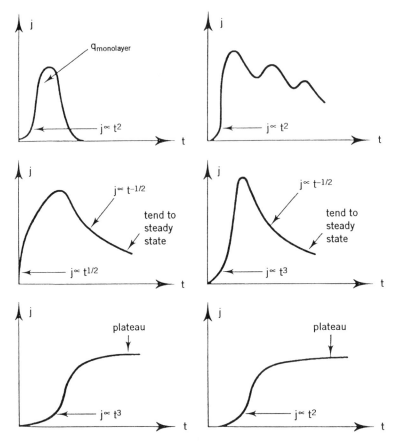

Figure 52. Potential step chronoamperometry as mechanistic indicator in electro-chemical nucleation/growth processes. Top right: monolayer deposition, progressive nucleation, followed by two-dimensional growth; top left: layer-by-layer deposition. Middle right: instantaneous nucleation followed by three-dimensional growth under diffusion control; middle left: progressive nucleation and three-dimensional nuclear growth under electron transfer (activation) control, layer growth under diffusion control. Bottom right: progressive nucleation and three-dimensional nuclear growth under electron transfer control, layer growth activation controlled. Bottom left: instantaneous nucleation and three-dimensional growth of nuclei under electron transfer control, layer growth under electron transfer control.

by N_0. For the case of progressive nucleation the following expression is used:

$$i(t) = \int_0^t i \, \frac{dN(t)}{dt} \, dt \qquad (4.17)$$

This idea may be illustrated by considering the growth of a single hemispheric center. In this case we can write

$$V = \frac{2\pi r^3}{3} \frac{dV}{dr} = 2\pi r^2 \qquad S = 2\pi r^2$$

Hence the current due to a single hemispheric center is given by

$$i = \frac{2zF\rho\pi r^2}{M} \frac{dr}{dt} = zFk_g 2\pi r^2$$

Simplification of the latter expression results in the assignment that $dr/dt = Mk_g/\rho$ and so $r(t) = (Mk_g/\rho)t$. Hence the current is given by

$$i(t) = 2\pi zFk_g r(t)^2 = \frac{2\pi zFM^2k_g^3}{\rho^2} t^2$$

If the nucleation is instantaneous, then at short times the current response to a potential step is given by

$$i(t) = \frac{2\pi zFN_0 M^2 k_g^3 t^2}{\rho^2}$$

Alternatively, if the nucleation occurs via a progressive mechanism, we note that $N(t) = k_n t$ and $dN/dt = k_n$, where k_n represents the nucleation rate constant discussed previously. Hence using Eq. (4.17), we see that the total current is given by

$$i(t) = \frac{2\pi zFN_0 k_n k_g^3 M^2}{\rho^2} \int_0^t t^2 \, dt = \frac{2\pi zFN_0 k_n k_g^3 M^2}{3\rho^2} t^3$$

A clear distinction may therefore be made between instantaneous nucleation/three dimensional growth and progressive nucleation/three dimensional growth by examination of the power law dependence of the current with respect to the time variable. For instantaneous nucleation $i \propto t^2$, whereas for progressive nucleation, $i \propto t^3$. A similar type of analysis may be made for two-dimensional growth processes. In this case a disc geometry is assumed for the growth center (radius r and height h). The lateral surface area $S = 2\pi rh$, whereas the volume $V = \pi r^2 h$. Proceeding as before we can show that the variation of the radius of the growth center with time is given by $r(t) = (Mk_g/\rho)t$, whence the current due to the growth of a single disclike center is given by

$$i(t) = 2\pi r F k_g hr(t) = \frac{2\pi z F k_g^2 h M}{\rho} t$$

Again, if the nucleation is instantaneous, then the net current response is given by

$$i(t) = \frac{2\pi z F k_g^2 N_0 h M}{\rho} t$$

If, on the other hand, the nucleation of growth centers is progressive, then the net current is given by

$$i(t) = \frac{2k_n N_0 \pi z F k_g^2 h M}{\rho} t^2$$

Hence for the two-dimensional growth situation $i \propto t$ if nucleation is instantaneous, whereas $i \propto t^2$ if nucleation proceeds in a progressive manner. In this way the expressions outlined in Table III are derived.

The important feature to note from Table III is that the power law relationship between i and t is of considerable diagnostic value. Let us consider the six cases presented in Table III for nucleation and growth under charge transfer control. In two of these six cases no ambiguity arises. The nucleation/growth mechanism may be readily quantified by examining the time dependence of the current alone. For the remaining four cases discrimination may be effected via the examination of the detailed nature of the potential dependence explicit in the k_g parameter. We assume that this is given by the Butler–Volmer equation:

$$k_g = A \exp\left(\pm\frac{\alpha z F \eta}{RT}\right) \tag{4.18}$$

where A is a constant preexponential factor and η, α denote the overpotential and the transfer coefficient respectively. We also note that evaluation of the constant factors preceding the $k_g^x t^y$ terms in Table III leads to the extraction of the number of nucleation sites N_0 in the instantaneous cases, but only the combined parameter $k_n N_0$ in progressive cases, The fundamental nucleation and growth parameters may be separated in principle using the technique of double-potential step chronoamperometry. This technique has been applied with profit by Gunawardena and co-workers [216]. In this experiment the first potential step is to the nucleation potential. A certain number of nuclei are formed. The second potential step is to a slightly lower potential, where the nuclei previously formed continue to grow, but no new nuclei are formed. Finally, returning to an examination of Table III, we note that if the nucleation/growth process is diffusion controlled, then half-integral current–time relations pertain, with $i \propto t^{1/2}$ if the nucleation is instantaneous and $i \propto t^{3/2}$ if the nucleation occurs progressively.

In many cases a high denity of nuclei is formed and overlap will rapidly occur. Hence when experimental data are subjected to analysis, one will observe a rapid departure from the simple expressions presented in Table III, since the latter were derived by assuming that the centers grow independently of each other. This assumption will be valid during the early stages of growth when the centers are widely spaced, but as radial growth proceeds, adjacent centers will come into contact, thereby reducing the edge area available for the incorporation of new material into the growth center. This effect is termed the *overlap problem* and has been discussed long ago in a number of classic papers by Avrami [217]. A particularly simple analysis of the overlap problem has been developed by the Southampton Electrochemistry Group [218], and we shall follow their development here. A more advanced discussion is presented in the reviews written by Thirsk and co-workers [205, 206]. Harrison and Rangarajan [219] have explored the consequences of the Avrami approach using simple computer simulation techniques.

A fundamental concept in the quantitation of overlap is that of *extended area*. The latter is defined as the notional area that would be covered by the centers if the overlap were not taken into account. This idea is illustrated in Fig. 53. In this diagram we illustrate how the overlap of growth centers presented in Fig. 53a gives rise to the real area S (Fig. 53b) and to the extended area S_e as outlined in Fig. 53c. Avrami [217] proved that the area S is related to the extended area S_e via an expression of the form

$$S = 1 - \exp(-S_e) \tag{4.19}$$

Clearly, as S_e becomes very large, $\exp(-S_e) \to 0$ and $S \to 1$, and the surface will be completely covered.

We can define a charge density q_e related to the extended area S_e using the expression

$$q_e = \int_0^t i_e \, dt \tag{4.20}$$

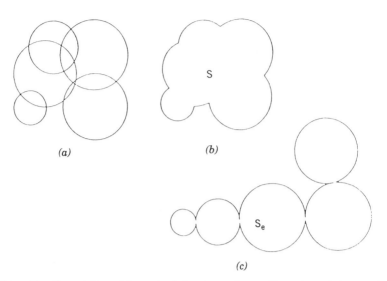

Figure 53. Presentation of Avrami solution to overlap problem. Avrami theorem [Eq. (4.19)] relates true surface area S to nominal extended area S_e. Figure illustrates way that overlap of growth centers outlined in (a) gives rise to real area (b) and extended area (c).

where i_e is given by

$$i_e = \frac{2N_0 zF\pi k_g^2 hM}{\rho} t$$

$$= \frac{2k_n N_0 zF\pi k_g^2 hM}{\rho} t^2 \tag{4.21}$$

depending on whether the nucleation is instantaneous or progressive. We consider two-dimensional growth here for simplicity. We can also relate q_e to the total charge density q_m required to form a monolayer. This may be done by noting that $q_e = S_e q_m$. The extended current density i_e is given by $i_e = q_m (dS_e/dt)$ and the total current density i is given by $i = q_m (dS/dt)$. We now recall Eq. (4.19) and note that

$$\frac{dS}{dt} = \exp(-S_e) \frac{dS_e}{dt}$$

Furthermore, $dS_e/dt = i_e/q_m$. Hence we may readily show that

$$i = i_e \exp(-S_e) \tag{4.22}$$

This expression may be used to compute the current transient when overlap is significant. To do this, we need to evaluate S_e, the extended area. We first compute q_e using Eqs. (4.20) and (4.21). If, for instance, the nucleation is instantaneous, then one may show by simple integration that

$$q_e = \frac{zF\pi MhN_0 k_g^2 t^2}{\rho}$$

Noting that $q_m = zFh\rho/M$ and $S_e = q_e/q_m$, for instantaneous nucleation

$$S_e = \frac{\pi N_0 k_g^2 M^2 t^2}{\rho^2} \tag{4.23}$$

Finally, in order to evaluate the transient current response via Eq. (4.22), we must determine i_e. We recall that

$$i_e = q_m \frac{dS_e}{dt} = \frac{zFh\rho}{M} \frac{2\pi N_0 k_g^2 M^2 t}{\rho^2} = \frac{2\pi z FN_0 Mh k_g^2 t}{\rho} \tag{4.24}$$

Hence, substituting the results obtained for S_e and i_e into Eq. (4.22), we finally obtain the following expression for the transient current response for instantaneous nucleation and two-dimensional growth:

$$i(t) = \frac{2\pi z FMhN_0 k_g^2}{\rho} t \exp\left(-\frac{\pi N_0 M^2 k_g^2}{\rho^2} t^2 \right)$$

$$= \alpha t \exp(-\beta t^2) \tag{4.25}$$

where

$$\alpha = \frac{2\pi z FMhN_0 k_g^2}{\rho} \qquad \beta = \frac{\pi N_0 M^2 k_g^2}{\rho^2} \tag{4.26}$$

An identical treatment for progressive nucleation and two-dimensional growth results in the following expression for the current transient response:

$$i(t) = \frac{zF\pi Mh k_n N_0 k_g^2}{\rho} t^2 \exp\left(-\frac{\pi M^2 k_n N_0 k_g^2}{3\rho^2} t^3 \right)$$

$$= \xi t^2 \exp(-\zeta t^3) \tag{4.27}$$

where

$$\xi = \frac{zF\pi Mhk_n N_0 k_g^2}{\rho} \qquad \zeta = \frac{\pi M^2 k_n N_0 k_g^2}{3\rho^2} \tag{4.28}$$

We see that inclusion of growth center overlap introduces a negative exponential factor into the expressions for the current transients. This will result in the current transient response exhibiting a maximum. Subsequent to the maximum, the current will decay and will eventually approach zero as the time becomes very long.

It is the established custom to present the chronoamperometric profiles in terms of a *reduced-variable representation*. This means in effect that the transients are normalized with respect to the maximum current i_m and the time corresponding to the onset of the current maximum t_m. Clearly $t = t_m$ when $di/dt = 0$. Hence from Eq. (4.25) we note that, on differentiation, we obtain

$$\alpha \exp(-\beta t_m^2)(1 - 2\beta t_m^2) = 0 \tag{4.29}$$

Clearly α must be nonzero, and so we note that $1 - 2\beta t_m^2 = 0$ or $t_m^2 = 1/2\beta$. Hence we may obtain the following expression for t_m:

$$t_m = \frac{1}{\sqrt{2\beta}} \tag{4.30}$$

Now $i = i_m$ when $t = t_m$, and so, from Eq. (4.25) we note that

$$i_m = \alpha t_m \exp(-\beta t_m^2) = \frac{\alpha}{\sqrt{2\beta}} \exp(-\tfrac{1}{2}) \tag{4.31}$$

We now evaluate the ratio i/i_m as follows:

$$\frac{i}{i_m} = \sqrt{2\beta}t \exp(\tfrac{1}{2}) \exp(-\beta t^2)$$

$$= \sqrt{2\beta}t \exp\left(-\frac{2\beta t^2 - 1}{2}\right) \tag{4.32}$$

We recall that $2\beta = 1/t_m^2$ and so

$$\frac{i}{i_m} = \frac{t}{t_m} \exp\left(-\frac{t^2 - t_m^2}{2t_m^2}\right) \tag{4.33}$$

This is the expression for the reduced-variable current transient for

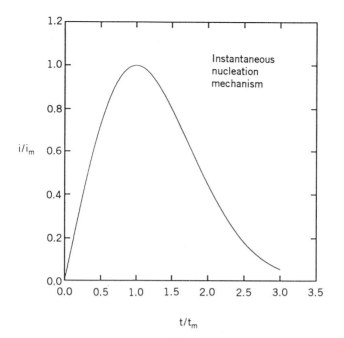

Figure 54. Normalized working curve for instantaneous nucleation followed by two-dimensional growth [Eq. (4.33)].

instantaneous nucleation and two-dimensional growth under charge transfer control. The normalized transient is presented in Fig. 54. In many respects this should be considered as a master curve. Experimental current transient data may be recast into nondimensional form and superimposed onto the master curve in order to check the extent of validity of an instantaneous nucleation/two-dimensional growth mechanism. A further point is of note here. We may readily show that

$$i_m t_m = \frac{\alpha \exp(-\tfrac{1}{2})}{2\beta} = \frac{0.61\alpha}{2\beta} = \frac{0.61zF\rho h}{M} = 0.61q_m \qquad (4.34)$$

This is an interesting result in that the product of i_m and t_m is independent of potential.

A similar type of analysis may be performed for progressive nucleation and two-dimensional growth. In this case we differentiate Eq. (4.27) wrt time and set the resulting expression equal to zero to obtain

$$\xi t_m \exp(-\zeta t_m^3)(2 - 3\zeta t_m^3) = 0 \qquad (4.35)$$

Again, since the product ξt_m is nonzero, we obtain that $2 - 3\zeta t_m^3 = 0$, and so

$$t_m = \left(\frac{2}{3\zeta}\right)^{1/3} \tag{4.36}$$

Furthermore $i = i_m$ when $t = t_m$, and so, from Eq. (4.27), we obtain

$$i_m = \xi t_m^2 \exp(-\zeta t_m^3) = \xi\left(\frac{2}{3\zeta}\right)^{2/3} \exp(-\tfrac{2}{3}) \tag{4.37}$$

Hence

$$\frac{i}{i_m} = \frac{t^2 \exp(-\zeta t^3)}{(2/3\zeta)^{2/3} \exp(-2/3)} = \frac{t^2}{t_m^2} \exp\left[-\frac{2}{3}\left(\frac{t^3}{t_m^3} - 1\right)\right]$$

$$= \frac{t^2}{t_m^2} \exp\left[-\frac{2(t^3 - t_m^3)}{3t_m}\right] \tag{4.38}$$

This is the expression for the normalized current transient response for progressive nucleation and two-dimensional growth. We illustrate this working curve in Fig. 55. Note that the master curve of i/i_m versus t/t_m is more symmetric for progressive nucleation. The normalized current response also decays to zero much more rapidly as t/t_m increases for this type of nucleation mechanism. We also note that

$$i_m t_m = \frac{2\xi}{3\zeta} \exp(-\tfrac{2}{3}) = 1.03 q_m \tag{4.39}$$

Again the product $i_m t_m$ is independent of potential. Hence the potential independence of the product $i_m t_m$ is a characteristic of two-dimensional growth under activation control regardless of the nature of the primary nucleation mechanism.

The parameters t_m and i_m vary with overpotential. For both instantaneous and progressive nucleation, t_m increases as the overpotential decreases. Converely, i_m increases as the overpotential increases.

We should note that a characterstic feature of a simple nucleation/growth model of the type just discussed is that the current transient observed under potentiostatic conditions starts from zero and exhibits a maximum. As previously noted, the origin of the maximum can be ascribed to growth center overlap. However, for many systems, most notably for the electrodeposition of electronically conducting hereroaromatic polymers, an initially falling current is often oberved at

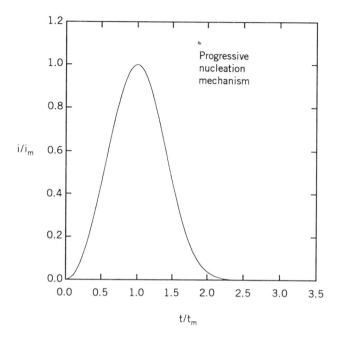

Figure 55. Normalized working curve for progressive nucleation followed by two-dimensional growth [Eq. (4.38)].

short times prior to the current peak. Hence both current minima and current maxima are observed. This observation of an initially falling current response can be ascribed to adsorption processes, or possibly by oligomer oxidation. Bosco and Rangaran [220] have examined adsorption–nucleation models and developed mathematical expressions for the current transitions expected under chronoamperometric conditions. We shall not discuss this approach here but merely refer the reader to the original paper [220].

The situation for three-dimensional nucleation/growth under charge transfer control is more complicated. The treatment of the overlap problem in this case presents formidable difficulties and has been considered by Armstrong et al. [221] and, more recently, by Bosco and Rangarajan [222, 223]. In these papers the growth centers are represented by hemispheres or right circular cones, or indeed by more complex shapes such as exponential profiles, error function–like geometries, or incomplete gamma function profiles. The papers by Bosco and Rangarajan [222, 223] are particularly useful in that the overlap problem is discussed in a careful and rigorous way.

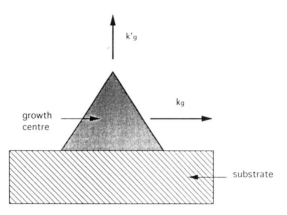

Figure 56. Growth of three-dimensional centers: right circular cone. Growth in directions normal and parallel to underlying substrate surface is quantified via rate constants k_g and k_g, respectively.

We can briefly consider the example of the growth of a three-dimensional center such as a right circular cone (Fig. 56). We see that growth in the normal and lateral directions is defined in terms of two rate constants k'_g and k_g. Note that the latter are expressed in units of $mol\,cm^{-2}\,s^{-1}$. Following Amstrong and co-workers [221] we can show that the current response to a potential step is given by

$$i = nFk'_g \left[1 - \exp\left(-\frac{\pi M^2 k_g^2 N_0 t^2}{\rho^2} \right) \right]$$

$$\approx \frac{nFk'_g \pi M^2 k_g^2 N_0}{\rho^2} t^2 \qquad (4.40)$$

for instantaneous nucleation and

$$i = nFk'_g \left[1 - \exp\left(-\frac{\pi M^2 k_g^2 k_n t^3}{3\rho^2} \right) \right]$$

$$\approx \frac{nFk'_g \pi M^2 k_g^2 k_n}{3\rho^2} t^3 \qquad (4.41)$$

for progressive nucleation. Note that the approximate expression for the current response transient at short times is also included in these expressions. Furthermore, in the long-time limit the exponential terms decay to zero and a limiting value of current given by nFk'_g is predicted.

Some computed transients are presented in Fig. 57. In contrast, the corresponding solution of the problem for hemispheric growth centers [222] results in the prediction of current transient curves, which exhibit maxima and minima in the approach toward the limiting current. This behavior is outlined in Fig. 58.

3. Application to Electronically Conducting Polymers

A recent application of nucleation/growth modeling has been in the area of the electrodeposition of electronically conducting heteroaromatic polymers. As previously noted, the polymers are deposited either via constant potential chronoamperometry or via cyclic voltammetry. This

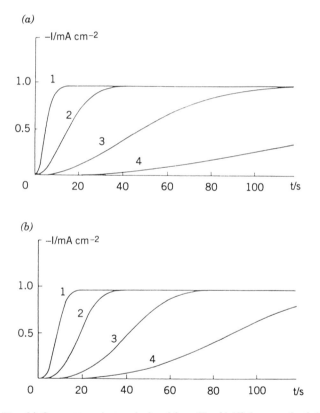

Figure 57. (a) Current transients calculated from Eq. (4.40) for growth of right circular cones formed by instantaneous nucleation: $k'_g = k_g = 10^{-8}\,\text{mol cm}^{-2}\,\text{s}^{-1}$; $M = 100\,\text{g mol}^{-1}$; $\rho = 10\,\text{g cm}^{-3}$; N_0 (in cm^{-2}): 10^{12} (1), 10^{11} (2), 10^{10} (3), and 10^9 (4). (b) Current transients calculated from Eq. (4.41) for growth of right circular cones formed via progressive nucleation. Nucleation rate constant k_n (in $\text{cm}^{-2}\,\text{s}^{-1}$): 10^{12} (1), 10^{11} (2), 10^{10} (3), and 10^9 (4). Other values are the same as those presented in (a).

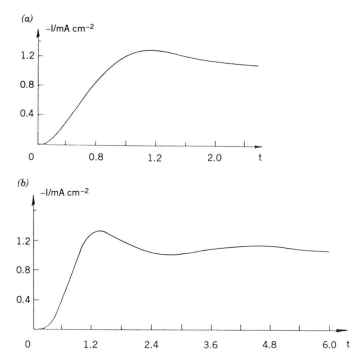

Figure 58. Current transients calculated for growth of hemispherical centers formed by (a) instantaneous nucleation and (b) progressive nucleation. Note presence of local maxima and minima at long times.

approach offers the advantage that the electrode can act simultaneously as an initiator, monitoring device, and deposition substrate.

Early work concentrated on the electrodeposition of poly(pyrole) and poly(N-methylpyrrole) [224, 225]. The electrodeposition of poly-(thiophene) was examined using a combination of electrochemical and spectroscopic techniques by Hillman and co-workers [226–229].

Conducting polymer film formation is a macrogrowth process and is three dimensional. One can visualize two possible mechanisms resulting in macrogrowth: three-dimensional nucleation/growth and two-dimensional layer-by-layer nucleation/growth. Most reports appearing in the literature to date suggest that instantaneous nucleation followed by three-dimensional growth is the preferred mechanistic avenue for electronically conducting polymer film formation [226, 230]. However, there are indications [231] that a two-dimensional layer by layer growth mechanism may also operate under certain circumstances. The differences between the two mechanisms can be of significance with respect to the morphology exhibited by the polymer film. For instance, Li and Albery [232] have pointed out that multinuclear growth of three-dimensional nuclei may form a packed-grain morphology with a substantial amount of empty space, while conducting polymer films formed layer by layer via lateral growth of two-dimensional nuclei may be expected to have a more uniform and compact arrangement of polymer chains. There are also indications [233] that one-dimensional nucleation/growth (via needle-type centers) may also be observed under certain conditions.

These rather general conclusions will now be examined by referring to recent work performed in the authors' laboratory on poly(pyrrole) depoistion.

Typical cyclic voltammograms recorded for the electropolymerization of pyrrole at a Pt electrode in aqueous KCl are outliend in Fig. 59. A number of interesting features may be noted. Figure 59a corresponds to the initial sweep. The oxidation reaction proceeds in an irreversible

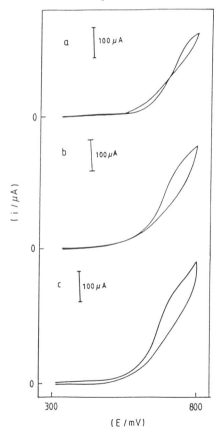

Figure 59. Typical cyclic voltammograms for electropolymerization of pyrrole at platinum electrode: (a) first cycle, (b) third cycle, and (c) fifth cycle. Synthesis solution contained 50 mM pyrrole in aqueous 0.1 M KCl. Sweep rate, 2 mV s^{-1}.

manner and only becomes significant once a potential of 600 mV (vs. SCE) is exceeded. Detailed kinetic analysis of the current–potential profile has indicated that the Tafel slope admits a value of 120 mV dec^{-1}, which implies that the initial electron transfer generating the radical cation is rate limiting. It is also significant that the current obtained on sweep reversal is higher than the corresponding value obtained during the forward sweep. A hysteresis loop can be observed, which is a characteristic feature of nucleation/growth phenomena. In Figs. 59b, c voltammetric profiles corresponding to the third and fifth potential sweeps are presented. In both cases, oxidation begins at slightly less positive potentials than that observed during the initial sweep, that is, the nucleation overpotential is lowered, and the oxidation currents at any given overpotential are higher than those obtained at the bare electrode during the initial sweep. Hence, we note that the pyrrole monomer oxidizes more readily on a polymer-coated electrode surface than on the bare-substrate surface. Kinetic analysis of the rising portion of the current–potential profiles presented in Figs. 59b, c indicate that dual-slope Tafel behavior is operative. At low potentials (~590–700 mV) the slope is ~132 mV dec^{-1}, whereas at more positive potentials the slope changes to ~240 mV dec^{-1}. We note that the ratio of the low-potential Tafel slope to that observed at higher potentials is ~0.45. This ratio is significant. It is well established that the Tafel slope for a Faraday process occurring at a rough or porous electrode surface is twice that recorded for the same Faraday reaction at a smooth planar surface [234, 235]. Furthermore, it is expected from theoretical considerations that the two Tafel slopes should be preent, the lower Tafel slope region preceding the region of high Tafel slope. If the first electron transfer step in the electropolymerization sequence is rate determining, then the low and high Tafel slopes should be given by $2.303RT/\alpha F$ and $2.303(2RT/\alpha F)$, respectively. The experimentally observed Tafel slopes of 132 and 240 mV dec^{-1} correspond to transfer coefficient α values of 0.45 and 0.44, repectively. These values are very close to that expected for a simple single-step rate-determining interfacial electron transfer reaction [236].

In Fig. 60 we present a series of chronoamperometric current transient response profiles corresponding to the formation of a poly(pyrrole) film on a Pt support surface by the oxidation of a dilute (50 mM monomer in 0.1 M KCl) pyrrole solution. Each curve corresponds to a different polymerization potential. The shape of these transitions are similar to those observed for other conducting polymers. Three regions are of note.

Region I, observed at short times, is characterized by a current spike of short (<200 ms) duraton followed by a subsequent decay in the current response to a local minimum. This feature of the chronoamperometric

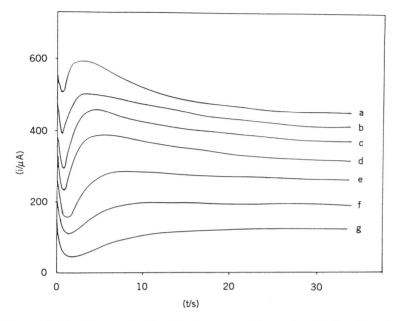

Figure 60.   Typical current vs. time transients recorded for polymerization of pyrrole at platinum electrode as function of polymerization potential. Lower potential value, 200 mV. Curves (a)–(g) correspond to upper potential values of 800, 780, 760, 740, 720, 700 and 680 mV, respectively. Synthesis solution similar to that presented in Fig. 59.

response may be ascribed to a number of possibilities such as double-layer charging, monomer oxidation, adsorption of intermediates, substrate dissolution, and most probably, polymer deposition. Hillman and Mallen [226] have proposed that in this region an initial monolayer of polymer forms and the current minimum observed just prior to the onset of the rising current response characterizing region II corresponds to the transition between initial monolayer and subsequent multilayer coverage. They have also defined a time τ_0, the time corresponding to the local current minimum, to be an estimate of the time required to produce a monolayer coverage of polymer and defined τ_0^{-1} as a rough estimate of the electrochemical rate constant for this electrodeposition process. Indeed, these workers have shown that a plot of $\ln \tau_0$ versus electrode potential E is linear (i.e., a Tafel-type behavior pertains). In their work on poly(thiophene) formation, Hillman and Mallen [226] have estimated a charge of $160 \pm 40 \ \mu C \, cm^{-2}$ for region I, which corresponds to a surface coverage of $0.69 \pm 0.20 \, nmol \, cm^{-2}$. Limiting monolayer coverages for monomeriac thiophene units adsorbed in parallel and perpendicular orientations were estimated as 0.46 and 0.75 $nmol \, cm^{-2}$, respectively, from examination of molecular models. These data tend to suggest that region I is associated with the formation of a layer of polymer one monomer unit thick, that is, a film of horizontally deposited polymer. Our studies [233] on poly(pyrrole) deposition at very short times also confirm the Hillman–Mallen hypothesis. It is quite possible that significant preadsorption of monomer may occur prior to electropolymerization that forms the initial monolayer.

Region II is characterized by an ascending current response. The rise in current becomes more rapid as the electropolymerization potential is made more positive. This feature is related to a second nucleation and growth process occurring on the previously deposited monolayer. Here we have nucleation/growth on an organic substrate. A problem arises when we wish to assign a temporal origin for this second nucleation process. A certain amount of overlap exists between region I and region II: There can be some nucleation of region II material on islands of region I material before full monolayer formation occurs. Analysis of the i-versus-t behavior in region II (Fig. 61) indicates that the current varies quadratically with time. Consequently, we can evaluate τ_0 by plotting $i^{1/2}$ versus t and noting the intercept on the time axis. This procedure is outlined in Fig. 61 for two polymerization potentials, where τ_0 is 0.2 s at $E = 760$ mV and increases to 0.5 s at $E = 720$ mV. At short times, the nuclei grow independently without overlapping. This free growth results in a rapid increase in current. At longer times the high density of nuclei generated will result in the rapid onset of overlap, which is manifested in the deviation from linear behavior, as shown in Fig. 61. Hence the duration of the linear $i^{1/2}$-versus-t behavior will depend on the overall rate of polymerization and hence on the polymerization potential.

The quadratic dependence of current on time can be accounted for by proposing either a progressive nucleation and two-dimensional growth mechanism or instantaneous nucleation followed by three-dimensional growth (see Table III). From Table III we note that the slope of the i-versus-t^2 plot should be proportional to k_g^2 for the two-dimensional progressive mechanism and to k_g^3 for the three-dimensional instantaneous mechanism. Since the growth parameter k_g obeys the Butler–Volmer equation [see Eq. (4.18)], the two alternative mechanistic possibilities predict a different value for the quantity

$$S = \frac{d \log[di/d(t - \tau_0)^2]}{dE}$$

For progressive nucleation followed by two-dimensional growth one can readily show that $S = 2\alpha F/2.303RT$, which reduces to 16.9 V^{-1} for $T =$

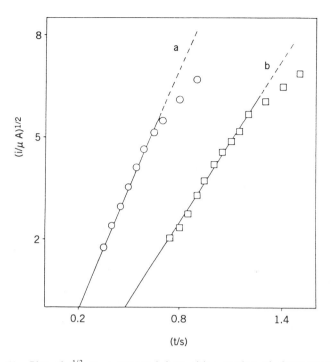

Figure 61. Plot of $i^{1/2}$ vs. t extracted from rising portion of chronoamperometric current transient response profile (of type presented in Fig. 60). Polymerization potentials of (*a*) 760 and (*b*) 720 mV.

298 K and $\alpha = 0.5$. On the other hand, for instantaneous nucleation followed by three-dimensional growth, $S = 3\alpha F/2.303RT$, which reduces to 25.3 V^{-1} at $T = 298$ K and $\alpha = 0.5$. The corresponding experimental data are presented in Fig. 62. It should be noted that the $\log[i/(t - \tau_0)^2]$-versus-E plot is linear with a slope S given by ~ 24 V^{-1} independent of monomer concentration. This corresponds very well with the prediction derived from the instantaneous nucleation/three-dimensional growth scenario. Hence under the condition studied, region II corresponds to the instantaneous nucleation of sites on the organic substrate produced in region I. These sites then grow three dimensionally until they overlap, at which point (the start of region III) continued growth is only possible perpendicular to the surface.

We now examine the transient response profile observed after the maximum, which we designate as region III. As noted in Fig. 60, when the applied potential is low, a steady-state response is obtained. As the polymerization potential is increased to more positive values, the current

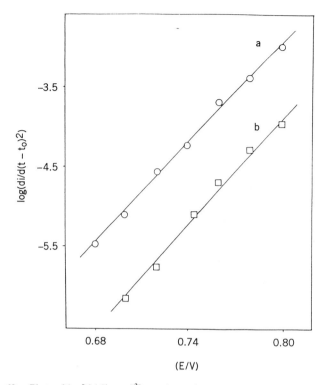

Figure 62. Plots of $\log[d_i/d(t - \tau_0)^2]$ vs. electrode potential E for two different pyrrole monomer concentrations: (a) 50 mM and (b) 30 mM pyrrole in 0.1 M KCl.

response is seen to decay slowly with increasing time, until for long times a steady-state response is observed. Also, the magnitude of the current response at long times depends markedly on the magnitude of the electrode potential. Indeed, we predict that at long times [see Eq. (4.40)] the current response reduces to $i = nFk_g$ regardless of whether nucleation is instantaneous or progressive, provided that the growth is three dimensional. This long-time response also does not depend on the geometry assumed for the growing centers (right circular conical or hemispheric). Since the growth parameter obeys the Butler–Volmer equation, we predict that the steady-state current response should exhibit Tafel-type behavior when plotted as a function of polymerization potential (i.e., $\log i$ versus E should be linear). A typical Tafel-type plot is presented in Fig. 63 for two different pyrrole monomer concentrations. In both cases a good linear plot is obtained. The theoretical Tafel slope expected is $\alpha F/2.303RT$, which reduces to 8.4 V^{-1} for $\alpha = 0.5$ and

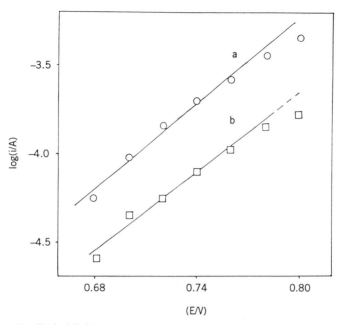

Figure 63. Typical Tafel plots recorded from quasi-steady-state section of chronoam-perometric current–time response curve at long times. Pyrrole monomer concentrations of (*a*) 50 and (*b*) 30 mM.

$T = 298$ K. The experimental slopes are 8.0 and 8.2 V^{-1} for the two pyrrole monomer concentrations employed. This observation provides further confirmation that three-dimensional, rather than layer-by-layer, growth pertains.

The slow current decay at intermediate times in region III warrants mention. The slow fall-off in current may be ascribed to the low electronic conductivity of the growing film. The resistance of the film changes with time. This assertion is supported by the observation that the fall-off in current becomes more marked as the polymerization potential is increased. The conductivity of oxidized poly(pyrrole) can be attributed to the presence of delocalized charged species such as polarons and bipolarons located on the polymer backbone. Since the growth of the polymer layer involves generation and subsequent oxidation of polaronic species, the overall concentration of polarons within the film will remain low, and consequently, the conductivity of the growing film will be low.

In this section we have discussed the nucleation and growth of poly(pyrrole) films doped with Cl⁻ ions. The results presented are specific only to this system. We have shown [233] that poly(pyrrole)

doped with surfactant counterions such as DBS⁻ exhibits an instantaneous nucleation/one-dimensional needle-type growth mechanism. This
is an unusual growth mechanism, and its occurrence is probably related to
the amphiphilic nature of the dodecylbenzenesulfonate ion. In contrast,
Albery and Li [232] have presented evidence for doped poly(pyrrole)
films growing via a two-dimensional layer-by-layer nucleation/growth
process. Hence, we can conclude that, even for a specific type of
polymer, the assignment of the pertinent nucleation/growth mechanism
must be done on a case-by-case basis. No a priori assignment may be
made.

V. COMPLEX IMPEDANCE SPECTROSCOPY: USEFUL TECHNIQUE FOR QUANTIFYING ELECTRONIC AND IONIC CONDUCTIVITY IN ELECTROACTIVE POLYMER FILMS

A. Introduction

The technique of complex-impedance spectroscopy (CIS) has been
extensively used in recent years to examine the electrochemical characteristics of surface-deposited electroactive polymer films. A useful summary of the fundamental principles may be found in the monograph by
MacDonald [237]. The application of impedance methods to electroactive
polymers has been recently reviewed by Musiani [238].

The fundamental principles of impedance spectroscopy are outlined in
a number of basic textbooks [239–243] and review articles [244–246], and
so only a very brief summary of the fundamental ideas will be presented
here. In essence, what one does is to examine the sinusoidal voltage
response of an electrochemical system to a small-amplitude sinusoidal
current perturbation. Let the perturbation have the form $\Delta I = I_m \sin \omega t$,
where I_m denotes the amplitude of the perturbation. Then the response
will have the form $\Delta V = V_m \sin(\omega t - \phi) = V_m(\cos \phi \sin \omega t - \sin \phi \cos \omega t)$,
where ϕ denotes the phase angle. The impedance is then calculated for
that specific frequency ω using the relation $Z(\omega) = \Delta V(\omega)/\Delta I(\omega)$. The
relationship between ΔV and ΔI is completely determined by the ratio of
the amplitudes V_m/I_m and the phase shift ϕ between current and voltage.
One can therefore define the impedance Z as a vector with a modulus $|Z|$
given by $|Z| = V_m/I_m$ and argument ϕ. This is presented in Fig. 64. It is
well known that a vector may in general be resolved into orthogonal
components, and so from Fig. 64 we note that the components of Z,
labeled as Z' and Z'' may be written as $Z' = |Z|\cos \phi$, $Z'' = |Z|\sin \phi$.
From the latter relationships we may obtain an expression for the phase
angle ϕ as follows: $\tan \phi = Z''/Z'$; hence $\phi = \tan^{-1}(Z''/Z')$. This pro-

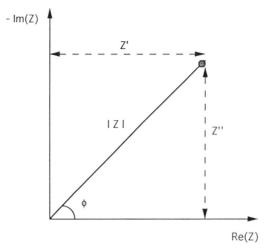

Figure 64. Schematic representation of Argand diagram illustrating modulus $|Z|$ and phase angle ϕ of complex number Z. Note that Z' and Z'' represent real and imaginary components of Z since $Z = Z' - jZ''$. Note also that $Z' = |Z|\cos\phi$ and $Z'' = |Z|\sin\phi$ and that $|Z| = (Z'^2 + Z''^2)^{1/2}$.

cedure is repeated over a very large frequency range, typically from 100 kHz to 0.1 mHz. The result is that one obtains an impedance profile or spectrum $Z(\omega)$ as a function of frequency ω.

It is convenient to recast our arguments in terms of a complex impedance $Z(j\omega)$, which is defined as the quotient of the Laplace transforms of ΔV and ΔI. We show this assertion as follows. We again assume that

$$\Delta I = I_m \sin \omega t$$
$$\Delta V = V_m \sin(\omega t - \phi) = I_m |Z|(\sin wt \cos \phi - \cos \omega t \sin \phi) \quad (5.1)$$

Hence we note that

$$\Delta V = I_m(Z' \sin \omega t - Z'' \cos \omega t)$$
$$= Z' \Delta I - Z'' \frac{1}{\omega} \frac{d\,\Delta I}{dt} \quad (5.2)$$

We now take Laplace transforms to obtain

$$\Delta \bar{V}(p) = \left(Z' - \frac{p}{\omega} Z''\right) \Delta \bar{I}(p) + \frac{Z''}{\omega} \Delta I(t = 0) \quad (5.3)$$

Now $\Delta I(t = 0) = 0$, and so the impedance in Laplace space $Z(p)$ is given by

$$Z(p) = \frac{\Delta \bar{V}(p)}{\Delta \bar{I}(p)} = Z' - \frac{p}{\omega} Z'' \qquad (5.4)$$

We can now replace the Laplace parameter p by $j\omega$ to obtain our final result:

$$Z(j\omega) = Z' - jZ'' \qquad (5.5)$$

This implies that the complex impedance vector may be represented in the Argang plane, where one plots the imaginary component $\text{Im}(Z) = Z''$ versus the real component $\text{Re}(Z) = Z'$. Hence we may write also that

$$Z(j\omega) = |Z|(\cos \phi - j \sin \phi) = |Z| \exp(-j\phi) \qquad (5.6)$$

where the modulus $|Z|$ is given by

$$|Z| = \sqrt{Z'^2 + Z''^2} \qquad (5.7)$$

The major point to note from this analysis is that an imnpedance spectrum in its most basic representation is a set of points in the complex plane. A plot of $-Z''$ versus Z' is termed a Nyquist plot (Fig. 65).

The general shape of this impedance spectrum may then be modeled in terms of an *equivalent circuit* consisting, for example, of resistance and capacitance circuit elements combined in definite ways (series or parallel combinations) that can range from relatively simple to rather complex combinations, depending on the degree of complexity of the system under study. This correspondence between an impedance spectrum and an equivalent circuit is well established in basic physics. These circuit elements reflect various physical features of the real electrochemical system under examination. For instance, resistive elements can correspond to interfacial electron transfer processes or may represent ionic or electronic transport rates; capacitors reflect charge separation at interfaces. Mass transport processes may also be described in terms of special circuit elements whose values depend on frequency. Much information may then be derived about the characteristics of an electroactive polymer material by examining the way in which the values of the characterstic circuit parameters vary with changes in quantities such as electrode potential, layer thickness, solution pH, solution composition, and temperature. The main difficulty is to devise an unambiguous equivalent circuit representation for a specific system. This can pose problems because of the fact that the system may be adequately described by a

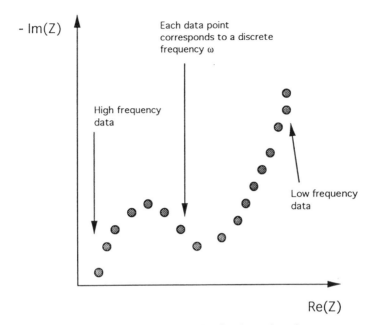

Figure 65.   Schematic representation of Nyquist plot or impedance spectrum.

number of different equivalent circuits. A more logical method of analysis is to start with a mathematical model containing the essential features of the electrochemical system of interest and to quantitatively develop a theoretical expression for the impedance response as a function of frequency in the context of this model. This type of approach is more complex and will be discussed in some detail in the succeeding sections of this chapter. The impedance spectrum predicted theoretically is then fitted to the experimental data using a nonlinear least-squares fitting protocol. An overall schematic of the various methods of approach is presented in Fig. 66.

There are various ways of presenting impedance data. One such method was mentioned previously, the Nyquist representation, where $-\mathrm{Im}(Z)$ is plotted versus $\mathrm{Re}(Z)$, with frequency as a parametric variable. Such complex-plane plots are used extensively in the electrochemical literature. In Fig. 67 we present some typical complex-plane plots for various simple combinations of typical circuit elements used in describing electrochemical systems. One should note, however, that because CIS experiments range over a very wide frequency range, it can sometimes be useful to introduce the frequency in an explicit manner. Hence in the Bode representation one plots both $\log|Z|$ versus $\log \omega$ and ϕ versus

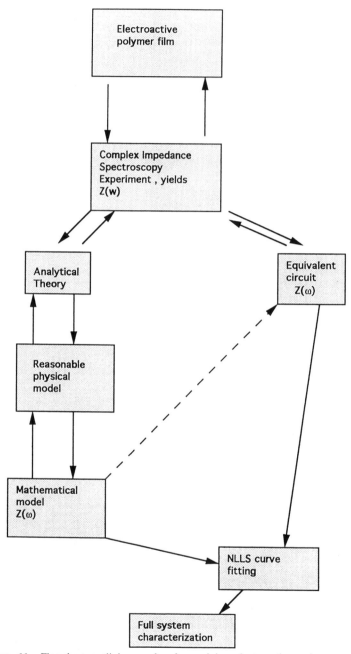

Figure 66. Flowchart outlining mode of examining electroactive polymer materials using complex impedance spectroscopy.

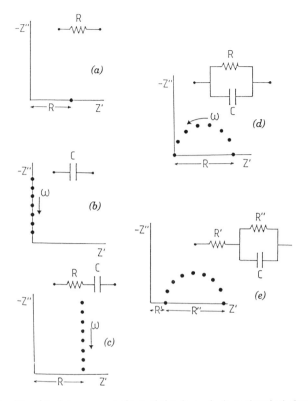

Figure 67. Nyquist plot representations of simple equivalent electrical circuits consisting of resistors and capacitors.

log ω. Some examples of Bode plots for simple equivalent circuits are illustrated in Fig. 68. Some workers choose to plot Z'' and Z' versus log ω. MacDonald [247] has advocated the use of a three-dimensional representation that affords the use of perspective. The three dimensions of the plot are usually the real and imaginary parts of the impedance and log ω. One can of course also make the Bode plot three dimensional, but this has a lesser diagnostic value. A three-dimensional MacDonald plot is illustrated in Fig. 69, the response being viewed from two different perspectives.

It is sometimes useful to display data in the form of admittance Y. This is defined as a vector with modulus $|Y| = I_m/V_m = 1/|Z|$ and argument ϕ. The components of this vector are given by $Y' = |Y| \cos \phi$, $Y'' = |Y| \sin \phi$.

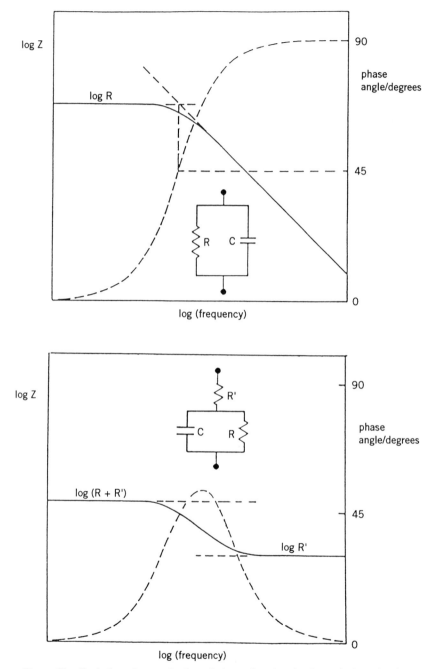

Figure 68. Typical mode magnitude and phase plots for simple equivalent circuit.

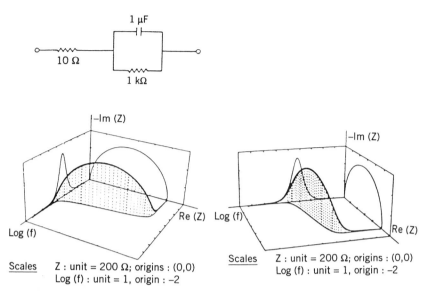

Figure 69. Schematic representation of three-dimensional MacDonald plot consisting of compilation of Bode and Nyquist representations.

The relationship between impedance and admittance is given by

$$\frac{Z'}{Y'} = \frac{Z''}{Y''} = Z'^2 + Z''^2 = \frac{1}{Y'^2 + Y''^2} \tag{5.8}$$

The admittance may also be defined as a complex number:

$$Y(j\omega) = \frac{\Delta \bar{I}(j\omega)}{\Delta \bar{V}(j\omega)} = Y' + jY'' = |Y|(\cos\phi + j\sin\phi) = |Y|\exp(j\phi) \tag{5.9}$$

and so admittance data may also be plotted in the complex plane (Y'' versus Y' with ω implicit). Some workers [248] choose to display data in terms of the complex capacitance $C(j\omega)$, where $C(j\omega) = Y(j\omega)/j\omega$. The latter type of representation can be useful when one examines the electrochemical response of electronically conducting polymer films. The low-frequency redox pseudocapacitance may be read directly from a plot of C'' versus C' at low frequency.

We now gather together some useful expressions for the most common equivalent circuit elements:

Resistance: $Z' = R$ $Y' = \dfrac{1}{R}$ $Z'' = Y'' = 0$

Capacitance: $Z' = Y' = 0$ $Z'' = \dfrac{1}{\omega C}$ $Y'' = \omega C$ (5.10)

Self-inductance: $Z' = Y' = 0$ $Z'' = -\omega L$ $Y'' = -\dfrac{1}{\omega L}$

This then concludes our brief survey of fundamental concepts. In the following sections we shall discuss the complex impedance response of surface-deposited electroactive polymer films. We shall consider redox conductors and conjugated organic conductors separately since the mathematical models used to describe the impedance response of these two classes of electroactive macromolecules can be quite different.

B. Complex Impedance Response of Redox Polymer Films

In this section we outline, in some detail, the salient features of the theory required to quantitatively interpret the complex impedance response of thin, surface-deposited redox polymer films. The analysis will also be valid for ionomer films loaded with electroactive species. The analysis presented is based on the work reported by Armstrong and co-workers [249, 250], Ho et al. [251], Gabrielli and co-workers [252–254], Mathias and Haas [255], Lang and Inzelt [256], Lindholm [257], Sharp and co-workers [258], and Lyons et al. [259]. The analysis discussed here not only is applicable to redox polymers but also may be applied to redox switching in metal–oxide films [251, 260] and membranes in which mobile electroactive ions are confined [261, 262]. The theoretical problem is simple in essence. To the first order of approximation, one has diffusion (of either charge-compensating counterions or electrons) to a blocking interface through a layer of finite thickness. This problem is essentially the same as that of a finite transmission line. We expand on the transmission line concept later in this section. This simple diffusional picture must, of course, be modified to take electromigration with the resultant coupling between ionic and electronic transport into account.* The further complicating factor of nonideality arising from the interaction between redox sites in the polymer must also be considered.†

---

*These modifications to the simple quasi-diffusional electron-hopping picture are discussed in Section II.A.4. In short, as will be shortly noted, the concept of the redox polymer as a simple binary electrolyte consisting of electrons and charge-compensating counterions arises, and a composite diffusion coefficient comprising contributions due to electron-hopping and counterion transport is defined.

† Intersite interaction effects are discussed in Section III.A.

In simple terms the impedance response of a redox polymer film may be modeled in terms of the general circuit illustrated in Fig. 70. The latter circuit has been proposed by Sharp and co-workers [258]. We note from this figure that the equivalent circuit contains two elements labeled I and II. The upper nonconducting branch of element I consists of the interfacial or double-layer capacitance C_{DL} at the support electrode–film interface connected in series with a parallel combination of a capacitance C_B and a resistance R_B that reflect the dielectric properties of the polymer. In the representation presented in Fig. 70 we have neglected a possible distribution of relaxation times linked with the underlying dipolar reorientations and short-range non-Faraday charge displacements. The lower branch of element I contains a resistance R_{CT} called the charge transfer resistance, which quantifies the rate of interfacial electron transfer between the underlying support electrode and the bound redox couple in the film. In effect, R_{CT} quantifies the rate of charge injection/ extraction at the support electrode–film interface. From this quantity one can obtain a value for the heterogeneous electron transfer rate constant k^0. The R_{CT} component is in series with a frequency-dependent impedance Z_D called the diffusional impedance, which reflects the kinetics of charge percolation (assumed to be diffusive) through the polymer film. The sum of R_{CT} and Z_D defines the Faraday impedance Z_F, which is the major parameter derived from any theoretical model describing the impedance response. Charge transfer across the film–solution interface (this is required to maintain electroneutrality) is described in terms of circuit element II, which consists of a capacitance C_{FS} and a resistance R_{FS} in a parallel configuration. The circuit is completed by a component R_u that quantifies the uncompensated solution resistance.

The circuit described in Fig. 70 is quite complex, but one may make some simplifications. For instance, one may suppose that charge-compensating counterions may freely partition into the film, and so we may set the resistance $R_{SF} = 0$. Hence circuit element II may be neglected. We are then left with element I, which is still rather complex. Now the bulk resistance R_B can be reasonably assigned [258] to the transport of unpinned counterions in the layer. These electroinactive species will respond to the alternating voltage in the absence of electron self-exchange between the fixed redox sites. When the free counterion concentration in the film is high, R_B will be low and vice versa. Sharp and co-workers [258] noted the following complicating factor. For loaded ionomer films such as Nafion containing mobile electroinactive cations and fixed SO_3^- sites, in addition to determining the magnitude of R_B, the latter mobile and fixed species serve as supporting electrolyte for the electrostatically incorporated redox couple. Hence Sharp and co-workers

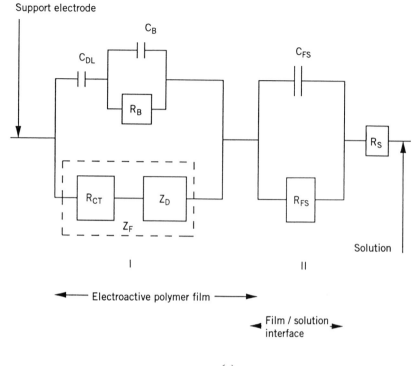

Support electrode

Electroactive polymer film

Film / solution
interface

(a)

Figure 70. (*a*) General equivalent circuit representing electroactive polymer film exhibiting redox conduction. Each circuit element corresponds to particular kinetic/transport/structural property of polymer. (*b*) Approximate equivalent circuits and associated Nyquist plots for various simplified limiting situations encountered in practice. Approximation (a) valid when R_{CT} is very low, $R_{FS} = R_B = 0$, the redox site concentration c_Σ is low and the charge compensating counterion concentration c_I is high. Approximation (b) valid when R_{CT} very low, R_{FS} near zero, redox site concentration, high and low charge-compensating counterion concentration. In approximation (c), R_{CT} is finite and exerts effect via charge injection at support electrode–polymer interface, $R_{FS} = R_B = 0$, c_Σ is low and counterion concentration is high. Magnitude of c_Σ will depend on applied electrode potential. (*c*) Simplified, yet general equivalent circuit used to discuss impedance response of electroactive polymer film. Corresponding impedance plot also presented. For this approximation R_{CT} is finite, R_{FS} is zero, R_B is finite, and both redox site concentration c_Σ and mobile charge-compensating counterion concentration c_I are high.

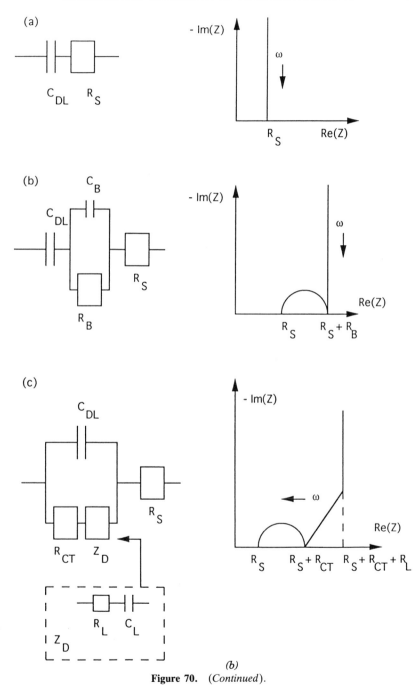

Figure 70. (*Continued*).

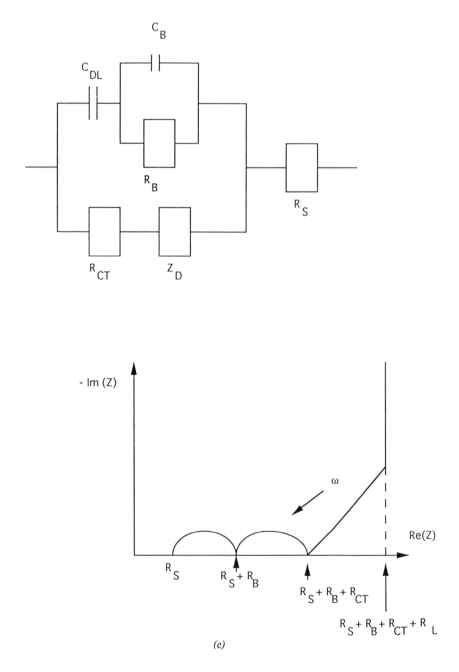

(c)

Figure 70. *(Continued)*.

[258] have delineated two situations: "supported" conditions, in which the total and incorporated redox species concentration c_Σ is low and the mobile counterion concentration in the film is high, and "unsupported" conditions, in which the opposite situation pertains. In the former case R_B will be small, but for the latter, R_B may well be appreciable, and migrational enhancement of charge propagation may occur. The latter situation has been discussed in Section II.A.4 of this chapter. The form of the complex impedance response for the various limiting cases discussed above is illustrated in Fig. 70b. In the following analysis we shall adopt the simplified equivalent outlined in Fig. 70c as the basis for our discussion.

This circuit is of the well-established Randles type. In the Randles approximation we neglect ion partitioning at the film–solution interface and also neglect the bulk polymer RC parallel network ascribed to dipole reorientation. Hence the major circuit elements are the double-layer capacitance C_{DL}, the Faraday impedance Z_F, and the uncompensated solution resistance R_u.

The total impedance Z of this network is given by

$$Z = R_u + Z_p = R_u + \frac{Z_F Z_C}{Z_F + Z_C} = R_u + \frac{1}{1/Z_F + 1/Z_c} \qquad (5.11)$$

From Eq. (5.10) we note that the impedance corresponding to the capacitance is given by $Z_C = -j/\omega C_{DL} = 1/j\omega C_{DL}$; hence Eq. (5.11) reduces to

$$Z = R_u + \frac{1}{1/Z_F + j\omega C_{DL}} \qquad (5.12)$$

In order to obtain a theoretical expression for the full interfacial impedance response we must evalaute Z_F. This involves treating the polymer film as a layer of finite thickness with a single blocking interface at $x = L$. The process of electron self-exchange between neighboring sites (labeled A and B) in the film is modeled in terms of a simple diffusional process (charge percolation via migrational enhancement is neglected) and is quantified via the apparent diffusion coefficient D_{CT}. The process of charge injection at the support electrode–film interface is described via the Butler–Volmer equation.

We assume that the potential of the electrode is perturbed by a small-amplitude sinusoidal potential given by $\Delta E = |\Delta E| \exp(j\omega t)$. Consequently the Faraday current will be perturbed by an amount $\Delta I = |\Delta I| \exp(j\omega t + \phi)$. This perturbation induces a concentration change Δc

in the layer that will be described by the time-dependent Fick diffusion equation as follows:

$$\frac{\partial \Delta C}{\partial t} = D_{CT} \frac{\partial^2 \Delta c}{\partial x^2} \tag{5.13}$$

where the quantity Δc denotes a change either in the reduced species concentration Δa or the oxidized species concentration Δb. In general the time dependence may be factored out, and we note that $\Delta c(x, t)$ is given by

$$\Delta c(x, t) = \Delta c(x) \exp(j\omega t) \tag{5.14}$$

where $j = (-1)^{1/2}$. Hence in order to derive an expression for Z_F, one must solve Eq. (5.13) for species A and B using the following boundary conditions:

$$x = 0: \quad \Delta I = nFAD_{CT} \left(\frac{\partial \Delta a}{\partial x} \right)_{x=0} = -nFAD_{CT} \left(\frac{\partial \Delta b}{\partial x} \right)_{x=0} \tag{5.15}$$

$$x = L: \quad \left(\frac{\partial \Delta a}{\partial x} \right)_{x=L} = \left(\frac{\partial \Delta b}{\partial x} \right)_{x=L} = 0$$

Integration of Eq. (5.13) results in the following expressions for Δa and Δb:

$$\Delta a(x, t) = \frac{2 \Delta a_0}{1 + \exp(-2\beta L)} \exp(-\beta L) \cosh[\beta(x - L)] \exp(j\omega t) \tag{5.16}$$

$$\Delta b(x, t) = \frac{2 \Delta b_0}{1 + \exp(-2\beta L)} \exp(-\beta L) \cosh[\beta(x - L)] \exp(j\omega t)$$

where the factor β is given by

$$\beta = \sqrt{\frac{j\omega}{D_{CT}}} \tag{5.17}$$

Using the Butler–Volmer equation at $x = 0$,

$$I = nFA(k'_E a_0 - k'_{-E} b_0) = nFAk^0 \{ a_0 \exp(\alpha\vartheta) \\ - b_0 \exp[-(1 - \alpha)\vartheta] \} \tag{5.18}$$

where α is the transfer coefficient and k^0 is the standard heterogeneous electrochemical rate constant for charge injection at the electrode–film interface, $\vartheta = nF(E - E^0_{A/B})/RT$, and a_0, b_0 denote the steady-state concentrations of reduced and oxidized centers at $x = 0$ ($a_0 + b_0 = c_\Sigma$). We may show, after some algebra, that the current ΔI resulting from the sinusoidal perturbation in potential ΔE is given by

$$\Delta I = nFA(k'_E a_0 [\exp(\alpha \, \Delta\vartheta) - 1] + k'_{-E} b_0 \{1 - \exp[-(1 - \alpha) \, \Delta\vartheta]\})$$
$$+ nFA\{k'_E \, \Delta a_0 \exp(\alpha \, \Delta\vartheta) - k'_{-E} \, \Delta b_0 \exp[-(1 - \alpha) \, \Delta\vartheta]\} \quad (5.19)$$

where Δa_0 and Δb_0 denote the perturbed concentrations of reduced and oxidized species at $x = 0$ and $\Delta\theta$ is a normalized potential given by $\Delta\vartheta = nF \, \Delta E / RT$. We now note that the amplitude ΔE of the potential perturbaton will be small, typically in the range 5–10 mV. If this is the case, then the exponential terms in $\Delta\theta$ may be linearized using a Taylor expansion, and so Eq. (5.19) reduces to

$$\Delta I = nFA[k'_E a_0 \alpha \, \Delta\vartheta + k'_{-E} b_0 (1 - \alpha) \, \Delta\vartheta]$$
$$+ nFA\{k'_E \, \Delta a_0 (1 + \alpha \, \Delta\vartheta) - k'_{-E} \, \Delta b_0 [1 - (1 - \alpha) \, \Delta\vartheta]\} \quad (5.20)$$

The latter expression may be simplified further if we neglect the products $\Delta a_0 \, \Delta E$ and $\Delta b_0 \, \Delta E$ to obtain

$$\Delta I = nFA \left(k'_E a_0 \frac{\alpha nF}{RT} + k'_{-E} b_0 \frac{(1 - \alpha)nF}{RT} \right) \Delta E$$
$$+ nFA(k'_E \, \Delta a_0 - k'_{-E} \, \Delta b_0) \quad (5.21)$$

We now note that the charge transfer resistance is given by

$$R_{CT} = \frac{\partial \, \Delta E}{\partial \, \Delta I} = \left[nFA \left(\frac{\alpha nFk'_E a_0}{RT} + \frac{(1 - \alpha)nFk'_{-E} b_0}{RT} \right) \right]^{-1} \quad (5.22)$$

Hence we finally note that

$$\Delta I = R_{CT}^{-1} \, \Delta E + nFA(k'_E \, \Delta a_0 - k'_{-E} \, \Delta b_0) \quad (5.23)$$

We now return to the solution of the finite diffusion problem given in Eq. (5.16) and use Eq. (5.15) to obtain

$$\Delta a_0 = -\frac{\Delta I}{nFAD_{CT}\sqrt{j\omega/D_{CT}}} \coth\left(\sqrt{\frac{j\omega}{D_{CT}}} \, L \right)$$

$$\Delta b_0 = \frac{\Delta I}{nFAD_{CT}\sqrt{j\omega/D_{CT}}} \coth\left(\sqrt{\frac{j\omega}{D_{CT}}} \, L \right) \quad (5.24)$$

Substituting these expressions into Eq. (5.23) and simplifying yields the following expression for the Faraday impedance:

$$Z_F = \frac{\Delta E}{\Delta I} = R_{CT}\left[1 + \frac{k'_E + k'_{-E}}{\sqrt{j\omega D_{CT}}}\coth\left(\sqrt{\frac{j\omega}{D_{CT}}}L\right)\right] \qquad (5.25)$$

This expression may be transformed still further if we note that R_{CT} may be expressed in the following manner:

$$R_{CT} = \frac{RT(k'_E + k'_{-E})}{n^2 F^2 A c_\Sigma k'_E k'_{-E}} \qquad (5.26)$$

The latter expression is readily derived from Eq. (5.22) by noting that the steady-state surface concentrations a_0 and b_0 are given by

$$b_0 = \frac{k'_E c_\Sigma}{k'_E + k'_{-E}} \qquad a_0 = \frac{k'_{-E} c_\Sigma}{k'_E + k'_{-E}} \qquad (5.27)$$

Hence Eq. (5.26) reduces to

$$Z_F = R_{CT} + \frac{RT(k'_E + k'_{-E})^2}{n^2 F^2 A c_\Sigma k'_E k'_{-E}}\frac{\coth(\sqrt{j\omega/D_{CT}}\,L)}{\sqrt{j\omega D_{CT}}} \qquad (5.28)$$

Note that the group of heterogeneous rate constants on the rhs of Eq. (5.28) may be expressed as

$$F(\vartheta) = \frac{(k'_E + k'_{-E})^2}{k'_E k'_{-E}} = \frac{[1 + \exp(-\vartheta)]^2}{\exp(-\vartheta)}$$

$$= 2 + \exp(\vartheta) + \exp(-\vartheta)$$

$$= 2[1 + \cosh(\vartheta)] = 4\cosh^2\left(\frac{\vartheta}{2}\right) \qquad (5.29)$$

We note therefore that the diffusional impedance Z_D is given by

$$Z_D = \frac{RT}{n^2 F^2 A D_{CT}^{1/2} c_\Sigma}F(\vartheta)\frac{\coth(\sqrt{j\omega/D_{CT}}\,L)}{\sqrt{j\omega}} \qquad (5.30)$$

We may show that

$$\frac{1}{\sqrt{j\omega}} = \frac{2}{\sqrt{2}}\frac{1}{1+j}\omega^{-1/2} = \frac{1}{\sqrt{2}}(1-j)\omega^{-1/2} \qquad (5.31)$$

Hence the Faraday impedance takes the form

$$
Z_F = R_{CT} + \frac{RTF(\vartheta)\omega^{-1/2}}{\sqrt{2}\,n^2F^2AD_{CT}^{1/2}c_\Sigma}(1-j)\coth\left(\sqrt{\frac{j\omega}{D_{CT}}}\,L\right)
$$

$$
= R_{CT} + (1-j)\sigma_w\omega^{-1/2}\coth\left(\sqrt{\frac{j\omega}{D_{CT}}}\,K\right) \qquad (5.32)
$$

where the parameter σ_W is termed the Warburg coefficient and is given by

$$
\sigma_w = \frac{RT}{\sqrt{2}\,n^2F^2AD_{CT}^{1/2}c_\Sigma}F(\vartheta) = \frac{4RT}{\sqrt{2}\,n^2F^2AD_{CT}^{1/2}c_\Sigma}\cosh^2\left(\frac{\vartheta}{2}\right) \quad (5.33)
$$

The expression outlined in Eq. (5.33) has been derived previously by Armstrong and co-workers [249, 250] and Gabrielli et al. [252]. Note that both the charge transfer resistance R_{CT} and the Warburg coefficient σ_W depend on potential because they contain terms in θ. We note from Eq. (5.26) that

$$
R_{CT} = \frac{RT(k_E' + k_{-E}')}{n^2F^2Ac_\Sigma k_E'k_{-E}'} = \frac{RT}{n^2F^zAc_\Sigma k^0}\exp(-\alpha\vartheta)(1+\exp(\vartheta)) \quad (5.34)
$$

Hence we see that R_{CT} should decrease rather rapidly with increasing θ due to the negative exponential multiplier in Eq. (5.34).

We now examine the behavior of Z_F in the limit of very high and very low frequencies. Let us first consider the high-frequency limit. This corresponds to the condition that $j\omega \gg D_{CT}/L^2$. Under such conditions we note that $\coth[\sqrt{j\omega/D_{CT}}\,L] \approx 1$ and Eq. (5.32) reduces to

$$
Z_F = R_{CT} + (1-j)\sigma_w\omega^{-1/2} \qquad (5.35)
$$

This is the well-established Randles expression for semi-infinite diffusion. We now look at the low-frequency limit. When $j\omega \ll D_{CT}/L^2$, we note that

$$
\coth\left(\sqrt{\frac{j\omega}{D_{CT}}}\,L\right) \approx \frac{D_{CT}^{1/2}}{\sqrt{j\omega}L} + \frac{\sqrt{j\omega}L}{3D_{CT}^{1/2}} \qquad (5.36)
$$

Hence the low-frequency representation for the Faraday impedance takes the form

$$Z_F = R_{CT} + \frac{\sqrt{2}\sigma_w}{\sqrt{j\omega}}\left(\frac{D_{CT}^{1/2}}{\sqrt{j\omega L}} + \frac{\sqrt{j\omega}L}{3D_{CT}^{1/2}}\right)$$

$$= R_{CT} + \frac{\sqrt{2}\sigma_w D_{CT}^{1/2}}{j\omega L} + \frac{\sqrt{2}\sigma_w L}{3D_{CT}^{1/2}}$$

$$= R_{CT} + \frac{1}{j\omega C_L} + R_L \qquad (5.37)$$

where we have introduced a low-frequency resistance R_L and a low-frequency capacitance C_L as follows:

$$R_L = \frac{\sqrt{2}\sigma_w L}{3D_{CT}^{1/2}} \qquad C_L = \frac{L}{\sqrt{2}\sigma_w D_{CT}^{1/2}} \qquad (5.38)$$

Hence we may represent the Faraday impedance component as a circuit consisting of two resistors and a capacitor in series (Fig. 71). The low-frequency capacitance C_L is simply the polymer redox capacitance previously discussed at some length. Note also from Eq. (5.38) that if R_L and C_L are known, then the fundamental parameter D_{CT}/L^2 may be determined according to the relation

$$R_L C_L = \frac{L^2}{3D_{CT}} = \frac{\tau_D}{3} \qquad (5.39)$$

where τ_D represents the diffusional time constant of the system.

Following from this idea we can write the diffusional impedance at any frequency as

$$Z_D = \frac{L^2/D_{CT}}{C_L} \frac{\coth(\sqrt{j\omega L^2/D_{CT}})}{\sqrt{j\omega L^2/D_{CT}}} = \frac{\tau_D}{C_L} \frac{\coth(\sqrt{j\omega\tau_D})}{\sqrt{j\omega\tau_D}} \qquad (5.40)$$

Letting $Z_{D,0} = \tau_D/C_L$, Eq. (5.40) reduces to

$$Z_D = Z_{D,0} \frac{\coth(\sqrt{j\omega\tau_D})}{\sqrt{j\omega\tau_D}} \qquad (5.41)$$

This is simply the impedance element describing diffusion through a finite medium where one boundary is blocking for the diffusing species. In

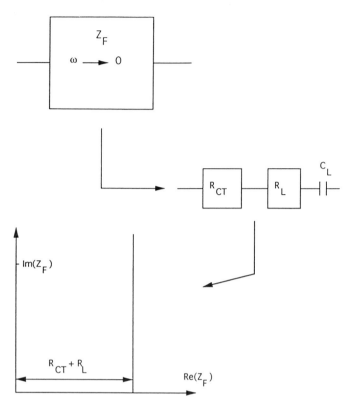

Figure 71. Equivalent circuit for Faraday impedance in low-frequency limit and corresponding impedance response.

terms of the admittance we write

$$Y_D = Y_{D,0}\sqrt{j\omega\tau_D}\,\tanh(\sqrt{j\omega\tau_D}) \tag{5.42}$$

These expressions for Z and Y are characteristic of a finite single-rail transmission line. These expressions have been derived many times in the literature of porous electrodes [263, 264] as well as that of oxide- and polymer-modified electrodes [251, 265]. For instance, in terms of a simple pore model Raistrick [265] has written that the impedance is given by

$$Z = \sqrt{\frac{R_D}{j\omega C_D}}\,\coth(\sqrt{j\omega R_D C_D}) \tag{5.43}$$

where R_D and C_D denote the total distributed (with respect to pore length) resistance and capacitance, respectively. This expression is very similar to that derived via the diffusion model [Eq. (5.41)]. Hence porous electrode models and diffusion models are similar since they both may be represented in terms of transmission lines. This transmission line approach has been applied extensively in the work reported by Pickup and co-workers [266–268].

The type of analysis presented here has been applied to both redox conducting and electronically conducting polymers. In both cases it is often assumed that the D_{CT} value corresponds to the process of ion transport through the polymer.

Let us now return to Eq. (5.12), where we have written an expression for the total impedance Z for the equivalent circuit in terms of the circuit elements R_u, Z_F, and C_{DL}. Consider the situation at low frequencies. We note from Eq. (5.37) that Z_F is given by

$$Z_F = R_{CT} + R_L + \frac{1}{j\omega C_{DL}} \qquad (5.44)$$

We now wish to obtain an expression for the total impedance at low frequency. The latter may be readily derived by noting that the interfacial capacitance C_{DL} becomes negligible at low frequencies, and so Eq. (5.44) reduces to

$$Z = R_u + Z_F = R_u + R_{CT} + R_L - \frac{j}{\omega C_L} = R_\Sigma - \frac{j}{\omega C_L}$$

$$= Z' - jZ'' \qquad (5.45)$$

Extracting the real and imaginary components of interfacial impedance, we obtain

$$Z' = R_\Sigma = R_u + R_{CT} + R_L$$
$$Z'' = \frac{1}{\omega C_L} \qquad (5.46)$$

Hence from Eq. (5.46) we note that R_L may be obtained from the numerical value of the real-axis intercept R_Σ of the Nyquist plot at low frequency provided both R_u and R_{CT} are known, where C_L may be determined from the slope of a plot of Z'' versus $1/\omega$. This situation is outlined in Fig. 72.

What about the situation at very high frequencies? In this case the diffusional impedance Z_D may be neglected and the Faraday impedance

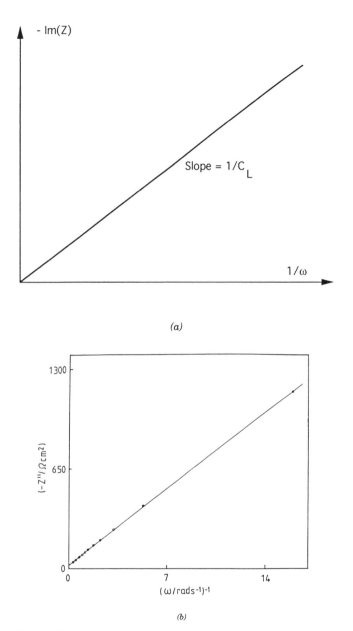

(a)

(b)

Figure 72. (*a*) Determination of redox capacity C_L via analysis of impedance data recorded at low frequency [Eq. (5.46)]. (*b*) Variation of imaginary component of total impedance recorded at low frequency, with inverse angular frequency for Cl^- ion doped poly(pyrrole) films ($L = 1.5\ \mu\text{m}$) in 0.1 M NaCl. Applied DC potential, 400 mV (Ag/AgCl).

Z_F is given by $Z_F = R_{CT}$. Hence Eq. (5.44) reduces to

$$Z = R_u + \frac{R_{CT}(1 - j\omega R_{CT}C_{DL})}{1 + (\omega R_{CT}C_{DL})^2} \qquad (5.47)$$

Extraction of the real and imaginary components of the impedance from Eq. (5.47) results in the assignations

$$Z' = R_u + \frac{R_{CT}}{1 + (\omega R_{CT}C_{DL})^2}$$
$$-Z'' = \frac{\omega R_{CT}C_{DL}}{1 + (\omega R_{CT}C_{DL})^2} \qquad (5.48)$$

Elimination of the angular frequency ω from the impedance expressions outlined in Eq. (5.48) results in

$$(Z' - R_u - \tfrac{1}{2}R_{CT})^2 + Z''^2 = (\tfrac{1}{2}R_{CT})^2 \qquad (5.49)$$

This is the equation of a circle with center on the real axis at $Z' = R_u + \tfrac{1}{2}R_{CT}$ and radius $\tfrac{1}{2}R_{CT}$. As ω becomes very large, the real-axis intercept is given by R_u, the uncompensated solution resistance. The second real-axis intercept at lower frequencies occurs at $Z' = R_u + R_{CT}$. As noted in Fig. 73, the maximum of the high-frequency semicircle gives the characteristic relaxation frequency for heterogeneous electron transfer at

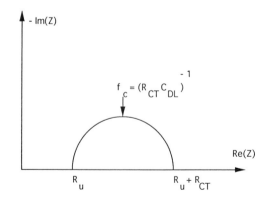

Figure 73.   Impedance response at high frequencies according to Eq. (5.47).

the support–polymer interface, and one notes that

$$f_c = \frac{1}{R_{CT}C_{DL}} = \frac{1}{\tau_{CT}} \qquad (5.50)$$

where τ_{CT} denotes the characteristic relaxation time for charge injection.

Finally, in the middle frequency range, the interfacial impedance becomes controlled by the diffusion of counterions into the polymer matrix, and thus the impedance response is here representative of material transport processes involved in the redox switching reaction. We note that in the Nyquist diagram one observes a linear behavior between Z'' and Z' with a frequency-independent phase angle of $\frac{1}{4}\pi$. In this frequency region we note that $\coth(\sqrt{j\omega/D_{CT}}\, L) \approx 1$, and substituting the result for the Faraday impedance obtained from Eq. (5.35) into Eq. (5.12) and simplifying, we obtain

$$Z' = R_u + R_{CT} - 2\sigma_w^2 C_{DL} + \sigma_w \omega^{-1/2}$$
$$Z'' = \sigma_w \omega^{-1/2} \qquad (5.51)$$

The expressions outlined in Eq. (5.51) define Randles plots in which either the real or imaginary components of the total interfacial impedance are predicted to vary in a linear manner with $\omega^{-1/2}$. The slope of either plot (termed the Randles slope) yields the Warburg coefficient σ_W (Fig. 74). Using Eq. (5.33), we can directly obtain an expression for D_{CT} from σ_W. Elimination of ω from Eq. (5.51) results in the following linear expression:

$$-Z'' = Z' - (R_u + R_{CT} - 2\sigma_w^2 C_{DL}) \qquad (5.52)$$

This result implies that a plot of $-Z''$ versus Z' should be linear, of unit slope, and exhibiting a real-axis intercept given by $Z' = R_u + R_{CT} - 2\sigma_w^2 C_{DL}$. This feature is clearly illustrated in Fig. 75.

Hence one may extract all of the relevant equivalent circuit parameters via a careful analysis of the impedance response provided that the data are collected over a very large frequency range, typically 100 kHz to 0.1 mHz.

In many situations it is preferable to obtain values for the fundamental equivalent circuit parameters via a nonlinear least-squares fitting protocol applied to the raw experimental impedance data. The latter may be accomplished if one can obtain an analytical expression for the real and imaginary components of the impedance that is valid over the entire range of frequency. Initial estimates of the circuit parameters may be

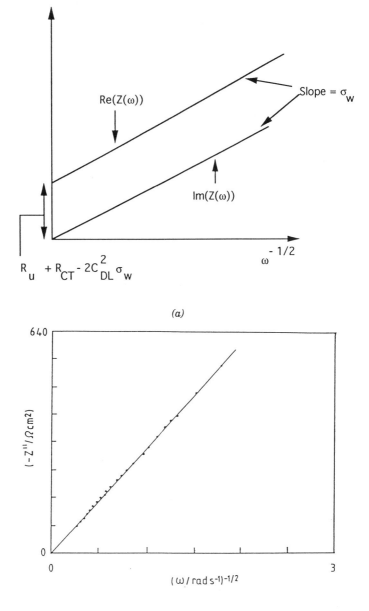

Figure 74. (a) Schematic representation of Randles plots used in data analysis where Warbourg region is evident in impedance spectrum. (b) Randles plot for poly(pyrrole) films doped with dodecylbenzenesulfonate (DBS) ion. Supporting electrolyte, 0.1 M NaDBS. Applied potential, −750 mV.

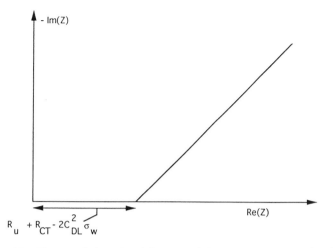

Figure 75.   Linear variation of Im(Z) with Re(Z) predicted from Eq. (5.52).

obtained via data analysis using the approximate expressions previously outlined.

Lang and Inzelt [256] have proposed the following method of analysis. We begin with Eq. (5.32) for the Faradaic impedance:

$$Z_F = R_{CT} + (1 - j)\sigma_w \omega^{-1/2} \coth(\beta L) \qquad (5.53)$$

where $\beta = \sqrt{j\omega/D_{CT}} = (1 + j)\omega^{1/2}/\sqrt{2D_{CT}}$. One now transforms the $\coth(\beta L)$ term into the form

$$\coth(\beta L) = \coth[\kappa(1 + j)] \qquad (5.54)$$

where we define κ to be

$$\kappa = \frac{\sqrt{2}}{2}\left(\frac{\omega}{D_{CT}}\right)^{1/2} L \qquad (5.55)$$

We now use the fundamental identities

$$\coth(x + y) = \frac{1 + \coth x \coth y}{\coth x + \coth y}$$

$$\coth x = j \coth jx \qquad (5.56)$$

in Eq. (5.54) to obtain, after some algebra,

$$\coth(\beta L) = \frac{\coth \kappa(1 + \cot^2 \kappa)}{\coth^2 \kappa + \cot^2 \kappa} + j \cdot \frac{\cot \kappa(1 - \coth^2 \kappa)}{\coth^2 \kappa + \cot^2 \kappa} \tag{5.57}$$

We now substitute this relation into Eq. (5.53) to obtain the following expression for Z_F:

$$Z_F = R_{CT} + F_1(\kappa)\sigma_w \omega^{-1/2} - jF_2(\kappa)\sigma_w \omega^{-1/2} \tag{5.58}$$

where the Lang–Inzelt functions $F_1(\kappa)$ and $S_2(\kappa)$ are given by [256]

$$F_1(\kappa) = \frac{\coth \kappa(1 + \cot^2 \kappa) - \cot \kappa(1 - \coth^2 \kappa)}{\coth^2 \kappa + \cot^2 \kappa}$$

$$F_2(\kappa) = \frac{\coth \kappa(1 + \cot^2 \kappa) - \cot \kappa(1 + \coth^2 \kappa)}{\coth^2 \kappa + \cot^2 \kappa} \tag{5.59}$$

If we now substitute Eq. (5.58) into Eq. (5.12) and simplify, then after some rather tedious algebra, we obtain

$$Z = R_u + \frac{[R_{CT}\omega^{1/2} + \sigma_w F_1(\kappa)]\omega^{-1/2}}{[1 + \sigma_w C_{DL} F_2(\kappa)\omega^{1/2}]^2 + \omega C_{DL}^2 [R_{CT}\omega^{1/2} + \sigma_w F_1(\kappa)]^2}$$

$$- j \frac{C_{DL}[R_{CT}\omega^{1/2} + \sigma_w F_1(\kappa)]^2 + \sigma_w F_2(\kappa)[\omega^{-1/2} + \sigma_w C_{DL} F_2(\kappa)]}{[1 + \sigma_w C_{DL} F_2(\kappa)\omega^{1/2}]^2 + \omega C_{DL}^2 [R_{CT}\omega^{1/2} + \sigma_w F_1(\kappa)]^2}$$

$$\tag{5.60}$$

This then is an analytical expression for the entire impedance response. It can be used to simulate the impedance spectrum and can be incorporated into a nonlinear least-squares fitting routine to obtain accurate estimates of R_{CT}, C_{DL}, C_L, L, σ_w, and D_{CT}.

Mathias and Hass [255a] have recently extended the analysis of the impedance response of a redox polymer film to consider electromigration processes and redox site interactions. They have recently published a second paper [255b] dealing with the application of their analysis to experimental impedance data. The analysis is rather complex, and therefore we will only present a summary of their results here. The analysis is, however, the most complete presented to date for the complex impedance response of an electroactive polymer film exhibiting redox conductivity. The Mathias–Haas theory is valid provided that Donnan exclusion (i.e., penetration of large quantities of co-ions from the bathing solution into the polymer film) is operative.

The Faradaic impedance is given by

$$Z_F = R_{CT} + \frac{RT}{F^2 A} \left(\frac{\beta}{n^2 \theta c_\Sigma} + \frac{1}{z_X^2 c_X} \right) \frac{2 t_E t_X + (t_E^2 + t_X^2) \cosh(\sqrt{j\omega/D_\Sigma} \, L)}{\sqrt{j\omega D_\Sigma} \sinh(\sqrt{j\omega/D_\Sigma} \, L)}$$

$$(5.61)$$

In this rather complex expression we note that

$$\theta = X_E (1 - X_E) \qquad (5.62)$$

where X_E denotes the mole fraction of reduced sites in the polymer (see Section III.A.5, the Chidsey–Murray model). Furthermore the parameter β is given by

$$\beta = 1 + \frac{\theta \varepsilon}{RT} \qquad (5.63)$$

where the parameter ε/RT quantifies the interaction between redox sites. If $\varepsilon/RT > 0$, then one has redox site repulsion effects and the transition potential range from oxidation to reduction is broadened. Alternatively, if $\varepsilon/RT < 0$, then one has attraction between neighboring redox sites and the transition potential range is narrowed. Note that ε represents the occupied-site interaction energy discussed previously in Section III.A.2. In the latter section we introduced, in Eq. (3.12), a dimensionless interaction parameter σ that is related to the Gibbs energy of interaction G via the relation $\sigma = G/RT$. Mathias and Haas [255] use a different notation, and we can reconcile the two notations by noting that $\varepsilon/RT = -\sigma$. This area is made quite confusing by a wide and rather arbitrary range of notations.

Note also that t_E and t_X represent the transference number of hopping electrons and mobile counterions X in the film, respectively, and are given by the expressions

$$t_E = \frac{n^2 \theta c_\Sigma D_E}{z_X^2 c_X D_X + n^2 \theta c_\Sigma D_E} \qquad t_X = \frac{z_X^2 c_X D_X}{z_X^2 c_X D_X + n^2 \theta c_\Sigma D_E} \qquad (5.64)$$

where z_X represents the signed valence of the mobile counterion and c_X denotes the counterion concentration in the film. Furthermore, the polymer phase is considered to be a binary electrolyte containing electrons and one mobile ionic species, and so one can introduce a binary diffusion coefficient D_Σ as follows:

$$D_\Sigma = \frac{D_E D_X (\beta z_X^2 c_X + n^2 \theta c_\Sigma)}{z_X^2 c_X D_X + n^2 \theta c_\Sigma D_E} \tag{5.65}$$

where n denotes the number of electrons transferred during redox switching and c_Σ denotes the total redox site concentration. The concept of a binary diffusion coefficient has been developed extensively by Buck in a series of papers [85–87]. It is related to the concept of a net resistance R_Σ of a dual-rail transmission line, as developed recently by Albery and co-workers [92–96]:

$$R_\Sigma = R_E + R_X$$

$$R_E = \frac{RT}{n^2 F} \frac{L}{A D_E} \frac{c_\Sigma}{a.b} \tag{5.66}$$

$$R_X = \frac{RT}{z_x^2 F^2} \frac{L}{A D_X b}$$

where a, b denote the concentrations of the reduced and oxidized forms of the redox sites and D_E, D_X represent the diffusion coefficient for electron hopping and counterion transport in the polymer, respectively.

The expression outlined in Eq. (5.61) for Z_F may be substituted into the following expression to obtain a master equation for the overall impedance response:

$$Z = R_u + R_F + \frac{1}{1/Z_F + j\omega C_{\text{DL}}} \tag{5.67}$$

where R_F denotes the film resistance and is given by

$$R_F = \frac{L}{\kappa A} \tag{5.68}$$

where the polymer conductivity κ is given by

$$\kappa = \frac{F^2}{RT} (n^2 \theta c_\Sigma D_E + z_X^2 c_X D_X) \tag{5.69}$$

One may show, via detailed analysis of the expression given in Eq. (5.61) for Z, that the low-frequency resistance R_L and redox capacitance C_L are given by

$$
R_L = \frac{RTL}{3F^2 A D_\Sigma} \left(\frac{\beta}{n^2 \theta c_\Sigma} + \frac{1}{z_X^2 c_X} \right)
$$

$$
C_L = \frac{F^2 A L}{RT} \left(\frac{\beta}{n^2 \theta c_\Sigma} + \frac{1}{z_X^2 c_X} \right)^{-1}
$$

(5.70)

Hence we note that

$$
R_L C_L = \frac{L^2}{3 D_\Sigma}
$$

(5.71)

Hence the binary diffusion coefficient D_Σ may be determined from the low-frequency data, as previously described in Eq. (5.39). Comparing Eqs. (5.71) and (5.39) we immediately note that D_{CT} is simply replaced by the binary diffusion coefficient D_Σ. Hence we see that the product of the low-frequency resistance R_L and the low-frequency capacitance C_L defines a fundamental quantity, which is independent of the complexity of the model assumed.

Furthermore, for higher frequencies one has a Warburg region, and one may show that the Warburg coefficient is given by

$$
\sigma_w = \frac{(t_E^2 + t_X^2) RT}{F^2 A \sqrt{2 D_\Sigma}} \left(\frac{\beta}{n^2 \theta c_\Sigma} + \frac{1}{z_X^2 c_X} \right) F(\vartheta)
$$

$$
= \frac{(t_E^2 + t_X^2) RT}{F^2 A \sqrt{2 D_\Sigma}} \left(\frac{\beta}{n^2 \theta c_\Sigma} + \frac{1}{z_X^2 c_X} \right) \cosh^2 \left(\frac{\vartheta}{2} \right)
$$

(5.72)

where, as previously noted, $\vartheta = (nF/RT)(E - E^0)$ denotes a normalized potential. In particular, at $E = E^0$, $F(\vartheta) = 1$, and Eq. (5.72) reduces to

$$
\sigma_w = \frac{(t_E^2 + t_X^2) RT}{F^2 A \sqrt{2 D_\Sigma}} \left(\frac{\beta}{n^2 \theta c_\Sigma} + \frac{1}{z_X^2 c_X} \right)
$$

(5.73)

For $E = E^0$ one may show that

$$
t_E^2 + t_X^2 = \sqrt{\frac{2 C_L}{3 R_L}} \, \sigma_w
$$

(5.74)

Now when either t_E or t_X approaches unity, the sum of the squares of the transference numbers will also approach unity. For the situation of mixed control, when $t_E = t_X = 0.5$, the latter value will approach 0.5.

Hence one can only distinguish between limiting behavior and intermediate control from analysis of a single spectrum. Equation (5.74) does
not enable differentiation between electron hopping and ion transport
control. Furthermore, Mathias and Haas [255] noted that when counterion transport is slow relative to electron hopping, the impedance spectrum contains a classical Warburg region just as when electron hopping
controls the rate of charge percolation. Hence these two tpes of possible
rate control cannot be distinguished from the analysis of a single-impe-
dance spectrum.

However, these workers proposed that one can make a precise
elucidation of the nature of the rate-controlling process by examining the
variation of characteristic quantities such as $(L/c_\Sigma D_\Sigma)$, $c_\Sigma D_\Sigma^{1/2}$, and $c_\Sigma L$
with electrode potential. The latter quantities may be derived from the
expressions

$$c_\Sigma L = \Gamma = \frac{RTC_L}{n^2 F^2 A} F(\vartheta)$$

$$\frac{L}{c_\Sigma D_\Sigma} = \frac{3n^2 F^2 AR_L}{RTF(\vartheta)} \tag{5.75}$$

$$c_\Sigma \sqrt{D_\Sigma} = \frac{RTF(\vartheta)}{\sqrt{2}\sigma_w n^2 F^2 A}$$

On the basis of the Mathias–Haas [255] model one may show that

$$c_\Sigma L = \Gamma = \frac{LF(\vartheta)}{n^2} \left(\frac{\beta}{n^2 \theta c_\Sigma} + \frac{1}{z_X^2 c_X} \right)^{-1}$$

$$\frac{L}{c_\Sigma D_\Sigma} = \frac{n^2 L}{D_\Sigma F(\vartheta)} \left(\frac{\beta}{n^2 \theta c_\Sigma} + \frac{1}{z_X^2 c_X} \right) \tag{5.76}$$

$$c_\Sigma \sqrt{D_\Sigma} = \frac{\sqrt{D_\Sigma} F(\vartheta)}{(t_E^2 + t_X^2)n^2} \left(\frac{\beta}{n^2 \theta c_\Sigma} + \frac{1}{z_X^2 c_X} \right)^{-1}$$

The theoretical quantities outlined in Eq. (5.76) are plotted in a
normalized format in Fig. 76 for various values of the parameter ε/RT
and various D_E/D_X ratios. Note that EH corresponds to electron hopping
control where $D_E/D_X \to 0$ and AM corresponds to anion transport
control, where $D_E/D_X \to \infty$. Well-defined and characteristic variations
with electrode potential for the characteristic quantities are predicted.
The shapes of the curves differ appreciably as one progresses from pure

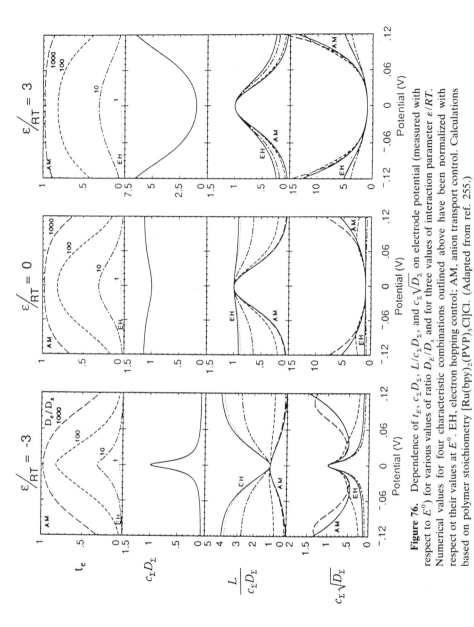

Figure 76. Dependence of t_E, $c_\Sigma D_\Sigma$, $L/c_\Sigma D_\Sigma$, and $c_\Sigma \sqrt{D_\Sigma}$ on electrode potential (measured with respect to E^0) for various values of ratio D_E/D_x and for three values of interaction parameter ε/RT. Numerical values for four characteristic combinations outlined above have been normalized with respect ot their values at E^0. EH, electron hopping control; AM, anion transport control. Calculations based on polymer stoichiometry [Ru(bpy)$_2$(PVP)$_5$Cl]Cl. (Adapted from ref. 255.)

electron-hopping control through mixed hopping/anion transport control to pure anion transport control. One can use the expressions outlined in Eq. (5.76) utilizing experimentally obtained data to make a direct comparison with theoretical prediction. In this way one can determine the extent of electron hopping/anion transport control manifested by a specific polymer system.

Very recently, Mathias and Haas [255b] have applied their analysis to an examination of charge transport in thin films consisting of redox active $Os(bpy)_2Cl$ centers coordinatively attached to protonated poly-(vinylpyridine) backbones. Analysis of the mobile ion concentrations in the film indicated that protons were rejected from the layer via the operation of Donnan exclusion, making anions and hopping electrons the only charge carriers. Detailed analysis of the complex impedance response indicated that even for the largest counterions studied, ion transport was much faster than electron hopping, in good agreement with previous work reported by Vos and co-workers [24–26]. However, it was reported that the counterion type affected the charge transport rate indirectly, by modifying the degree of film swelling and the extent of repulsions between electroactive sites.

C. Complex Impedance Response of Electronically Conducting Polymer Films

1. Introduction

The dynamics of charge transport in surface-deposited electronically conducting polymer films may also be probed using complex impedance spectroscopy. The advantages in using the latter method include the fact that slow kinetic processes in the polymer may be observed if impedance measurements are made over an extended frequency range and that the redox state of the polymer (and hence its morphology and chemical composition) may be fixed via application of a controlled DC potential, and the film is only perturbed slightly from equilibrium via application of the superimposed AC voltage. The latter point is especially important for electronically conductive polymers, because both the electrical properties and chemical structures of the polymers are changed by large-amplitude voltage perturbations. This is the so-called apples-to-oranges problem [269]. For example in the reduced state poly(pyrrole) is a nonconductive hydrophobic organic polymer, whereas in the oxidized form it is a hydrophilic polyelectrolyte that is electronically conductive. The redox kinetics of conducting polymers such as poly(pyrrole) [270], poly-(thiophene) [271], and poly(aniline) [272] have been examined using complex impedance spectroscopy, and a number of equivalent circuits have been proposed to account for the impedance responses observed. In many cases the theoretical formulation adopted has been derived from that developed for redox polymers as discussed in the previous section. This approach can provide a certain amount of information but, in general, is unsatisfactory, in that redox polymers are quite different materials than electronically conducting organic polymers. No totally satisfactory theoretical model exists at the present time that accounts in a truly fundamental way for all the observed behavior exhibited by doped conductive polymers in response to a small-amplitude AC perturbation.

2. Fletcher Dual-Rail Transmission Line Approach

In the present section we present a novel approach recently developed by Fletcher [96, 97] that provides some indication of the type of theory needed in order to acquire a full understanding of the complex impedance response of surface-deposited electronically conducting polymer films in contact with aqueous electrolyte solutions.

Fletcher proposed [96, 97] that a porous electrode model be adopted, the essential concept being that the conductive polymer film in contact with an aqueous electrolyte solution be regarded as consisting of a large number of identical, noninterconnected pores. The electrolyte solution is contained within the pores. The analysis then considered a single pore of uniform cross section. Three general impedance elements were considered: the solution impedance x within the pore, the interfacial impedance y between the solution within the pore and the pore wall, and z, the internal impedance of the polymer. The latter quantities are assumed not to vary with distance inside the pore. The model circuit is illustrated in Fig. 77. It takes the form of a diagonally connected discrete ladder network, or in simple terms, a dual-rail transmission line of finite dimension. The essential problem is to replace the general impedance elements x, y, and z by suitable arrangements of passive circuit elements such as resistors and capacitors that will adequately represent the microscopic physics occurring within an electronically conducting polymer.

One can apply the theory of electrical networks [273] to show that for a ladder having m steps there are $N = M - 1$ "squares", and the overall impedance Z is given by

$$Z = N \frac{xz}{x + z} + \frac{x^2 + y^2}{2(x + z)} \alpha + (x + z)\beta \qquad (5.77)$$

where

$$\alpha = \frac{\sinh(N\mu)}{\sinh(\mu)\cosh[(N + 1)\mu]} \qquad \beta = \frac{\cosh(\mu)}{2\sinh(\mu)\sinh[2(N + 1)\mu]} \qquad (5.78)$$

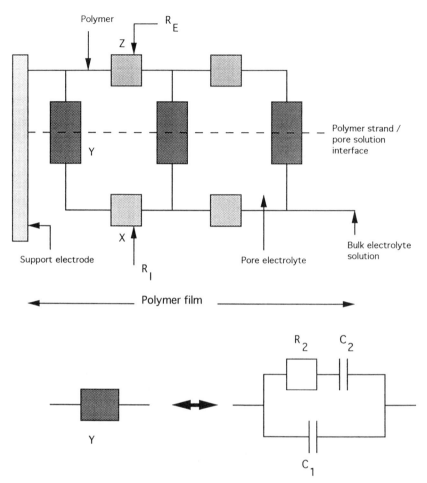

Figure 77. Schematic representation of equivalent circuit ladder network corresponding to Fletcher porous electrode model for electronically conducting polymers. Specific equivalent circuit representation of interfacial impedance element also illustrated.

and

$$\mu = \sinh^{-1}\left(\sqrt{\frac{x+z}{4y}}\right) = \ln\left(\sqrt{\frac{x+z}{4y}} + \sqrt{1 + \frac{x+z}{4y}}\right) \qquad (5.79)$$

Equation (5.77) may be simplifed to yield

$$Z = \frac{xz}{x+z} A(N, \mu) + y^{1/2}(x+z)^{1/2}B(N, \mu) \qquad (5.80)$$

where

$$A(N, \mu) = N - \frac{\sinh(N\mu)}{\sinh(\mu)\cosh[(N+1)\mu]}$$

$$B(N, \mu) = \frac{\cosh[(2N+1)\mu]}{\sinh[2(N+1)\mu]}$$

$$(5.81)$$

It is reasonable to suppose that the microstructure of the polymer is on a much finer scale than the thickness of the film, and so unit increments of N have little effect on the total impedance. Hence for large N, we set $N + 1 \approx N$, and Eq. (5.81) reduces to

$$A(N, \mu) = N - \frac{\tanh(N\mu)}{\sinh(\mu)} \qquad B(N, \mu) = \coth(2N\mu) \qquad (5.82)$$

Hence the impedance is given by

$$Z = \frac{xz}{x+z}\left(N - \frac{\tanh(N\mu)}{\sinh(\mu)}\right) + \sqrt{y}(x+z)^{1/2}\coth(2N\mu) \qquad (5.83)$$

We now assume that $|\mu| \ll 1$; then we note that $\mu \approx \sqrt{x + z/4y}$, and we can write that $\sinh(\mu) \approx \mu$ and $\tanh(N\mu) \approx N\mu$ and $\coth(2N\mu) \approx 1/2N\mu + 2N\mu/3$, and so the impedance becomes

$$Z \approx \sqrt{y}(x+z)^{1/2}\left(\frac{1}{2N\mu} + \frac{2N\mu}{3}\right) = \frac{y}{N} + \frac{N(x+z)}{3} \qquad (5.84)$$

Equation (5.84) will be valid only under conditions where $N \gg 1$ and $|\mu| \ll 1$ simultaneously. We can write Eq. (5.83) in another way. Again for $|\mu| \ll 1$ we have

$$Z \approx (x+z)^{1/2}y^{1/2}\coth\left(2N\sqrt{\frac{x+z}{4y}}\right) = (x+z)^{1/2}y^{1/2}\coth\left(N\sqrt{\frac{x+z}{y}}\right)$$

$$(5.85)$$

If we now assume that x and z are pure resistances and set $x = R_I$ the ionic resistance and $z = R_P$ the polymer resistance, then we let $N(x + z) = R_\Sigma = R_I + R_P$. Furthermore, if we assume that y may be represented as a pure capacitance, then we note that $y/N = -j/\omega NC = -j/\omega C_\Sigma = 1/j\omega C_\Sigma$, and so the total impedance is given by

$$Z = \sqrt{N(x + z)\frac{y}{N}}\, \coth\!\left(\sqrt{N(x + z)\frac{N}{y}}\right)$$

$$= \sqrt{\frac{R_\Sigma}{j\omega C_\Sigma}}\, \coth(\sqrt{j\omega R_\Sigma C_\Sigma}) \tag{5.86}$$

This is simply the expression for the impedance of a dual-rail transmission line of finite length and is very similar in form to that already presented in Eq. (5.43) for a simple single-rail transmission line model. We have simply introduced the composite resistance R_Σ into the expression.*

In terms of the RC model we note that the low-frequency expression outlined in Eq. (5.86) reduces to

$$Z \approx \frac{R_\Sigma}{3} - \frac{j}{\omega C_\Sigma} \tag{5.87}$$

This expression has been previously derived by Albery and co-workers [92–95] in their work on dual-transmission-line models.

Let us now return to Eq. (5.80) and examine the situation for $|\mu| \gg 1$. In this case $1/\sinh \mu \to 0$ and the overall impedance is given by

$$Z = N\!\left(\frac{xz}{x + z}\right) \tag{5.88}$$

Taking the inverse of Eq. (5.88) we obtain

$$\frac{1}{Z} = \frac{1}{Nx} + \frac{1}{Nz} \tag{5.89}$$

The latter is just the textbook formula for a series of n pairs of impedances in parallel.

The overall expression for the impedance is generally given by Eq. (5.80). The behavior exhibited by the impedance of the transmission line is governed by the absolute size of the dimensionless product $2N|\mu|$. If $2N|\mu| \ll 1$ a condition guaranteed by the joint existence of a thin film and a large impedance y, then the impedance reduces to

$$Z \approx \frac{y}{N} \tag{5.90}$$

---

*This composite resistance R_Σ is equivalent to the binary diffusion coefficient D_Σ introduced earlier.

and the overall impedance will be governed mainly by the interfacial impedance element y. Alternatively, if $2N|\mu| \gg 1$, the full expression in Eq. (5.80) must be used. However, for $2N[\text{Re } \mu] \gg 1$, we note from Eq. (5.80) that Z is given by

$$Z \approx \frac{Nxz}{x+z} + (x+z)^{1/2}y^{1/2} \qquad (5.91)$$

This "root product" behavior is characteristic of a Warburg impedance response since Z varies as $\omega^{-1/2}$ assuming that y is represented as a simple capacitor. A characteristic feature of the impedance response of many electronically conducting polymers is that Warburg behavior is not often observed experimentally. This is probably due to the fact that the deposited layers must be very thick in order that semi-infinite Warburg behavior be observed. Furthermore, it should be noted that Warburg-type behavior will be restricted to the range $Nxz/(x+z) < Z < \frac{1}{3}N(x+z)$.

At low frequencies the total impedance will be dominated by the impedance element y that deals with interfacial processes at the pore solution–pore wall interface. As noted in Fig. 77, this may be represented by a capacitance C_1 in parallel with a series combination of a resistance R_2 and a pseudocapacitance C_2. The capacitance C_1 is the double-layer capacitance and R_2 is the charge transfer resistance quantifying the redox kinetics of polarons and bipolarons located on the polymer chains. Fletcher has shown that the total impedance for this combination of circuit elements is given by

$$Z(\phi) = Z'(\phi) - jZ''(\phi)$$
$$= \frac{R_2 C_2^2}{(C_1 + C_2)^2 + \phi^2 C_1^2} - \frac{jR_2 C_2}{\phi} \frac{C_1 + C_2 + \phi^2 C_1}{(C_1 + C_2)^2 + \phi^2 C_1^2} \qquad (5.92)$$

where ϕ denotes a dimensionless frequency defined by the relation

$$\phi = \frac{\omega}{\omega_2} = R_2 C_2 \omega \qquad (5.93)$$

The expression defined by Eq. (5.92) is illustrated in Fig. 78. We note from this figure that a well-defined semicircular feature followed by a vertical line at low frequencies will be observed only when one has the condition that $C_2 \gg C_1$. This type of impedance response has been observed many times for electronically conducting polymer films. Note that the low-frequency intercept on the real axis is given by $R_2 C_2^2/(C_1 +$

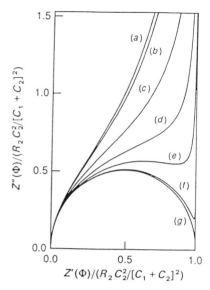

Figure 78. Complex plane representation of Eq. (5.92) that represents impedance response expected for interfacial impedance element Y in Fig. 77 for various values of capacitance ratio C_1/C_2; C_1/C_2 is 100 (a), 10 (b), 1 (c), 0.3 (d), 0.1 (e), 0.01 (f), and 0.001 (g). If polymer film is thick or there is charge leakage across pore walls, vertical response observed at low frequencies will be skewed to right.

$C_2)^2$. If $C_2 \gg C_1$, then this low-frequency intercept on the real axis reduces to R_2. Note also that at high frequencies one has that $Z''(\omega) \rightarrow 1/\omega C_1$, while at low frequencies $Z''(\omega) \rightarrow 1/\omega(C_1 + C_2)$. Hence values of C_1 and C_2 may be obtained by examining the variation of Z'' with frequency ω at low and high frequencies, and R_2 may be determined from the low-frequency real-axis intercept.

We can extend the analysis and consider the entire ladder network in terms of distinct R and C circuit elements. The impedance x may be represented by a resistance R_I that defined the resistane of counterions in the pore electrolyte. Furthermore, the impedance element z, which is that of the solid polymer, is replaced by a Randles equivalent circuit (Fig. 79), where one has a parallel arrangement of a resistor R_z and a capacitor C_z in series with a resistor R_Ω. Hence we see that the pore solution is modeled in terms of a simple resistor, whereas the solid polymer is described as a binary composite medium. The latter assumption may be justified as follows. From a macroscopic viewpoint (and this has been borne out experimentally), the electronic resistance of the polymer can be due to two contributions: the first, R_Ω, being due to the presence of regions of *high structural order*, and the second, R_z, associated with regions of *low structural order*. Hence R_Ω will be smaller than R_z. From a microscopic point of view, the polymer may exhibit two fundamentally different types of conduction. As previously noted in Section II.B.3, one has *intrachain conduction* along the backbone of the polymer and *interchain conduction* between adjacent polymer backbones. Clearly, the intrachain conduction is facile and is associated with the low-resistance

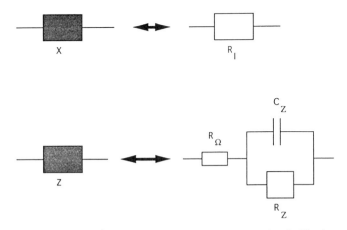

Figure 79. Equivalent circuit representation of elements X and Z in Fletcher model.

component R_Ω, whereas the interchain conduction is less kinetically facile and is associated with the high-resistance component R_z. What about the C_z component? The origin of C_z arises from the polarizability of the solid polymer. When an electric field is applied across the polymer film, electric dipole moments are induced in all poorly conducting regions that become polarized. Charge carriers are displaced from their field-free positions by distances of the order of an atomic diameter, with the result that there is a buildup of charge at the interfaces between poorly conducting and highly conducting regions. This situation may be represented via a capacitance C_z, and rapidly varying AC currents will flow through these elements with low impedance.

The general equivalent circuit incorporating the RC approximations to the general impedance elements x, y, and z is illustrated in Fig. 80. The overall impedance response in the low to very low frequency range is described by the expression

$$Z = \tfrac{1}{3}N(R_\Omega + R_z + R_x) + \frac{1}{N}\frac{R_2 C_2^2}{(C_1 + C_2)^2 + (\omega C_1 R_2 C_2)^2}$$

$$-\frac{j}{\omega N}\frac{C_1 + C_2 + (\omega R_2 C_2)^2 C_1}{(C_1 + C_2)^2 + (\omega C_1 R_2 C_2)^2} \tag{5.94}$$

Hence at low and very low frequencies, the impedance response of the ladder network is the same as that illustrated in Fig. 79 except that the impedance is divided by a factor N (related to the thickness of the film) and the locus of impedance points is sifted along the real axis of the first quadrant of the complex plane by a factor $(N/3)(R_\Omega + R_z + R_x)$. A

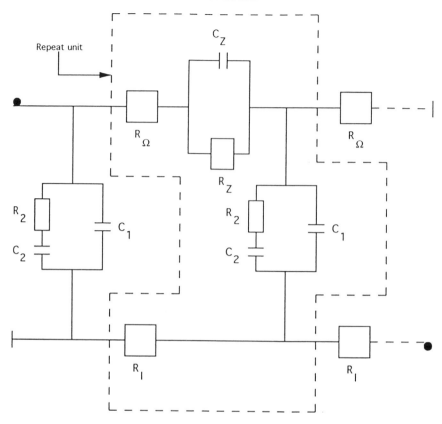

Figure 80. Overall equivalent circuit incorporating specifically all circuit elements.

further point should be noted. At high frequencies a second semicircle may also be observed that is due to the intrinsic conduction properties of the solid polymer. At these frequencies the interfacial behavior may be neglected. One can show that the network may be described in terms of a Randles equivalent circuit with parallel components C_z^*/N and NR_z^* and a series component NR_Ω^*. The latter quantities are defined as

$$R_\Omega^* = R_\Omega \frac{R}{R_\Omega + R_x}$$

$$R_z^* = R_z \frac{R_x^2}{R_z(R_\Omega + R_x) + (R_\Omega + R_x)^2} \qquad (5.95)$$

$$C_z^* = C_z \left(\frac{R_x + R_\Omega}{R_x}\right)^2$$

The diameter of the semicircle is NR_z^* and the high-frequency intercept on the real axis is NR_Ω^*. This semicircle will appear when both R_x and R_z are large and will be readily noted experimentally when the polymer is reduced. The semicircle will decrease in magnitude as the polymer film becomes oxidized and therefore more conducting.

We recall that the interfacial impedance element y also gives rise to a semicircle that becomes visible at lower frequencies. How may one distinguish between the two semicircles experimentally? This is accomplished by examining the impedance response of the polymer film as a function of layer thickness at a fixed value of electrode potential. If the diameter of the semicircle *increases* as the film thickness increases, then the semicircle is due to the R_z^* and C_z^* components since the diameter $D = NR_z^*$ in this case, where N plays the role of film thickness. In contrast, if the diameter D of the semicircle *decreases* as the layer thickness is decreased, then the semicircular feature in the spectrum is due to R_2 and C_1 the interfacial quantities. The latter statement follows from the fact that when the interfacial properties of the porous electrode predominate, the diameter D of the semicircle is given by

$$D = \frac{R_2 C_2^2}{N(C_1 + C_2)^2} \tag{5.96}$$

We note from Eq. (5.94) that for low frequencies we can write

$$Z'' = \frac{1}{N(C_1 + C_2)\omega}\left(1 + \frac{(\omega R_2 C_2 C_1)^2}{(C_1 + C_2)^2}\right) \approx \frac{1}{N(C_1 + C_2)\omega} \tag{5.97}$$

Hence at low frequencies for an oxidized film we note that the total low-frequency capacitance is given by

$$C_L = N(C_1 + C_2) \approx NC_2 \tag{5.98}$$

Furthermore, if the polymer is reduced, then at high frequencies the total impedance is given by

$$Z \approx \frac{NR_x(R_\Omega + R_z)}{R_\Omega + R_z + R_x} \tag{5.99}$$

Hence the high-frequency impedance of a reduced nonconducting polymer and the low-frequency capacitance of an oxidized polymer both depend linearly on the layer thickness (via the N factor). The linear dependence of capacitance on thickness is of considerable diagnostic importance in model verification, since one usually expects that the total capacitance of a thin film be inversely proportional to film thickness. We can explain the prediction in Eq. (5.98) by noting that the low-frequency capacitance is confined to the pore wall, which has its major axis oriented at right angles to the electrode surface. Thus the deeper the pore, the more capacitors there are in parallel in the pore wall, and consequently, the greater the total capacitance.

Experimental evidence for the Fletcher model has been recently obtained in the author's laboratory. In Fig. 81 the Nyquist plots for a poly(pyrrole) electrode doped with Cl^- ions in $0.1\,M$ NaCl solution is presented as a function of applied potential. Special attention should be focused on the first of these plots, which was recorded at a potential of $-550\,mV$ (vs. SCE) where the polymer is reduced and essentially nonconductive. Two semicircles are observed, one of smaller dimension at high frequencies and the other, much more pronounced, at mid to low frequencies. One notes that the high-frequency semicircle decreases in magnitude as the polymer becomes oxidized and cannot be discerned at potentials greater than $-200\,mV$. This semicircle is clearly associated with the R_z, C_z circuit elements. The second semicircle, in the mid- to low-frequency range, is present over the entire range of potentials examined (-500–$400\,mV$) and decreases in magnitude with increasing potential. This spectral feature is due to the redox chemistry of the polarons and bipolarons located on the polymer chains. The vertical line at very low frequencies corresponding to capacitive behavior is also evident, except for very negative potentials when the polymer is totally reduced. A plot of low-frequency capacitance as a function of layer thickness (the latter recorded at a potential of $400\,mV$ where the polymer is conductive and oxidized) is presented in Fig. 82. Note that a linear relationship between C_L and L is observed, in agreement with the Fletcher model. One would expect that R_2, the charge transfer resistance at the pore wall should decrease with increasing electrode potential and approach zero when the layer is oxidized and conducting. This prediction is confirmed in Fig. 83. We note from Eq. (5.94) that, on setting $R_2 = 0$ and $\omega \to 0$, when the polymer is conducting, the low-frequency resistance R_L should depend in an inverse manner with layer thickness since $1/(R_\Omega + R_z + R_x) = 1/R_L = N/3$. This prediction is confirmed in Fig. 84, where $1/R_L$ varies linearly with layer thickness L. This result was obtained for an oxidized film recorded at a potential of $400\,mV$ in $0.1\,M$ NaCl. It is also interesting to note from Fig. 83 that the low-frequency resistance R_L decreases quite rapidly with increasing electrode potential to a final steady value when the polymer is completely oxidized. Clearly both R_Ω and R_z decrease with increasing degree of oxidation. Also, R_x, the major component of the low-frequency resistance, will decrease. The

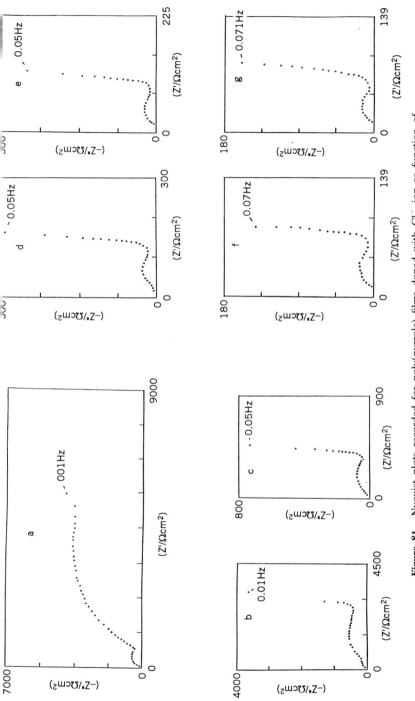

Figure 81. Nyquist plots recorded for poly(pyrrole) films doped with Cl⁻ ion as function of applied electrode potential. Supporting electrolyte, 0.1 M NaCl; $E = -550$ mV (a), -300 mV (b), -200 mV (c), 0 mV (d), 100 mV (e), 200 mV (f), and 400 mV (g). Frequency range used was 100 kHz down to lower limit indicated in figures.

521

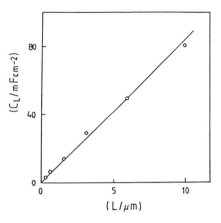

Figure 82. Variation of low-frequency capacitance C_L with layer thickness for Cl⁻ doped poly(pyrrole) film in 0.1 M BNaCl. Applied potential, 400 mV.

results in Fig. 85 indicate that the morphology of the polymer film changes when it is transformed from an insulating to a conductive state. Counterion transport in the pore solution (the latter manifested as R_x) is facilitated to a greater degree when the polymer is oxidized. Similar results to those outlined for poly(pyrrole) have been obtained for Cl⁻-

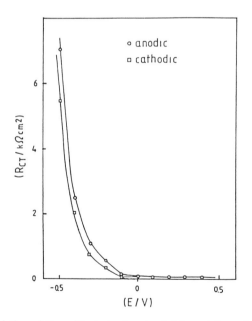

Figure 83. Variation of R_{CT} with electrode potential for poly(pyrrole) films (thickness 1.5 μm) examined in 0.1 M NaCl.

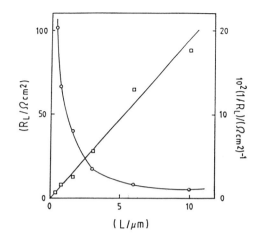

Figure 84. Variation of low-frequency resistance R_L with layer thickness. Same experimental conditions as in Fig. 82. Note that $1/R_L$ varies linearly with layer thickness L.

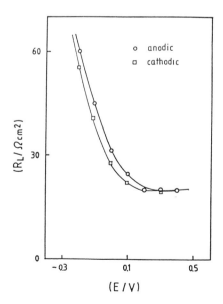

Figure 85. Variation of low-frequency resistance with electrode potential. Same experimental conditions as in Fig. 83.

doped poly(aniline) films in $1.0\,M$ HCl. Hence we see that the Fletcher model is confirmed by experiment in many important aspects. However, much work is still required before the model can be regarded as providing a definitive explanation for the complex impedance response of electronically conducting polymers.

VI. MEDIATED ELECTROCATALYSIS USING POLYMER-MODIFIED ELECTRODES

A. Introduction and Overview

In this and subsequent sections of the chapter we shall present a discussion of the use of electroactive polymer films in the important area of electrocatalysis. The material presented will also be relevant with respect to the quantitative description of the mechanism of operation of amperometric chemical and biological sensors. In the latter context, the efficient operation of the amperometric sensor will depend largely on how readily the polymer layer can enhance the rate of substrate oxidation or reduction. This, of course, is related to the electrocatalytic properties of the polymer film.

Electrocatalysis of electrode reactions at macromolecular layers involves the direct participation of the polymer material. Instead of a direct electron transfer between the Fermi level of the metallic electrode and the redox active substance in solution (which is the classical electrocatalytic situation), the electron transfer is *mediated* by the surface immobilized film. Furthermore, the overpotential at which a given substrate reaction occurs at an appreciable rate may be appreciably lowered at a polymer-modified electrode, compared to that obtained at an uncoated electrode. Consequently, the electroactive polymer layer plays a central role. The inert support electrode in such chemically modified electrode systems merely serves as a connector for electron flow, appropriate electronic circuitry providing the electrical driving force for redox transformation within the immobilized film. Compared to conventional unmodified electrodes, greater control of electrode characteristics and reactivity is achieved on surface modification.

This, therefore, is the attractive feature of chemically modified electrodes. The deposited chemical microstructure may be the subject of a *rational design* and be *tailormade* to perform a specific task. In order to be able to achieve this, however, we must be able to suggest a basis for a rational design of these systems. We must understand the fundamental principles underlying the way that the deposited layer mediates the oxidation or reduction of the substrate of interest. This may be accomplished by formulating very simple mathematical models of electrocatalytically active systems and by stating and solving (in an analytical manner if possible) the differential equations governing the diffusional transport of substrate through, as well as substrate reaction within, the polymer matrix. If an analytical solution to the problem is not possible, then one can attempt a digitial simulation of the problem. The analytically derived models, or indeed the results of the digital simulation, can then be compared with experiment and used as a basis for the subsequent rational design of the microstructure.

In this and subsequent sections of the chapter, we will present, in general terms, a survey of the various generations of electrocatalytic systems utilizing electroactive polymer films that have been developed since the late 1970s. Keeping the idea presented at the end of the last paragraph in mind, we shall be concerned in particular with presenting a detailed discussion of *approximate analytical models* that have been recently developed for these systems and we shall indicate how the predictions arising from these models may be compared with experimental data. We shall also examine a number of specific examples of recent experimental work in this area and indicate the direction of future developments in the field.

There have been considerable advances made in the development of electrocatalytic systems based on chemically modified electrodes in recent years. A particularly good summary has been provided by Hillman [3]. This review contains a wealth of experimental data and is written at a level that makes it a useful introduction to the area. A somewhat more general and more recent review written by Lyons [9] is also useful. Early work in the area has been well covered by Murray [1] and Albery and Hillman [2]. Specific applications to analytical chemistry have been covered in a brief, but topical, review written by Wring and Hart [274]. Interesting reviews have also been written by Abruna [4] and by Kaneko and Wohrle [275]. Lyons [276] has recently summarized current theoretical models in a brief but comprehensive manner. The recent chapter by Mortimer [277] is also of interest.

One can classify the development of electrocatalytic systems based on chemically modified electrodes into three generations. The first-generation systems were the so-called *monolayer modified electrodes*. These are the most simple systems. The chemical microstructures deposited on support electrode surfaces are two dimensional and very thin, usually a monolayer thick. These systems are quite easy to model, both for steady-state and transient conditions. However, as electrocatalytic systems they are limited utility since a three-dimensional dispersion of catalytically active centers cannot be realized. The early work of Murray [278] and Wrighton [279] is of note in this area.

Subsequent generations of electrocatalytically active systems utilize deposited three-dimensional chemical microstructures. These are the *polymer-modified electrodes*, multilayer polymeric structures containing redox active groups. Typical examples of the latter include *redox polymers, electronically conducting polymers*, and ionomer films loaded with redox active species, or *loaded ionomers*. Much emphasis has been palced on these second-generation systems. Their operational characteristics under steady-state conditions have been elucidated, mainly by Andrieux and Saveant [14] and by Albery and Hillman [280].

The fundamental characteristic of these second generation of polymer-based electrocatalytic systems is that the electrochemically active centers contained within the polymer matrix exhibit a dual purpose. They must be efficient shuttlers of electrons (the layer must have reasonable electronic or redox conductivity) as well as display good inherent electrocatalytic activity.

In many cases this dual requirement is significantly restrictive. Consequently, in the latest third-generation electrocatalytic systems, this condition has been dispensed with, and the concept of the *polymer-based integrated system* has been developed [17, 276]. These have also been termed *microheterogeneous systems*. In an integrated system the functions of charge transport through the film to the catalytic site and catalytic activity are carried out by *different* components within the layer. A number of different types of integrated or microheterogeneous systems have been developed. The first are those based on the use of microscopic metal particles [281–290] or enzymes [291–295] dispersed uniformly within electronically conducting polymer matrices. In another variant microparticles have been dispersed within ionically conductive ionomer matrices [296–301]. Systems containing dispersed catalytic microparticles within a redox polymer matrix have also been described [302–306]. The theory of microheterogeneous systems has been recently developed by Lyons and Bartlett in a number of papers [307–311].

Polymer-bound microheterogeneous catalytic systems have a number of catalytic advantages. First, they are relatively easy to fabricate. Second, the functions of catalysis and electron transport between the support surface and the catalytic site are distinct. Third, there is a three-dimensional dispersion of active sites throughout the polymer host matrix. A high local concentration of catalytic sites may be achieved, even though the total quantity of active material is small. This *dispersion* of catalytic material offers important catalytic advantage. Indeed, highly dispersed platinum catalysts have been widely employed in the field of fuel cell technology. Fourth, the microscopic particles can act as catalytic sites for multielectron transfer reactions. Finally, the polymer matrix appears to stabilize the ensemble of dispersed microparticles. This, of course, leads to an increased operational lifetime of the catalytic system.

B. Heterogeneous Mediated Electrocatalysis: General Considerations

In this section we consider the process of mediated electron transfer via a surface-immobilized redox couple. The latter either is assumed to be covalently attached to a support electrode surface to form a two-dimensional chemically derivatized electrode or is located in a three-dimensional polymeric matrix to form a polymer-modified electrode. In the absence of the surface or polymer matrix immobilized redox couple, the solution-phase substrate is assumed to display rather sluggist electron transfer kinetics. The situation is outlined in schematic form in Fig. 86.

We now consider the catalytic process in general terms. Consider the reaction

$$A \mp ne^- \Rightarrow B$$
$$S + B \Rightarrow P + A \tag{6.1}$$

In this reaction sequence A is terms the *precatalyst* and B is the catalytically active form of the surface-immobilized redox couple. The thermodynamics is encapsulated in the standard potential $E^0(A-B)$, whereas the electrochemical kinetics governing the A–B transformation is described via the heterogeneous rate constant k'_E (units: centimeters per second) and the physical chemistry of the electron transfer is governed by the symmetry factor α (usually $0 < \alpha < 1$, with typically $\alpha = \frac{1}{2}$). The catalytic process involves the heterogeneous reaction between the catalytically active form B and the substrate S to form the product P. The latter heterogeneous reaction results in the regeneration of the precatalyst A. The process is therefore cyclical. Therein lies the utility of mediated electrocatalysis: One has constant regeneration of the precatalytic form, which may subsequently be transformed back into the catalytically active form, which can then react further with substrate. The rate of the heterogeneous step (termed the cross-exchange reaction) is quantified via the second-order rate constant k (units: $cm^3 mol^{-1} s^{-1}$).

If the rate of the cross-exchange reaction is fast and the rate of redox transformation of the immobilized species A is faster (i.e., $k'_E \gg k$), then the substrate S will be transformed into product P at a potential near that of the standard potential $E^0(A/B)$ of the immobilized redox couple. This is what we mean by mediated electrocatalysis.

The equilibrium constant K for the exchange reaction is related to the

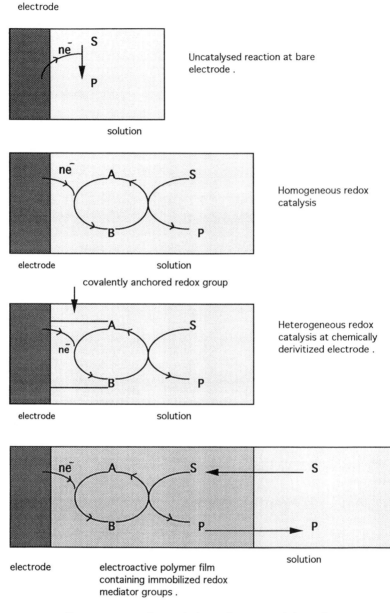

Figure 86. Schematic representation of various catalytic potentialities. Uncatalyzed reaction at bare electrode, homogeneous redox catalysis, and heterogeneous redox catalysis both at monolayer and multilayer hemically modified electrodes are shown.

standard potentials of the two couples A/B and S/P according to

$$K = \exp\left\{ \mp \frac{nF}{RT} [E^0(S/P) - E^0(A/B)] \right\}$$ (6.2)

where the positive sign in the exponential refers to the process of substrate reduction and the negative sign refers to substrate oxidation. Hence in thermodynamic terms, for the most efficient catalysis, the precatalyst–mediator and substrate–product formal potentials should be similar.

One further aspect must be considered at this stage. The logistics of substrate supply to the electrode surface must be specifically considered. Consequently, the kinetics of the overall process is assumed to be conjointly controlled by the rate of reaction between the active form of the bound catalyst centers and the substrate and by the diffusional transport of the substrate from the bulk of the solution to the interfacial region. Hence the phenomenon of concentration polarization in the solution must also be taken into account. The rate of diffusional transport of substrate (we neglect electromigration effects here by assuming a large excess of electroinactive supporting electrolyte) is quantified in terms of a rate constant k_D (units: centimeters per second).

Hence only two processes need to be considered when modeling very simple two-dimensional catalytic systems: the heterogeneous cross-exchange reaction and diffusional transport in the solution. The situation becomes more complex when three-dimensional polymer-modified electrodes are used, as we shall now see.

The currently accepted model for mediated electron transfer at polymer-coated electrodes may be presented in the following way. A distinction needs to be made at this point between mediated electrocatalysis at two-dimensional chemically derivatized electrodes and three-dimensional polymer-modified electrodes. In both cases matter transport in the diffusion layer and electron exchange between the substrate S and the catalytically active form B must be considered. However, *in addition*, for three-dimensional catalytic systems, the rate of charge propagation through the polymer matrix and the permeability of the substrate through the polymer film will, in many cases, be very important and may even be rate determining. The former will depend on the intrinsic redix or electronic conductivity of the layer, whereas the latter will depend on the morphology of the film.

The fundamental reaction scheme is outlined in Fig. 87. Reduction or oxidation of the substrate S by the catalytically active species B involves the following processes: injection of charge and substrate at the elec-

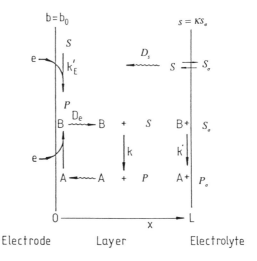

Figure 87. Schematic representation of processes of transport and kinetics within electrocatalytically active multilayered polymer film according to Albery–Hillman reaction scheme. Various symbols are outlined in text.

trode–polymer and polymer–solution interfaces, respectively, subsequent diffusion of substrate and electrons through the polymer matrix, and chemical reaction involving electron cross exchange between substrate- and polymer-bound catalyst group in some region of the layer.

The substrate enters into the polymer film at the polymer–solution interface. The permeability is quantified in terms of a partition coefficient κ and a substrate diffusion coefficient D_S. It is to be expected that the substrate diffusion coefficient within the film will be rather less than that exhibited by the substrate when it diffuses in the bulk solution phase. Typically D_S is in the range 10^{-13}–$10^{-7}\,\mathrm{cm}^2\,\mathrm{s}^{-1}$, compared to a solution-phase diffusion coefficient of $\sim 10^{-6}\,\mathrm{cm}^2\,\mathrm{s}^{-1}$.

Charge injection at the electrode–polymer interface will be to bound redox species in close proximity to the electrode and will normally be kinetically facile. This means that the surface concentration of the mediator species B, labelled b_0, is then fixed by the electrode potential via the Nernst equation. The A/B transformation at this inner interface is potential gradient driven. As we have previously noted, in contrast, the propagation of charge throughout the layer is, to a first approximation, concentration gradient driven. The facility of charge percolation may be quantified via an electron-hopping diffusion coefficient D_E. In the following discussion we assume that charge injection will be more rapid than charge percolation via electron hopping between neighboring redox sites.

Finally, S and B will react in some region of the film to produce A and product P. This region of the film is called the *reaction zone*. The proces of reaction between substrate and mediator is quantified in terms of a second-order rate constant k. It should be noted that depending on the realtive rates of the kinetic and transport parameters, reaction may occur throughout or only in a restricted region of the layer. As well as reaction occurring in the layer between substrate and mediator quantified by the rate constant k, one can envisage two further possibilities. The first involves reaction at the outer surface of the film without the requirement osf substrate partitioning. This is essentially a surface process and is quantified using a rate constant k'. The second possibility involves a direct unmediated reaction at the underlying support electrode. This is the electrode case and is quantified by a heterogeneous rate constant k_E. We see therefore that the location of the reaction zone, and indeed its thickness, will be determined by the relative rates of transport of the two species in the layer (substrate and electrons) and the rate of the catalytic reaction. Hence, for example, if the diffusion of electrons in the layer is much faster than the diffusion of substrate, then the reaction is likely to occur close to the layer–solution interface. If, on the other hand, the substrate diffuses much more rapidly through the layer than electrons, then the reaction is likely to occur close to the electrode–layer interface.

C. Theoretical Analysis of Mediated Electrocatalysis at Polymer-Modified Electrodes

1. Introduction

We now consider a theoretical analysis of mediated electrocatalysis at electroactive polymer-modified electrodes. In this case the deposited layer is rather thick, and one has a three-dimensional dispersion of catalytic material throughout the polymer matrix. This three-dimensional catalytic system is intrinsically more efficient than the chemically derivatized electrode consisting of a monolayer of deposited material. However, the fact that one has a three-dimensional multilayer means that the theoretical analysis is somewhat complicated.

Fundamental contributions to the theory of mediated electrocatalysis at polymer-modified electrodes have been made by a number of workers, most notably by Andrieux, Saveant and co-workers [312–318], Murray and Rocklin [319, 320], Anson [321], Laviron [322], and Albery and Hillman [280]. A numerical analysis of the problem has been presented by Daul and co-workers [323, 324], Cassidy and Vos [325], and Turner-

Jones and Faulkner [326]. Recent extensions to the theory have been reported by Sharp [327], Dong and Che [328], and Lyons et al. [329, 330]. In all of this work the steady-state response was only considered. The theory was cast in the framework of rotating disc voltammetry [331, 332]. The extension to the transient domain has not yet received a large amount of attention. The analysis of mediated electrocatalysis using cyclic voltammetry has been presented by Aoki and co-workers [333], whereas an analysis of the complex impedance response has been presented recently by Deslouis and co-workers [334]. The mathematics describing mediated electrocatalysis under transient conditions is quite complex, and the reader is referred to the original papers for further details.

We shall only consider the steady-state theory in this chapter. Our discussion shall concentrate on the model developed by Albery and Hillman [280]. We do this because we feel that this particular approach has not received the attention it deserves in the literature. Many workers have tended to adopt the Andrieux–Saveant approach [312–318], which is based on so-called characteristic currents. The reader is referred to the recent review by Andrieux and Saveant [14] for a full discussion of this completely equivalent approach. Albery and Hillman [280] also compare the two approaches in their original paper.

2. Albery–Hillman Model for Mediated Electrocatalysis

The mathematics can be stated in the following way [280]. In the steady state, the transport and kinetics of mediator and substrate may be quantified in terms of the following pair of reaction–diffusion expressions:

$$D_E \frac{\partial^2 b}{\partial x^2} - kbs = 0 \qquad D_S \frac{\partial^2 s}{\partial x^2} - kbs = 0 \qquad (6.3)$$

where b denotes the mediator concentration and s the substrate concentration. At the electrode surface where $x = 0$, $b = b_0$, where b_0 is determined by the electrode potential. The flux of electrons at the electrode–polymer interface for the A/B transformation is given by

$$j_E = -D_E \left(\frac{\partial b}{\partial x}\right)_{x=0} \qquad (6.4)$$

The boundary condition at $x = L$ at the layer–solution interface is obtained from the relation $s_L = \kappa s_\sigma$, where s_σ denotes the interfacial concentration of the substrate. The flux of electrons at $x = L$ is given by the rate of reaction between B and S at the layer–solution interface and is $j_E = k' b_L s_\sigma$.

One must also consider the direct or unmediated reaction of S at the electrode–polymer interface. This is given by

$$j_S = D_S \left(\frac{\partial s}{\partial x} \right)_{x=0} = k_E s_0 \qquad (6.5)$$

The total flux at $x = 0$ is given by

$$j = j_B + j_S = -D_E \left(\frac{\partial b}{\partial x} \right)_{x=0} + D_S \left(\frac{\partial s}{\partial x} \right)_{x=0} = \frac{i}{nFA} = k'_{ME} s_\sigma \qquad (6.6)$$

where k'_{ME} is the modified electrode rate constant, which serves to quantify the kinetics at the polymer-coated electrode.

We can relate the interfacial concentration s_σ to the bulk concentration s^∞ by considering the following reaction scheme:

$$S(\infty) \underset{k_D}{\overset{k_D}{\rightleftharpoons}} S(\sigma) \overset{k'_{ME}}{\longrightarrow} P(\sigma) \qquad (6.7)$$

Hence under steady-state conditions we note that

$$s_\sigma = \frac{k_D s^\infty}{k'_{ME} + k_D} \qquad (6.8)$$

where k_D represents the diffusional rate constant for substrate transport through the Nernst diffusion layer. If a rotating disc electrode is used, then this quantity is well defined, as is a simple function of electrode rotation speed ω. Therefore, concentration polarization in the solution poses no problem and can be readily calculated. The problem is to evaluate the modified electrode rate constant k'_{ME}.

To proceed further in the analysis it is useful to introduce the concept of a characteristic length, called a *reaction layer thickness*. The first reaction layer thickness X_L describes the consumption of substrate S as it enters the layer from the solution. This is given by

$$X_L = \left(\frac{D_S}{kb_L} \right)^{1/2} = \left(\frac{D_S}{kb_0} \right)^{1/2} \qquad (6.9)$$

Note that we have used either b_L or b_0 in Eq. (6.9) because under conditions where X_L enters expressions for k'_{ME}, species B is not concentration polarized, that is, the concentration of mediator is uniform throughout the layer. This reaction layer thickness will be important when the electrons penetrate easily across the layer and react with the substrate before it can diffuse very far across the layer. Under such conditions $X_L < L$, where L is the layer thickness.

We can also introduce a second reaction layer thickness X_0 that describes the consumption of an electron or, equivalently, a mediator group as it enters the layer from the electrode. This reaction layer thickness is defined as

$$X_0 = \left(\frac{D_E}{ks_0}\right)^{1/2} = \left(\frac{D_E}{k\kappa s_0}\right)^{1/2} \qquad (6.10)$$

In this case similar arguments as to concentration polarization of substrate S in the layer apply as previously delineated above for Eq. (6.9). The reaction layer X_0 will be important when substrate S moves readily through the layer and reacts with mediator species before they can diffuse very far from the inner interface via electron self-exchange processes, that is, electron "hopping".

We can use Eqs. (6.9) and (6.10) to qualitatively identify the various reaction zones. We first examine a rather extreme case. If the catalytic rate constant k is zero, then S has to diffuse across the film to react directly at the electrode. When the kinetics between B and S are slow, Eqs. (6.9) and (6.10) indicate that the reacton layers will be thick and mediated catalysis will take place throughout the bulk of the film. However, if k is quite large, then B and S will react as soon as they meet. They cannot coexist in the same region of space. In such a situation separate regions dominated by each will be found in the film. The relative sizes of these regions will of course depend on the respective diffusional fluxes given by $D_E b_0/L$ and $\kappa D_S s_\sigma/L$. Finally, when k is sufficiently large or, indeed, κD_S sufficiently small, the reaction between B and S will be confined to a region of molecular dimensions at the outer film–solution interface and S will not partition into the film. Hence we see that the concept of reaction layer thickness is very useful. It conveys a concise picture of the interplay between transport and kinetics in the modified electrode.

To proceed on a more quantitative manner, we have to solve the steady-state reaction–diffusion equations presented in Eq. (6.3) and hence obtain expressions for k'_{ME}.

This aim is accomplished by first recasting the reaction diffusion equations into a nondimensional form by introducing the following normalized parameters:

$$u = \frac{b}{b_0} \qquad v = \frac{s}{\kappa s_\sigma} \qquad \Phi_0 = \mu_0^{-1} = \frac{L}{X_0} \qquad \Phi_L = \mu_L^{-1} = \frac{L}{X_L}$$

$$\chi = \frac{x}{L} \qquad \lambda_B = \frac{k's_\sigma L}{D_E} \qquad \lambda_S = \frac{k'_E L}{D_S}$$

(6.11)

Note that u and v denote normalized concentrations of mediator and substrate, respectively, χ is a normalized thickness parameter ($0 < \chi < 1$), λ_B compares the rate of the surface reaction with the rate of the elelctron-hopping process across the film, λ_S compares the rate of the direct reaction of the substrate at the support electrode–film interface with the rate of substrate transport through the film, and Φ_0, Φ_L are Thiele moduli that compare the reaction layer thicknesses X_0 and X_L to the overall layer thickness L.* We note that Φ_0^2 defines the ratio of the pseudo-first-order rate constant for substrate reaction in the layer near the electrode surface (ks_0) to the diffusional frequency for eletron hopping (D_E/L^2), whereas Φ_L^2 defines the ratio of the pseudo-first-order rate constant for reaction of mediator species B in the outer regions of the film (kb_L) to the pseudo-first-order rate constant for substrate transport in the film (D_S/L^2). These definitions are in accord with the usual concept of Thiele modulus, which always relates a chemical reaction rate to a transport rate.

Using Eq. (6.11) we note that the reaction–diffusion equations presented in Eq. (6.3) may be transformed to

$$\frac{d^2u}{d\chi^2} - \frac{\Phi_0^2 uv}{v_0} = 0$$

(6.12)

and

$$\frac{d^2v}{d\chi^2} - \frac{\Phi_L^2 uv}{u_1} = 0$$

(6.13)

* We have also introduced the notation μ_0 and μ_L, which define the reciprocal Thiele moduli. This notation was used by Albery and Hillman in their original paper.

Also the boundary conditions for the problem transform to

$$\chi = 0 \qquad u = 1 \qquad v = v_0 \qquad \left(\frac{dv}{d\chi}\right)_{\chi=0} = \lambda_2 v_0$$

$$\chi = 1 \qquad v = 1 \qquad u = u_1 \qquad \left(\frac{du}{d\chi}\right)_{\chi=1} = -\lambda_B u_1 \tag{6.14}$$

Now subtracting Eqs. (6.12) and (6.13) and integrating (also noting that $s_L = \kappa s_\sigma$), we obtain

$$D_E b_0 \left[\left(\frac{du}{d\chi}\right)_{\chi=1} - \left(\frac{du}{d\chi}\right)_{\chi=0}\right] = D_S \kappa s_\sigma \left[\left(\frac{dv}{d\chi}\right)_{\chi=1} - \left(\frac{dv}{d\chi}\right)_{\chi=0}\right] \tag{6.15}$$

whereas a second integration yields

$$D_e b_0 \left[u - 1 - \chi\left(\frac{du}{d\chi}\right)_{\chi=0}\right] = D_S \kappa s_\sigma \left[v - v_0 - \chi\left(\frac{dv}{d\chi}\right)_{\chi=0}\right] \tag{6.16}$$

We also note that Eq. (6.6) reduces to

$$k'_{ME} = -\frac{D_E b_0}{s_\sigma L}\left(\frac{du}{d\chi}\right)_{\chi=0} + \frac{\kappa D_S}{L}\left(\frac{dv}{d\chi}\right)_{\chi=0} \tag{6.17}$$

Hence the theoretical problem reduces to determining $(du/d\chi)_0$ and $(dv/d\chi)_0$.

The essential point noted by Albery and Hillman [280] is the balance between the quantities $D_E b_0$ and $D_S \kappa s_\sigma$. If $D_E b_0 \gg D_S \kappa s_\sigma$, then electrons will penetrate across the layer more readily than substrate species S penetrates to the electrode. The layer will exhibit good redox conductivity. Under such conditions, we note, from Eq. (6.16), that

$$u \approx 1 + \left(\frac{du}{d\chi}\right)_0 \chi \tag{6.18}$$

and to complete the analysis, we must solve Eq. (6.13) for v. If, on the other hand, we have $D_E b_0 \ll D_S \kappa s_\sigma$, then substrate S penetrates to the electrode more readily than electrons percolate via electron hopping across the layer, and so from Eq. (6.16), we note that

$$v \approx v_0 + \left(\frac{dv}{d\chi}\right)_0 \chi \tag{6.19}$$

and we are required to solve Eq. (6.12) for u.

We first examine the situation when the redox conductivity within the layer is large and/or the morphology of the layer is compact. We must integrate Eq. (6.13). We can assume that under such conditions the substrate S will react with B sites in a thin reaction layer of dimension X_L, which is very much less than the layer thickness L. Hence, to a good approximation, we write that $u \approx u_1$, and so Eq. (6.13) reduces to

$$\frac{d^2 v}{d\chi^2} - \Phi_L^2 v = 0 \tag{6.20}$$

This differential equation admits the following solution*:

$$v(\chi) = \frac{\cosh(\Phi_L \chi)}{\cosh(\Phi_L)} + \frac{\lambda_S \Phi_L [\sinh(\Phi_{L-1}\chi) - \tanh(\Phi_L)\cosh(\Phi_L \chi)]}{\cosh(\Phi_L) + \lambda_S \Phi_L^{-1} \sinh(\Phi_L)} \tag{6.21}$$

Furthermore the concentration gradient at $\chi = 0$ is given by

$$\left(\frac{dv}{d\chi}\right)_0 = \frac{\lambda_S}{\cosh(\Phi_L) + \lambda_S \Phi_L^{-1} \sinh(\Phi_L)} \tag{6.22}$$

We can also show that

$$\left(\frac{dv}{d\chi}\right)_1 - \left(\frac{dv}{d\chi}\right)_0 = \Phi_L \tanh(\Phi_L) - \frac{\lambda_S \tanh^2(\Phi_L)}{1 + \lambda_S \Phi_L^{-1} \tanh(\Phi_L)}$$

$$= \frac{\Phi_L \tanh(\Phi_L)}{1 + \lambda_S \Phi_L^{-1} \tanh(\Phi_L)} \tag{6.23}$$

We now use Eq. (6.15) to note that

$$\left(\frac{du}{d\chi}\right)_1 = \left(\frac{du}{d\chi}\right)_0 + \frac{D_S \kappa s_\sigma}{D_E b_0}\left[\left(\frac{dv}{d\chi}\right)_1 - \left(\frac{dv}{d\chi}\right)_0\right]$$

$$= -\lambda_B u_1 = -\lambda_B - \lambda_B \left(\frac{du}{d\chi}\right)_0 \tag{6.24}$$

where we have used Eq. (6.18) to evaluate u_1. From Eq. (6.23) we can

---

* The solution for the concentration profile of substrate has been derived for the purposes of this review by the present author. It differs from the expression originally presented by Albery and Hillman in thier original paper [280]. The expression presented by Albery and Hillman [given in Eq. (43) of their 1984 paper] does not obey the boundary conditions set for the problem. It may readily be shown that the expression presented in Eq. (6.21) satisfies the boundary conditions defining the problem.

write

$$\frac{D_S \kappa s_\sigma}{D_E b_0}\left[\left(\frac{dv}{d\chi}\right)_1 - \left(\frac{dv}{d\chi}\right)_0\right] = \frac{k\kappa s_\sigma L^2}{D_E} u_1 \frac{\Phi_L^{-1} \tanh(\Phi_L)}{1 + \lambda_S \Phi_L^{-1} \tanh(\Phi_L)}$$

$$= \frac{k\kappa s_\sigma L^2}{D_E} \frac{\Phi_L^{-1} \tanh(\Phi_L)}{1 + \lambda_S \Phi_L^{-1} \tanh(\Phi_L)}\left[1 + \left(\frac{du}{d\chi}\right)_0\right] \qquad (6.25)$$

Substituting Eq. (6.25) into Eq. (6.24) and solving for $(du/d\chi)_0$ yield

$$\left(\frac{du}{d\chi}\right)_0 = \left[1 + \left(\lambda_B + k\kappa s_\sigma \frac{L^2}{D_E} \frac{\Phi_L^{-1} \tanh(\Phi_L)}{1 + \lambda_S \Phi_L^{-1} \tanh(\Phi_L)}\right)^{-1}\right]^{-1} \qquad (6.26)$$

We are now in a position to evaluate the modified electrode rate constant k'_{ME}. To do this, we simply substitute Eqs. (6.22) and (6.26) into Eq. (6.17) to obtain (after retransforming from dimensionless to dimensioned parameters)

$$k'_{ME} = \left[\frac{s_\sigma L}{D_E b_0} + \left(k'b_0 + k\kappa b_0 \frac{X_L \tanh(L/X_L)}{1 + (k'_E X_L/D_S)\tanh(L/X_L)}\right)^{-1}\right]^{-1}$$

$$+ \frac{\kappa \operatorname{sech}(L/X_L)}{\dfrac{1}{k'_E} + \dfrac{X_L}{D_S}\tanh(L/X_L)} \qquad (6.27)$$

This is the most general expression for the modified electrode rate constant when the rate of redox conduction is much greater than the rate of substrate transport through the polymer. We follow Albery and Hillman [280] and propose that $1 + (k'_E X_L/D_S)\tanh[(L/X_L)] \approx 1$. The reader is referred to the original publication for details of the logic behind this choice.

Let us now examine Eq. (6.27) in more detail. There are two terms in this expression for k'_{ME}. The denominator for the first term on the rhs of Eq. (6.27) contains, first, a term describing the transport of electrons across the layer and, second, a composite term quantifying the mediation at the surface or in the reaction layer X_L. The second major term on the rhs of Eq. (6.27) describes the reaction of substrate S directly at the electrode surface. The denominator consists of two contributions, one corresponding to the electrochemical kinetics of substrate at the electrode–polymer interface and the other a transport term due to the diffusion of S across the film to this inner interface. The numerator describes the partition of S into the layer via the κ coefficient. The latter is modified by the $\operatorname{sech}(L/X_L)$ factor. If $L \ll X_L$, then very little S is lost in its passage across the layer. Under such conditions $\operatorname{sech}(L/X_L) \approx 1$. On the other hand, if $L \gg X_L$, then $\operatorname{sech}(L/X_L)$ becomes very small and the second term in Eq. (6.27) may be neglected. In this case one considers only the first major term due to the mediation process.

Hence keeping the latter approximations in mind we can write that if $D_E b_0 \gg D_S \kappa s_\sigma$, then the modified electrode rate constant when $L \ll X_L$ is

$$k'_{\mathrm{ME}} = \left(\frac{s_\sigma L}{D_E b_0} + \frac{1}{k' b_0 + k\kappa b_0 L} \right)^{-1} + \left(\frac{L}{D_S} + \frac{1}{k'_E} \right)^{-1} \qquad (6.28)$$

We can invert Eq. (6.28) to obtain

$$\frac{1}{k'_{\mathrm{ME}}} = \frac{s_\sigma L}{D_E b_0} + \frac{1}{k' b_0 + k\kappa b_0 L} + \frac{1}{k'_E} + \frac{1}{D_S / L} \qquad (6.29)$$

The reciprocal form of the expression for k'_{ME} means that the slowest term (transport or kinetic) will be rate determining. Now we consider the case where $X_L \gg L$, and the kinetics are so slow that little substrate is lost in its passage across the layer. Hence the entire layer is used in the reaction. We note four terms on the rhs of Eq. (6.29). The first corresponds to electron hopping across the polymer. The second is a kinetic term and consists of a term ascribed to the kinetics of the surface reaction and a term due to the reaction between S and B in the bulk of the film. The third component is due to the direct, unmediated electrode reaction, whereas the final term on the rhs of Eq. (6.29) is due to substrate diffusion through the polymer matrix.

When $X_L \ll L$, $\operatorname{sech}(L/X_L) \to 0$, and Eq. (6.27) reduces to

$$k'_{\mathrm{ME}} = \frac{1}{s_\sigma L / D_E b_0 + 1/(k' b_0 + k\kappa b_0 X_L)} \qquad (6.30)$$

This corresponds to the situation where the kinetics are so fast that S only travels a distance X_L into the layer before it is destroyed by reaction with mediator species B. Again inverting Eq. (6.30), we note

$$\frac{1}{k'_{\mathrm{ME}}} = \frac{s_\sigma L}{D_E b_0} + \frac{1}{k' b_0 + k\kappa b_0 X_L} \qquad (6.31)$$

We are left with two possible rate-determining situations: electron transport and chemical reaction (either at the surface or in a thin reaction layer).

We now turn to the situation where the redox conductivity of the layer is poor and/or the layer exhibits a porous morphology enabling facile transport of substrate through the film. This corresponds to the situation where $D_E b_0 \ll D_S \kappa s_\sigma$. We solve Eq. (6.12) for u to obtain

$$u(\chi) = \cosh(\Phi_0 \chi) - \Phi_0^{-1} \frac{\lambda_B + \Phi_0 \tanh(\Phi_0)}{1 + \lambda_B \Phi_0^{-1} \tanh(\Phi_0)} \sinh(\Phi_0 \chi) \quad (6.32)$$

Also we note that

$$\left(\frac{du}{d\chi}\right)_0 = -\frac{\lambda_B + \Phi_0 \tanh(\Phi_0)}{1 + \lambda_B \Phi_0^{-1} \tanh(\Phi_0)} \quad (6.33)$$

We recall from Eq. (6.14) that $(dv/d\chi)_0 = \lambda_S v_0$. Hence we must evaluate v_0. We note that the concentration gradient of v in a region of the film (at $\chi = \chi^*$) close to the electrode but outside the reaction layer X_0 is given by

$$\left(\frac{dv}{d\chi}\right)_* = 1 - v_* \quad (6.34)$$

where $v^* \approx v_0$ denotes the dimensionless concentration of substrate close to the electrode surface but outside the reaction layer X_0. We can also show that

$$\left(\frac{dv}{d\chi}\right)_* = v_* \frac{k_E' L}{D_S} + \frac{k X_0 b_0 L}{D_S} \approx v_0 \frac{L}{D_S} (k_E' + k X_0 b_0) \quad (6.35)$$

where the two terms describe consumption of substrate S by direct reaction on the electrode and in the reaction layer of thickness X_0. From Eqs. (6.34) and (6.35) we obtain

$$v_0 \approx \frac{1}{1 + (L/D_S)(k X_0 b_0 + k_E')} \quad (6.36)$$

We recall from Eq. (6.14) that

$$\left(\frac{dv}{d\chi}\right)_0 = \lambda_S v_0 = \frac{k_E' L}{D_S} \frac{1}{1 + L(k X_0 b_0 + k_E')/D_S} \quad (6.37)$$

We now substitute the results obained in Eqs. (6.33) and (6.37) into Eq. (6.17) to obtain, after some algebraic manipulation, the following expression for the modified electrode rate constant*:

$$k'_{ME} = \frac{k\kappa b_0 X_0 \dfrac{k' + k\kappa X_0 \tanh(L/X_0)}{k' \tanh(L/X_0) + k\kappa X_0} + \kappa k'_E}{1 + (L/D_S)(k'_E + kb_0 X_0)} \tag{6.38}$$

Again this expression is rather complex. There are two terms in the numerator on the rhs of Eq. (6.38). The first of these describes mediated electron transfer and the second the kinetics for direct reaction at the electrode surface. The denominator on the rhs of Eq. (6.38) describes the concentration polarization of S in the layer where it may be consumed at the electrode surface by direct unmediated reaction described by the heterogeneous rate constant k'_E, or in a homogeneous reaction layer of dimension X_0. Let us now assume that the direct unmediated process may be neglected. If this is true, then one may simplify Eq. (6.38) as

$$\frac{1}{k'_{ME}} \approx \frac{L}{\kappa D_S} + \frac{k' \tanh(L/X_0) + k\kappa X_0}{k\kappa X_0 b_0 [k' + k\kappa X_0 \tanh(L/X_0)]} \tag{6.39}$$

Again we have transport and kinetic terms. The first term on the rhs of Eq. (6.39) defines the transport of substrate S across the layer. If this is rate determining, then we write that $k'_{ME} \approx \kappa D_S / L$. The second kinetic term is more complex and requires further analysis. If $L/X_0 \gg 1$ or $L \gg X_0$, then $\tanh(L/X_0) \approx 1$ and the expression for k'_{ME} reduces to $k'_{ME} \approx k\kappa X_0 b_0$, and we have mediation in a reaction layer of thickness X_0 near the support electrode surface. On the other hand, the situation becomes more complex when we start to decrease the value of the ratio L/X_0. When this ratio becomes small, if $L/X_0 \approx 1$, we obtain, for $k' > k\kappa X_0$,

$$\frac{1}{k'_{ME}} \approx \frac{\tanh(L/X_0)}{k\kappa X_0 b_0} = \frac{L/X_0}{k\kappa X_0 b_0} = \frac{L}{k\kappa X_0^2 b_0} = \frac{k\kappa s_\sigma L}{k\kappa D_E b_0} = \frac{s_\sigma L}{D_E b_0}$$

$$k'_{ME} \approx \frac{D_E b_0}{L y_\sigma}$$

and the flux will be limited by the transport of electrons. On the other

---

* The derivation of Eq. (6.38) requires some approximations that have been discussed in detail in the original paper by Albery and Hillman [280]. The reader is referred to this paper for full details.

hand, if $k' < k\kappa X_0$, then

$$\frac{1}{k'_{\mathrm{ME}}} \approx \frac{k\kappa X_0}{(k\kappa)^2 X_0^2 b_0 \tanh(L/X_0)} = \frac{1}{k\kappa X_0 b_0 (L/X_0)} = \frac{1}{k\kappa b_0 L}$$

$$k'_{\mathrm{ME}} \approx k\kappa b_0 L$$

and the reaction occurs throughout the layer. Finally, when $L/X_0 \ll 1$, and again if $k' > k\kappa X_0$, then we obtain $1/k'_{\mathrm{ME}} \approx (L/X_0)/k\kappa X_0 b_0$, and so, if we assume that $L/X_0 \approx k\kappa X_0/k'$, then the modified electrode rate constant is given by $k'_{\mathrm{ME}} \approx k'b_0$. This is the expression for a simple surface reaction. On the other hand, if $k' < k\kappa X_0$, then as before, we have $1/k'_{\mathrm{ME}} \approx 1/k\kappa X_0 b_0 (L/X_0)$, and if we further assume that $L/X_0 \approx k'/k\kappa X_0$, then we obtain $k'_{\mathrm{ME}} \approx k'b_0$, which is again the expression for a surface reaction. Hence for very low values of L/X_0 the reaction takes place on the surface rather than in a very thin layer.

There is one further possibility that we have not considered. This is the situation when $D_E b_0/\kappa D_S s_\sigma = X_0^2/X_L^2 \cong 1$. Under such conditions, if X_0 or X_L is less than the layer thickness L, then the reaction will take place in a reaction zone somewhere in the middle of the film, with a transport-controlled supply of electrons from the electrode and a transport-controlled supply of substrate from the solution. In this case one may show that

$$\frac{1}{k'_{\mathrm{ME}}} = \frac{L}{\kappa D_S + D_E b_0/s_\sigma} \tag{6.40}$$

In this case the modified electrode rate constant is given by $k'_{\mathrm{ME}} = \kappa D_S/L + D_E b_0/L s_\sigma$, and we see that we have joint rate control via transport of electrons and substrate. This case is called the layer reaction zone situation.

3. Albery–Hillman Kinetic Case Diagram

We can now summarize the situation. We have identified *six* different locations for the reaction of B and S. First, the reaction may take place at the film–solution interface with a rate constant k'. This is the *surface (S) case*. Second, the reaction may take place throughout the layer over the entire distance L. This is the *layer (L) case*. Third, the reaction may take place in a layer close to the support electrode surface over the distance of the reaction length X_0. This is the *layer–electrode (LE) case*. Fourth, the reaction may occur in a region close to the film–solution interface defined by a characteristic length X_L. This is termed the *layer–solution (LS) case*.

Fifth, the reaction may take place in some region near the middle of the film. This is called the *layer–reaction zone (LRZ) case*. Finally, the reaction of S can take place at the support electrode. This is the unmediated situation and we call this the *electrode (E) case*.

These six situations are outlined in a pictorial manner in Fig. 88. We can proceed still further. There are *ten* different possible cases involving the location of the reaction and the nature of the rate-determining step regardless of whether reaction kinetics or transport of electrons or substrate is used. Albery and Hillman [280] suggest the following notation. One can indicate the reaction zone via the designation L, LS, LE, S, E, or LRZ. The rate can then depend on the kinetics of the reaction in the layer k, the reaction at the polymer–solution interface k', the reaction at the electrode surface k'_E, or indeed the transport across the layer of electrons, t_E, or substrate, t_S. The 10 possible situations are outlined in Fig. 88. In this figure the location of the reaction zone is shaded. We also gather together the various expressions for k'_{ME} discussed previously in Table IV.

Albery and Hillman [280] have also introduced the following modified definitions of the reaction layer thicknesses:

$$X'_L = \left(\frac{D_S}{kb_0}\right)^{1/2} = \left(\frac{b_L}{b_0}\right)^{1/2} X_L$$

$$X'_0 = \left(\frac{D_E}{k\kappa s_\sigma}\right)^{1/2} = \left(\frac{s_0}{s_L}\right)^{1/2} X_0 \tag{6.41}$$

Clearly, when there is no concentration polarization of S or B in the film, then the modified definitions of the reaction layer thicknesses revert to the original definitions.

We now introduce the very important concept of a *kinetic case diagram*. This is a very useful device that enables one to present in a very concise manner the essential kinetic behavior of a complex system. We recall that the quantities X'_L/L and X'_0/L define the behavior of the six layer cases. We can then define and pictorially present the various limiting kinetic situations by constructing a diagram with X'_L/L and X'_0/L serving as axes. This type of representation is illustrated in Fig. 89. Also included in this figure are the various concentration profiles of mediator and substrate in the layer. A number of features may be noted. The 45° line $X'_L = X'_0$ divides the two major limiting cases defined via the definition of the parameter $\Psi = D_E b_0 / D_S \kappa s_\sigma$. When $\Psi > 1$ corresponding to the situation where $X'^2_0 > X'^2_L$, one is located in the southeast corner of the case diagram. Note that Eq. (6.27) is valid here. Alternatively, when

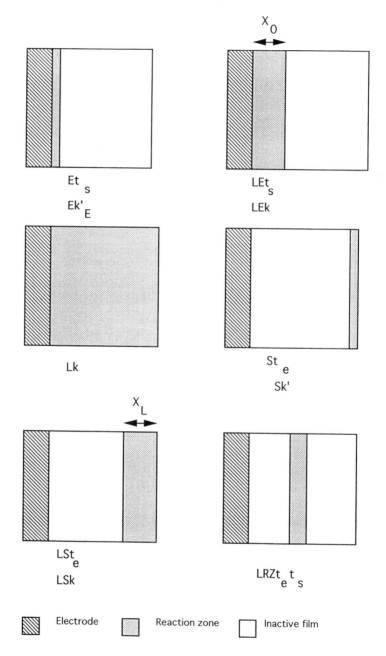

Figure 88. Pictorial representation of 10 cases considered by Albery and Hillman illustrating pertinent reaction zones.

TABLE IV
Albery–Hillman Case Notation for Electroactive Polymers

Location of Reaction Zone

L	Layer
LS	Layer/surface
LE	Layer/electrode
S	Surface
E	Electrode
LRZ	Layer reaction zone

Rate-Limiting Kinetics

t_S	Transport of substrate
t_e	Transport of electrons
k'	Surface reaction
k	Layer reaction
k'_E	Direct electrode reaction

Case Notation	k'_{ME}	Location

Limiting Expression for K'_{ME}

Case Notation	K'_{ME}	Location
Sk'	$k''b_0$	Surface reaction at
St_e	$D_E b_0 / Ls_\sigma$	layer–electrolyte interface
$LSk(SR)$	$\kappa\sqrt{kb_0 D_s}$	Reaction layer
$LSt_e(E)$	$D_E b_0 / Ls_\sigma$	close to film–electrolyte interface
$Lk(R)$	$\kappa k b_0 L$	Throughout layer
$LRZt_e t_s(S+E)$	$D_E b_0 / Ls_\sigma + \kappa D_s / L$	Narrow reaction zone in layer
$LEt_s(S)$	$\kappa D_s / L$	Reaction layer
$LEk(ER)$	$\kappa_{b0}\sqrt{D_E k / s_\sigma}$	close to electrode
Ek'_E	$\kappa k'_E$	Direct reaction on
Et_s	$\kappa D_s / L$	electrode

$\Psi < 1$ corresponding to ${X'_0}^2 < {X'_L}^2$, one moves to the northwest corner of the case diagram. In this region Eq. (6.38) pertains. The layer case Lk is located in the northeast quadrant of the case diagram, defined by the lines $X'_L = L$ and $X'_0 = L$. Let us now proceed to the southwest corner of the diagram. This part of the diagram is quite complex. The LS and LE cases are further subdivided by broken lines into a situation where the kinetics (k) are rate limiting and a case where either the transport of electrons (t_E) or substrate (t_S) is rate limiting. For the situation presented

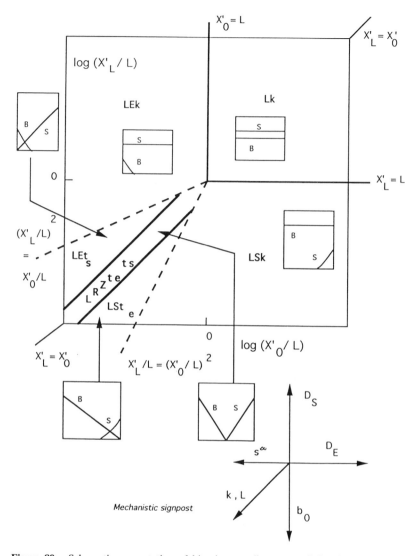

Figure 89. Schematic presentation of kinetic case diagram outlining interrelationship between six layer cases. Axes are log (X'_L/L) and log(X'_0/L), where characteristic lengths X'_L and X'_0 are defined in Eq. (6.41) in text. Solid lines separate cases for different reaction locations: L, throughout layer; LE near electrode–layer interface; LS, near layer–solution interface; LRZ, reaction zone located in middle of layer. Dashed lines subdivide LE and LS cases into regions where there is kinetic (k) or transport (t_e or t_s) control. Typical concentration profiles of substrate S and mediator B are shown as insets. Also included is mechanistic signpost showing how one moves through case diagram as one varies indicated parameters.

in Eqs. (6.27) and (6.29) the broken line will be located in a region determined by $D_E/s_\sigma L = k\kappa X_L$ or when $(X_0'/L)^2 = X_L'/L$. We have noted that in the southwest corner of the case diagram the reaction will be limited by the transport of either electrons or substrate. In this regon it will be the larger of the two transport fluxes that will be rate limiting. This is because the reaction zone locates itself in the layer in such a way that the slower moving species do not have as far to travel. The LRZ case is found in the middle of the southwest corner. It is bounded on either side by the t_E or the t_S cases. Note that in regions very close to the case boundaries one usually has to use the full expressions for the modified electrode rate constant [Eqs. (6.27) and (6.38)]. One can use the approximate expressions (such as those outlined in Table IV) when one moves away from the boundaries.

The case diagram presented in Fig. 89 is two dimensional. It deals with all of the layer cases. However, one must also consider the surface reactions at the layer–electrode and layer–solution interfaces. This will introduce a third dimension into the case diagram. In this case one can introduce the parameters

$$\phi_0 = \frac{k_E'}{kb_0 L} \qquad \phi_\sigma = \frac{k'}{\kappa k L} \tag{6.42}$$

The third axis is defiend by the ratio

$$\vartheta = \frac{\phi_\sigma}{\phi_0} = \frac{k'b_0}{\kappa k_E'} \tag{6.43}$$

Clearly $\vartheta > 1$ or $\vartheta < 1$. The corresponding three-dimensional case diagrams are illustrated in Figs. 90 and 91. These diagrams consist of the axes defining values of $\log \vartheta$, $\log(X_0'/L)$, and $\log(X_L'/L)$. The last four cases are covered when the third dimension is added, namely the Sk′, St$_E$, Ek$_E'$, and Et$_S$ situations. We note from these figures that the layer cases are located at the bottom of the three-dimensional case diagram. However, as one moves up the block diagram, surface processes become important. At the very top of the figure one has surface cases entirely. The surface case S appears on the south side of the block, whereas the electrode case E is located on the west side of the block. This arrangement is obtained because the surface reaction defined as S requires transport of electrons through the layer. This process is facile in the southeast quadrant. The electrode E case requires rapid transport of substrate S through the layer. This process will be kinetically facile in the northwest quadrant of the case diagram. At the very top of the block the

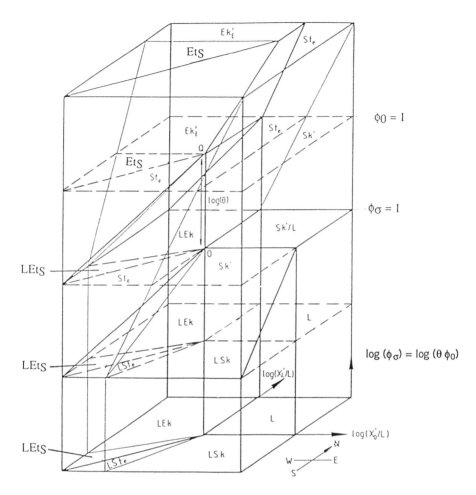

Figure 90. Three-dimensional case diagram indicating existence of layer cases (L), two electrode cases (ER), and two surface cases (S) as function of X_L'/L, X_0'/L, and ϕ_σ, where ϕ_σ is defined in Eq. (6.42). For diagram $\theta = \phi_\sigma/\phi_0 > 1$. Planes with thick solid lines separate cases with different reaction location: layer cases are the same as in Fig. 89 except that LRZ has been omitted for clarity. In electrode (E) cases reaction takes place by direct electron transfer to underlying support electrode while for surface (S) cases reaction takes place at the layer–solution interface. Planes with light solid lines subdivide cases into those with rate-limiting kinetics (k, k_E', or k') and those with rate-controlling transport (t_e or t_S).

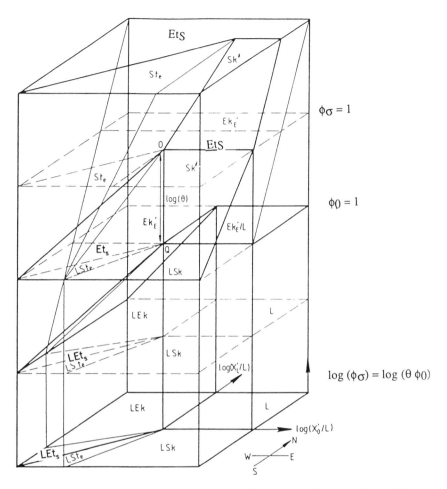

Figure 91. Three-dimensional kinetic case diagram according to Albery–Hillman approach. Block diagram is similar to that presented in Fig. 90 but now $\theta = \phi_\sigma/\phi_0 < 1$.

kinetics are so fast that the mediation reaction will be controlled by either the transport of electrons to the outer surface region (St_E) or the transport of substrate to the inner interface (ET_S). Consider the points O and Q in the block diagrams in Figs. 90 and 91. When $\vartheta > 1$ we see that Q is located above O, whereas when $\vartheta < 1$ the opposite pertains and Q is located below O. The point Q is associated wih the E regions and cases whereas the point O is associated with the S regions and cases. Hence the ϑ parameter quantifies the distance between O and Q and so defines the relation between the two classes of surface processes, those located at the inner and outer interfaces, respectively.

4. Experimental Determination of Diagnostic Parameters

One immediately notes that the parameters κ, k, k', D_S, and D_E are constant for any given polymer. However, the parameters s_σ, b_0, k'_E, and L may be experimentally varied by changing the rotation speed, electrode potential, and coating procedure, respectively. We can move through the case diagram by varying each of these parameters. Albery and Hillman [280] have proposed the following simple diagnostic scheme to label and identify the various rate-limiting possibilities for mediated electrocatalysis using electroactive polymer films. The procedure is based on the use of the rotating disc electrode and the Koutecky–Levich equation. The latter expression may be written as

$$\frac{nFAs^\infty}{i_L} = \frac{1}{B\omega^{1/2}} + \frac{1}{k'_{ME}} \qquad (6.44)$$

where B is the Levich factor given by

$$B = 1.55 D_S^{2/3} \nu^{-1/6} \qquad (6.45)$$

where ω represents the rotation speed in hertz and ν denotes the kinematic viscosity. As previously noted, the heterogeneous modified electrode rate constant k'_{ME} is obtained from the Koutecky–Levich intercept. Various expressions for k'_{ME} are outlined in Table IV, and the Albery–Hillman flow diagram for mechanistic diagnosis is outlined in Table V. The experimental quantities examined using this protocol are the rotation speed, the layer thickness L, and the mediator concentration b_0 at the electrode–polymer interface. The application of the flow diagram presented in Table V is apparent. For some specific circumstances, notably the LEk case, the Koutecky–Levich plot will be nonlinear, and so one must examine the following expression, termed the Albery–Hillman plot, for linearity:

$$\left(\frac{nFAs^\infty}{i_L}\right)^2 = \frac{nFAs^\infty}{i_L B\omega^{1/2}} + \left(\frac{1}{k'_{ME}}\right)^2 \qquad (6.46)$$

Hence we see that a plot of i_L^{-2} versus $i_L^{-1}\omega^{-1/2}$ is linear, with the intercept given by the quantity $k'_{ME}{}^{-2}$, where, from Table IV, we note that $k'_{ME} = \kappa b_0 (D_E k/s^\infty)^{1/2}$.

TABLE V
Albery–Hillman Diagnostic Scheme

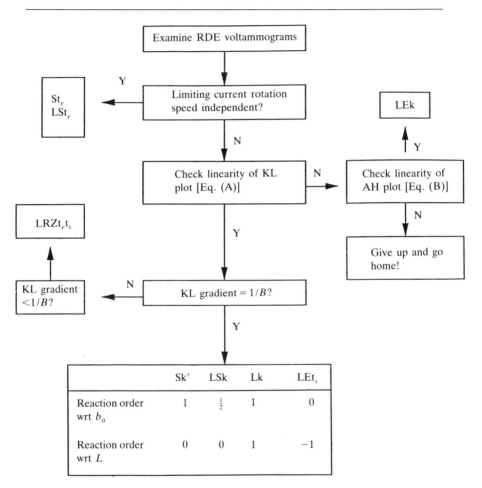

(A) Koutecky–Levich equation: $nFAs/i_L = B^{-1}\omega^{-1/2} + 1k'_{ME}$

(B) Albery–Hillman equation: $(nFAs/i_L)^2 = nFAs/o_L B\omega^{1/2} + (1/k'_{ME})^2$

5. *Optimizing Electrocatalysis*

Since we are considering electrocatalysis using multilayered films, it is reasonable to discuss the conditions for an optimum layer thickness. Let us assume that both substrate and charge may percolate through the film. Under such circumstances the value of k'_{ME} will initially increase with L (all sites mediate under kinetic control), then pass through a maximum, before finally decreasing with still larger L. In the latter situation only some sites mediate under transport control. Dependin on the ratio $\Psi = D_E b_0 / \kappa D_S s_\sigma$, this will correspond to the sequence of cases $Lk \rightarrow LEk \rightarrow LEt_S$ or $Lk \rightarrow LSk \rightarrow LSt_E$. Albery and Hillman [280] have shown that the optimum cases are LEk or LSk with $L \cong 3X_0$ or $3X_L$. The bottom line is that these conditions imply that the layer is thick enough to provide sufficient mediator sites for the consumption of S without constituting a barrier to substrate or charge diffusion. What about the situation where charge or substrate is unable to penetrate through the film? In this case the film should be thin. One should therefore aim for a surface reaction and to reduce the layer thickness L such that the transport of electrons or substrate through the layer is efficient. If the layer is thin enough, then one will have a direct competition between the rate of the surface reaction $k'b_0$ and the electrode reaction $\kappa k'_E$. The optimum cases here are the electrode or surface reaction under kinetic control (Ek'_E or Sk' cases depending on the species that is able to penetrate the film). The former corresponds to the zero mediation or direct reaction process, whereas the latter corresponds to mediation without partition. One should therefore aim for a surface reaction and to reduce the layer thickness L such that the transport of electrons or substrate through the layer is efficient.

Outlined in Table VI are the results of a calculation carried out by Albery and Hillman [280], who outline the optimum values of the layer thickness for most efficient mediated electrocatalysis. One can perform some very simple calculations to obtain quantitative estimates of the catalytic advantage in using a polymer-modified electrode using the results outlined in this table.

Albery and Hillman [280] have assumed that typically, for efficient mediation, the modified electrode rate constant k'_{ME} must be $\sim 10^{-2} \, \text{cm s}^{-1}$. Furthermore the rate constant for the mediated process at the polymer-coated electrode k'_{ME} must be greater than that for the direct unmediated process k'_E. We see from Table VI that when $\Psi \gg 1$ the optimum case is LSk, with $k'_{ME} = k\kappa X_L b_0$ and $L \approx 3X_L$. When $\Psi = 1$, $L \approx X_0 \approx X_L$ and $k'_{ME} = k\kappa (X_0 X_L)^{1/2} b_0$. Furthermore, when $\Psi \ll 1$, the optimum case is LEk, with $L \approx 3X_0$ and $k'_{ME} = k\kappa X_0 b_0$. For the surface

TABLE VI
Criteria for Layer Thickness L for Optimum Performance of Modified Electrode

Conditions	Optimum case	Condition for L	k'_{ME}
$D_e b_0 \gg \kappa D_s s_\sigma$ and $k\kappa X_L > k' + \kappa k'_E/b_0$	LSk	$L \approx 3X_L$	$k\kappa X_L b_0$
$D_e b \cong \kappa D_s s_\sigma$ $\kappa X_0 \cong \kappa X_L > k' + \kappa k'_E/b_0$	Not known	$L \approx X_0 \approx X_L$	$k\kappa X_0^{1/2} X_L^{1/2} b_0$
$D_e b_0 \ll \kappa D_s s_\sigma$ and $k\kappa X_0 > k' + \kappa k'_E/b_0$	LEk	$L \approx 3X_0$	$k\kappa X_0 b_0$
$k' > \kappa k'_E/b_0$	Sk'	$L < D_e/k's_\sigma$	$k'b_0$
$k' < \kappa k'_E/b_0$	Ek'$_E$	$L < D_s/k'_E$	$\kappa k'_E$

reactions (Sk' and Ek'$_E$) the reaction takes place at the film–solution and electrode–film interfaces, respectively. The Sk' case occurs for $L < D_E/k's_\sigma$ and for $k' > \kappa k'_E/b_0$, whereas the Ek'$_E$ case occurs for an optimal L value given by $L < D_S/k'_E$ and for $k' < \kappa k'_E/b_0$. In the former we note that $k'_{ME} = k'b_0$, whereas in the latter case we have that $k'_{ME} = \kappa k'_E$. In these cases one can show that the catalytic advantage is given by

$$\frac{k'_{ME}}{k'_E} = \kappa + \frac{k'b_0}{k'_E} \qquad (6.47)$$

Albery and Hillman [280] have shown that very little catalytic advantage is obained for these surface reactions. This of course is what one would expect. They came to the rather gloomy conclusion that for surface reactions a reasonable current density would only be obtained for reactions that are already rapid under homogeneous conditions. Hence the interfacial reactions are unlikely to yield much catalytic advantage unless the partition coefficient $\kappa > 1$ or if the bound mediator species B is a specific catalyst for the S-to-P conversion. A reasonable turnover requires that the second-order homogeneous rate constant between B and S be greater than $\sim 10^4 \, dm^3 \, mol^{-1} \, s^{-1}$.

Considering the optimum layer case LSk, we note that $k'_{ME} = \kappa(D_s k b_0)^{1/2}$, and hence the catlytic advantage is

$$\frac{k'_{ME}}{k'_E} \cong \frac{\kappa X_L}{\delta} \qquad (6.48)$$

where δ is a characteristic distance of magnitude $\sim 1 \, nm$. Hence we see that the ratio X_L/δ may be quite large, and so one can obtain quite a significant catalytic advantage if one used a multilayer system. Note that the bottom line in this situation is the value of D_S, the substrate diffusion coefficient in the film. If we assume that $\kappa = 1$ and $b_0 = 10\,M$, then for $k'_{ME} = 10^{-2}\,\text{cm s}^{-1}$ one can determine the catalytic activity and efficiency of a modified electrode for various values of the substrate diffusion coefficient D_S. This type of calculation is illustrated in Table VII. The point to note from this table is that D_S should be as large as possible. Hence the electroactive polymer should have a very open morphology that affords the facile transit of substrate through the matrix. A ballpark figure for the lowest allowed value for D_S is $10^{-9}\,\text{cm}^2\,\text{s}^{-1}$. Furthermore it is necessary that D_E should be similar to D_S, otherwise the rate will be limited by electron percolation through the layer.

6. Comparison with Experiment

Forster and Vos [335] examined the mediated reduction of $[\text{Fe}(\text{H}_2\text{O})_6]^{3+}$ at an $[\text{Os}(\text{bpy})_2\text{PVP}_{10}\text{Cl}]\text{Cl}$ film in 0.1 M H_2SO_4. The mediated reduction proceeds as follows:

$$\text{Fe(III)} + \text{Os(II)} \rightarrow \text{Fe(II)} + \text{Os(III)}$$

The standard potential of the Os(III/II) redox couple is 250 mV versus SCE (Fig. 92a). The formal potential for the FE(III/II) redox couple is 460 mV. Hence the driving force for the reaction is some 210 mV. Typical RDE voltammograms for the mediated reduction of 0.2 mM Fe^{3+} (aq) at the osmium-containing polyvinylpyrrolidone (PVP) polymer film are outlined in Fig. 92b. A series of Koutecky–Levich (KL) plots as a function of surface coverage Γ are illustrated in Fig. 92c. Also illustrated in this figure as an inset is the voltammetric response of the polymer layer (surface coverage $\Gamma = 1.7 \times 10^{-8}\,\text{mol cm}^{-2}$) in supporting electrolyte (0.1 M H_2SO_4) in the absence of any Fe^{3+} (aq) substrate. It is clear from these data that mediation occurs in the potential region where the

TABLE VII
Performance of an optimum layer electrode (LSk case)

D_s (cm^2 s^{-1})	k (dm^3 mol^{-1} s^{-1})^a	X_L (cm)^b	k'_{ME}/k'_E ^{c}
10^{-9}	10^4	10^{-7}	1
10^{-7}	10^2	10^{-5}	10^2
10^{-5}	1	10^{-3}	10^4

^{a} Value of k calculated from the expression $k'_{ME} = \kappa\sqrt{D_s k b_0}$ assuming that $k'_{ME} = 10^{-2}\,\text{cm s}^{-1}$ and setting $\kappa = 1$ and $b_0 = 10\,\text{mol dm}^{-3}$.
^{b} Optimum thickness of layer.
^{c} Catalytic advantage calculated using Eq. (6.48).

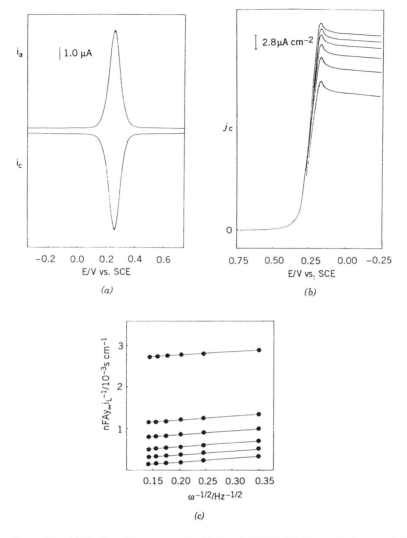

Figure 92. (a) Cyclic voltammogram for [Os(bpy)$_2$(PVP)$_{10}$Cl] Cl modified electrode in 0.1 M H$_2$SO$_4$ as supporting electrolyte. Sweep rate, 5 mV s^{-1}; surface coverage $\Gamma = 1.7 \times 10^{-8}$ mol cm^{-2}. (b) RDE voltammograms for reduction of [Fe(OH$_2$)$_6$]$^{3+}$ ion (0.2 mM) in 0.1 M H$_2$SO$_4$ at Pt electrode coated with redox active metallopolymer. Surface coverage, 5.0×10^{-9} mol cm^{-2}. Rotation speeds from bottom to top are 500–3000 rpm at 500-rpm increments. (c) Typical Koutecky–Levich plots for reduction of [Fe(OH$_2$)$_6$]$^{3+}$ ion (0.2 mM) in 0.1 M H$_2$SO$_4$ at Pt electrode coated with redox active metallopolymer as function of surface coverage of latter. Surface coverages (expressed in mol cm^{-2}) are, from top to bottom, 7.0×10^{-10}, 1.8×10^{-9}, 2.7×10^{-9}, 5.0×10^{-9}, and 1.1×10^{-8}. Lowest curve corresponds to that for Fe^{3+} (aq) ion reduction at bare Pt. (Adapted from ref. 335.)

Os(III/II) redox transition occurs. Furthermore, the KL behavior is markedly thickness dependent. The KL intercept *decreases* with *increasing* surface coverage; hence the rate of the catalytic reaction increases with increasing layer thickness. This effect is quantified in Fig. 93 where the modified electrode rate constant k'_{ME} is plotted as a function of L for a given Fe^{3+} (aq) concentration (0.2 mM). We see that the reaction order of unity is obtained. The slopes of the KL plots were constant for all layer thicknesses studied, and the mean value, $9.4 \pm 0.8 \times 10^{-4}$ cm s$^{-1/2}$, corresponded well with the value obained for Fe^{3+} reduction at a bright unmodified Pt disc electrode in the same solution. We now turn to Table V, where one can apply the Albery–Hillman flowchart [280] for mechanism diagnosis. Proceeding down this chart we see that the KL gradient is indeed given by B^{-1} and that a first-order dependence on layer thickness is observed. Hence the St$_E$, LSt$_E$, LEk, and LRZt$_E$t$_S$ cases can all be liminated from consideration. The first-order dependence on L observed eliminates the Sk', LSk, and LEt$_S$ cases. Hence the kinetics are described in terms of a Lk mechanism.

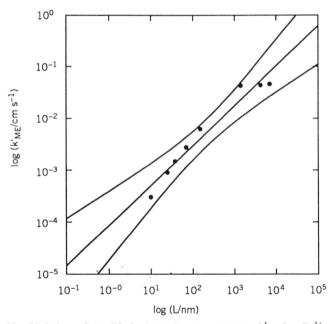

Figure 93. Variation of modified electrode rate constant k'_{ME} for Fe^{3+} (aq) ion reduction with layer thickness L for Os-loaded redox active metallopolymer described in Fig. 92. Least-squares linear regression line drawn through data. Also included are computer calculated 98% confidence limits corresponding to experimental data. Supporting electrolyte, 0.1 M H_2SO_4.

This assignment may be confirmed by examining the variation of k'_{ME} with the mediator concentration b_0. For the Lk mechanism a first-order dependence of k'_{ME} on b_0 should be observed. Now b_0 may be estimated by determining the concentration of Os(II) within the layer as a function of potential. This in general can be rather difficult. However, we note from the voltammetric response for the metallopolymer that the film exhibits a voltammetric profile characteristic of so-called surface behavior. This means that under the experimental conditions employed by Forster and Vos [335], there is no concentration polarization of B in the layer: All the redix centers are oxidized or reduced during the time scale of the voltammetric experiment. Consequently it is very likely that the redox switching process governing the Os(III/II) transformation is well described by the Nernst equation. One can write that the fraction x of the film that is oxidized [Os(III)] is given by

$$\ln\left(\frac{x}{1-x}\right) = \frac{F}{RT}(E - E^0) \qquad (6.49)$$

The ratio [Os(III)]/[Os(II)] = $x/(1-x)$ was determined via controlled potential coulometry. The latter protocol involved incrementing the electrode potential negatively from 1000 mV, where the film is fully oxidized, to -400 mV, where the film is fully reduced. The value of each increment was 50 mV and each potential increment was held for 5 min during which time the cathodic current was integrated. A correction for background charging was also applied. The results of this work are illustrated in Fig. 94. The slope of this Nernst plot is 58 ± 2 mV dec^{-1}, which is in excellent agreement with the theoretical Nernst slope of 59.6 mV dec^{-1}. The quantity $1 - x$ may be obtained from Eq. (6.49). A plot of k'_{ME} versus $1 - x$ is illustrated in double logarithmic form in Fig. 95. The slope of this plot is unity, thus confirming that the kinetics are

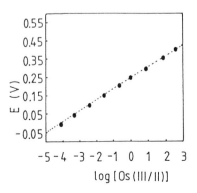

Figure 94. Nernst plot illustrating variation of redox potential with log[Os(III)/Os(II)] in 0.1 M H$_2$SO$_4$ supporting electrolyte. Dashed line indicates theoretical Nernst response.

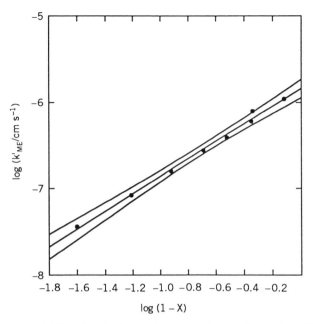

Figure 95. Double-logarithmic plot used to determine reaction order with respect to mediator concentration b_0. Mole fraction of Os(II) sites, $1 - X$, determined from Nernst plot illustrated in Fig. 94. Refer to text for further details of analysis. Supporting electrolyte is $0.1 M$ H_2SO_4. Computer-fitted least-squares linear regression line along with 98% confidence limits are illustrated.

Lk. Hence the entire layear is being utilized in the reaction. The second-order rate constant k for the reaction between B and S may be obtained from $k'_{ME} = \kappa k L b_0$. Hence one obtains that $\kappa k = 5.6 \times 10^2 M^{-1} s^{-1}$. For larger surface coverages $\Gamma > 10^{-7} mol\, cm^{-2}$, the kinetic zone passes from the Lk case and the limitin current becomes independent of layer thickness. In such a situation rate control passes to that of simple diffusional transport of the substrate in the solution to the polymer surface.

Forster and Vos [335] also examined the Fe^{3+} (aq) mediation process in 1.0 $HClO_4$. Previous studies [336] have indicated that the presence of perchlorate salts serve to dehydrate the metallopolymer and make the morphology of the layer considerably more compact. Hence one might expect that one would have inhibited penetration of FE^{3+} (aq) into the polymer matrix. Hence a decrease in permeation would serve to decrease the reaction layer thickness X_L and lead to a changeover in kinetic zone from Lk to that of a surface type. Koutecky–Levich plots for this system are illustrated in Fig. 96. These are all linear and have the same slope as

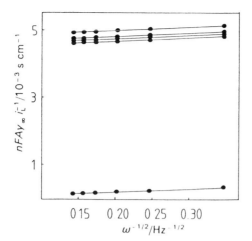

Figure 96. Typical Koutecky–Levich plots for reduction of 0.2 mM Fe^{3+} (aq) ion in 1.0 M $HClO_4$ supporting electrolyte at Os-loaded redox metallopolymer. Surface coverages (in mol cm^{-2}) are, from top to bottom, 7.0×10^{-10}, 1.8×10^{-9}, 2.7×10^{-9}, 5.0×10^{-9}, and 1.1×10^{-8} and bare Pt. (Adapted from ref. 335.)

that observed for a bare electrode. However, when one examines the dependence of k'_{ME} on layer thickness L (Fig. 97), one notes that the reaction order is zero with respect to L. From the diagnostic scheme in Table IV we see that the two possibilities are Sk' or LSk. Both of these are surface cases. To distinguish between these two possiblities, one must, as before, examine the dependence of k'_{ME} on mediator concentration b_0. From Table IV we see for the Sk' case a reaction order of $\frac{1}{2}$ is expected, whereas for the LSk situation the reaction order is unity. A typical Nernst-type plot obtained via potential step coulometry is illustrated in Fig. 98. In this case the plot deviates significantly from linearity when we have very reduced or very oxidized layers. Hence the thermodynamics of the Os(III/II) transformation in perchlorate media is rather complex. This may be due to the operation of repulsive interactive effects between the redox centers in the polymer layer. The slope of the Nernst plot is approximately 59 mV dec^{-1} only in the region near the formal potential E^0. One oberves deviations from this behavior for small and large Os(III)/Os(II) ratios. One can therefore utilize the Nernst part of the plot to determine b_0. By examining the rising portion of the RDE voltammogram in conjunction with this Nernst plot data, the dependence of k'_{ME} on b_0 may be determined. The results are outlined in Fig. 99. A good linear plot is obtained with slope 1.05 ± 0.2. Hence to a good approximation the reaction order with respect to b_0 is unity and so the kinetic case is of the Sk' type. For a fully reduced layer where $x = 0$ one

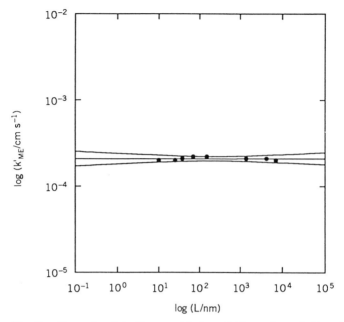

Figure 97. Reaction order plot with respect to layer thickness for reduction of 0.2 mM Fe^{3+} (aq) ion in 1.0 M $HClO_4$ supporting electrolyte at Os-loaded redox metallopolymer. Linear regression line and 98% confidence limits are indicated. Note that k'_{ME} is independent of layer thickness in this case.

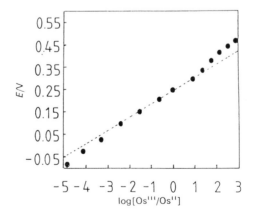

Figure 98. Nernst plot obtained via potential step coulometry for metallopolymer in 1.0 M $HClO_4$. Dashed line in theoretically predicted Nernst response. [335].

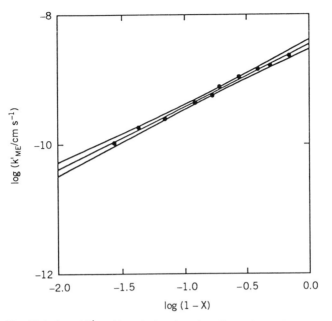

Figure 99. Variation of k'_{ME} with mole fraction of mediator, $1 - X$. Data extracted from analysis of Fig. 98 and Koutecky–Levich plots. Regression line and 98% confidence limits are indicated.

can see from Fig. 99 that $k'_{ME} = 2.9 \pm 0.2 \times 10^{-4} \, \text{cm s}^{-1}$. For the Sk' case $k'_{ME} = k'b_0$, where k' is the second-order rate onstant for the reaction between the surface-bound mediator and the Fe^{3+} (aq) species in solution. If one assumes that the fixed-site Os concentration within the layer is $0.7 \, M$, then the bimolecular surface rate constant $k' = 0.31 \, \text{cm}^4 \, \text{mol}^{-1} \, \text{s}^{-1}$. This value of the homogeneous rate constant is similar to that obtained by Albery and co-workers [337] for the reaction of Fe^{3+} (aq) at a polythionine-modified electrode.

VII. CATALYTIC SYSTEMS UTILIZING ELECTROACTIVE POLYMER FILMS THAT EXHIBIT COMPLEX MICHAELIS–MENTEN KINETIC BEHAVIOR

A. Introduction

In this section of the review we discuss a situation where the substrate reaction is complex and may be described in terms of Michaelis–Menten kinetics. Again, the major emphasis will be on the development of reasonably simple analytical models to describe the catalytic current response under steady-state conditions.

Consider the situation where one has catalytic particles dispersed in a reasonably thin polymeric film, where the substrate–product reaction occurs via Michaelis–Menten kinetics [338]. This problem is directly relevant to the associated problems of immobilied enzyme catalysis [339, 340] and diffusion–chemical reaction processes in chemical engineering [341]. Aspects of the theory presented in this section have recently been described by Albery and co-workers [342] for enzyme electrodes.

B. Lyons–Bartlett Model

A schematic of the model is outlined in Fig. 100. A physical example is the distribution of RuO_2 particles in a Nafion film for the electrooxidation of catechol [310, 330]. In the following analysis, the effect of substrate concentration polarization in solution will be ignored, and we shall concentrate on the processes of matter transport and chemical reaction within the polymer film. We further assume that electronic transport between the dispersed catalytic particles is rapid.

The differential equation describing the transport and kinetics in the

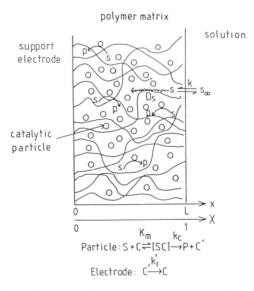

Figure 100. Schematic representation of composite electrocatalytic system consisting of dispersion of catalytic microparticles in polymeric matrix.

layer is given by

$$D_S \frac{d^2s}{dx^2} - ks = 0 \qquad (7.1)$$

where s denotes the concentration of substrate and D_S is the substrate diffusion coefficient in the layer. We assume that the substrate reacts with the catalyst particle via Michaelis–Menten (or in the parlance of surface chemistry, Langmuir–Hinshelwood) kinetics, according to the scheme

Catalytic particle: $\quad S + C \xrightarrow{K_M} [SC] \xrightarrow{k_c} P + C'$

Electrode: $\qquad\qquad C' \xrightarrow{k'_E} C$

where C and C' represent the catalytically active form and the pre-catalytic form, respectively. We are again considering heterogeneous redox catalysis. Note that [SC] denotes the substrate–catalyst complex. In the Michaelis–Menten scheme the second-order rate constant k quantifying the reacton between S and C is concentration dependent and takes the form

$$k = k(s) = \frac{k_c c_\Sigma}{K_M + s} \qquad (7.2)$$

In this expression k_c denotes the catalytic rate constant (units: reciprocal seconds), K_M is the Michaelis constant (units: moles per cubic centimeters), and c_Σ denotes the total catalyst concentration in the layer. If this expression is substituted into Eq. (7.1), we obtain

$$D_S \frac{d^2s}{dx^2} - \frac{k_c c_\Sigma s}{K_M + s} = 0 \qquad (7.3)$$

This is a nonlinear differential equation due to the presence of Michaelis–Menten term. In order to obtain expressions for the substrate concentration profiles in the film and the steady-state current response, one must integrate Eq. (7.4) subject to the boundary conditions

$$x = 0 \ \frac{ds}{dx} = 0 \qquad x = L \ s = \kappa s^\infty \qquad (7.4)$$

where s^∞ denotes the bulk concentration of the substrate and κ denotes the partition coefficient.

In order to make further progress, it is useful to transform the differential equation outlined in Eq. (7.3) and the boundary conditions presented in Eq. (7.4) into nondimensional form. To do this, we introduce the normalized variables

$$u = \frac{s}{\kappa s^{\infty}} \qquad \alpha = \frac{\kappa s^{\infty}}{K_M} \qquad \chi = \frac{x}{L} \qquad (7.5)$$

where L denotes the layer thickness. We also introduce a parameter called the Thiele modulus, given by

$$\Phi^2 = \frac{k_c c_{\Sigma} L^2}{K_M D_S} \qquad (7.6)$$

It is instructiave to note that the first-order rate constant for the chemical transofrmation of substrate to product, k', is given by

$$k' = \frac{k_c c_{\Sigma}}{K_M} \qquad (7.7)$$

Hence, we note that the Thiele modulus may be recast in the form

$$\Phi^2 = \frac{k' L^2}{D_S} = \frac{k' L}{k'_D} = \frac{k' L s^{\infty}}{D_S s^{\infty}/L} = \frac{j_R}{j_D} \qquad (7.8)$$

In this expression k'_D represents the diffusional rate constant for the substrate in the layer. We note from Eq. (7.8) that Φ^2 is simply the ratio of the maximum chemical conversion rate in the layer given by the reaction flux j_R to the maximum diffusional transport rate in the layer given by the flux j_D. We can obtain another useful relationship from Eq. (7.8) as follows:

$$\Phi = L\left(\frac{k'}{D_S}\right)^{1/2} = \frac{L}{X_K} \qquad X_K = \left(\frac{D_S}{k'}\right)^{1/2} \qquad (7.9)$$

where the reaction layer thickness X_K is simply the distance into the layer the substrate travels before undergoing reaction with the catalyst. This concept was met before in Section VI when the Albery–Hillman theory [280] was discussed. Hence we note that the Thiele modulus is simply the ratio of the layer thickness to the reaction layer thickness.

The important parameters defining the behavior of the system are α and Φ. We see that α defines the ratio of the substrate concentration in the layer to the Michaelis constant and Φ quantifies the extent of the reaction zone in the film. We shall shortly show that these two parameters may be used as axes defining a kinetic case diagram for the system.

It is easily shown that the master equation describing transport and kinetics in the polymer matrix may be transformed to the nondimensional form

$$\frac{d^2u}{d\chi^2} - \frac{\Phi^2 u}{1 + \alpha u} = 0 \qquad (7.10)$$

The boundary conditions transform to

$$\chi = 0 \qquad \frac{du}{d\chi} = 0 \qquad \chi = 1 \qquad u = 1 \qquad (7.11)$$

We may also introduce a dimensionless flux parameter y given by

$$y = \alpha \left(\frac{du}{d\chi} \right)_{\chi = 1} \qquad (7.12)$$

This dimensionless parameter y is related to the flux j (units: mol cm^2 s^{-1}) via the expression

$$j = \frac{i}{nFA} = \frac{K_M D_S y}{L} \qquad (7.13)$$

Note that the expression outlined in Eq. (7.10) is nonlinear, and so in order to facilitate discussion, we will consider a number of *approximate analytical solutions* to this expression subject to the boundary conditions previously presented. For instance, when $\alpha u \ll 1$ corresponding to $\kappa s^\infty \ll K_M$, we have the situation of *unsaturation*. In this case the system exhibits simple first-order kinetics with respect to the bulk concentration of substrate. In contrast, when $\alpha u \gg 1$ corresponding to $\kappa s^\infty \gg K_M$, *saturation* conditions apply and the observed kinetics are zero order with respect to substrate concentration. Furthermore, the finite dimension of the layer must be taken into account. This is accomplished by examination of the L/X_K ratio. Consequently $L/X_K \ll 1$ represents the thin-film situation where reaction takes place within the entire film. The reaction zone is quite thick. In this case no concentration polarization of substrate exists within the film. In contrast, when $L/X_k \gg 1$, we are working with thick films and the reaction occurs in a thin zone near the polymer–solution interface. We note therefore that the system may be characterized by a case diagram that consists of a plot of $\log \Phi$ versus $\log \alpha$.

C. Thin-Film Limit: Neglect of Concentration Polarization of Substrate in Layer

Let us initially examine the situation of a very thin film. In this case the concentration of substrate is uniform throughout the layer, and we may set $u = 1$ in Eq. (7.10). Hence we obtain

$$\frac{d^2u}{d\chi^2} = \frac{\Phi^2}{1+\alpha} \tag{7.14}$$

Integration of this expression between the limits of $\chi = 0$ and $\chi = 1$ results in the assignment

$$\frac{du}{d\chi} = \frac{\Phi^2\chi}{1+\alpha} \tag{7.15}$$

The normalized flux y is therefore given by

$$y = \alpha\left(\frac{du}{d\chi}\right)_{\chi=1} = \frac{\alpha\Phi^2}{1+\alpha} \tag{7.16}$$

Retransforming back into dimensioned parameters, we obtain that the reaction flux j is given by

$$j = \frac{i}{nFA} = \frac{k_C c_\Sigma L\kappa s^\infty}{K_M + \kappa s^\infty} \tag{7.17}$$

This of course is the simple Michaelis–Menten equation that is well known in enzyme kinetics. This expression describes the flux for the reaction of the substrate at the catalyst particle when there is no concentration polarization of substrate in the film. The expression presented is valid for all values of substrate concentration.

D. Low-Substrate-Concentration Limit

We now examine the situation when there is concentration polarization of subtrate in the layer. We first look at the situation of low substrate concentration, the unsaturated situation where $\kappa s^\infty \ll K_M$. Hence $\alpha u \ll 1$, and we have that $1 + \alpha u \approx 1$, and so the master expression given in Eq. (7.10) reduces to

$$\frac{d^2u}{d\chi^2} - \Phi^2 u = 0 \tag{7.18}$$

Hence we see that the problem reduces to that of diffusion–reaction with first-order kinetics. The solution to this equation is given by

$$y = \alpha \Phi \tanh \Phi \tag{7.19}$$

and the concentration profile for substrate in the layer is given by

$$u(\chi) = \frac{\cosh(\Phi\chi)}{\cosh \Phi} = \cosh(\Phi\chi)\mathrm{sech}(\Phi) \tag{7.20}$$

This concentration profile is illustrated in Fig. 101 as a function of the Thiele modulus Φ. We note from this diagram that when Φ is small the concentration profiles are fairly shallow, and the entire layer is being utilized. We can see this by noting that

$$\mathrm{sech}(\Phi)\cosh(\Phi\chi) \approx (1 - \tfrac{1}{2}\Phi^2)(1 + \tfrac{1}{2}\Phi^2\chi^2) \approx 1 \tag{7.21}$$

There is little concentration polarization of substraate within the film. The unsaturated catalytic kinetics need only be considered, and we can

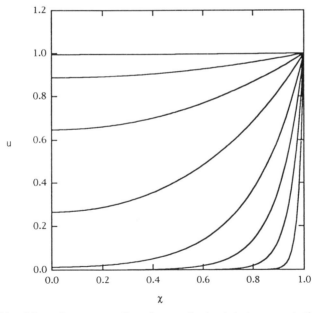

Figure 101. Schematic representation of normalized substrate concentration profiles $u(\chi)$ computed form Eq. (7.20) as function of normalized distance $\chi = x/L$ for various values of Thiele modulus $\Phi = L/X_K$. From top to bottom: $\Phi = 0.1$, 0.5, 1.0, 2.0, 5.0, 10.0, 20.0, 50.0. Valid for unsaturated conditions where $\alpha \ll 1$.

neglect substrate diffusion effects due to the fact that the latter are rapid. However, when Φ is large the concentration of substrate falls rapidly with distance into the polymer film. The reaction is essentially complete within a thin reaction zone (thickness X_K) near the polymer–solution interface. This statement may be readily illustrated by examining the form of the concentration profile when Φ is large. Under such conditions we can make the following approximations and write $\cosh(\Phi\chi) = \frac{1}{2}\exp(\Phi\chi)$ and $\operatorname{sech}(\Phi) \approx 2\exp(-\Phi)$, and so we write that $u \approx \exp[-\Phi(1-\chi)]$. Physically, this expression corresponds to an exponential decay in substrate concentration from an initial value $u = 1$ at $\chi = 1$, with a time constant given by the Thiele modulus Φ.

This discussion may be amplified by examining Eq. (7.19) in more detail. When $\Phi < 0.3$, one notes that $\tanh\Phi \approx \Phi$. In this case assuming that $\Phi \ll 1$ (i.e., $L \ll X_K$), the expression for the flux outlined in Eq. (7.19) reduces to

$$y = \alpha\Phi^2 \tag{7.22}$$

We can retransform into a flux as follows:

$$j = \frac{i}{nFA} = k_c c_\Sigma K_M^{-1} L \kappa s^\infty \tag{7.23}$$

Hence, provided that $s^\infty < K_M$, and we use thin films, the steady-state current response will depend in a linear manner on the substrate concentration, the catalyst concentration in the layer, and the layer thickness. The reaction zone will encompass the entire layer, and the rate of reaction at the catalyst surface will be much lower than the rate of substrate transport. In this sense, when $\Phi \ll 1$, the rate is reaction controlled.

Alternatively, for large Φ ($\Phi > 2$) we may show that $\tanh\Phi \approx 1$. This is the case for thick films when $L \gg X_K$. In this case the flux expression reduces to

$$y = \alpha\Phi \tag{7.24}$$

which may again be transformed into a flux expression:

$$j = \frac{i}{nFA} = \left(\frac{k_c c_\Sigma D_S}{K_M}\right)^{1/2} \kappa s^\infty \tag{7.25}$$

In the case where $s^\infty \ll K_M$ and if the layer thickness is large, the current will be independent of layer thickness and will be linearly dependent on substrate concentration. It will also vary as the square root of the catalyst loading. In this case there is a thin reaction layer, and the reaction kinetics at the particle surface will be much more rapid than the diffusional transport of the substrate through the film. Therefore the reaction will be diffusion controlled, and there will be considerable concentration polarization of substrate through the film. This fact is clearly indicated in the concentration profiles in Fig. 101.

E. High-Substrate-Concentration Limit

The situation at high substrate concentrations is now examined when $\alpha u \gg 1$ or $s^\infty \gg K_M$, in which the catalyst is saturated by substrate. This saturation condition results in the observation of zero-order kinetics. When $\alpha u \gg 1$, $1 + \alpha u \approx \alpha u$, and the master equation reduces to

$$\frac{d^2 u}{d\chi^2} - \frac{\Phi^2}{\alpha} = 0 \tag{7.26}$$

This expression may be readily integrated with respect to χ to obtain

$$\frac{du}{d\chi} = \frac{\Phi^2 \chi}{\alpha} \tag{7.27}$$

Further integraton of Eq. (7.27) results in the concentration profile

$$u(\chi) = 1 + \frac{\Phi^2}{2\alpha}(\chi^2 - 1) \tag{7.28}$$

The latter expression for the concentration profile will be valid for $\Phi^2 < 2\alpha$. Typical concentration profiles are presented in Fig. 102 for various Φ values.

The normalized flux is readily obtained from Eq. (7.27) as follows:

$$y = \Phi^2 \tag{7.29}$$

Introduction of the definition of Φ^2 from Eq. (7.9) obtains the following expression for the steady-state current response:

$$j = \frac{i}{nFA} = k_c c_\Sigma L \tag{7.30}$$

Thus, under conditions of reactant saturation, the current response is independent of substrate concentration but is first order with respect to layer thickness and catalyst loading. The rate-determining step involves breakdown of the catalyst–substrate complex and is quantified by the rate constant k_c.

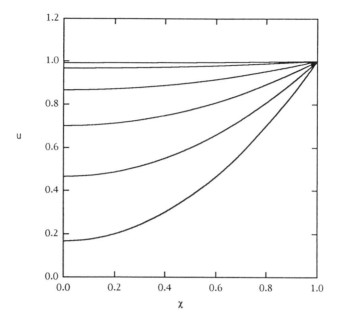

Figure 102. Schematic representation of normalized substrate concentration profiles $u(\chi)$ computed from Eq. (7.28) as function of normalized distance $\chi = x/L$ for various values of Thiele modulus $\Phi = L/X_K$. From top to bottom: $\Phi = 0.1$, 1.0, 2.0, 3.0, 4.0, 5.0. Diagram constructed using $\alpha = 15$ and noting restriction that $\Phi^2 < 2\alpha$.

F.   General Situation

We now return to the master differential equation (7.10). We require an analytical solution to this nonlinear problem in order to derive an expression for the flux and hence an expression for the current. Much of the previous work described in the literature has employed a numerical finite difference approah to this problem [339, 340], although Albery and co-workers [342] and Lyons et al. [310, 330] have presented an analytical solution. In the following discussion the approach developed by Lyons et al. [310, 330] and Albery et al. [342] will be presented. The following analysis will be valid for thick films where $\Phi \gg 1$.

We return to Eq. (7.10) and multiply both sides of the expression by $du/d\chi$ and note the identity

$$\frac{d}{d\chi}\left(\frac{du}{d\chi}\right)^2 = 2\left(\frac{du}{d\chi}\right)\frac{d^2u}{d\chi^2} \qquad (7.31)$$

to obtain

$$\frac{d}{d\chi}\left(\frac{du}{d\chi}\right)^2 d\chi = \frac{2\Phi^2 u}{1 + \alpha u}\, du \tag{7.32}$$

Integrating this expression and noting that when $u = 0$, $du/d\chi = 0$, we obtain

$$\left(\frac{du}{d\chi}\right)^2 = 2\left(\frac{\Phi}{\alpha}\right)^2 [\alpha u - \ln(1 + \alpha u)] \tag{7.33}$$

Hence

$$\frac{du}{d\chi} = \frac{\Phi}{\alpha}\sqrt{2[\alpha u - \ln(1 + \alpha u)]} \tag{7.34}$$

We recall that the normalized flux is given by $y = \alpha(du/d\chi)_{\chi=1}$. Also we note that $u = 1$ when $\chi = 1$. Hence from Eq. (7.34) we obtain

$$y = \sqrt{2\Phi^2[\alpha - \ln(1 + \alpha)]} \tag{7.35}$$

This is a rather complex expression for the flux and is valid for thick films and for all values of the substrate concentration. When $\alpha \ll 1$, $\ln(1 + \alpha) \cong \alpha - \frac{1}{2}\alpha^2$ and Eq. (7.35) reduces to $y \cong \alpha\Phi$ as obtained previously. For $\Phi > 1$ and $\alpha \gg 1$ we note that $\ln(1 + \alpha) \cong \ln\alpha$ and the flux y reduces to

$$y \cong \sqrt{2\Phi^2(\alpha - \ln\alpha)} \cong \sqrt{2\alpha}\,\Phi \tag{7.36}$$

The expression outlined above in Eq. (7.36) is valid for thick films when part of the layer is saturated and the other part unsaturated. We note that Eq. (7.36) will be valid up to $\Phi^2 = 2\alpha$. Transforming Eq. (7.36) into a current, we obtain

$$j = \frac{i}{nFA} = \sqrt{2\kappa k_c c_\Sigma D_S s^\infty} \tag{7.37}$$

Hence we see that the current response is half order in substrate concentration and catalyst loading and independent of layer thickness.

G. Kinetic Case Diagram

We now present the analysis in terms of a kinetic case diagram. We first note that a natural set of cases to choose are $\log\Phi$ as ordinate and $\log\alpha$ as abscissa. Hence movement along the ordinate takes us from thin to thick films, and movement along the abscissa takes us from a condition of reactant unsaturation to one of saturation. The case diagram is presented in Fig. 103. We see that four cases must be presented. The main approximate expressions for the flux or current are those given by Eqs. (7.22) (case I), (7.24) (case II), (7.29) (case III), and (7.36) (case IV). The approximate expressions for the flux and the reaction orders with respect to catalyst loading, substrate concentration, and layer thickness are gathered together in Table VIII. Case I defined by Eq. (7.22) is located in the quadrant bounded by $\Phi \leq 1$ and $\alpha \leq 1$. Case II is defined by Eq. (7.24) and is located in the quadrant bounded by $\Phi \geq 1$ and $\alpha \leq 1$. Case III [quantified via Eq. (7.29)] is located in the lower right-hand region of the case diagram and is defined by the boundary lines $\alpha = 1$ and $\alpha = \frac{1}{2}\Phi^2$. Case IV [quantified via Eq. (7.36)] is located in the upper right hand region where $\Phi \gg 1$. The boundary lines are $\alpha = 1$ and $\alpha = \frac{1}{2}\Phi^2$.

Albery and co-workers [342] have proposed the following rather complex expression for the general case located near the origin in the case diagram:

$$y = \sqrt{2\Phi^2[\alpha - \ln(1 + \alpha)]} \tanh\left[\frac{\alpha\Phi}{(1 + \alpha)\sqrt{2[\alpha - \ln(1 + \alpha)]}}\right] \quad (7.38)$$

When $\alpha \ll 1$, we have $\ln(1 + \alpha) \cong \alpha - \frac{1}{2}\alpha^2$ and $1 + \alpha \cong 1$; hence Eq. (7.38) reduces to

$$y = \sqrt{2\Phi^2(\alpha - \alpha + \frac{1}{2}\alpha^2)} \tanh\left(\frac{\alpha\Phi}{\sqrt{2(\alpha - \alpha + \frac{1}{2}\alpha^2)}}\right)$$

$$= \alpha\Phi \tanh(\Phi) \quad (7.39)$$

This expression provides a link between cases I and II and will therefore be valid for all values of the layer thickness under conditions where $s^\infty \ll K_M$. Alternatively, when $\alpha \gg 1$, we have $\ln(1 + \alpha) \cong \ln \alpha$ and $1 + \alpha \cong \alpha$. Also $\alpha \gg \ln \alpha$. Hence Eq. (7.38) reduces to

$$y = \sqrt{2\Phi^2(\alpha - \ln \alpha)} \tanh\left(\frac{\alpha\Phi}{\alpha\sqrt{2(\alpha - \ln \alpha)}}\right)$$

$$= \sqrt{2\Phi^2\alpha} \tanh\left(\frac{\alpha\Phi}{\alpha\sqrt{2\alpha}}\right)$$

$$= \sqrt{2\Phi^2\alpha} \tanh\left(\sqrt{\frac{\Phi^2}{2\alpha}}\right) \quad (7.40)$$

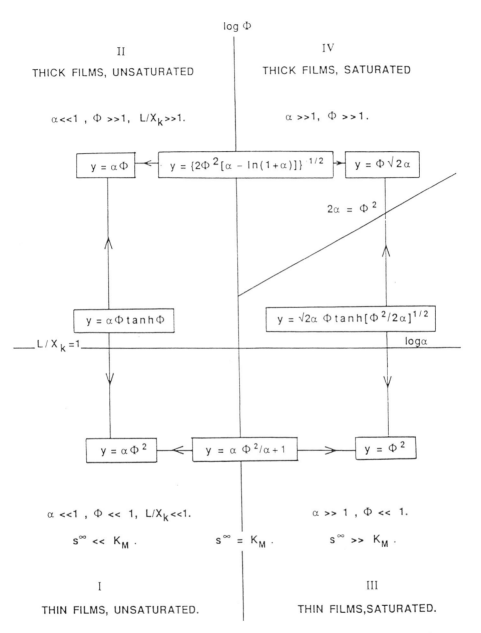

Figure 103. Kinetic case diagram ($\log \Phi$ vs. $\log \alpha$) for electrocatalysis in polymer film exhibiting Michaelis–Menten kinetics. Four major cases discussed in text are illustrated. Also included as insets are approximate analytical expressions for normalized flux for various limiting values of Φ and α.

TABLE VIII
Mechanistic indicators for multicomponent composite systems exhibiting Michaelis-Menten kinetics

Case Designation	Case/Flux Expression Correlations, Flux j
I	$\dfrac{k_c c_\Sigma L \kappa s^\pi}{K_M}$
II	$\sqrt{\dfrac{k_c c_\Sigma D_S}{K_M}}\,\kappa s^\infty$
III	$k_c c_\Sigma L$
IV	$\sqrt{2\kappa k_c c_\Sigma D_s}\,s^\infty$

	Case/Reaction Order Correlation		
Case Designation	c_Σ	L	s^∞
I	1	1	1
II	$\frac{1}{2}$	0	1
III	1	1	0
IV	$\frac{1}{2}$	0	$\frac{1}{2}$

This expression provides a link between cases III and IV in the kinetic case diagram. It is valid for all values of Φ and for $s^\infty \gg K_M$.

Turning back to Eq. (7.38) we let $\beta = \alpha - \ln(1 + \alpha)$ and obtain

$$y = \sqrt{2\beta\Phi^2}\,\tanh\left(\frac{\alpha\Phi}{\sqrt{2\beta}(1+\alpha)}\right) \qquad (7.41)$$

When $\Phi^2 \ll 1$,

$$\tanh\left(\frac{\alpha\Phi}{\sqrt{2\beta}(1+\alpha)}\right) \cong \frac{\alpha\Phi}{\sqrt{2\beta}(1+\alpha)}$$

and so the normalized flux reduces to

$$y = \frac{\sqrt{2\beta\Phi^2}\,\alpha\Phi}{\sqrt{2\beta}(1+\alpha)} = \frac{\alpha\Phi^2}{1+\alpha} \qquad (7.42)$$

This is the normalized form of the Michaelis–Menten equation derived previously. We see that this expression provides a link between cases I and III.

When $\Phi^2 \gg 1$, we note that $\tanh[\alpha\Phi/\sqrt{2\beta}(1+\alpha)] \cong 1$ and so the flux reduces to

$$y = \sqrt{2\beta}\,\Phi = \sqrt{2\Phi^2[\alpha - \ln(1 + \alpha)]} \qquad (7.43)$$

which has been derived previously. This expression links up cases II and IV.

Thus the case diagram presents a summary of the kinetics in a very convenient manner. The diagnostic criteria illustrated in Table VIII enable the identification of any particular case by suitable variation of the experimental quantities s^{∞}, L, and c_{Σ}.

H.   Comparison with Experiment

We now examine a specific example, that of the electrooxidation of catechol at a Nafion-coated electrode containing dispersed RuO_2 particles. This work has been published recently by Lyons and co-workers [310, 330].

Typical cyclic voltammograms recorded for the RuO_2–Nafion composite material in $0.2\,M$ H_2SO_4 are outlined in Fig. 104. The broad nature of the current response across the entire potential window

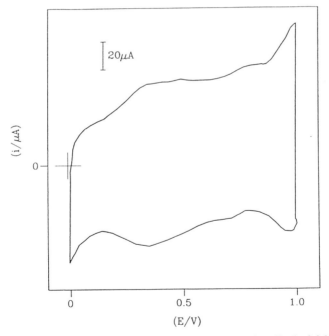

Figure 104.   Typical cyclic voltammogram for RuO_2-loaded Nafion film in $0.2\,M$ H_2SO_4.

examined suggests that the changes in oxidation state for the redox active oxide particles can occur over a wide potential range. Previous work [343] has indicated that the peaks in the voltammogram involve the Ru(III/IV) and Ru(IV/VI) redox transitions. Owing to the rather ill-defined peaks in the voltammogram, it is difficult to estimate the standard redox potentials of the oxyruthenium species, but an approximate assignment in aqueous acid solution is 440 ± 20 mV for Ru(III/IV) and 820 ± 20 mV for Ru(IV/VI). It has been shown [343] that the voltammetric peaks may be assigned to a series of surface redox transitions involving tightly bound oxyruthenium groups located on the surface of the RuO_2 particle. The oxide particle surface will be hydrated even if the particle is located in a Nafion polymer matrix. These surface groups are assumed to exhibit a low degree of bridging oxygen coordination to the bulk oxide particle.

The redox chemistry may be described using the following equations, the bridging oxygens being represented as $-O-$:

Ru(III/IV): $\quad (-O-)_2 RuOH(OH_2)_3 \rightarrow (-O-)_2 Ru(OH)_2(OH_2)_2 + H^+ + e^-$

Ru(IV/VI): $\quad (-O-)_2 Ru(OH)_2(OH_2)_2 \rightarrow (-O-)_2 Ru(OH)_4 + 2H^+ + 2e^-$

Note that the oxyruthenium surface groups maintain an octahedral geometry in all cases. Confirmation that oxidation involves the injection of protons from the hydrated oxide layer into the solution phase (with proton injection from the solution phase into the oxide on the reverse sweep) has been demonstrated [344] using a suitable proton-sensitive dye, bromocresol blue. Thus each oxide microparticle has an outer hydrated, catalytically active layer that consists of "dangling" oxyruthenium surface groups.

Catechol undergoes a $2e^-$, $2H^+$ oxidation reaction, as is typical for many organic compounds. An immediate point to note is that a good match exists between the stoichiometry exhibited by the transformation involving the Ru(IV/VI) surface groups and the catechol oxidation stoichiometry. Both involve the transfer of $2H^+$ and $2e^-$. The voltammetric response for catechol oxidation at the RuO_2–Nafion composite along with a typical series of RDE profiles are illustrated in Fig. 105. It is clear from this diagram that catechol oxidizes in the potential region where the Ru(IV/VI) redox chemistry predominates. It is therefore reasonable to suppose that there is mediation of the substrate oxidation via the surface-bound oxyruthenium groups. The mediation occurs via participation of the Ru(VI) group as follows:

$$Ru(VI) + S \rightarrow Ru(IV) + P$$

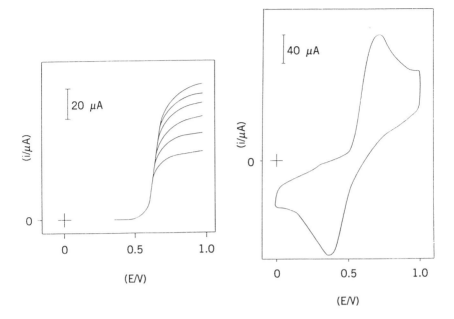

Figure 105. (*a*) Typical RDE voltammograms recorded for oxidation of catechol (1 m*M*) in 0.2 *M* H$_2$So$_4$ at metal oxide–ionomer composite (RuO$_2$–Nafion) electrode. Lowest curve corresponds to rotation speed of 500 rpm, whereas highest curve obtained at 3000 rpm. Rotation speed increment was 500 rpm. Sweep rate, 2 mV s^{-1}. (*b*) Cyclic voltammetric response for catechol (1 m*M*) in 0.2 *M* H$_2$SO$_4$ at metal oxide–ionomer composite electrode. Sweep rate, 20 mV s^{-1}.

The fact that catechol oxidation occurs via a Michaelis–Menten mechanism is confirmed by obtaining the RDE response for a range of catechol concentrations and subsequently subjecting the data to a Koutecky–Levich analysis (Fig. 106*a*). The Koutecky–Levich intercept I_{KL} is then obtained and a Lineweaver–Burk analysis applied, which predicts that I_{KL} should vary in an inverse manner with bulk substrate concentration according to

$$I_{KL} = \frac{K_M}{nFAk_cL\kappa c_\Sigma} \frac{1}{s^\infty} + \frac{1}{nFAk_cLc_\Sigma} \qquad (7.44)$$

The results of such an analysis performed with a thin composite film are presented in Fig. 106*b*. The Lineweaver–Burk plot is linear, thereby confirming Eq. (7.44) and that we are working in cases I and III of the kinetic diagram. Furthermore, we note from Eq. (7.44) that the fun-

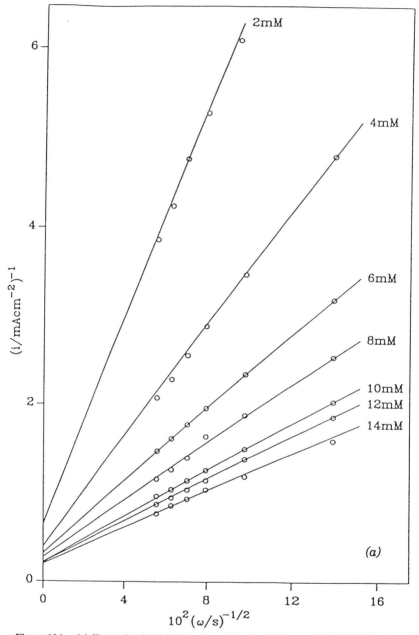

Figure 106. (*a*) Koutecky–Levich plots for catechol oxidation in 0.2 *M* H$_2$SO$_4$ at metal oxide–ionomer composite electrode as function of catechol concentration. (*b*) Lineweaver–Burk analysis of data presented in Fig. 106*a*.

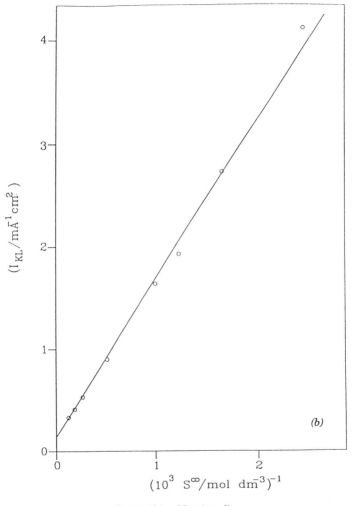

Figure 106. (*Continued*).

damental parameters K_M and k_c may be obtained from the Lineweaver–Burk plot by noting that these parameters are related to the slope S_{LB} and intercept I_{LB} as follows:

$$k_c = \frac{1}{nFAI_{LB}Lc_\Sigma} = \frac{1}{nFAI_{LB}\Gamma}$$

$$K_M = nFAk_cLc_\Sigma S_{LB} = nFAk_c\Gamma S_{LB}$$

where Γ represents the surface coverage of oxyruthenium species. Typically, $\Gamma = 2 \times 10^{-7}$ mol cm^{-2} and $n = 2$, and we find that $k_c = 2.3$ s^{-1} and $K_M = 9.75$ mM. Therefore the case I situation where there is a linear first-order relation between current response and substrate concentration is valid up to concentrations close to 10 mM catechol. This prediction is confirmed in Fig. 107, where the changeover from case I to case III occurs at 10 mM.

The variation of steady-state current for the oxidation of a dilute (0.8 mM) catechol solution recorded at a high rotation speed in order to minimize concentration polarization effects in the solution with layer thickness is illustrated in Fig. 108a. In these experiments the condition $s^\infty \ll K_M$ pertains since $K_M = 10$ mM. A good linear response is observed in the thin-film region ($2 < L < 10 \ \mu$m). Hence as predicted for the case I situation the rate is first order in layer thickness. For thicker Nafion

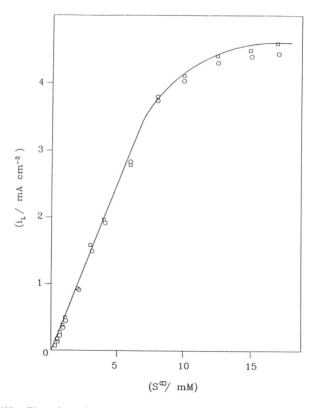

Figure 107. Plot of steady-state amperometric response vs. catechol concentration. First-order (case I) and zero-order regions (case III) clearly indicated.

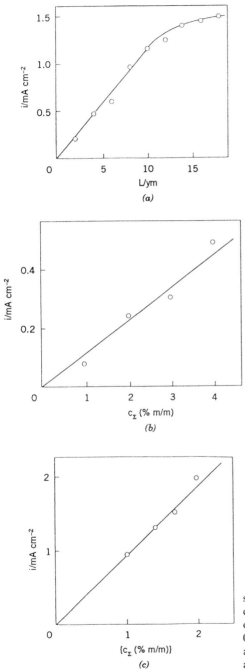

Figure 108. Dependence of steady-state amperometric current response recorded for catechol oxidation at metal oxide–ionomer composite electrode in $0.2\,M$ H_2SO_4 with layer thickness L (a) and with catalyst loading c_Σ for thin (b) and thick (c) films.

deposits ($L > 10 \ \mu$m) a turn-off in the response curve is observed, indicating a tendency toward zero-order behavior. This is predicted since one moves into case II in the kinetic diagram when the layer thickness increases. We can also examine the reaction order with respect to catalyst loading. The variation of steady-state current with c_Σ for a composite film 2 μm thick is outlined in Fig. 108b. A good linear plot is observed over the range 1–4% RuO$_2$. Hence we have a first-order dependence on c_Σ as predicted for case I. The situation corresponding to thicker Nafion layers ($L = 18 \ \mu$m) is illustrated in Fig. 108c. In this case the current response depends in a linear manner with $c_\Sigma^{1/2}$, as predicted for a case II situation.

We see therefore that many of the predictions of the simple analytical model have been confirmed experimentally for the RuO$_2$–Nafion system.

We recall finally that the magnitude of the Koutecky–Levich intercept can be used to obtain an estimate of the heterogeneous modified electrode rate constant k'_{ME}. We note that

$$I_{KL} = \frac{1}{nFAs^\infty k'_{ME}} \tag{7.45}$$

Comparing Eqs. (7.44) and (7.45), we obtain

$$k'_{ME} = \frac{\kappa k_c L c_\Sigma}{\kappa s^\infty + K_M} \tag{7.46}$$

Hence even though k'_{ME} is a comosite quantity, it can still be used to estimate the degree of reversibility of the mediated electron transfer reaction. Typically for catechol oxidation at low pH we have $k'_{ME} \approx 2 \times 10^{-2}$ cm s^{-1}. This is a reasonable value and indicates that the mediation occurs at an efficient rate.

A number of other systems appear to exhibit Michaelis–Menten kinetic behavior. For example, glucose (G) oxidation to form gluconolactone (GL) at RuO$_2$-loaded carbon paste electrodes in aqueous alkaline solution occurs via a mediated mechanism involving catalytically active oxyruthenium surface groups:

$$Ru(VI) + G \rightarrow Ru(IV) + GL$$

$$2Ru(VII) + G \rightarrow 2Ru(VI) + GL$$

A detailed kinetic analysis [311] indicates that the oxidation current is first order in glucose concentration for low values of the latter and is zero order in glucose concentration at high values of the latter. Indeed, the steady-state amperometric current response may be well fitted to the

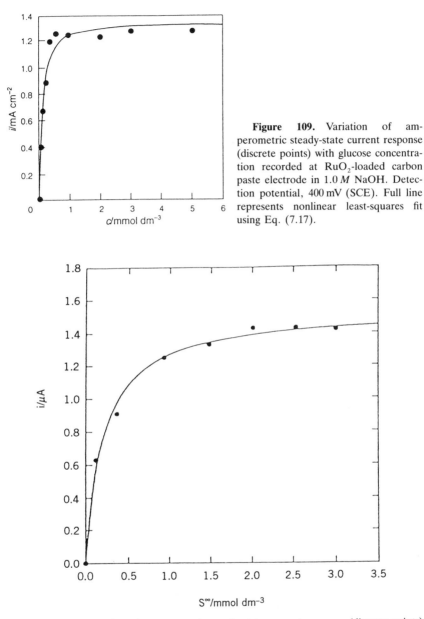

Figure 109. Variation of amperometric steady-state current response (discrete points) with glucose concentration recorded at RuO_2-loaded carbon paste electrode in $1.0\,M$ NaOH. Detection potential, $400\,mV$ (SCE). Full line represents nonlinear least-squares fit using Eq. (7.17).

Figure 110. Variation of amperometric steady-state current response (discrete points) with ascorbate concentration recorded at PPy/DBS$^-$-coated glassy carbon electrode in $0.1\,M$ KCl solution. Detection potential, $500\,mV$ (SCE). Full line represents nonlinear least-squares fit using Eq. (7.17).

Michaelis–Menten equation [Eq. (7.17)] as illustrated in Fig. 109. These observations may be accounted for by assuming that some type of tightly bound intermediate forms between the mediator and the glucose. Finally, recent, as yet unpublished work conducted in the author's laboratory [345] has indicated that the kinetics of ascorbate oxidation at doped poly(pyrrole)-modified electrodes exhibits Michaelis–Menten tight binding kinetics. A typical kinetic plot is presented in Fig. 108. For the data presented in Figs. 109 and 110, rather than utilizing the Lineweaver–Burk approximation, one can instead apply a nonlinear least-square fitting protocol to obtain values for the fundamental kinetic parameters k_c and K_M. One can show that the heterogeneous electrochemical rate constant is approximately related to the latter parameters via $k'_{ME} = k_c \Gamma / K_M$, where Γ denotes the surface coverage of the immobilized catalyst species. For example, for glucose oxidation at dispersed RuO_2 particles [311], $k_c = 6.35 \pm 0.23 \times 10^{-2}\,s^{-1}$, $K_m = 0.116 \pm 0.017\,mmol\,dm^{-3}$, and with $\Gamma = 1.65 \times 10^{-6}\,mol\,cm^{-2}$, the modified electrode rate constant $k'_{ME} = 0.92 \pm 0.11\,cm\,s^{-1}$. This is quite a large value and is suggestive of efficient catalysis at the dispersed particle surface.

VIII.   MICROHETEROGENEOUS ELECTROCATALYTIC SYSTEMS

A.   Introduction

In this section of the review we discuss electrocatalytic systems consisting of a dispersion of catalytic metal microparticles in an electroactive polymer matrix. These composite films are called *microheterogeneous systems*. This type of system is attractive since the functions of catalysis and conductivity are distinct: Catalysis occurs at the microparticles and charge percolation is taken care of via the electroactive polymer. We note that electrons either shuttle along the polymer backbone if the polymer is intrinsically electronically conductive or hop from redox site to redox site if the polymer exhibits redox conduction. Poly(pyrrole) (PPy) and poly(aniline) (PAN) seem to be the host polymeric matrices most often used in this field. It is also possible to incorporate metallic microparticles in conducting polymer–ionomer compositie films. In the latter systems the mechanical integrity of the polymer composite is excellent. These composite polymers are quite porous, and it is reasonably easy to obtain a homogeneous distribution of catalytically active microparticles into the composite film. For instance poly(pyrrole) may be modified with a wide variety of metals including Pd, Pt, Pb, Cu, Ni, Sn, and Au [346, 347]. The PPy–Pt composite films have been examined as electrocatalysts for the electroreduction of oxygen by a number of workers [282, 348, 349]. The oxidation of hydrogen [348] and hydrogen peroxide [350] has also been exmined using these composite systems. The oxidation of methanol at PPy–Pt composite films has been examined recently [351]. It is well known that methanol oxidation causes considerable poisoning of metal electrodes during operation due to the formation of adsorbed CO species. It appears that the poisoning reaction is inhibited if the metallic particles are incorporated into a polymeric matrix. Very little current decay is observed at PPy–Pt composite electrodes under conditions of methanol oxidation when the latter are operated under potentiostatic polarization conditions for periods of up to 12 h.

Some numerical results may be of interest at this stage. For methanol oxidation, Kost and co-workers [283] report current densities of ~ 1 mA cm^{-2} for 2–5-μm-thick PAN films loaded with 30 μg cm^{-2} of Pt. This oxidation rate corresponds to a constant polarization potential of 600 mV versus Ag–AgCl in 1.0 M H$_2$SO$_4$. The methanol concentration was 1.0 M. Aramata [352] reports current densities of ~ 200 μA cm^{-2} for methanol oxidation at Pt-loaded Nafion films 200 μm thick. In this work the catalyst loading was in the range 4–8 mg cm^{-2}. The reaction was carried out in 1.0 M H$_3$PO$_4$. Again the methanol concentration was 1.0 M. The polarization potential was 600 mV (vs. RHE). Very recently Strike and co-workers [351] obtained current densities of ~ 2 mA cm^{-2} for methanol oxidation at PPy–Pt films (PPy thickness, 300 nm; Pt loading, 200 μg cm^{-2}) after prolonged polarization at 500 mV (vs. SCE) in 0.1 M H$_2$SO$_4$ or 1.0 M HClO$_4$. The bottom line here is that appreciable oxidation currents may be obtained using these microheterogenous systems and the problem of electrode deactivation due to poisoning may be considerably reduced.

The *mechanism* of catalysis and stability is still uncertain [353, 354]. The observed catalytic effect observed may be due to that of an increased surface area being available using a dispersion of catalytic microparticles. It is very difficult to estimate the surface area in these polymer–particle systems due to the fact that well-defined hydrogen adsorption waves (integration of the hydrogen adsorption peaks in the cyclic voltammogram is a classical method for determination of the real surface area) are not observed for the composite electrodes. One might expect such an observation given the fact that if the microparticles are well dispersed, it will be improbable that Pt islands will form in the film. The formation of such clusters would lead to the formation of hydrogen adsorption peaks in the voltammogram since the islands would exhibit similar adsorption characteristics to that displayed by bulk Pt surfaces. The question of catalytic activity and prolonged stability under practical operating conditions can be addressed by postulating some, as still undefined, metal–

polymer interaction, which leads to an inhibition in the formation of strongly chemisorbed intermediates on the microparticle surface. These intermediates can lead to surface poisoning and ultimately to a degradation in the current output under prolonged polarization. It should also be noted that polymers such as PPy act as anion exchange membranes. Such materials can stabilize surface redox couples such as $Pt–Pt^{2+}$, which are active in mediating the methanol oxidation reaction [355].

The fundamental physics of these microparticulate systems is still relatively undeveloped. Consequently, in this section, as in previous ones, we shall concentate on the development of simple theoretical models based on the solution of diffusion–reaction differential equations to describe the steady-state amperometric response of these microparticulate systems. The simple model developed on the next sections can be used to predict the steady-state current response of the microheterogeneous catalytic system as a function of particle size, particle loading, and polymer film thickness.

B. Model

We now present an analysis of the transport and kinetics of a substrate species S in an electroactive polymer matrix containing a homogeneous dispersion of catalytically active microparticles. We assume for simplicity that the distribution of microparticles is monodisperse*: Each particle has a radius R. We also assume that the microparticles exhibit a spherical geometry. These assumptions may, of course be difficult to achieve in practice. If the polymer matrix is electronically conducting, then one can assume that the microparticle is in intimate electrical contact with the conducting polymer and, hence, with the support electrode. Thus we can assume that each particle is held at the same potential regardless of its location in the film. If, however, the microparticles are dispersed either in a redox polymer or in an ionomer (such as Nafion) containing an inorganic mediator species, then the situation becomes a little more complicated. The mediator species, which is either covalently attached to

---

* This can be quite difficult to achieve in practice. Many workers choose to incorporate metallic microparticles into electroactive polymer matrices via a controlled electrodeposition proces from a plating solution containing the metal salt. Although this method is easy to perform, there is no real control over the particle size obtained, and there can be problems regarding the extent of homogeneity of the resulting particle dispersion throughout the host matrix. A better approach is to chemically synthesize a monodisperse sample of colloidal metal microparticles and then incorporate the latter into a host matrix. The latter procedure is not without its difficulties however.

an electrically insulating polymer backbone or electrostatically anchored in the polymer matrix, acts as an electron relay unit between support electrode and the catalytic site. Hence on each particle the rate of reaction will be determined by the balance between the reaction of mediator and the reaction of substrate. This leads to the consequence that the electrochemical potential of each particle throughout the film need not be the same. The systems to be considered are illustrated schematically in Fig. 111. We assume that the polymer film has a uniform thickness L. We also assume that concentration polarization in the solution phase may be neglected. This simplifies the algebra, but it is quite easy to incorporate solution-phase concentration polarization effects into the anlysis using the flux balance condition at $x = L$. The substrate concentration within the layer is given by κs^{∞}, where κ is the partition coefficient and s^{∞} denotes the bulk concentration of the substrate S.

The analysis of the transport and kinetics must be approached on two levels. The first is essentially *macroscopic*. The steady-state Ficksian diffusion–reaction equation must be solved for the substrate in the bounded diffusion space of the film of extent L. This type of analysis has been discussed in previous sections of this chapter. From this analysis the pseudo-first-order rate constant k for substrate reaction may be derived. However, the analysis must be taken a step further. One also has to adopt a *microscopic* approach. In this case the spherical geometry of the microparticle must be considered. The steady-state spherical diffusion of

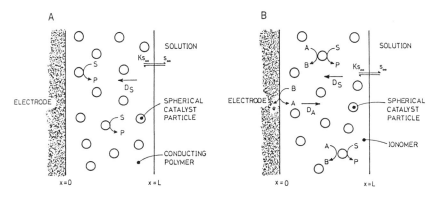

Figure 111. Schematic representation of electrocatalytic system comprising of dispersed catalytically active microparticles in polymeric matrix. (*a*) Microparticles dispersed in electronically conducting polymer matrix. (*b*) Electron relay complex/catalytic microparticle dispersion in ionically conducting polymer matrix.

the substrate to the microparticle must be examined. One must then relate the macroscopic rate constant k for substrate reaction with the spherical diffusion and reaction at each microparticle.

C. Conducting Polymer–Microparticulate Metal Composite Systems

We begin by considering the microparticle–electronically conducting polymer composite system.

In the steady state the diffusion–reaction equation takes the form

$$D_S \frac{d^2s}{dx^2} - ks = 0 \qquad (8.1)$$

The boundary conditions are

$$x = 0 \quad \frac{ds}{dx} = 0 \qquad x = L \qquad s = \kappa s^{\infty} \qquad (8.2)$$

Equation (8.1) may be readily integrated using the boundary conditions outlined in Eq. (8.2) to obtain expressions for the substrate concentration profile,

$$s(x) = \frac{\kappa s^{\infty} \cosh(x/X_K)}{\cosh(L/X_K)} \qquad (8.3)$$

and reaction flux,

$$j = D_S \left(\frac{ds}{dx}\right)_{x=L} = \frac{D_S \kappa s^{\infty}}{X_K} \tanh\left(\frac{L}{X_K}\right) \qquad (8.4)$$

where as before the reaction layer thickness X_K is given by

$$X_K = \sqrt{\frac{D_S}{k}} \qquad (8.5)$$

We now examine the substrate reaction at a microparticle of radius R. Substrate diffusion to the latter is described by the spherical diffusion equation given by

$$D_S \left(\frac{d^2s}{dr^2} + \frac{2}{r}\frac{ds}{dr}\right) = 0 \qquad (8.6)$$

where r denotes the radial coordinate. This expression may be solved most readily if we set $u = sr$ and substitute into Eq. (8.7) to obtain

$$D_S \frac{d^2 u}{dr^2} = 0 \tag{8.7}$$

Note that Eq. (8.7) may be twice integrated to obtain

$$u(r) = \alpha r + \beta \quad \text{or} \quad s(r) = \alpha + \frac{\beta}{r} \tag{8.8}$$

where α and β are integration constants to be determined. The latter are evaluated using the following boundary conditions. When $r = R$ at the surface of the particle, we note that $s = s_0$ and $D_S(ds/dr) = k'_E s_0$, where k'_E is the heterogeneous electrochemical rate constant given by the Butler–Volmer equation. One must also define a distance R_A that describes a sphere of action associated with the catalytic particle. This concept has been developed by Albery and Bartlett for microheterogeneous photocatalytic systems [356]. When $r = R_A$, $s = s(x)$. Note that the relationship between this critical distance R_A and the number density of catalytic particles N is $R_A = (3/4\pi N)^{1/3}$.

Application of these microscopic boundary condition to Eq. (8.8) leads to the assignment that

$$\alpha = s(x) + \frac{k'_E s_0 R^2}{D_S R_A} \qquad \beta = -\frac{k'_E s_0 R^2}{D_S} \tag{8.9}$$

Hence from Eqs. (8.8) and (8.9) and the boundary condition that $s = s(x)$ at $r = R_A$ we obtain that the substrate concentration at the surface of the particle is related to the substrate concentration at the outer edge of the zone of action via

$$s_0 = \frac{s(x)}{1 + k'_E R/D_S - k'_E R^2/D_S R_A} \tag{8.10}$$

Hence we obtain

$$\frac{s_0}{s(x)} = \frac{1}{1 + k'_E R/D_S - k'_E R^2/D_S R_A} \tag{8.11}$$

The substrate–catalyst reaction can be formally considered as a second-order reaction with the rate of consumption of S being given by

$$-\frac{ds}{dt} = k_2 s(x) c_\Sigma \tag{8.12}$$

where c_Σ denotes the concentration of catalyst particles in the layer. Alternatively, we may write an expression for the rate of reaction of S that contains the heterogeneous electrochemical rate constant k_E' as follows:

$$-\frac{ds}{dt} = k_E' N A s_0 \tag{8.13}$$

where $A = 4\pi R^2$ denotes the surface area of the spherical particle and N represents the number density of catalytic particles in the film. Comparing Eqs. (8.12) and (8.13), we obtain

$$k_2 = \frac{k_E' N A}{c_\Sigma} \frac{s_0}{s(x)} \tag{8.14}$$

Assuming that the quantity c_Σ is a constant, which will be the case if the catalytic particles are uniformly distributed throughout the layer, we may write that the pseudo-first-order rate constant $k = k_2 c_\Sigma$, and so we obtain

$$\frac{1}{k} = \frac{1}{4\pi R^2 N} \left(\frac{1}{k_E'} + \frac{R}{D_S} - \frac{R^2}{D_S R_A} \right)$$

$$\cong \frac{1}{4\pi R^2 N} \left(\frac{1}{k_E'} + \frac{R}{D_S} \right) \tag{8.15}$$

In deriving the latter expression we have assumed that $R_A \gg R$. Equation (8.15) is a very fundamental equation and provides a relationship between the macroscopic rate constant k and the microscopic processes of substrate electrode kinetics and spherical diffusion at the catalytic particle. On the rhs of Eq. (8.15) the first term describes the effect of heterogeneous kinetics of the substrate at the particle surface. This rate constant k_E' is a strong function of electrode potential. The second term describes the spherical diffusion of substrate to the catalytic particle. When $k_E' \gg D_S/R$, mass transport of S to the particle surface is rate determining. Alternatively, when $k_E' \ll D_S/R$, the electron transfer process at the surface of the particle will be rate determining.

We now are in a position to obtain an expression for the flux. We recall Eqs. (8.4) and (8.5) and use the result obtained in Eq. (8.15) to obtain

$$j = \kappa D_S s^\infty \left(4\pi R^2 N \frac{k_E'/R}{k_E' + D_S/R} \right)^{1/2} \tanh \left(L \sqrt{4\pi R^2 N \frac{k_E'/R}{k_E' + D_S/R}} \right) \tag{8.16}$$

We can identify four limiting cases from this expression. We begin by taking $L\{4\pi R^2 N[(k'_E/R)/(k'_E + D_S/R)]\}^{1/2} \ll 1$. Under such conditions Eq. (8.16) reduces to

$$j \cong \kappa D_S s^\infty L 4\pi R^2 N \frac{k'_E/R}{k'_E + D_S/R} \qquad (8.17)$$

We note that Eq. (8.17) consists of two parts, the first involving diffusional substrate transport through the layer and the other involving the electrochemical kinetics at the particle surface. As previously noted, when $k'_E \gg D_S/R$ corresponding to case I, the reaction at the particle surface is diffusion controlled and Eq. (8.17) reduces to

$$j \cong 4\pi R N \kappa D_S L s^\infty \qquad (8.18)$$

On the other hand, if $k'_E \ll D_S/R$, then we have case II and the electrode kinetics at the particle surface are rate determining. In this case the flux expression reduces to

$$j \cong 4\pi R^2 N k'_E \kappa L s^\infty$$
$$= 4\pi R^2 N \kappa L s^\infty k'_{E,0} \exp\left[\pm\frac{\alpha F(E - E^0)}{RT}\right] \qquad (8.19)$$

where α denotes the transfer coefficient and $k'_{E,0}$ is the standard rate constant.

Turning to the remaining pair of cases, when $L\{4\pi R^2 N[(k'_E/R)/(k'_E + D_S/R)]\}^{1/2} \gg 1$, Eq. (8.16) reduces to

$$j \cong \kappa D_S s^\infty \left(4\pi R^2 N \frac{k'_E/R}{k'_E + k'_E/R}\right)^{1/2} \qquad (8.20)$$

Again we can identify two further cases depending on the relative values of k'_E and D_S/R. If $k'_E \gg D_S/R$, then we have case III and Eq. (8.20) reduces to

$$j \cong \kappa D_S s^\infty \sqrt{4\pi R N} \qquad (8.21)$$

whereas if $k'_E \ll D_S/R$, we have case IV, and Eq. (8.20) reduces to

$$j \cong \kappa s^\infty \sqrt{4\pi R^2 D_S N k'_E}$$
$$= \kappa s^\infty \sqrt{4\pi R^2 D_S N k'_{E0}} \exp\left[\pm\frac{\alpha F(E - E^0)}{2RT}\right] \qquad (8.22)$$

Each of these four cases corresponds to a distinct physical situation. In case I the film is much thinner than the reaction layer thickness X_K. Consequently the substrate penetrates throughout the film. Hence increasing the layer thickness L leads to an increase in the current. The reaction on the particles in this case is very rapid and the overall kinetics will be limited by spherical diffusion within the film to each particle. Note also that k'_E does not appear in the flux expression [Eq. (8.18)] and so the flux and therefore the current response are independent of the applied electrode potential. This will give rise to a very large Tafel slope or a current plateau region in the current–voltage curve. The flux is first order in particle radius and also first order with respect to microcatalyst loading. In case II, the substrate still penetrates throughout the film, but now the reaction at the surface of the catalytic particle is rate determining. The pertinent flux expression is outlined in Eq. (8.19). Again the flux is first order in layer thickness, as it should be for a layer reaction. The current response is second order with respect to particle radius and first order in catalyst loading. In this case the flux will depend on electrode potential. The Tafel slope obtained is given by $b = dE/(d \log i) = \pm 2.303 RT/\alpha F$. This is the standard Tafel slope. We now turn to cases III and IV. In this regime, the film is sufficiently reactive that the substrate is consumed in a reaction layer of extent X_K located near the film–solution interface. For case III [Eq. (8.21)] the reaction at the catalyst particles is diffusion controlled and so the thickness of this reaction layer is determined by diffusion of substrate to the microparticles. The flux is independent of layer thickness and is half order with respect to particle radius and catalyst loading. In case IV [Eq. (8.22)] the reaction at the surface of the particles is kinetically controlled, and hence the reaction layer thickness depends upon k'_E and the value of the applied electrode potential. This fact leads to the prediction that the Tafel slope b will be twice that expected for case II and is given by $b = \pm 2.303(2RT/\alpha F)$.

The relation between these four cases is illustrated in the case diagram outlined in Fig. 112. The axes are defined as

$$\log\left(\frac{k'_E}{k'_D}\right) = \log\left(\frac{k'_E}{D_S/R}\right) \qquad \log\left(\frac{L}{X_K}\right) = \log\left(l\sqrt{4\pi NR^2 \frac{k'_E/R}{D_S/R + k'_E}}\right)$$

Case I is located in the upper left hand region of the case diagram, whereas case II is in the lower left hand quadrant. Case III is located in the upper right hand zone. The boundary between cases I and III is defined by the line $L = X_K$. Finally case IV is located in the lower right hand quadrant. The concentration profiles of substrate are also included as insets in the diagram, and we also show a signpost indicating how one

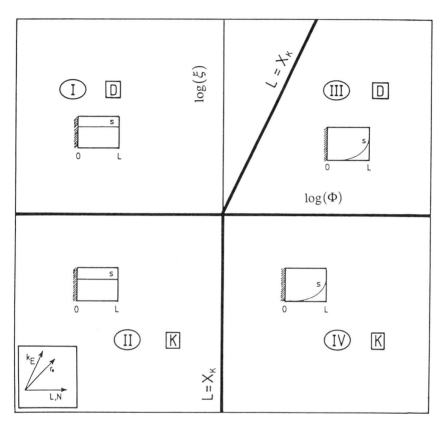

Figure 112. Kinetic case diagram for catalytic microparticles immobilized in electronically conducting polymer matrix. Axes of case diagram are $\log(\xi) = \log(k_E/k_D)$ with $k_D = D_s/R$ and $\log(\Phi) = L/X_K$, where kinetic length X_K is given by $X_K = \sqrt{D_s/k} = \sqrt{(D_s/4\pi R^2 N)(1/k_E + R/D_s)}$. Four possible rate-limiting cases are outlined along with concentration profiles of substrate within layer; K and D represent kinetic and diffusion control, respectively.

moves between the cases as the experimental quantities E, R, L, and N are changed. The pertinent mechanistic indicators as well as the flux expressions are outlined in Table IX.

We now consider the conditions under which it is valid to treat the macroscopic planar diffusion of the reactant S independently of the microscopic spherical diffusion to the microcatalyst particles. In Fig. 113, a single catalyst particle of radius R and its associated sphere of action of

TABLE IX

Expressions for Flux j and Mechanistic Indicators for Microheterogeneous Catalysis at Conducting Polymer–Dispersed Microparticle Composite Systems

Case Designation	Case/Flux Expression Correlation, Flux j
I	$4\pi r_0 N\kappa D_s s^\infty L$
II	$4\pi r_0^2 N k_E \kappa s^\infty L$
III	$\kappa D_s\sqrt{4\pi r_0 N s^\infty}$
IV	$\kappa\sqrt{4\pi r_0^2 D_s N k_E s^\infty}$

	Case/Reaction Order Correlation				
Case	s^∞	L	r_0	N	k_E
I	1	1	1	1	0
II	1	1	2	1	1
III	1	0	$\frac{1}{2}$	$\frac{1}{2}$	0
IV	1	0	1	$\frac{1}{2}$	$\frac{1}{2}$

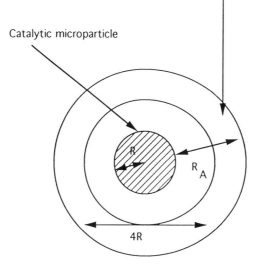

Figure 113. Schematic representation of sphere of action concept for catalytic microparticle.

radius R_A are outlined. Now, for cases I and II, the concentration of S in the film is uniform, and so the microparticles see the same concentration of S in all directions. Under these circumstances we can clearly separate the planar and spherical components. In cases III and IV, however, the concentration of S varies throughout the film. The treatment developed so far is most likely to break down when the reaction of S at the microcatalyst surface is diffusion controlled. This will be case III. Under these conditions the particle perturbs the concentration of S over a distance R from its surface. Our approximation of separability is therefore valid provided that the concentration of S does not vary significantly over a distance $4R$, as illustrated in Fig. 113.

Thus the approximation is valid provided that

$$4R \frac{ds}{dx} \ll s(x) \tag{8.23}$$

If we differentiate Eq. (8.3), we obtain

$$\frac{ds}{dx} = \frac{\kappa s^\infty}{X_K \cosh(L/X_K)} \sinh\left(\frac{x}{X_K}\right) \tag{8.24}$$

Substituting the expression for $s(x)$ obtained in Eqs. (8.3) and (8.24) into Eq. (8.23), we obtain

$$\frac{4R}{X_K} \sinh\left(\frac{x}{X_K}\right) \ll \cosh\left(\frac{x}{X_K}\right) \quad \text{or} \quad \frac{4R}{X_K} \tanh\left(\frac{x}{X_K}\right) \ll 1 \tag{8.25}$$

We can set the tanh term equal to unity in Eq. (8.25) to obtain

$$4R \ll X_K \tag{8.26}$$

This means that the particles must be small and much less than X_K. Furthermore, we can introduce the expression for X_K outlined in Eq. (8.5) into Eq. (8.26) and note the definition of k from Eq. (8.15) to obtain

$$4R \ll \sqrt{\frac{D_S}{4\pi R^2 N}\left(\frac{1}{k'_E} + \frac{R}{D_S}\right)}$$

$$\cong \sqrt{\frac{1}{4\pi RN}} \tag{8.27}$$

since the particles only perturb the concentration of substrate significantly when $R/D_S \gg 1/k'_E$. We recall that $1/N = 4\pi R_A^3/3$; hence introducing this relation into Eq. (8.27), one obtains

$$3.3R \ll R_A \tag{8.28}$$

which represents the condition for the particle dimension under which one can separate the macroscopic and microscopic aspects of the problem.

Thus we see that the model presented is valid up to quite high catalyst loadings where 3% of the film by volume is made up of catalyst particles.

D. Ionomer–Mediator–Microparticle and Redox Polymer–Microparticle Composite Systems

We now consider the situation where one utilizes an ionomer film or a redox polymer film loaded with microscopic catalytic particles. In this situation a mediator is used that is either electrostatically incorporated into the ionomer or covalently bound to the polymer backbone. The function of this mediator species is to shuttle charge between the support electrode and the catalytic particle. This situation is outlined in Fig. 111b. In this case, in addition to the diffusion–reaction equation for the substrate S given by the expression

$$D_S \frac{d^2 s}{dx^2} - k_S s = 0 \tag{8.29}$$

we must also consider the diffusion–reaction equation for the mediator species A:

$$D_A \frac{d^2 a}{dx^2} - k_A a = 0 \tag{8.30}$$

The pseudo-first-order rate constant k_j with $j = S$, A, is given by

$$k_j = \frac{4\pi R^2 N k_j' D_j}{D_j + k_j' R} \tag{8.31}$$

where the heterogeneous electrochemical rate constant k_j' is given by the Butler–Volmer equation:

$$k_j' = k_{j,0}' \exp\left(\pm \frac{\alpha F(E - E^0)}{RT}\right) \tag{8.32}$$

Under steady-state conditions the rates of reaction of A and S on each individual particle must balance, and so we write

$$k'_S s_0 = k'_A a_0 \qquad (8.33)$$

where s_0 and a_0 denote the substrate and mediator concentrations at the surface of the catalyst. One can readily show that

$$k'_A = \sqrt{\frac{k'_{A,0} k'_{S,0} s}{a}} \qquad k'_S = \sqrt{\frac{k'_{A,0} k'_{S,0} a}{s}} \qquad (8.34)$$

If we substitute this result into Eqs. (8.29) and (8.30), we immediately obtain

$$D_A \frac{d^2 a}{dx^2} - \frac{4\pi R^2 N D_A \sqrt{k'_{A,0} k'_{S,0}} \sqrt{as}}{D_A + R\sqrt{k'_{A,0} k'_{S,0}} \sqrt{s/a}} = 0 \qquad (8.35)$$

and

$$D_S \frac{d^2 s}{dx^2} - \frac{4\pi R^2 N D_S \sqrt{k'_{A,0} k'_{S,0}} \sqrt{as}}{D_S + R\sqrt{k'_{A,0} k'_{S,0}} \sqrt{a/s}} = 0 \qquad (8.36)$$

These nonlinear differential equations have to be solved subject to the boundary conditions

$$
\begin{aligned}
x = 0 \qquad & \frac{ds}{dx} = 0 \qquad a = a^\infty \\
x = L \qquad & s = \kappa s^\infty \qquad \frac{da}{dx} = 0
\end{aligned}
\qquad (8.37)
$$

We now introduce a dimensionless substrate concentration u, mediator concentration v, and distance χ,

$$u = \frac{s}{\kappa s^\infty} \qquad v = \frac{a}{a^\infty} \qquad \chi = \frac{x}{X_K} \qquad (8.38)$$

where

$$X_K = \sqrt{\frac{1}{4\pi R N}}$$

and also introduce two further dimensionless parameters β and γ:

$$\beta = \sqrt{\frac{a^\infty D_A}{\kappa s^\infty D_S}} \qquad \gamma = R\sqrt{\frac{k'_{A,0}k'_{S,0}}{D_A D_S}} \qquad (8.39)$$

Substitution of Eqs. (8.38) and (8.39) into the differential equations outlined in Eqs. (8.35) and (8.36) results in the generation of the following dimensionless expressions:

$$\frac{d^2 u}{d\chi^2} - \frac{\beta\gamma\sqrt{uv}}{1 + \beta\gamma\sqrt{v/u}} = 0 \qquad (8.40)$$

for the substrate and

$$\frac{d^2 v}{d\chi^2} - \frac{\gamma\sqrt{uv}}{\beta + \gamma\sqrt{u/v}} = 0 \qquad (8.41)$$

for the mediator. These expressions obey the boundary conditions

$$\chi = 0 \qquad v = 1 \qquad \frac{du}{d\chi} = 0$$

$$\chi = \frac{L}{X_K} \qquad \frac{dv}{d\chi} = 0 \qquad u = 1 \qquad (8.42)$$

The flux j is given by either of the following:

$$j = -D_A\left(\frac{da}{dx}\right)_{x=0} = D_S\left(\frac{ds}{dx}\right)_{x=L}$$

$$j = -D_A a^\infty \sqrt{4\pi RN}\left(\frac{dv}{d\chi}\right)_{\chi=0} = \kappa D_S s^\infty \sqrt{4\pi RN}\left(\frac{du}{d\chi}\right)_{x=L/X_K} \qquad (8.43)$$

The system of equations must be solved in order to obtain analytical expressions for the flux. To do this, the differential equations may be simplfied as follows.

First, we consider the master equation describing the transport and kinetics of the mediator within the layer. When $\gamma \ll \beta$, Eq. (8.41) takes the form

$$\frac{d^2 v}{d\chi^2} - \frac{\gamma\sqrt{uv}}{\beta} = 0 \qquad (8.44)$$

whereas when $\gamma \gg \beta$, Eq. (8.41) reduces to

$$\frac{d^2v}{d\chi^2} - v = 0 \tag{8.45}$$

Second, we consider the master equation describing the transport and kinetics of the substrate in the layer given by Eq. (8.40). When $\beta\gamma \ll 1$, we have

$$\frac{d^2u}{d\chi^2} - \beta\gamma\sqrt{uv} = 0 \tag{8.46}$$

whereas if $\beta\gamma \gg 1$, we obtain

$$\frac{d^2u}{d\chi^2} - u = 0 \tag{8.47}$$

We note that Eqs. (8.45) and (8.47) describe the diffusion of mediator and diffusion of substrate to the spherical microparticle, whereas Eqs. (8.44) and (8.46) are kinetic equations. We see therefore that the parameters β and γ make natural axes for a case diagram.

We first consider Eq. (8.45). This expression may be readily integrated to yield

$$v(\chi) = \frac{a}{a^\infty} = \cosh\chi - \sinh\chi \tanh\left(\frac{L}{X_K}\right) \tag{8.48}$$

and

$$j = D_A a^\infty \sqrt{4\pi RN} \tanh\left(\frac{L}{X_K}\right) \tag{8.49}$$

For thin films, when $L \ll X_K$, Eq. (8.49) reduces to

$$j \cong 4\pi RND_A a^\infty L \tag{8.50}$$

Hence, the reaction takes place throughout the entire layer and the rate-determining step is the spherical diffusion of mediator to the microcatalytic particles. Alternatively when $L \gg X_K$ we have

$$j \cong \sqrt{4\pi RND_A} a^\infty \tag{8.51}$$

In this case the mediator is consumed in a first-order reaction layer at the inner electrode surface and the reaction of mediator is diffusion controlled at the particle.

Equation (8.47) which describes the transport of substrate in the layer, may also be directly integrated to yield

$$u(\chi) = \frac{s}{\kappa s^{\infty}} = \cosh \chi \, \text{sech}\left(\frac{L}{X_K}\right) \tag{8.52}$$

and

$$j = D_S \kappa s^{\infty} \sqrt{4\pi RN} \, \tanh\left(\frac{L}{X_K}\right) \tag{8.53}$$

Again this expresesion decomposes into two special cases when the film is thin and the layer is thick. When $L \ll X_K$, we have

$$j \cong 4\pi RND_S \kappa s^{\infty} L \tag{8.54}$$

and reaction occurs through the entire layer. The rate determining step (rds) is the spherical diffusion of substrate S to the microparticles. On the other hand, when $L \gg X_K$, we have

$$j \cong \sqrt{4\pi RN} D_S \kappa s^{\infty} \tag{8.55}$$

In this case the substrate is consumed in a first-order reaction layer at the polymer–solution interface. The reaction of substrate is diffusion controlled at the microparticle.

We then consider the interesting situation where the substrate and mediator reaction somewhere in the middle of the film. This *titration situation* is analogous to the *layer reaction zone* (LRZ) case developed by Albery and Hillman [280] and described in Section VI of this chapter. We recall that two reaction layers X_0 and X_L were described that were used to quantify the consumption of mediator or substrate in the layer. Under conditions where either of these kinetic lengths are much less than the layer thickness L, the reaction will take place in a reaction zone somewhere in the middle of the layer with a transport-controlled supply of electrons from the electrode and a transport-controlled supply of substrate from the bulk solution. In the present case the titration situation corresponds to the case when $X_K \ll L$. The concentration profiles for mediator and substrate are linear, meeting at some point x^* in the middle of the layer. Since the flux is transport controlled, we write

$$j = \frac{D_A a^{\infty}}{x^*} = \frac{D_S \kappa s^{\infty}}{L - x^*} \tag{8.56}$$

Noting that $\beta = \sqrt{D_A a^\infty / D_S \kappa s^\infty}$, we have

$$\beta^2(L - x^*) = x^* \qquad x^* = \frac{\beta^2 L}{1 + \beta^2} \qquad (8.57)$$

Hence substituting Eq. (8.57) into Eq. (8.56), we obtain

$$j = \frac{D_A a^\infty (1 + \beta^2)}{\beta^2 L} = \frac{D_S \kappa s^\infty}{L} + \frac{D_A a^\infty}{L} \qquad (8.58)$$

Hence the total flux is simply the sum of the fluxes due to the mediator and the substrate. We note that when $\beta^2 \gg 1$, $D_A a^\infty \gg D_S \kappa s^\infty$ and Eq. (8.58) reduces to

$$j \cong \frac{D_A a^\infty}{L} \qquad (8.59)$$

whereas if $\beta^2 \ll 1$, then $D_A a^\infty \ll D_S \kappa s^\infty$, and the flux reduces to

$$j \cong \frac{D_S \kappa s^\infty}{L} \qquad (8.60)$$

We now turn to the kinetic equations given by Eqs. (8.44) and (8.46). Looking at Eq. (8.44) we have that if $\beta \gg \gamma$, then $\gamma/\beta \ll 1$ and $d^2 v / d\chi^2 = 0$, and so $v = 1$. We therefore solve

$$\frac{d^2 u}{d\chi^2} - \beta\gamma \sqrt{u} = 0 \qquad (8.61)$$

Turning to Eq. (8.46) we note that if $\beta\gamma \ll 1$, then $d^2 u / d\chi^2 = 0$, and so $u = 1$, and we need to solve the simplified expression

$$\frac{d^2 v}{d\chi^2} - \frac{\gamma}{\beta} \sqrt{v} = 0 \qquad (8.62)$$

The expressions outlined in Eqs. (8.61) and (8.62) must be integrated to obtain the flux and the concentration profiles. These equations look rather simple but they are not.

An approximate solution of Eq. (8.61), valid for regions of the film near the layer–solution interface, is given by

$$u(\chi) \cong \left[1 - \frac{\sqrt{\beta\gamma}}{2\sqrt{3}} \left(\frac{L}{X_K} - \chi\right)\right]^4 \qquad (8.63)$$

This solution will cease to be valid when the parameter $\xi = \sqrt{\beta\gamma}(L/X_K) > 5$. One can show that the flux is approximately given by

$$j \cong \frac{4D_S\kappa s^\infty\sqrt{\beta\gamma}}{2\sqrt{3}X_K} = \sqrt{4\pi RND_S\kappa s^\infty\sqrt{k'_{A,0}k'_{S,0}\kappa s^\infty a^\infty}} \qquad (8.64)$$

This expression for the flux is valid for the situation where $X_K/\sqrt{\beta\gamma} < L$. This corresponds to the situation where the substrate is consumed in a first-order reaction layer at the polymer–solution interface, and the heterogeneous kinetics at the particle surface are rate determining. Note the unusual reaction order with respect to substrate concentration of $\frac{3}{4}$. If one considers the alternative situation where $X_K/\sqrt{\beta\gamma} > L$, then the substrate concentation is uniform throughout the layer and $u = 1$. In this case Eq. (8.61) reduces to

$$\frac{d^2u}{d\chi^2} - \beta\gamma = 0 \qquad (8.65)$$

which integrates to

$$u = \frac{\beta\gamma\chi^2}{2} + 1 - \left(\frac{\sqrt{\beta\gamma}L}{4X_K}\right)^2 \cong 1 + \frac{\beta\gamma\chi^2}{2} \qquad (8.66)$$

since we have assumed that $X_K/\sqrt{\beta\gamma}L \gg 1$. The flux is obtained by differentiating Eq. (8.66) to obtain

$$j = \frac{D_S\kappa s^\infty\beta\gamma L}{X_K^2} = 4\pi\sqrt{k'_{A,0}k'_{S,0}a^\infty\kappa s^\infty}RNL \qquad (8.67)$$

Equation (8.67) corresponds to the situation where the film is thinner than the reaction layer thickness and the reaction takes place throughout the film. The rate is limited by the heterogeneous kinetics at the microparticle surface. Note that in this case the flux is half order in substrate concentration.

We now finally examine the kinetics of the mediator turnover at the catalytic particle. We need to solve Eq. (8.62). This expression is very similar to that outlined in Eq. (8.61) and a similar expression for $v(\chi)$ applies in this case. We obtain

$$v(\chi) \cong \left(1 + \frac{1}{2\sqrt{3}}\sqrt{\frac{\gamma}{\beta}}\chi\right)^4 \qquad (8.68)$$

This solution is only approximate and pertains when $\sqrt{\beta/\gamma}X_K < L$. The region of validity is near the support electrode–polymer film interface. One can also show that the flux is given by

$$j \cong \frac{4}{2\sqrt{3}} \sqrt{\frac{\gamma}{\beta}} \sqrt{4\pi RND_A a^\infty}$$

$$= \sqrt{4\pi R^2 ND_A a^\infty} \sqrt{k'_{A,0} k'_{S,0} \kappa s^\infty a^\infty} \tag{8.69}$$

where again we have set $4/2\sqrt{3} = 1.15 \approx 1$. Note that in this case the reacton order with respect to mediator is $\frac{3}{4}$.

We finally consider the case when $\sqrt{\beta/\gamma} X_K > L$. In this case we set $v = 1$ in Eq. (8.62) to obtain

$$\frac{d^2 v}{d\chi^2} - \frac{\gamma}{\beta} = 0 \tag{8.70}$$

This expression is exactly similar to that already solved in Eq. (8.65) for the substrate. Proceeding by a similar argument one may show that the flux is given by

$$j = \frac{D_A a^\infty \gamma L}{\beta X_K^2} = 4\pi \sqrt{k'_{A,0} k'_{S,0} \kappa s^\infty a^\infty} R^2 NL \tag{8.71}$$

This expression corresponds to the situation where the mediator species A is consumed in a reaction layer close to the inner electrode–polymer interface and the reaction at the surface of the microcatalyst is kinetically controlled. It will be valid for the case where X_K is large and the film is wholly reactive.

The analysis developed so far may be summarized in terms of a number of cases labeled I–VIII. We note that each case is distinguished by a unique set of dependencies on the experimental parameters a^∞, s^∞, L, R, and N. Thus it is possible to identify a specific case by careful systematic variation of these parameters. A summary of the approximate expressions for the flux, the various case designations, and the mechanistic indicators or reaction orders are presented in Table X. The relationships between the various cases are outliend in the case diagrams in Figs. 114–116. We recall that the axes of the case diagrams are cast in terms of the dimensionless parameters $\log(\beta) = \log(\sqrt{D_A a^\infty / \kappa D_S s^\infty})$, $\log(\gamma) = \log(R\sqrt{k'_{A,0} k'_{S,0} / D_A D_S})$, and L/X_K.

For cases I–III [defined in Eqs. (8.50), (8.54), (8.67), and (8.71), where the latter two equations correspond to case III], outlined in Fig. 114, the film is thinner than the reaction layer thickness, that is, $L < X_K$,

TABLE X

Expressions for Flux j and Mechanistic Indicators for Microheterogeneous Catalysis at Microparticle–Loaded Ionomer or Microparticle–Redox Polymer Composite Films

Case Designation	Case/Flux Expression Correlation, Flux j
I	$4\pi r_0 N\kappa D_s Ls^\infty$
II	$4\pi r_0 ND_A La^\infty$
III	$4\pi r_0^2 NL\sqrt{k'_{A,0}k'_{S,0}a^\infty \kappa s^\infty}$
IV	$\sqrt{4\pi r_0^2 N\kappa D_s s^\infty}\left(\sqrt{k'_{A,0}k'_{S,0}a^\infty \kappa s^\infty}\right)$
V	$\sqrt{4\pi r_0 N\kappa D_s s^\infty}$
VI(A)	$\dfrac{\kappa D_s s^\infty}{L}$
VI(B)	$\dfrac{D_A a^\infty}{L}$
VII	$\sqrt{4\pi r_0 ND_A a^\infty}$
VIII	$\sqrt{4\pi r_0^2 ND_A a^\infty}\left(\sqrt{k'_{A,0}k'_{S,0}a^\infty \kappa s^\infty}\right)$

	Case/Reaction Order Correlation				
Case	a^∞	s^∞	L	N	r_0
I	0	1	1	1	1
II	1	0	1	1	1
III	$\frac{1}{2}$	$\frac{1}{2}$	1	1	2
IV	$\frac{1}{4}$	$\frac{3}{4}$	0	$\frac{1}{2}$	1
V	0	1	0	$\frac{1}{2}$	$\frac{1}{2}$
VI(A)	0	1	-1	0	0
VI(B)	1	0	-1	0	0
VII	1	0	0	$\frac{1}{2}$	$\frac{1}{2}$
VIII	$\frac{3}{4}$	$\frac{1}{4}$	0	$\frac{1}{2}$	1

and the reaction takes place throughout the film. Thus for all three cases the concentrations of A and S are constant across the film and the flux, and hence the current increases with increasing film thickness. In case I the kinetics are controlled by spherical diffusion of substrate to the microparticles and for case II by the spherical diffusion of mediator to the particles. For case III, the balance between the kinetics for the reaction of mediator and substrate on the microparticle surface determines the flux. From Fig. 114 we see that starting from case II, if $\gamma < 1$, we can move into case III and then into case I by either increasing the concentration of the mediator A or decreasing the concentration of the

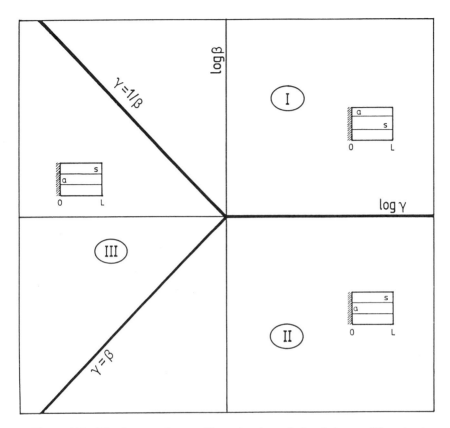

Figure 114. Kinetic case diagram illustrating interrelation between different rate-determining situations for catalytic microparticles immobilized in an ionomer of redox polymer film. Diagram is drawn for $\Phi = L/X_K < 1$. System is characterized by two dimensionless parameters $\beta = \sqrt{k_{Dm,A} a^{\infty}/k_{D,S} s^{\infty}} = \sqrt{D_A a^{\infty}/kD_S s^{\infty}}$ and $\gamma = \sqrt{k'_{A,0} k'_{S,0}/k_{D,A} k_{D,S}} = R\sqrt{k'_{A,0} k'_{S,0}/\kappa D_S D_A}$. Parameter β compares diffusional flux of mediator and diffusional flux of substrate, whereas parameter γ compares electrode kinetics of mediator and substrate at catalytic microparticle with corresponding spherical diffusional kinetics. Concentration profiles for mediator A and substrate S shown in each case.

substrate S. Furthermore if $\gamma > 1$ we move straight from case II to case I without passing through case III. This is because $\gamma > 1$ corresponds to the situation where spherical diffusion represented by the term $\sqrt{D_A D_S}/R$ is slower than the heterogeneous surface kinetics, which is described by the term $\sqrt{k'_{A,0} k'_{S,0}}$.

Turning now to Fig. 115, which is drawn for $L > X_K$, we find that the

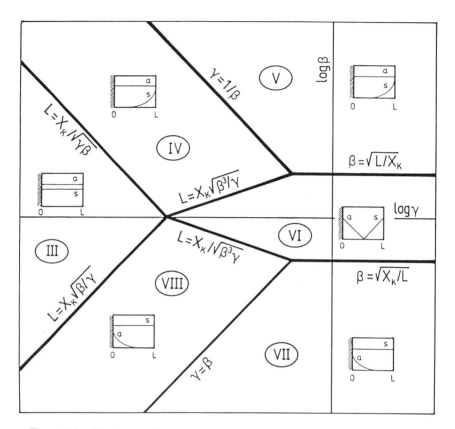

Figure 115. Kinetic case diagram. Parameters as in Fig. 114 but with $\Phi = L/X_K > 1$.

description of the transport and kinetics is more complex and that one has six distinct cases. The reaction layer is narrower than the film thickness, and so concentration polarization of A or S or indeed both within the film is important. In cases IV and V [Eqs. (8.64) and (8.65), respectively], the substrate S is consumed in a reaction layer of dimension X_K at the polymer–solution interface. For case IV the heterogeneous kinetics at the surface of the microparticle are rate determining. As the heterogeneous surface kinetics become more facile (increasing γ), we move into case V, where the spherical diffusion of the substrate to the particles is rate determining. Cases VII and VIII [given by Eqs. (8.51) and (8.69), respectively] are similar, but now it is the reaction of the mediator A at the microparticle surface, that is considered. Again this can be controlled

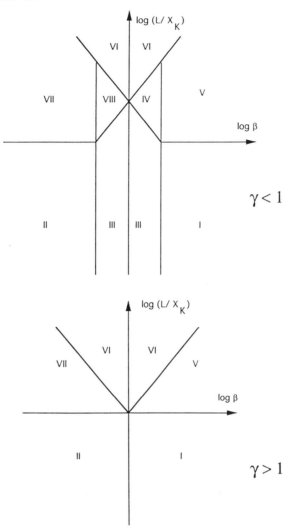

Figure 116.   Different perspective on kinetic case diagram. Here rate-determining cases indicated using $\log \Phi$ and $\log \beta$ as axes. Situations for $\gamma < 1$ and $\gamma > 1$ shown separately.

either by the heterogeneous kinetics (case VIII) or by spherical diffusion (case VII). Finally we have case VI [defined via Eqs. (8.58)–(8.60)]. This corresponds to the titration situation. Mediator diffusing from the electrode meets substrate diffusing in from the solution, and the two react together at some point within the layer. Note that the titration case is only found when spherical diffusion is rate determining. This is becuause when the heterogeneous kinetics are rate limiting, the condition of balancing the reaction rates for A and S at each microparticle means that we obtain case III again. The concentration profiles for each case are outlined as insets in Figs. 114 and 115.

Finally, we note that the case diagrams presented in Figs. 114 and 115 represent horizonal slices of the general three-dimensional case diagram in the $\log \beta$–$\log \gamma$ plane. The parameter L/X_K provides the third dimension. Vertical slices in the $\log(L/X_K)$–$\log \beta$ plane are presented in Fig. 116 for the situations where $\gamma < 1$ and $\gamma > 1$. The various cases viewed from this perspective are also presented in the diagrams. Finally a three-dimensional signpost illustrating the manner in which varying the experimental parameters takes one from one case to another is illustrated in Fig. 117.

E. Microheterogeneous Systems: Development of Optimal Strategy for Electrocatalysis

We complete this section by discussing some implications for electro-catalysis and attempt to determine which case represents the optimum strategy for the utilization of polymer-bound dispersions of microparticles for electrocatalytic applications.

First, let us consider the microcatalytic particles immobilized in a conducting polymer matrix. In cases III and IV (Fig. 112) only the outer region of the film takes part in the reaction since the substrate does not penetrate throughout the layer. This is inefficient in the use of the microcatalytic particles because not all the particles are being utilized. In case I the current is limited by spherical diffusion to the surface of each particle and not by the electrode kinetics on the particle surface. Consequently, this is also an inefficient strategy, because the overpoten-tial applies is now excessive. The only efficient strategy is case II, where all of the catalyst particles are being used and the rate of reaction is determiend by the heterogeneous kinetics on the particle surface. Under these condition the current is greater by a factor of $4\pi R^2 NL$ over that which would be found at a macroscopic planar electrode of the same projected geometric area.

Second, we consider the microcatalyst-loaded ionomer or redox polymer system. We can again identify an optimum strategy. One can immediately reject all cases in which the current response depends upon the concentration of the mediator, since, unless the solubility of the mediator were limited, we could presumably increase the mediator loading. This leaves cases I, V, and possibly VI (Figs. 114 and 115). In cases V and VI only a fraction of the catalyst particles are being used since the substrate does not penetrate the entire film. We therefore reject these

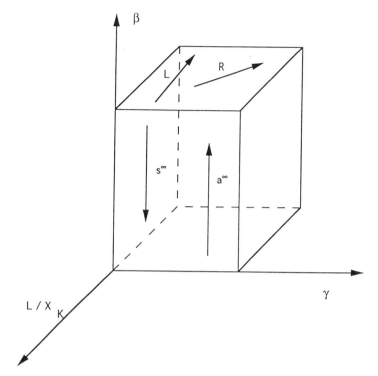

Figure 117. Three-dimensional block diagram indicating how system (and thus rate-determining case) responds to changes in various parameters that may be experimentally varied.

two cases. Thus we are left with case I, where the reaction occurs throughout the entire layer at a rate controlled by the spherical diffusion of reactant to each catalytic particle. This represents the optimum strategy for this class of microheterogeneous systems. In this situation the rate will be ehanced by decreasing the particle radius, increasing the particle loading, or increasing the layer thickness. If the solubility of the mediator is limiting, then this will correspond to a fixed maximum possible value of β for any given substrate concentration.

IX.   CONCLUDING COMMENTS

In this chapter we have presented a fairly comprehensive unified discussion of charge percolation, transport, and kientics in electroactive polymer thin-film-modified electrodes. We have shown that charge transport and substrate kinetics in electroactive polymers are well founded in basic physics and can be quantitatively formulated in terms of simple models.

With respect to charge percolation, we have discussed the mechanism of redox conduction in redox and loaded ionomer materials and have also described delocalized conduction processes in conjugated organic polymer materials. We have also examined ionic and solvent transport processes in these polymeric materials and presented a detailed description of the important technique of complex impedance spectroscopy, which can be used to experimentally probe the mixed-conduction behavior of electroactive polymers. The complex impedance approach results in the formulation of dual-fail transmission line models that to a first approximation well describe the conductance behavior of electroactive polymers.

We have also presented an analysis of the nucleation and growth behavior of electronically conducting polymer materials. These systems exhibit a novel oxidative polycondensation mechanism.

We finally discussed current theoretical approaches used to quantitatively describe mediated electrocatalysis at electroactive polymer thin films. The steady-state amperometric response was derived for a number of different systems by solving the pertinent nonlinear reaction–diffusion equations in a bounded spatial region. Approximate analytical solutions were obtained, which yielded, on analysis, simple expressions for the current response involving experimentally quantifiable parameters such as layer thickness, catalyst loading, and electrode potential. The determination of reaction orders with respect to the latter parameters enables one to unambiguously identify the kientic case. Furthermore, we have shown that the kinetics could be conveniently represented in terms of pictorial case diagrams, which could be used as an aid toward mechanistic diagnosis.

We initially discussed systems where the electroactive polymer exhibited both good redox conductivity and electrocatalytic activity. The discussion was then extended to more interesting and complex situations where the functions of catalysis and conductivity were separated. We discussed microheterogeneous systems, in which distinct catalytic species were incorporated into a polymeric matrix. The cases of incorporated oxide particles in ionomer matrices and metallic microparticles in electroactive polymeric supports were discussed in some detail.

ACKNOWLEDGMENTS

The author acknowledges grant support from the EC Science Programme, the Strategic Research Programme in Advanced Materials (EOLAS), Materials Ireland, and Bioresearch Ireland.

REFERENCES

1. (a) R. W. Murray, in *Electroanalytical Chemistry*, Vol. 13, A. J. Baird, ed., Marcel Dekker, New York (1984) pp. 191–368. (b) R. W. Murray, *An. Rev. Mater. Sci.* **14**, 145 (1984). (c) R. W. Murray, Acc. Chem. Res. **13**, 135 (1980).
2. W. J. Albery and A. R. Hillman in *Ann. Repts. Progr. Chem.*, Vol. 78, Section C, Royal Society of Chemistry, London, 1981, pp. 377–437.
3. A. R. Hillman, in *Electrochemical Science and Technology of Polymers*, Vol. 1, R. G. Linford, ed., Elsevier Applied Science, 1987, pp. 103–239, 241–291.
4. (a) H. D. Abruna, *Coord. Chem. Rev.*, **86**, 135 (1988). (b) H. D. Abruna, in *Electroresponsive Molecular and Polymeric Systems*, Vol. 1, T. A. Skotheim, ed., Marcel Dekker, New York, 1988, pp. 98–160.
5. G. P. Evans, in *Advances in Electrochemical Science and Engineering*, Vol. 1, H. Gerisher and C. W. Tobias, eds., VCH, Weinheim, 1990, pp. 1–74.
6. W. H. Smyrl and M. Lien, in *Applications of Electroactive Polymers*, B. Scrosati, ed., Chapman and Hall, London, 1993, pp. 29–74.
7. R. B. Kaner, in *Electrochemical Science and Technology of Polymers*, Vol. 2, R. G. Linford, ed., Elsevier Applied Science, London, 1990, pp. 97–147.
8. R. A. Pethrick, in *Electrochemical Science and Technology of Polymers*, Vol. 2, R. G. Linford, ed., Elsevier Applied Science, London, 1990, pp. 149–199.
9. M. E. G. Lyons, *Ann. Rep. C., R. Soc. Chem.* **87**, 119 (1990).
10. R. W. Murray, ed., *Molecular Design of Electrode Surfaces*, Techniques of Chemistry Series, Vol. XXII, Wiley Interscience, New York, 1992.
11. M. Majda, in *Molecular Design of Electrode Surfaces*, R. W. Murray, ed., Techniques of Chemistry Series, Vol. XXII, Wiley Interscience, New York, 1992, pp. 159–206.
12. N. Oyama and T. Ohsaka, in *Molecular Design of Electrode Surfaces*, R. W. Murray, ed., Techniques of Chemistry Series, Vol. XXII, Wiley Interscience, New York, 1992, pp. 333–402. This reference contains a wealth of experimental data on ionic diffusion coefficients in electroactive polymers.
13. R. W. Murray in *Molecular Design of Electrode Surfaces*, R. W. Murray, ed., Techniques of Chemistry Series, Vol. XXII, Wiley Interscience, New York, 1992, pp. 1–48.
14. C. P. Andrieux and J. M. Saveant, in *Molecular Design of Electrode Surfaces*, R. W. Murray, ed., Techniques of Chemistry Series, Vol. XXII, Wiley Interscience, New York, 1992, pp. 207–270.
15. *Charge Transfer in Polymeric Systems*, Faraday Discuss. Chem. Soc., R. Soc. Chem., London, Vol. 88, 1989. This volume considers both polymer ionics and modified electrodes based on electroactive polymers.
16. M. E. G. Lyons, ed., *Electroactive Polymer Electrochemistry: Part 1: Fundamentals*, Plenum Press, New York, 1994.
17. M. E. G. Lyons, *Analyst* **119** (1994) 805.
18. B. Scrosati, ed., *Applications of Electroactive Polymers*, Chapman and Hall, London, 1993.
19. C. A. Vincent, in *Electrochemical Science and Technology of Polymers*, Vol. 2, R. G. Linford, ed., Elsevier Applied Science, London, 1990, pp. 47–96.
20. R. G. Linford in *Applications of Electroactive Polymers*, B. Scrosati, ed., Chapman and Hall, London, 1993, pp. 1–28.
21. J. R. Owen, in *Electrochemical Science and Technology of Polymers*, Vol. 1, R. G. Linford, ed., Elsevier Applied Science, London, 1987, pp. 45–66.

22. J. R. MacCallum and C. A. Vincent, eds., *Polymer Electrolyte Reviews*, Vol. 1, Elsevier Applied Science, London, 1988.

23. E. F. Dalton, N. A. Surridge, J. C. Jernigan, K. O. Wilbourn, J. S. Facci, and R. W. Murray, *Chem. Phys.* **141**, 143 (1990).

24. (a) C. M. Clear, J. M. Kelly, D. C. Pepper, and J. G. Vos, *Inorg. Chim. Acta* **33**, 139 (1979). (b) C. M. Clear, J. M. Kelly, C. M. O'Connell, and J. G. Vos, *J. Chem. Res. (M)*, 3037 (1981).

25. (a) O. Haas and J. G. Vos., *J. Electroanal. Chem.* **113**, 139 (1980). (b) O. Haas, M. Kriens, and J. G. Vos, *J. Am. Chem. Soc.* **103**, 1318 (1981). (c) O. Haas, H. R. Zumbrunnen, and J. G. Vos, *Electrochim. Acta* **30**, 1551 (1985). (d) S. M. Geraty and J. G. Vos, *J. Electroanal. Chem.* **176**, 389 (1984); *J. Chem. Soc. Dalton Trans.* 3073 (1987). (e) M. E. G. Lyons, H. G. Fay, J. G. Vos, and A. J. Kelly, *J. Electroanal. Chem.* **250**, 207 (1988). (f) R. J. Forster, A. J. Kelly, J. G. Vos, and M. E. G. Lyons, *J. Electroanal. Chem.* **270**, 365 (1989). (g) R. J. Forster and J. G. Vos, *Macromolecules*, **23**, 4372; (1990); *J. Inorg. Organomet. Polym.* **1**, 67 (1991). (h) R. J. Forster and J. G. Vos, *J. Electroanal. Chem.* **314**, 135 (1991); *Electrochim. Acta* **37**, 159 (1992). (i) R. J. Forster, J. G. Vos, and M. E. G. Lyons, *J. Chem. Soc. Faraday Trans.* **87**, 3761, 3769 (1991).

26. A. P. Clarke and J. G. Vos, *Trends Electrochem.* **1**, 167 (1992).

27. R. J. Gale, ed., *Spectroelectrochemistry: Theory and Practice*, Plenum Press, New York, 1988.

28. V. G. Levich, *Adv. Electrochem. Electrochem. Eng.* **4**, 314 (1966).

29. I. Ruff, *Electrochim. Acta* **15**, 1059 (1970).

30. (a) F. B. Kaufman and M. B. Engler, *J. Am. Chem. Soc.* **101**, 547 (1979). (b) F. B. Kaufmann, A. M. Schroeder, E. M. Engler, S. R. Kramer, and J. Q. Chambers, *J. Am. Chem. Soc.* **102**, 483 (1980).

31. C. P. Andrieux and J. M. Saveant, *J. Electroanal. Chem.* **111**, 377 (1980).

32. E. Laviron, *J. Electroanal. Chem.* **112**, 1 (1980).

33. I. Ruff and L. Botar, *J. Chem. Phys.* **83**, 1292 (1985).

34. L. Botar and I. Ruff, *Chem. Phys. Lett.* **126**, 348 (1986).

35. H. Dahms, *J. Phys. Chem.* **72**, 362 (1968).

36. (a) I. Ruff, V. J. Friedrich, K. Demeter, and K. Csillag, *J. Phys. Chem.* **75**, 3303 (1971). (b) I. Ruff and V. J. Friedrich, *J. Phys. Chem.* **75** 3297 (1971).

37. M. Sharp, B. Lindholm, and E. L. Lind, *J. Electroanal. Chem.* **274**, 35 (1989).

38. D. N. Blauch and J. M. Saveant, *J. Am. Chem. Soc.* **114**, 3323 (1992).

39. I. Fritsch–Faules and L. R. Faulkner, *J. Electroanal. Chem.* **263**, 237 (1989).

40. L. R. Faulkner, *Electrochim. Acta* **34**, 1699 (1989).

41. P. He and X. Chen, *J. Electroanal. Chem.* **256**, 353 (1988).

42. S. R. Broadbent and J. M. Hammersley, *Proc. Cambridge Phil. Soc.* **53**, 629 (1957).

43. V. K. S. Shante and S. Kirkpatrick, *Adv. Phys.* **20**, 325 (1971).

44. S. Kirkpatrick, *Rev. Mod. Phys.* **45**, 574 (1973).

45. R. Zallen, *The Physics of Amorphous Solids*, Wiley, New York, 1983, Chapter 4.

46. G. S. Grest, I. Webman, S. A. Safran, and A. L. R. Bug, *Phys. Rev. A* **33**, 2842 (1986).

47. A. L. R. Bug and Y. Gefen, *Phys. Rev. A* **35**, 1301 (1987).

48. S. D. Druger, A. Nitzan, and M. A. Ratner, *J. Chem. Phys.* **79**, 3133 (1983).

49. S. D. Druger, M. A. Ratner, and A. Nitzan, *Phys. Rev. B* **31**, 3939 (1985).

50. M. A. Ratner and A. Nitzan, *Faraday Discuss. Chem. Soc.* **88**, 19 (1989).

51. R. Granek and A. Nitzan, *J. Chem. Phys.* **90**, 3784 (1989).

52. R. Granek, A. Nitzan, S. D. Druger, and M. A. Ratner, *Solid State Ionics* **28/30**, 128 (1988).

53. S. D. Druger, M. A. Ratner, and A. Nitzan, *Mol. Cryst. Liq. Cryst.* **190**, 171 (1990).

54. J. H. Noggle, *Physical Chemistry*, Little Brown, Boston, 1985, pp. 430–439.

55 K. W. Kehr, R. Kutner, and K. Binder, *Phys. Rev. B* **23**, 4931 (1981).

56. R. Kutner and K. W. Kehr, *Philos. Mag. A* **48**, 199 (1983).

57. (a) K. Nakazato and K. Kitahara, *Prog. Theor. Phys.* **64**, 2261 (1980). (b) M. Koiwa and S. Ishioka, *J. Stat. Phys.* **30**, 477 (1983).

58. G. E. Murch and R. J. Thorn, *Philos, Mag. A* **39**, 673 (1979).

59. P. A. Fedders and O. F. Sankey, *Phys. Rev. B* **18**, 18 (1978).

60. S. Chandrasekhar, *Rev. Mod. Phys.* **15**, 1 (1943).

61. (a) J. R. Miller, *J. Chem. Phys.* **56**, 5173 (1972). (b) J. R. Miller, *Science* **189**, 221 (1975). (c) J. V. Bietz and J. R. Miller, *J. Chem. Phys.* **71**, 4579 (1979); **74**, 6746 (1981). (d) J. R. Miller, K. W. Hartman, and S. Abrash, *J. Am. Chem. Soc.* **104**, 4296 (1982). (e) J. R. Miller, J. V. Beitz, and R. K. Huddleston, *J. Am. Chem. Soc.* **106**, 5057 (1984). (f) J. R. Miller, *N. J. Chem.* **11**, 83 (1987).

62. S. W. Feldberg, *J. Electroanal. Chem.* **198**, 1 (1986).

63. (a) J. J. Hopfield, *Proc. Nat. Acad. Sci., U.S.A.* **71**, 3640 (1974). (b) J. J. Jortner, *J. Chem. Phys.* **64**, 4860 (1976). (c) B. Chance, D. DeVault, H. Frauenfelder, R. A. Marcus, J. R. Schrieffer, and N. Sutin, eds., *Tunnelling in Biological Systems*, Academic Press, New York, 1979. (d) S. J. Larsson, *J. Chem. Soc., Faraday Trans. II.* **79**, 1375 (1983). (e) A. A. S. daGama, *Theor. Chim. Acta* **68**, 159 (1985).

64. T. J. Lewis, *Faraday Discuss. Chem. Soc.* **88**, 189 (1989).

65. The literature on electron transfer in metalloproteins is extensive. For a list of pertinent references see ref. 8 on the Fritsch-Faules/Faulkner study (ref. 39) and also J. R. Bolton, N. Mataga, and G. McLendon, eds., *Electron Transfer in Inorganic, Organic and Biological Systems*, Advances in Chemistry Series 228, ACS, Washington, DC, 1991, especially pp. 71–90, pp. 191–200, pp. 201–214, and pp. 229–246.

66. (a) J. S. Facci, R. H. Schmehl, and R. W. Murray, *J. Am. Chem. Soc.* **104**, 4959 (1982). (b) M. Majda and L. R. Faulkner, *J. Electroanal. Chem.* **169**, 77 (1984).

67. (a) M. Majda and L. R. Faulkner, *J. Electroanal. Chem.* **137**, 149 (1982). (b) M. Majda and L. R. Faulkner, *J. Electroanal. Chem.* **169**, 97 (1984).

68. (a) D. A. Buttry and F. C. Anson, *J. Electroanal. Chem.* **130**, 333 (1982). (b) D. A. Buttry and F. C. Anson, *J. Am. Chem. Soc.* **105**, 685 (1983).

69. M. Von Smoluchowski, *Z. Phys. Chem.* **92**, 129 (1917).

70. (a) R. A. Marcus, *J. Chem. Phys.* **24**, 966 (1956), **26**, 867, 872 (1957); *Can. J. Chem.*, **37**, 155 (1959). (b) R. A. Marcus in *Special Topics in Electrochemistry*, P. A. Rock, ed., Elsevier, New York, 1977, pp. 161, 180. (c) R. A. Marcus, *Trans. N.Y. Acad. Sci.* **19**, 423 (1957). (d) R. A. Marcus, *J. Phys. Chem.* **67**, 853 (1963). (e) R. A. Marcus, *Ann. Rev. Phys. Chem.* **15**, 155 (1964). (f) R. A. Marcus, *J. Chem. Phys.* **43**, 679 (1965). (g) R. A. Marcus, *Electrochim. Acta* **13**, 997 (1968). (h) R. A. Marcus, *J.*

Phys. Chem. **72**, 891 (1968). (i) R. A. Marcus, *Int. J. Chem. Kin.* **13**, 865 (1981). (j) R. A. Marcus and N. Sutin, *Inorg. Chem.* **14**, 21 (1975). (k) R. A. Marcus, *Faraday Discuss. Chem. Soc.* **29**, 21 (1960); **74**, 7 (1982).

71. (a) H. Heitele, *Angew. Chem. Int. Ed. Engl.* **32**, 359 (1993). (b) K. V. Mikkelsen and M. A. Ratner, *Chem. Rev.* **87**, 113 (1987).

72. J. R. Bolton and M. D. Archer in *Electron Transfer in Inorganic, Organic and Biological Systems*, J. R. Bolton, N. Mataga, and G. McLendon, eds., Advances in Chemistry Series 228, ACS, Washington DC, 1991, pp. 7–23.

73. (a) R. A. Marcus and N. Sutin, *Biochim. Biophys. Acta* **811**, 265 (1985). (b) R. A. Marcus and N. Sutin, *Comments Inorg. Chem.* **5**, 119 (1986).

74. (a) M. J. Weaver in *Comprehensive Chemical Kinetics*, R. G. Compton, ed., Elsevier, Amsterdam, 1987, Vol. 27, *Electrode Kinetics: Reactions*, pp. 1–60 and references therein. (b) J. T. Hupp and M. J. Weaver, *J. Electroanal. Chem.* **152**, 1 (1983). (c) J. T. Hupp and M. J. Weaver, *J. Electroanal. Chem.* **145**, 43 (1983). (d) J. T. Hupp and M. J. Weaver, *J. Phys. Chem.* **89**, 2795 (1985) and references therein. (e) M. J. Weaver and G. E. McManis III, *Acc. Chem. Res.* **23**, 294 (1990). (f) M. J. Weaver, *Chem. Rev.* **92**, 463 (1992).

75. (a) N. Sutin, *Prog. Inorg. Chem.*, vol. 30, S. J. Lippard, ed., Wiley Interscience, New York, 1983, pp. 441–498 and references therein. (b) N. Sutin, in *Electron Transfer in Inorganic, Organic and Biological Systems*, J. R. Bolton, N. Mataga, and G. McLendon, eds., Advances in Chemistry Series 228, ACS, Washington DC 1991, pp. 25–43. (c) N. Sutin, *Acc. Chem. Res.* **15**, 275 (1982).

76. M. D. Newton and N. Sutin, *Ann. Rev. Phys. Chem.* **35**, 437 (1984).

77. (a) W. J. Albery, *Ann. Rev. Phys. Chem.* **31**, 227 (1980). (b) W. J. Albery, *Electrode Kinetics*, Clarendon Press, Oxford, 1975, pp. 92–124.

78. J. M. Saveant, *J. Electroanal. Chem.* **201**, 211 (1986).

79. J. M. Saveant, *J. Electroanal. Chem.* **238**, 1 (1987).

80. J. M. Saveant, *J. Electroanal. Chem.* **242**, 1 (1988).

81. J. M. Saveant, *J. Phys. Chem.* **92**, 4526 (1988).

82. J. M. Saveant, *J. Phys. Chem.* **92**, 1011 (1988).

83. C. P. Andrieux and J. M. Saveant, *J. Phys. Chem.* **92**, 6761 (1988).

84. R. P. Buck, *J. Electroanal. Chem.* **219**, 23 (1987).

85. (a) R. P. Buck, *J. Phys. Chem.* **92**, 4196 (1988). (b) R. P. Buck, *J. Phys. Chem.* **92**, 6445 (1988).

86. R. P. Buck, *J. Electroanal. Chem.* **243**, 279 (1988).

87. (a) R. P. Buck, *J. Electroanal. Chem.* **258**, 1 (1989). (b) R. P. Buck, *J. Phys. Chem.* **93**, 6212 (1989).

88. C. J. Baldy, C. M. Elliott, and S. W. Feldberg, *J. Electroanal. Chem.* **283**, 53 (1990).

89. W. J. Albery, Z. Chen, B. R. Horrocks, A. R. Mount, P. J. Wilson, D. Bloor, A. T. Monkman, and C. M. Elliott, *Faraday Discuss. Chem. Soc.* **88**, 247 (1989).

90. I. Rubenstein, E. Sabatani, and J. Rishpon, *J. Electrochem. Soc.* **134**, 3078 (1987).

91. R. P. Buck, *J. Electroanal. Chem.* **210**, 1 (1986).

92. W. J. Albery, C. M. Elliott, and A. R. Mount, *J. Electroanal. Chem.* **288**, 15 (1990).

93. W. J. Albery and A. R. Mount, *J. Electroanal. Chem.* **305**, 3 (1991).

94. W. J. Albery and A. R. Mount, *J. Chem. Soc., Faraday Trans.* **89** 327 (1993).

95. W. J. Albery and A. R. Mount in *Electroactive Polymer Electrochemistry, Part 1: Fundamentals*, M. E. G. Lyons, ed., Plenum Press, New York, 1994, pp. 443–483.

96. S. Fletcher, *J. Electroanal. Chem.* **337**, 127 (1992).

97. S. Fletcher, *J. Chem. Soc., Faraday Trans.* **89**, 311 (1993).

98. Th. M. Nieuwenhuizen and J. M. Luck, cited in *Adv. Phys.* **39**, 191 (1990).

99. (a) K. Doblhofer and M. Vorotyntsev in *Electroactive Polymer Electrochemistry, Part 1: Fundamentals*, M. E. G. Lyons, ed., Plenum Press, New York, 1994, pp. 375–442. (b) A. Eisenberg, *Macromolecules*, **3**, 147 (1970). (c) K. A. Mauritz and A. J. Hopfinger in *Modern Aspects of Electrochemistry*, Vol. 16, J. O'M. Bockris, B. E. Conway, and R. E. White, eds., Plenum Press, New York, 1982, pp. 425–508.

100. (a) S. Roth and H. Bleier, *Adv. Phys.* **36**, 385 (1987). (b) S. Roth, H. Bleier, and W. Pukacki, *Faraday Discuss. Chem. Soc.* **88**, 223 (1989).

101. A. J. Heeger, S. Kivelson, J. R. Schrieffer, and W. P. Su, *Rev. Mod. Phys.* **60**, 781 (1988).

102. W. P. Su, J. R. Schrieffer, and A. J. Heeger, *Phys. Rev.*, B **22**, 2209 (1980); erratum, **28**, 1138 (1983).

103. S. Kivelson and A. J. Epstein, *Phys. Rev.* B **29**, 336 (1984).

104. J. C. Scott, P. Pfluger, M. Krombi, and G. B. Street, *Phys. Rev.* B **28**, 40 (1983).

105. (a) N. F. Mott, *Conduction in Noncrystalline Materials*, Oxford University Press, Oxford, 1987, p. 28. (b) N. F. Mott and E. A. Davis, *Electronic Processes in Noncrystalline Materials*, 2nd ed., Clarendon Press, Oxford, 1979.

106. A. J. Epstein in *Handbook of Conducting Polymers*, Vol. 2, T. Skotheim, ed., Marcel Dekker, New York, 1986, p. 1041.

107. R. Zallen, *The Physics of Amorphous Solids*, Wiley Interscience, New York, 1983, pp. 135–204.

108. D. Stauffer, *Introduction of Percolation Theory*, Taylor and Francis, London, 1985.

109. D. Kim, H. Reiss, and H. M. Rabeony, *J. Phys. Chem.* **92**, 2673 (1988).

110. W. D. Murphy, H. M. Rabeony, and H. Reiss, *J. Phys. Chem.* **92**, 7007 (1988).

111. A. Prock and W. P. Giering, *J. Phys. Chem.* **93**, 2192 (1989).

112. A. Prock and W. P. Giering, *J. Phys. Chem.* **93**, 8382 (1989).

113. (a) P. N. Bartlett and J. W. Gardner, *Phil. Trans. Roy. Soc.*, in press. (b) J. W. Gardner and P. N. Bartlett, *Synth. Met.*, in press.

114. J. J. Hermans, *J. Colloid Sci.* **2**, 387 (1947).

115. H. S. Carlsaw and J. C. Jaeger, *Conduction of Heat in Solids*, Oxford University Press, Oxford, 1959, p. 284.

116. J. Callaway, *Quantum Theory of the Solid State*, Academic Press, New York, 1974.

117. W. Jones and N. H. March, *Theoretical Solid State Physics*, Dover, New York, 1973.

118. (a) J. L. Bredas and R. R. Chance, *Phys. Rev. B.* **26**, 2886 (1982). (b) J. L. Bredas, R. R. Chance, and R. Silbey, *Phys. Rev. B.* **26**, 5843 (1982). (c) R. R. Chance, D. S. Boudreaux, J. F. Wolf, L. W. Shacklette, R. Silbey, B. Themans, J. M. Andre, and J. L. Bredas, *Synth. Met.* **15**, 105 (1986) and references therein.

119. J. L. Bredas, B. Themans, J. M. Andre, R. R. Chance, D. S. Boudreaux, and R. Silbey, *J. Phys. Colloq.* **C3**, 44, 374 (1983).

120. A. R. Bishop, D. K. Campbell, and K. Fesser, *Mol. Cryst. Liq. Cryst.* **77**, 253 (1980).

121. S. Kivelson, *Phys. Rev.* B **25**, 3798 (1982).

122. P. D. Townsend and R. H. Friend, *Synth. Met.* **28**, 735 (1989); *Phys. Rev. B* **40**, 3112 (1989).

123. R. R. Chance, J. L. Bredas, and R. Sibley, *Phys. Rev. B* **29**, 4491 (1984).

124. R. H. Friend and J. H. Burroughs, *Faraday Discuss. Chem. Soc.* **88**, 213 (1989).

125. J. L. Bredas and G. B. Street, *Acc. Chem. Res.* **18**, 309 (1985).

126. J. L. Bredas, B. Themans, and J. M. Andre, *Phys. Rev. B* **27**, 7827 (1983).

127. J. L. Bredas, B. Themans, J. G. Fripiat, J. M. Andre, and R. R. Chance, *Phys. Rev. B* **29**, 6761 (1984).

128. J. L. Bredas, *Mol. Cryst. Liq. Cryst.* **118**, 49 (1985).

129. A. J. Heeger, *Faraday Discuss. Chem. Soc.* **88**, 203 (1989).

130. A. J. Heeger, S. Kivelson, J. R. Schrieffer, and W. P. Su, *Rev. Mod. Phys.* **60**, 781 (1988).

131. S. Kivelson and A. J. Heeger, *Synth. Met.* **22**, 371 (1988).

132. D. S. Pearson, P. A. Pincus, G. W. Heffner, and S. J. Dahman, *Macromolecules* **26**, 1570 (1993).

133. P. G. de Gennes, *C. R. Acad. Sci. Paris, Ser. 2* **302**, 1 (1986).

134. H. S. Carslaw and J. C. Jaeger, *Conduction of Heat in Solids*, 2nd ed., Oxford University Press, Oxford, 1959. Chapter XIV.

135. J. Moulton and P. Smith, *Synth. Met.* **40**, 13 (1991).

136. (a) S. Tokito, P. Smith, and A. J. Heeger, *Polymer* **32**, 464 (1991). (b) Y. Cao, P. Smith, and A. J. Heeger, *Polymer* **32**, 1210 (1991).

137. (a) P. Pincus, G. Rossi, and M. E. Cates, *Europhys. Lett.* **4**, 41 (1987). (b) D. R. Spiegel, P. Pincus, and A. J. Heeger, *Polym. Commun.* **29**, 264 (1988). (c) A. Viallat and P. A. Pincus, *Polymer* **30**, 1997 (1989).

138. (a) K. Y. Jen, R. Oboodi, and R. L. Elsenbaumer, *Polym. Mater. Sci.* **53**, 79 (1985). (b) M. Sato, S. Tanaka, and K. Kaeriyama, *Chem. Commun.* 873 (1986). (c) A. O. Patil, Y. Ikenoue, N. Basescu, J. Chem, F. Wudl, and A. J. Heeger, *Synth. Met.* **20**, 151 (1987). (d) D. Spiegel and A. J. Heeger, *Polym. Commun.* **29**, 266 (1988).

139. P. J. Flory, *Statistical Mechanics of Chain Molecules*, Wiley, Interscience, New York, 1969.

140. (a) K. Aoki and Y. Tezuka, *J. Electroanal. Chem.* **267**, 55 (1989). (b) Y. Tezuka, K. Aoki, and K. Shinozaki, *Synth. Met.* **30**, 369 (1989). (c) Y. Tezuka and K. Aoki, *J. Electroanal. Chem.* **273**, 161 (1989).

141. (a) K. Aoki, Y. Tezuka, K. Shinozaki, and H. Sato, *J. Electrochem. Soc. Jpn.* **57**, 397 (1989). (b) K. Aoki, *J. Electroanal. Chem.* **292**, 53 (1990).

142. K. Aoki, *J. Electroanal. Chem.* **310**, 1 (1991).

143. K. Aoki, T. Aramoto, and Y. Hoshino, *J. Electroanl. Chem.* **340**, 127 (1992).

144. M. Kalaji, L. M. Peter, L. M. Abrautes, and J. C. Mesquita, *J. Electroanal. Chem.* **274**, 289 (1989).

145. A. Bunde and S. Havlin, eds., *Fractals and Disordered Systems*, Springer Verlag, Berlin, 1993.

146. K. Aoki, J. Cao, and Y. Hoshino, *Electrochim. Acta* **38**, 1711 (1993).

147. K. Aoki, J. Cao, and Y. Hoshino, *Electrochim. Acta* **39**, 2291 (1994).

148. T. Koyayashi, H. Yoneyama, and H. Tamura, *J. Electroanal. Chem.* **177**, 281 (1984).

149. C. Barbero, M. C. Miras, O. Haas, and R. Kotz, *J. Electrochem. Soc.* **138**, 669 (1991).

150. (a) B. J. Feldmann, P. Burgmayer, and R. W. Murray, *J. Am. Chem. Soc.* **107**, 872 (1985). (b) W. W. Focke and G. E. Wnek, *J. Electroanal. Chem.* **256**, 343 (1988).

151. A. G. Macdiarmid, J. C. Chiang, A. F. Richer, and A. J. Epstein, *Synth. Met.* **18**, 285 (1986).

152. A. B. Brown and F. C. Anson, *Anal. Chem.* **49**, 1589 (1977).

153. W. J. Albery, M. G. Boutelle, P. J. Colby, and A. R. Hillman, *J. Elecroanal. Chem.* **133**, 135 (1982).

154. E. Laviron, *J. Electroanal. Chem.* **63**, 245 (1975).

155. H. Matsuda, K. Aoki, and K. Tokuda, *J. Electroanal. Chem.* **217**, 1 (1987); **217**, 15 (1987).

156. H. Daifuku, K. Aoki, K. Tokuda, and H. Matsuda, *J. Electroanal. Chem.* **183**, 1 (1985).

157. C. E. D. Chidsey and R. W. Murray, *J. Phys. Chem.* **90**, 1479 (1986).

158. D. Ellis, M. Eckhoff, and V. D. Neff, *J. Phys. Chem.* **85**, 1225 (1981).

159. S. T. Coleman, W. R. McKinnon, and J. R. Dahn, *Phys. Rev. B* **29**, 4147 (1984).

160. R. J. Gale, ed., *Spectroelectrochemistry: Theory and Practice*, Plenum Press, New York, 1988, and references therein.

161. D. A. Buttry and M. D. Ward, *Chem. Rev.* **92**, 1355 (1992).

162. J. H. Kaufman, K. K. Kanazawa, and G. B. Street, *Phys. Rev. Lett.* **53**, 2461 (1984).

163. S. Burckenstein and M. Shay, *J. Electroanal. Chem.* **188**, 131 (1985).

164. S. Bruckenstein and S. Swathirajan, *Electrochim. Acta* **30**, 851 (1985).

165. P. T. Varineau and D. A. Buttry, *J. Phys. Chem.* **91**, 1292 (1987).

166. D. Orata and D. A. Buttry, *J. Am. Chem. Soc.* **109**, 3574 (1987).

167. M. D. Ward, *J. Phys. Chem.* **92**, 2049 (1988).

168. A. R. Hillman, D. C. Loveday, M. J. Swann, R. M. Eales, A. Mamnett, S. J. Higgins, S. Bruckenstein, and C. P. Wilde, *Faraday Discuss. Chem. Soc.* **88**, 151 (1989).

169. A. R. Hillman, D. C. Loveday, S. Bruckenstein, and C. P. Wilde, *J. Chem. Soc. Faraday Trans.* **86**, 437 (1990).

170. G. Suaerbrey, *Z. Phys.* **155**, 206 (1959).

171. S. Bruckenstein and A. R. Hillman, *J. Phys. Chem.* **92**, 4837 (1988).

172. C. K. Baker and J. P. Reynolds, *J. Electroanal. Chem.* **251**, 307 (1988).

173. A. J. Kelly and N. Oyama, *J. Phys. Chem.* **95**, 9579 (1991).

174. S. Bruckenstein, C. P. Wilde, M. Shay, A. R. Hillman, and D. C. Loveday, *J. Electroanal. Chem.* **258**, 457 (1989).

175. A. R. Hillman, M. J. Swann, and S. Bruckenstein, *J. Phys. Chem.* **95**, 3271 (1991).

176. A. R. Hillman, D. C. Loveday, and S. Bruckenstein, *J. Electroanal. Chem.* **300**, 67 (1991).

177. A. R. Hillman and S. Bruckenstein, *J. Chem. Soc. Faraday Trans.* **89**, 339 (1993).

178. B. S. H. Royce, D. Voss, and A. Bocarsly, *J. Phys. Chem.* **88**, 325 (1983).

179. J. Pawliszyn, M. F. Weber, M. J. Dignam, A. Mandelis, R. D. Venter, and S. M. Park, *Anal. Chem.* **58**, 236 (1986).

180. J. Pawliszyn, M. F. Weber, M. J. Dignam, A. Mandelis, R. D. Venter, and S. M. Park, *Anal. Chem.* **58**, 239 (1986).

181. J. Pawliszyn, *Anal. Chem.* **60**, 1751 (1988).

182. R. E. Russon, F. R. McLarnon, J. D. Spear, and E. J. Cairns, *J. Electrochem. Soc.* **134**, 2783 (1987).

183. A. Mandelis and B. S. H. Royce, *Appl. Opt.* **23**, 2892 (1984).

184. E. M. Genies, M. Lapkowski, and C. Tsintavis, *N. J. Chem.* **12**, 181 (1988).

185. A. Kitiani, I. Izumi, J. Yano, Y. Hiromoto, and K. Sasaki, *Bull. Chem. Soc. Jpn.* **57**, 2254 (1984).

186. G. Horanyi and G. Inzelt, *Electrochim. Acta* **33**, 947 (1988).

187. W. W. Focke, G. E. Wnek, and Y. Wei, *J. Phys. Chem.* **91**, 5813 (1987).

188. T. Kobayashi, H. Yoneyama, and H. Tamura, *J. Electroanal. Chem.* **177** 281 (1984).

189. W. S. Huang, B. D. Humphrey, and A. G. MacDiarmid, *J. Chem. Soc. Faraday Trans. I* **82**, 2385 (1986).

190. K. Shimazu, K. Murakoshi, and K. Kita, *J. Electroanal. Chem.* **277**, 347 (1990).

191. C. Barbero, M. C. Miras, O. Haas, and R. Kotz, *J. Electrochem. Soc.* **138**, 669 (1991).

192. O. Haas, *Faraday Discuiss. Chem. Soc.* **88**, 123 (1989).

193. (a) J. H. Edwards and W. J. Feast, *Polymer* **21**, 595 (1980). (b) J. H. Edwards, W. J. Feast, and D. C. Bott., *Polymer* **25**, 395 (1984).

194. E. M. Genies, G. Bidan, and A. F. Diaz, *J. Electroanal. Chem.* **149**, 101 (1983).

195. C. K. Baker and J. R. Reynolds, *J. Electroanal. Chem.* **251**, 307 (1988).

196. A. F. Diaz, J. Crowley, J. Bargon, G. P. Gardini, and J. B. Torrance, *J. Electroanal. Chem.* **121**, 355 (1981).

197. R. J. Waltman and J. Bargon, *Tetrahedron* **40** 3963 (1984).

198. C. K. Baker and J. R. Reynolds, *Polym. Prepr.* **28**, 284 (1987).

199. R. J. Waltman, A. F. Diaz, and J. Bargon, *J. Phys. Chem.* **88**, 4343 (1984).

200. R. J. Waltman and J. Bargon, *Can. J. Chem.* **64**, 76 (1986).

201. C. P. Andrieux, P. Adebert, P. Hapiot, and J. M. Saveant, *J. Phys. Chem.* **95**, 10158 (1991).

202. P. Lang, F. Chao, M. Costa, and F. Garnier, *Polymer* **28**, 668 (1987).

203. K. Tanaka, T. Shichiri, S. Wang, and T. Yamabe, *Synth. Met.* **24**, 203 (1988).

204. A. Hamnett, P. A. Christensen, and S. J. Higgins, *Analyst* **119**, 735 (1994).

205. M. Fleischmann and H. R. Thirsk, *Adv. Electrochem. Electrochem. Eng.* **3**, 123 (1963).

206. J. A. Harrison and H. R. Thirsk, *Electroanal. Chem.* **5**, 67 (1971).

207. R. De Levie, *Adv. Electrochem. Electrochem. Eng.* **13**, 1 (1984).

208. *Faraday Symp. Chem. Soc.* **12** (1977) and references therein.

209. E. Budevski and W. J. Lorenz, eds., Electrocrystallization, *Electrochim. Acta* **28**, 863–1010 (1983) and references therein.

210. D. Pletcher, *A First Course in Electrode Processes*, Electrochemical Consultancy, 1991, pp. 37–44.

211. J. B. Zeldovich, *Acta Phys. URSS* **18**, 1 (1943).

212. Southampton Electrochemistry Group, *Instrumental Methods in Electrochemistry*, Ellis Horwood, Chichester, 1985, pp. 283–316.

213. (a) S. K. Rangarajan, *Faraday Symp. Chem. Soc.* **12**, 101 (1977). (b) H. Angerstein-Kozlowska, B. E. Conway, and J. Klinger, *J. Electroanal. Chem.* **87**, 301, 321 (1978).

214. (a) I. Epelboin, M. Ksouri, and R. Wiart, *Faraday Symp. Chem. Soc.* **12** 115 (1977). (b) C. Cachet, I. Epelboin, M. Keddam, and R. Wiart, *J. Electroanal. Chem.* **100**, 745 (1979). (c) C. Cachet, M. Froment, and R. Wiart, *Electrochim. Acta* **24**, 713 (1979).

(d) C. Cachet, C. Gabrielli, F. Huet, M. Keddam, and R. Wiart, *Electrochim. Acta* **28**, 899 (1983). (e) R. Wiart, *Electrochim. Acta* **35**, 1587 (1990).

215. B. Scharifker and G. Hills, *Electrochim. Acta* **28**, 879 (1983).

216. (a) G. Gunawardena, G. J. Hills, I. Montengegro, and B. Scharifker, *J. Electroanal. Chem.* **138**, 225 (1982). (b). G. J. Hills, A. K. Pour, and B. Scharifker, *Electrochim. Acta* **28**, 891 (1983).

217. M. Avrami, *J. Chem. Phys.* **7**, 1130 (1939); **8**, 212 (1940); **9**, 177 (1941).

218. Southampton Electrochemistry Group, *Instrumential Methods in Electrochemistry*, Ellis Horwood, Chichester, 1985, pp. 300–302.

219. J. A. Harrison and S. K. Rangarajan, *Faraday Symp. Chem. Soc.* **12**, 70 (1977).

220. E. Bosco and S. K. Rangarajan, *J. Chem. Soc. Faraday Trans. I* **77**, 1673 (1981).

221. R. D. Armstrong, M. Fleischmann, and H. R. Thirsk, *J. Electroanal. Chem.* **11**, 208 (1966).

222. E. Bosco and S. K. Rangarajan, *J. Electroanal. Chem.* **134**, 213 (1982).

223. E. Bosco and S. K. Rangarajan, *J. Electroanal. Chem.* **134**, 225 (1982).

224. (a) S. Asavapiriyanont, G. K. Chandler, G. A. Gunawardena, and D. Pletcher, *J. Electroanal. Chem.* **177**, 229 (1984); (b) G. K. Chandler and D. Pletcher, *Specialist Periodical Reports. Roy. Soc. Chem.* **10**, 117 (1986).

225. A. J. Downard and D. Pletcher, *J. Electroanal. Chem.* **206**, 139 (1986).

226. A. R. Hillman and E. F. Mallen, *J. Electroanal. Chem.* **220**, 351 (1987).

227. A. R. Hillman and M. J. Swann, *Electrochim. Acta* **33**, 1303 (1988).

228. A. R. Hillman and E. F. Mallen, *J. Electroanal. Chem.* **243**, 403 (1988).

229. A. Hamnett and A. R. Hillman, *J. Electrochem. Soc.* **135**, 2517 (1988).

230. (a) M. L. Marcos, I. Rodrigues, and J. G. Velasco, *Electrochim. Acta* **32**, 1453 (1987). (b) G. Zotti, S. Cattarin, and N. Comisso, *J. Electroanal. Chem.*, **235** (1987) 259; (c) G. Zotti, S. Cattarin and N. Comisso, *J. Electroanal. Chem.* **239**, 387 (1988). (d) B. R. Scharifker, E. Garcia-Pastoriza, and W. Marino, *J. Electroanal. Chem.* **300**, 85 (1991). (e) L. L. Miller, B. Zinger, and Q. X. Zhou, *J. Am. Chem. Soc.* **109**, 2267 (1987). (f) U. Konig and J. W. Schultze, *J. Electroanal. Chem.* **242**, 243 (1988). (g) Y. Wei, Y. Sun, and X. Tang, *J. Phys. Chem.* **93**, 4878 (1989). (h) G. Dian, N. Merlet, G. Barkey, F. Outurquin, and C. Paulmier, *J. Electroanal. Chem* **238**, 225 (1987). (i) R. Yang, D. F. Evans, K. M. Dalsin, L. Christensen, and W. A. Hendrickson, *J. Phys. Chem.* **93**, 511 (1989). (j) R. Yang, D. F. Evans, L. Christensen, and W. A. Hendrickson, *J. Phys. Chem.* **94**, 6117 (1990). (k) R. John and G. G. Wallace, *J. Electroanal. Chem.* **306**, 157 (1991). (l) F. Beck, *Electrochim. Acta* **33**, 839 (1988). (m). I. Rubenstein, J. Rishpon, E. Sabatani, A. Redondo, and S. Gottsefeld, *J. Am. Chem. Soc.* **112**, 6135 (1990).

231. T. McCabe, Ph.D. Thesis, University of Dublin, 1992.

232. F. Li and W. J. Albery, *Electrochim. Acta.* **37**, 393 (1992).

233. C. H. Lyons, Ph.D. Thesis, University of Dublin, 1993.

234. J. O'M. Bockris and S. Srinivasan, *J. Electroanal. Chem.* **11**, 350 (1966).

235. L. G. Austin and H. Lerner, *Electrochim. Acta* **9**, 1469 (1964).

236. J. O'M. Bockris and S. U. M. Khan, *Surface Electrochemistry: a Molecular Level Approach*, Plenum Press, New York, 1993, Chapter 3.

237. J. R. McDonald, *Impedance Spectroscopy—Emphasising Solid Materials and Systems*, Wiley, New York, 1987.

238. M. M. Musiani, *Electrochim. Acta* **35**, 1665 (1990).

239. A. J. Bard and L. R. Faulkner, *Electrochemical Methods*, Wiley, New York, 1980, pp. 316–369.

240. Southampton Electrochemistry Group, *Instrumental Methods in Electrochemistry*, Ellis Horwood, Chichester, pp. 251–282.

241. D. D. Macdonald, *Transient Techniques in Electrochemistry*, Plenum Press, New York, 1977, pp. 229–272.

242. (a). E. Gileadi, *Electrode Kinetics*, VCH, Weinheim, 1993, pp. 428–443. (b) P. H. Rieger, *Electrochemistry*, Prentice-Hall, Englewood Cliffs, NJ, 1987, pp. 310–323. (c) C. M. A. Brett and A. M. Oliveira Brett, *Electrochemistry: Principles, Methods and Applications,* Oxford Science Publications, Oxford, 1993, pp. 224–252.

243. (a) C. Gabrielli, Identification of Electrochemical Processes by Frequency Response Analysis, Solartron Schlumberger Technical Report, 1983. (b) C. Gabrielli, Use and Applications of Electrochemical Impedance Techniques, Schlumberger Technical Report, 1990.

244. (a) M. Sluyters-Rehbach and J. H. Sluyters, *Electroanalytical Chemistry*, Vol. 4, A. J. Bard, ed., Marcel Dekker, New York, 1970, pp. 1–128. (b) M. Sluyters-Rehbach and J. J. Sluyters, *Comprehensive Treatise of Electrochemistry*, Vol. 9, J. O'M. Bockris, B. E. Conway, and E. Yeager, S. Sarangapani, eds., Plenum Press, New York, 1984, pp. 177–292.

245. M. Sluyters-Rehbach and J. H. Sluyters, *Comprehensive Chemical Kinetics*, Vol. 26, C. H. Bamford and R. G. Compton, eds., Elsevier, Amsterdam, 1986, pp. 203–354.

246. D. D. Macdonald and M. C. H. McKubre, *Modern Aspects of Electrochemistry*, Vol. 14, J. O'M. Bockris, B. E. Conway, and R. E. White, Plenum Press, New York, 1982, pp. 61–150.

247. J. R. Macdonald, *J. Electroanal. Chem.* **223**, 25 (1987).

248. I. Rubinstein, E. Sabatani, and J. Rishpon, *J. Electrochem. Soc.* **134**, 1467 (1987).

249. R. D. Armstrong, B. Lindholm, and M. Sharp, *J. Electroanal. Chem.* **202**, 69 (1986).

250. R. D. Armstrong, *J. Electroanal. Chem.* **198**, 177 (1986).

251. C. Ho, I. D. Raistrick, and R. A. Huggins, *J. Electrochem. Soc.* **127**, 343 (1980).

252. C. Gabrielli, O. Haas, and H. Takenouti, *J. Appl. Electrochem.* **17**, 82 (1987).

253. C. Gabrielli, O. Haas, and H. Takenouti, *Proc. Journie d'Etudes SEE*, Electro-catalyse, Electrodes modifiees, 1986, paper C5, pp. 117–129.

254. C. Gabrielli, H. Takenouti, O. Haas, and A. Tsukada, *J. Electroanal. Chem.* **302**, 59 (1991).

255. (a) M. F. Mathias and O. Haas, *J. Phys. Chem.* **96**, 3174 (1992). (b) M. F. Mathias and O. Haas, *J. Phys. Chem.* **97**, 9217 (1993).

256. G. Lang and G. Inzelt, *Electrochim. Acta* **36**, 847 (1991).

257. B. Lindholm, *J. Electroanal. Chem.* **289**, 85 (1990).

258. M. Sharp, B. Lindholm-Sethson, and E. Lotta Lind, *J. Electroanal. Chem.* **345**, 223 (1993).

259. M. E. G. Lyons, H. G. Fay, and T. McCabe, *Key Eng. Mat.* **72–74**, 381 (1992).

260. S. H. Glarum and J. H. Marshall, *J. Electrochem. Soc.* **127**, 1467 (1980).

261. I. Rubenstein, J. Rishpon, and S. Gottesfeld, *J. Electrochem. Soc.* **133**, 729 (1986).

262. (a) K. Doblhofer, *Electrochim. Acta* **25**, 871 (1980). (b) K. Doblhofer and R. D. Armstrong, Electrochim. Acta **33**, 453 (1988).

263. F. A. Posey, *J. Electrochem. Soc.* **111**, 1173 (1964). (b) F. A. Posey and T. Morozumi, *J. Electrochem. Soc.* **113**, 176 (1966).

264. R. De Levie, *Electrochim. Acta* **8**, 751 (1963); **9**, 1231 (1964).

265. I. D. Raistrick, *Electrochim. Acta* **35**, 1579 (1990).

266. (a) C. D. Paulse and P. G. Pickup, *J. Phys. Chem.* **92**, 7002 (1988). (b) H. Mao, J. Ochmanska, C. D. Paulse, and P. G. Pickup, *Faraday Discuss. Chem. Soc.* **88**, 165 (1989). (c) J. Ochmanska and P. G. Pickup, *J. Electroanal. Chem.* **271**, 83 (1989).

267. P. G. Pickup, *J. Chem. Soc. Faraday Trans.* **86**, 3631 (1990).

268. (a). X. Ren and P. G. Pickup, *J. Electrochem. Soc.* **139**, 2097 (1992). (b) G. L. Duffit and P. G. Pickup, *J. Chem. Soc. Faraday Trans.* **88**, 1417 (1992). (c) X. Ren and P. G. Pickup, *J. Chem. Soc. Faraday Trans.* **89**, 321 (1993).

269. (a) C. R. Martin and L. S. van Dyke in *Molecular Design of Electrode Surfaces*, R. W. Murray, ed., *Techniques of Chemistry Series, Vol. 22*, Wiley Interscience, New York, 1992, pp. 403–424. (b) R. M. Penner, L. S. van Dyke, and C. R. Martin, *J. Phys. Chem.* **92**, 5274 (1988). (c) R. M. Penner and C. R. Martin, *J. Phys. Chem.* **93**, 984 (1989). (d) Z. Cai and C. R. Martin, *J. Electroanal. Chem.* **300**, 35 (1991).

270. Some representative references include: (a) R. A. Bull, F. R. F. Fan, and A. J. Bard, *J. Electrochem. Soc.* **129**, 1009 (1982). (b) N. Mermilliod, J. Tanguy, and F. Petiot, *J. Electrochem. Soc.* **133**, 1073 (1986). (c) J. Tanguy, N. Mermilliod, and M. Hoclet, *J. Electrochem. Soc.* **134**, 795 (1987). (d) J. Tanguy and N. Mermilliod, *Synth. Met.* **21**, 129 (1987). (e) A. M. Waller, A. N. S. Hampton, and R. G. Compton, *J. Chem. Soc. Faraday Trans. I* **85**, 773 (1989). (f) P. Burgmayer and R. W. Murray, *J. Phys. Chem.* **88**, 2515 (1984). (g) S. Panero, P. Prosperi, F. Bonino, B. Scrosati, A. Corradini, and M. Mastragostino, *Electrochim. Acta* **32**, 1007 (1987). (h) S. Panero, P. Prosperi, and B. Scrosati, *Electrochim. Acta* **32**, 1461 (1987). (i) H. C. Lyons, Ph.D. Thesis, University of Dublin, 1993. (j) T. Amemiya, K. Hashimoto, and A. Fujihima, *J. Phys. Chem.* **97**, 4187, 4192, 9736 (1993). (k) T. B. Hunter, P. S. Tyler, W. H. Smyrl, and H. S. White, *J. Electrochem. Soc.* **134**, 2198 (1987). This list is not exhaustive.

271. Some representative references include (this list is not exhaustive): (a) T. F. Otero and E. De Laretta, *J. Electroanal. Chem.* **244**, 311 (1988). (b) R. K. Yuan, D. Peramunage, and M. Tomkiewicz, *J. Electrochem. Soc.* **134**, 886 (1987). (c). S. Sunde, G. Hagen, and R. Odegard, *J. Eletroanal. Chem.* **345**, 43, 59 (1993).

272. Some representative references include: (a) S. H. Glarum and J. H. Marshall, *J. Electrochem. Soc.* **134**, 142 (1987). (b) I. Rubenstein, E. Sabatani, and J. Rishpon, *J. Electrochem. Soc.* **134**, 3078 (1987). (c) C. Delouis, M. M. Musiani, and B. Tribollet, *J. Electroanal. Chem.* **264**, 57 (1989). (d) R. S. Hutton, M. Kalaji, and L. M. Peter, *J. Electroanal. Chem.* **270**, 429 (1989). (e) M. Kalaji and L. M. Peter, *J. Chem. Soc. Faraday Trans.* **87**, 853 (1991). (f) T. McCabe, Ph.D. Thesis, University of Dublin, 1992.

273. B. I. Bleaney and B. Bleaney, *Electricity and Magnetism*, 2nd ed., Clarendon Press, Oxford, 1965, pp. 291–328.

274. S. A. Wring and J. A. Hart, *Analyst* **117**, 1215 (1992).

275. M. Kaneko and D. Wohrle, *Adv. Polym. Sci.* **84**, 143 (1988).

276. M. E. G. Lyons in *Electroactive Polymer Electrochemstry*, M. E. G. Lyons, ed., Plenum Press, New York, 1994, pp. 237–374.

277. R. J. Mortimer in R. G. Compton and G. Hancock, eds., *Research in Chemical Kinetics*, Vol. 2, Elsevier, Amsterdam, 1994.

278. R. W. Murray, *Acc. Chem. Res.* **13**, 135 (1980).

279. M. W. Wrighton, *Acc. Chem. Res.* **12**, 303 (1979).

280. W. J. Albery and A. R. Hillman, *J. Electroanal. Chem.* **170**, 27 (1984).

281. R. A. Bull, F. R, Fan, and A. J. Bard, *J. Electrochem. Soc.* **130**, 1636 (1983).

282. S. Holdcroft and B. L. Funt, *J. Electroanal. Chem.* **240**, 89 (1988).

283. K. M. Kost, D. E. Bartak, B. Kazee, and Y. Kuwana, *Anal. Chem.* **60**, 2379 (1988).

284. A. Yassar, J. Roncali, and F. Garnier, *J. Electroanal. Chem.* **225**, 53 (1988).

285. F. T. A. Vork, L. J. J. Janssen, and E. Barendrecht, *Electrochim. Acta* **31**, 1569 (1986).

286. E. W. Paul, A. G. Ricco, and M. W. Wrighton, *J. Phys. Chem.* **89**, 1441 (1985).

287. G. Tourillion and F. Garnier, *J. Phys. Chem.* **88**, 5281 (1984); G. Tourillion, E. Dartyge, H. Dexpert, A. Fountaine, A. Jucha, P. Lagarde, and D. E. Sayers, *J. Electroanal. Chem.* **178**, 366 (1984).

288. P. Ocon Esteban, J. M. Leger, C. Lamy, and E. Genies, *J. Appl. Electrochem.* **19**, 462 (1989).

289. Y. Takasu, Y. Fujii, K. Yasuda, Y. Iwanaga, and Y. Matsuda, *Electrochim. Acta* **34**, 453 (1989).

290. R. Noufi, *J. Electrochem. Soc.* **130**, 2126 (1983).

291. P. N. Bartlett and R. G. Whitaker, *J. Electroanal. Chem.* **224**, 27 (1987).

292. P. N. Bartlett and R. G. Whitaker, *J. Electroanal. Chem.* **224**, 37 (1987).

293. P. N. Bartlett, P. Tebbutt, and C. H. Tyrrell, *Anal. Chem.* **64**, 138 (1992).

294. N. C. Foulds and C. R. Lowe, *Anal. Chem.* **60**, 2473 (1988); N. C. Foulds and C. R. Lowe, *J.C.S. Faraday Trans. I* **82**, 1259 (1986).

295. (a) P. N. Bartlett, P. Tebbutt, and R. G. Whitaker *Prog. Reaction Kinetics* **16**, 55 (1991). (b) P. N. Bartlett, Z. Ali, and V. Eastwick-Field, *J.C.S. Faraday Trans.* **88**, 2677 (1992). (c) K. Yokoyama, E. Tamiya, and I. Karube, *J. Electroanal. Chem.* **273**, 107 (1989).

296. A. Michas, J. M. Kelly, R. Durand, M. Pineri, and J. M. D. Coey, *J. Membr. Sci.* **29**, 239 (1986).

297. K. Itaya, H. Takahashi, and J. Uchida, *J. Electroanal. Chem.* **208**, 373 (1986).

298. W. H. Kao and T. Kuwana, *J. Am. Chem. Soc.* **106**, 473 (1984).

299. D. E. Bartak, B. Kazee, K. Shimazu, and T. Kuwana, *Anal. Chem.* **58**, 2756 (1986).

300. H. Y. Liu and F. C. Anson, *J. Electroanal. Chem.* **158**, 181 (1983).

301. R. N. Dominey, N. S. Lewis, J. A. Bruce, D. C. Bookbinder, and M. S. Wrighton, *J. Am. Chem. Soc.* **104**, 476 (1982).

302. J. A. Bruce, T. Murashi, and M. S. Wrighton, *J. Phys. Chem.* **86**, 1552 (1982).

303. R. A. Simon, T. E. Mallouk, K. A. Daube, and M. S. Wrighton, *Inorg. Chem.* **24**, 3119 (1985).

304. D. J. Harrison and M. S. Wrighton, *J. Phys. Chem.* **88**, 3932 (1984).

305. C. J. Stalder, S. Chao, and M. S. Wrighton, *J. Am. Chem. Soc.* **106**, 3673 (1984).

306. J. F. Andre and M. S. Wrighton, *Inorg. Chem.* **24**, 4288 (1985).

307. M. E. G. Lyons, D. E. McCormack, and P. N. Bartlett, *J. Electroanal. Chem.* **261**, 51 (1989).

308. M. E. G. Lyons, D. E. McCormack, O. Smyth, and P. N. Bartlett, *Faraday Discuss. Chem. Soc.* **88**, 139 (1989).

309. M. E. G. Lyons and P. N. Bartlett, *J. Electroanal. Chem.* **316**, 1 (1991).

310. M. E. G. Lyons, C. H. Lyons, A. Michas, and P. N. Bartlett, *Analyst* **117**, 1271 (1992).

311. M. E. G. Lyons, C. A. Fitzgerald, and M. R. Smyth, *Analyst* **119**, 855 (1994).

312. C. P. Andrieux and J. M. Saveant, *J. Electroanal. Chem.* **93**, 163 (1978).

313. C. P. Andrieux, J. M. Dumas-Bouchiat, and J. M. Saveant, *J. Electroanal. Chem.* **114**, 159 (1980).

314. C. P. Andrieux, J. M. Dumas-Bouchiat, and J. M. Saveant, *J. Electroanal. Chem.* **131**, 1 (1982).

315. C. P. Andrieux and J. M. Saveant, *J. Electroanal. Chem.* **139**, 163 (1982).

316. C. P. Andrieux, J. M. Dumas-Bouchiat, and J. M. Savant, *J. Electroanal. Chem.* **169**, 9 (1984).

317. C. P. Andrieux and J. M. Saveant, *J. Electroanal. Chem.* **171**, 65 (1984).

318. (a) C. P. Andrieux in *Electrochemistry, Sensors and Analysis*, M. R. Smyth and J. G. Vos, eds., Elsevier, Amsterdam, 1986, pp. 235–245. (b) J. Leddy, A. J. Bard, J. T. Maloy, and J. M. Saveant, *J. Electroanal. Chem.* **187**, 205 (1985).

319. R. D. Rocklin and R. W. Murray, *J. Phys. Chem.* **85**, 2104 (1981).

320. R. W. Murray, *Phil. Trans. Roy. Soc.* **302**, 253 (1981).

321. F. C. Anson, *J. Phys. Chem.* **84**, 3336 (1980).

322. E. Laviron, *J. Electroanal. Chem.* **112**, 1 (1980).

323. C. Daul and O. Haas in *Electrochemistry Sensors and Analysis*, M. R. Smyth and J. G. Vos, eds., Elsevier, Amsterdam, 1986, pp. 277–284.

324. E. Deiss, O. Haas, and C. Daul, *J. Electroanal. Chem.* **337**, 299 (1992).

325. J. F. Cassidy and J. G. Vos, *J. Electroanal. Chem.* **235**, 41 (1987).

326. E. T. Turner-Jones and L. R. Faulkner, *J. Electroanal. Chem.* **222**, 201 (1987).

327. M. Sharp, *J. Electroanal. Chem.* **230**, 109 (1987).

328. S. Dong and G. Che, *J. Electroanal. Chem.* **309**, 103 (1991).

329. M. E. G. Lyons, P. N. Bartlett, C. H. Lyons, W. Breen, and J. F. CAssidy, *J. Electroanal. Chem.* **304**, 1 (1991).

330. M. E. G. Lyons, D. E. McCormack, A. Michas, C. H. Lyons, and P. N. Bartlett, *Key Eng. Mater.* **72/74**, 477 (1992).

331. W. J. Albery, *Electrode Kinetics*, Clarendon Press, Oxford, 1975, pp. 49–67.

332. A. J. Bard and L. R. Faulkner, *Electrochemical Methods*, Wiley, New York, 1980, pp. 280–315.

333. K. Aoki, K. Tokuda, and H. Matsuda, *J. Electroanal. Chem.* **199**, 69 (1986).

334. C. Deslouis, M. M. Musiani, and B. Tribollet, *J. Electroanal. Chem.* **264**, 37 (1989).

335. R. J. Forster and J. G. Vos, *J.C.S. Faraday Trans.* **87**, 1863 (1991).

336. (a) R. J. Forster, J. G. Vos, and M. E. G. Lyons, *J.C.S. Faraday Trans.* **87**, 3761 (1991). (b) R. J. Forster, J. G. Vos, A. J. Kelly, and M. E. G. Lyons, *J. Electroanal. Chem.* **270**, 365 (1989).

337. W. J. Albery, M. G. Bouteille, P. J. Colby, and A. R. Hillman, *J. Electroanal. Chem.* **133**, 135 (1982).

338. K. J. Laidler, *Chemical Kinetics*, 3rd ed., Harper & Row, New York, 1987, pp. 399–412.

339. J. M. Engasser and C. Horvath in *Applied Biochemistry and Bioengineering: Immobilized Enzyme Principles*, Vol. 1, L. Wingrad, E. Katchalski-Katzir, and L. Goldstein, eds., Academic Press, New York, 1976, pp. 127–221.

340. J. M. Engasser and C. Horvath, *Biotechnol. Bioeng.* **16**, 909 (1974).

341. R. Aris, *The Mathematical Theory of Diffusion and Reaction in Permeable Catalysts*, Clarendon Press, Oxford, 1975.

342. W. J. Albery, A. E. G. Cass, and Z. H. Shu, *Biosens. Bioelectron.* **5**, 367 (1990).

343. M. E. G. Lyons and L. D. Burke, *J.C.S. Faraday Trans.* **83**, 299 (1987).

344. G. W. Jang, E. W. Tsai, and K. Rajeshwar, *J. Electrochem. Soc.* **134**, 2377 (1987).

345. M. E. G. Lyons and C. F. Fitzgerald, unpublished work.

346. G. K. Chandler and D. Pletcher, *J. Appl. Electrochem.* **16**, 62 (1986).

347. J. Y. Lee and T. C. Tan, *J. Electrochem. Soc.* **137**, 1402 (1990).

348. F. T. A. Vork and B. Barendrecht, *Synth. Met.* **28**, C121 (1989).

349. F. T. A. Vork and B. Barendrecht, *Electrochim. Acta* **35**, 135 (1990).

350. D. Belanger, E. Brassard, and G. Fortier, *Anal. Chim. Acta* **228**, 311 (1990).

351. D. J. Strike, N. F. De Rooij, M. Koudelka-Hep, M. Ulmann, and J. Augustynski, *J. Appl. Electrochem.* **22**, 922 (1992).

352. A. Aramata, T. Kodera, and M. Matsuda, *J. Appl. Electrochem.* **18**, 577 (1988).

353. H. Laborde, J. M. Leger, C. Lamy, F. Garnier, and A. Yassar, *J. Appl. Electrochem.* **20**, 524 (1990).

354. M. Gholamirn and A. Q. Contractor, *J. Electroanal. Chem.* **281**, 69 (1990).

355. A. Aramata and R. Ohnishi, *J. Electroanal. Chem.* **162**, 153 (1984).

356. W. J. Albery and P. N. Bartlett, *J. Electroanal. Chem.* **131**, 137 (1982).

POLYMERS IN DISORDERED MEDIA

A. BAUMGÄRTNER

Institut für Festkörperforschung Forschungszentrum, Jülich, Germany

M. MUTHUKUMAR

Department of Polymer Science University of Massachusetts Amherst, Massachusetts

CONTENTS

I. INTRODUCTION

The statistical mechanics of flexible-chain molecules [1–4], and in particular our knowledge about their phase transitions and critical phenom-

Advances in Chemical Physics, Volume XCIV, Edited by I. Prigogine and Stuart A. Rice.
ISBN 0-471-14324-3 © 1996 John Wiley & Sons, Inc.

ena, have now settled beyond fundamentals. Modern concepts of critical phenomena such as scaling and renormalization group theory have provided powerful tools in exploring the physics of macromolecules [4, 5] and will be the key to future progress.

One particular area in polymer research, however, is still limping behind this general pace of progress. Although of considerable importance, the statistical mechanics of polymers with intrinsic or external quenched disorder, such as proteins or chains in porous media, respectively, are not well understood. The intention of the present work is to review the current status of this research. Although necessarily preliminary, it may well stimulate progress. The physics of proteins and similar macromolecules is certainly one of the great challenges now and in the future. In particular, from a theoretical point of view, the possible relation [6] between proteins and spin glasses [6, 7] is a formidable task. Since there exist several reviews on protein structures and dynamics [e.g., 6, 8], this subject is omitted in the present work. Hence we will focus our discussions on polymers in disordered media only. The problem of a polymer trapped in a quenched random environment has been investigated for more than one decade. One may consider this class of problems to be divided in two parts. One is concerned with the problem of how the structures of disordered porous solids affect the conformational and thermodynamic properties of macromolecules. This is addressed in the first section. In the second section, the transport properties of polymers in a porous environment are discussed.

A special and very interesting problem, which is the behavior of *directed polymers* in disordered media, will not be discussed in detail in the present review. Directed polymers are chains with a local stiffness sufficiently strong to prohibit local self-interactions, such as ordinary excluded volume repulsion. Directed polymers may be of relevance to polyelectrolytes, liquid crystal polymers, and interfacial problems. From a theoretical point of view, they are easier to treat in comparison with flexible chains. (However, from a conceptual point of view, directed polymers do not provide basically new features as compared to flexible polymers.) The interested reader is suggested to consult the relevant articles and reviews [9–12]. Another special case, the influence of environments with *annealed disorder* on the conformations and thermodynamics of polymers [13–17], will not be considered in the present review, but rather is discussed at the appropriate places. Annealed disorder is encountered in such cases, where the polymer's environment is in thermal equilibrium with the chain itself, or if the structure of the disordered media relaxes on a much faster time scale than the chain. Of course, there are interesting limiting cases between media of quenched and annealed disorder, which are discussed below.

II. CONFORMATIONS AND CRITICAL PHENOMENA

Obviously, the number of possible conformations of a flexible polymer trapped in a porous solid is more restricted than in free space. This restriction may lead to reductions of some amplitudes, however independent of chain length, but also, and in addition, to a qualitative change of conformational statistics eventually revealed by changes of some critical exponents. Therefore, the very first question, which is addressed in the following two sections, is related to the entropy reduction of a single chain in a random environment and the relations between typical correlation lengths of the porous media and the polymer chain. Since randomness requires some special concepts, mathematical techniques, and computational methods, particular attention is paid to various Flory-type mean-field theories, replica calculations, Monte Carlo methods, and finite-size scaling laws.

Since randomness-induced entropic effects may produce dramatic deviations from classical polymer behavior, it is conceivable that thermodynamic properties of polymers in disordered media may change as well. Special phenomena are discussed in the third part of section II. The first two examples contain single-chain adsorption phenomena modified, or even induced, by a disordered environment. The third example is a demonstration of how the critical behavior of short liquid crystal polymers trapped in a porous matrix is changed as compared to the pure case.

A. Self-Avoiding Walks on Fractal Lattices

1. Scaling Laws and Mean-Field Theories

Self-avoiding walks on regular lattices as a model for polymers have a long and successful history [2]. The statistics of lattice self-avoiding walks (SAWs) is now well understood [4]. For example, the total number of SAWs, Z_N, the total number of loops G_N, and the average end-to-end distance $\langle R^2 \rangle$ of N steps are given by

$$Z_N \sim \mu^N N^{\gamma - 1} \tag{1}$$

$$G_N \sim \mu^N N^{-(2-\alpha)} \tag{2}$$

$$\langle R^2 \rangle \sim N^{2\nu} \tag{3}$$

where μ is the connectivity constant, which depends on the type of lattice and α, ν, γ are universal exponents depending only on the dimensionality d. Since the Heisenberg n-vector model corresponds in the limit $n \rightarrow 0$ to SAWs [18], the exponents α, ν, γ correspond the well-known exponents of specific heat, correlation length, and susceptibility in this limit: $\nu = \frac{3}{4}$, 0.59, $\frac{1}{2}$; $\gamma = \frac{43}{32}$, 1.16, 1; and $\alpha = 2 - \nu$ in dimensions $d = 2$, 3, 4, respectively [19–22].

More recently the effects of dilution of lattice connectivity have been addressed. Consider a lattice with randomly occupied bonds of concentration p upon which SAWs are allowed to conform. In this case the connectivity constant μ would change such that $\mu(p)$ is expected to decrease with decreasing p. The critical exponents are expected to be nontrivially affected by either annealed [23] or quenched [24] disorder [25] (see the end of this chapter). However, arguments based upon a modified Harris criterion [26] applied to the $n \to 0$ vector model (see the end of this chapter) provides no indication in favor of critical disorder effects on SAWs. This has also been supported by field theoretical renormalization group calculations [27] close to the percolation threshold p_c. However, real space renormalization group studies [28–30] have indicated that the renormalization group flow is toward the pure SAW fixed point for $p < p_c$, indicating the same critical behavior for all $p < p_c$ but a new fixed point with slightly larger values of ν at $p = p_c$. Numerical estimates in $d = 2$ dimensions and at $p = p_c$ [29, 30] gave $\nu \simeq 0.77$ as compared to the pure case where $\nu_0 = \frac{3}{4}$. Moreover, field theoretical estimates [31] indicate $\nu = \frac{1}{2} + \varepsilon/42$, with $\varepsilon = 6 - d$ indicating a new fixed point for SAW on percolation clusters at $p = p_c$ with a new upper critical dimension, where $\nu = \frac{1}{2}$, of $d_c = 6$. The same conclusions can be obtained by a simple scaling theory [32]. Assuming that, for $p > p_c$, SAW still have their classical fixed point as in the pure case, then

$$R \sim (p - p_c)^{-\sigma} N^{\nu_0} \tag{4}$$

where the first factor, corresponds to the swelling with increasing lattice dilution with $\sigma \geq 0$ [33–35]. Below the percolation threshold $p < p_c$, the size of the SAW is given by

$$R \sim \xi_p \sim (p - p_c)^{-\nu_p} \tag{5}$$

where ξ_p is the correlation length of the percolation cluster [36] and ν_p the corresponding exponent. Assuming a crossover scaling function $f(y)$ for $p \to p_c$ and

$$R \sim N^\nu f[N^x(p - p_c)] \tag{6}$$

provides by comparison of Eqs. (4)–(6) that $x = \nu/\nu_p$ and the scaling relation

$$\sigma = \frac{(\nu - \nu_0)\nu_p}{\nu} \tag{7}$$

Assuming further that $\sigma \geq 0$, which is also supported by computer simulations [33–35], one must conclude that at the percolation threshold $\nu \geq \nu_0$.

A useful approach to estimate the exponent ν at $p = p_c$ is the application of mean-field arguments of the Flory type. Variations of Flory formulas have been proposed [37–40]. An interesting derivation of the general Flory formula proposed by Aharony and Harris [39] has been given by Roy and Blumen [40] based on simple considerations about the distribution of the radius of gyration.

To derive the correlation exponent ν, one has to use three different geometric dimensionalities: the fractal dimension of the backbone, D_B, the minimal (or chemical) path on the fractal, d_{min}, and the anomalous fractal dimension of a regular random walk on the backbone, d_{wB}. The Flory formula is the free energy consisting of a potential energy and the entropic or "kinetic" energy term

$$F \sim a\frac{N^2}{R^{D_B}} + b\left(\frac{R^{d_{wB}}}{N}\right)^\alpha \tag{8}$$

Minimizing with respect to R yields

$$\nu = \frac{2 + \alpha}{D_B + d_{wB}\alpha} \tag{9}$$

In the potential energy part, $R^{d_{wB}}$ represents the number of sites on the backbone, within the radius R, in which the self-avoiding walk might self-intersect. The entropic part arises from the probability of a regular random walk to reach a distance R after N steps on the backbone, $P(R)$ [37, 39]. Since the dependence on R is expected to arise via the scaled variable $(R^{d_{wB}}/N)$, and since for large R one expects an exponential decay, it is reasonable to assume that

$$P(R, N) \sim \exp\left[-b\left(\frac{R^{d_{wB}}}{N}\right)^\alpha\right] \tag{10}$$

which in fact yields the free-energy form given above. However, the value of α has been the subject of some controversy but now is commonly

TABLE I

Connectivity constant μ for Lattices without Dilution and with Dilution at p_c

Lattice	μ_0	Dilution	p_c	$\mu(p_c)$
Square	2.638	Site	0.5928	1.53
Square	—	Bond	0.5	1.31
Triangular	4.151	Site	0.5	
Triangular	—	Bond	0.3473	
Cubic	4.684	Site	0.3117	
Cubic	—	Bond	0.2492	

Note: Data obtained for μ_0 from ref. [22], p_c from ref. 36, and $\mu(p_c)$ from ref. 42.

accepted to be of the form

$$\alpha = \frac{1}{d_{wB}/d_{\min} - 1} \tag{11}$$

This has been obtained [39, 41] by the supper and lower bounds of $P(R)$:

$$\frac{1}{d_{wB} - 1} < \alpha < \frac{d_{\min}}{d_{wB} - d_{\min}} \tag{12}$$

where the upper bound applies to typical walks or to the quenched average over $\ln P(R)$, whereas the lower bound is found if $P(R)$ is averaged over all possible starting points and local geometries, that is, the annealed average of $P(R)$.

In Tables I and II we have listed some current estimates of the various exponents and connectivity constants.

Similar considerations as outlined above for self-avoiding walks on fractals have been carried out for self-avoiding walks near the theta point on fractal structures [43, 44]. Using a Flory approximant for the size

TABLE II

Theoretical Estimates of Self-avoiding Walk Exponent ν on Percolation Clusters

d	ν_0	ν	ν_{MC}	ν_{EN}	$1/d_B$
2	$\frac{3}{4}$	0.89	0.75	0.81	0.62
3	0.592	0.72	0.61	0.67	0.57
4	$\frac{1}{2}$	0.62	—	0.63	0.52
5	$\frac{1}{2}$	0.59	—	0.54	0.52
6	$\frac{1}{2}$	0.50	—	—	$\frac{1}{2}$

Note: The exponent ν_0 corresponds to the undiluted case, and d is the dimensionality; ν_{MC} and ν_{EN} are estimates obtained by Monte Carlo and exact enumeration, respectively; d_B is the dimensionality of the backbone of the cluster.

exponent and the crossover exponent, the approximant involves the three fractal dimensionalities for the backbone, the minimal path, and the resistance of the fractal structures.

2. Fisher–Harris Criterion

Fisher and Harris gave heuristic arguments of how the critical behavior of a system would be affected by the presence of annealed [23] and quenched [24] disorder, that is, if the specific heat of the pure system diverges with positive exponent α. For quenched disorder the Harris criterion only provides evidence for a crossover of the pure system but cannot describe the new critical behavior. In the case of annealed disorder, the arguments by Fisher also give the nature of the critical behavior.

a. Harris Criterion for Quenched Disorder. If ξ denotes the thermal correlation length of the system, the mean-square fluctuation on the concentration p of impurities in a typical volume element is

$$(\Delta p)^2 \equiv \sum_i [\langle p_i^2 \rangle - \langle p_i \rangle^2] \sim \xi^d p(1-p) \tag{13}$$

The corresponding induced fluctuations in the internal energy leads to a shift of the transition temperature T_c,

$$\xi^d \, \Delta T_c \sim \Delta p \sim \xi^{d/2} \tag{14}$$

or equivalently

$$\Delta T_c \sim \xi^{-d/2} \tag{15}$$

On the other hand, denoting $\Delta T \equiv |T - T_c|/T_c$, we have the classical relation $\Delta T \sim \xi^{-1/\nu}$. For small shifts of the transition temperature such that $\Delta T_c \ll \Delta T$, one has $d/2 > 1/\nu$, or equivalently $2 - d\nu < 0$. Comparing this with the hyperscaling relation $\alpha = 2 - d\nu$ implies that the presence of impurities leads to $\alpha < 0$. This means that the disorder induces a different transition with negative α as compared to the pure system with $\alpha > 0$.

In the particular situation of a self-avoiding walk described by the n-vector model in the limit $n \to 0$, a similar calculation gives

$$\xi^d \, \Delta T_c \sim n \, \Delta p \tag{16}$$

where the prefactor n comes from the sum over all n components of the Hamiltonian. Hence, for self-avoiding walks $\Delta T_c \ll \Delta T$ is always fulfilled, independent of the specific heat exponent $\alpha = 2 - d\nu$. Consequently, any disorder effects coming from the impurities on the lattice imposed on self-avoiding walks should not be detectable [26]. The disorder-induced shift of the critical temperature is always smaller as compared to the finite-size-induced broadening of the transition regime.

b. Fisher Renormalization for Annealed Disorder. In the annealed case the impurities are in the thermodynamic equilibrium with the "magnetic" system. Given the coupling constant λ for the impurities, Fisher proposed, for the free energy,

$$F(\Delta T, \lambda) \sim F[\Delta T^*(\Delta T, \lambda)] \qquad (17)$$

where $F(\Delta T) \sim |\Delta T|^{2-\alpha}$ and α are the free energy and the specific heat exponent of the pure system. Here, ΔT^* is an analytic function of λ. The concentration of impurities is related by

$$p = \frac{\partial F}{\partial \lambda} = \frac{\partial F}{\partial \Delta T^*} \frac{\partial \Delta T^*}{\partial / \partial \Lambda} \qquad (18)$$

Since both ΔT and ΔT^* are smaller than unity, both sides can be expanded and the lowest order terms can be compared to

$$|\Delta T^*| \sim |\Delta T|^{1/(1-\alpha)} \quad \text{for } \alpha > 0 \qquad (19)$$

and $|\Delta T^*| = |\Delta T|$ if $\alpha < 0$. Therefore, one obtains the Fisher renormalized exponents $\alpha_R = -\alpha/(1-\alpha)$, $\beta_R = \beta/(1-\alpha)$, $\gamma_R = \gamma/(1-\alpha)$, and $\nu_R = \nu/(1-\alpha)$ of the impure system if $\alpha > 0$ for the pure one. These renormalized exponents leave the scaling and hyperscaling relations $\alpha + 2\beta + \gamma = 2$ and $\alpha = 2 - d\nu$, respectively, unchanged.

In particular, for self-avoiding walks, where $\alpha = (4-d)/(2+d)$ and $\nu \approx 3/(2+d)$, the renormalized correlation length exponent is

$$\nu_R = \frac{3}{2(d-1)} \qquad (20)$$

However, since for $d = 2$ the exponent is $\nu > 1$, the self-avoiding walk seems to violate the Fisher renormalization [13–16].

B. Polymers on Diluted Lattices and in Porous Media

Random walks on fractal lattices where self-intersections of the chain are strictly forbidden may serve in a first approximation as a reasonable model for polymers in porous media. However, from a more general point of view the porous media is not necessarily fractal, at least not on length scales that are large compared to the chain length. In addition, the lattice bonds may in fact represent the channels of the porous medium in which the polymer may fold back and forth, which then spoils the dominance of the backbone of the percolation cluster for the polymer dimension (compare previous section). Therefore, in the following section, we consider the general case of diluted lattices far from the percolation threshold and the competing effect between the strength of self-avoidance and the restricting constraints of the random environment. The considerations are not restricted to lattices but are more general and pertain also to continuum models. In particular it has been found that for sufficiently weak self-avoidance the chain may collapse to a dense globule due to the mutual repulsion from the random environment. After the discovery of this "localization" effect by means of computer simulations [45], many numerical and analytical studies have been devoted to the elucidation of this phenomenon [46–62].

1. Mean-Field Theory: Replica Method

Consider a single chain of contour length L in a volume V containing a collection of N obstacles that are fixed in space. The obstacles are randomly distributed, and their concentration is low and below the percolation threshold. The obstacles interact with the chain segments with either repulsive or attractive two-body interactions. The segments of the polymer chain interact among themselves with a prescribed model interaction, such as the excluded-volume interaction, screened Coulomb interaction, and so on. The system is described by the generalized Edwards Hamiltonian,

$$\beta H = \frac{3}{2l} \int_0^L ds \left(\frac{\partial \mathbf{R}(s)}{\partial s} \right)^2 + \int_0^L ds \int_0^L ds' \, W[\mathbf{R}(s) - \mathbf{R}(s')]$$

$$+ \sum_{i=1}^N \int_0^L ds \, X[\mathbf{R}(s) - \mathbf{r}_i] \tag{21}$$

where $\mathbf{R}(s)$ is the position vector of the chain at the arc length s, l is the Kuhn length, and $\mathbf{r}$ is the position vector of the ith obstacle; $\beta = 1/k_B T$, where k_B is Boltzmann's constant and T is the absolute temperature; X is some arbitrary potential describing the interaction between the obstacles and the polymer; and W describes the intrachain interaction between two segments and can be taken to be any of the model interactions studied in the preceding chapters.

The simplest situation corresponds to a Gaussian chain where $W = 0$, which is discussed in the following two sections.

In this section we are interested in the calculation of the mean-square end-to-end distance of a chain described by Eq. (21) with $w = 0$. In this calculation two averages have to be performed: first with respect to the possible chain configurations and the second with respect to the possible realizations of the frozen obstacles. In a particular realization of the configuration of the frozen obstacles, the mean-square end-to-end distance of the chain is obtained by summing over all possible configurations of the chain in that realization and is given by

$$R^2(\{\mathbf{r}_i\}) = \frac{\int_{\mathbf{R}(0)}^{\mathbf{R}(L)} D[\mathbf{R}(s)][\mathbf{R}(L) - \mathbf{R}(0)]^2 e^{-\beta H}}{\int_{\mathbf{R}(0)}^{\mathbf{R}(L)} D[\mathbf{R}(s)] e^{-\beta H}} \tag{22}$$

The averaging over all possible configurations of the frozen obstacles is then performed to obtain the average of the mean-square end-to-end distance of the chain,

$$\langle R^2 \rangle = \int \prod_{i=1}^{N} \frac{d\mathbf{r}_i}{V} R^2(\{\mathbf{r}_i\}) = \left\langle \frac{\int D\mathbf{R}[\mathbf{R}(L) - \mathbf{R}(0)]^2 e^{-\beta H}}{\int D\mathbf{R} e^{-\beta H}} \right\rangle \tag{23}$$

where the angular brackets indicate the averaging over the different configurations of the random medium. It is to be noted that $\langle R^2 \rangle$ is not equal to the ratio of averages,

$$\langle R^2 \rangle \neq \frac{\left\langle \int D\mathbf{R}[\mathbf{R}(L) - \mathbf{R}(0)]^2 e^{-\beta H} \right\rangle}{\left\langle \int D\mathbf{R} e^{-\beta H} \right\rangle} \tag{24}$$

The right-hand side of Eq. (24) corresponds to the annealed situation where the obstacles are mobile. This situation is a different problem from the "quenched" situation of frozen obstacles studied in this chapter.

The procedure for the calculation of $\langle R^2 \rangle$ given by Eq. (23) involves two key steps that are now described. First is the replica formalism to correctly deal with the quenched average. The second is the use of the Feynman variational procedure in which the parameters of a trial Hamiltonian are so chosen to extremize the free energy of the chain.

a. Replica Formalism. Since

$$[\mathbf{R}(L) - \mathbf{R}(0)] = \int_0^L ds \frac{\partial \mathbf{R}(s)}{\partial s} \tag{25}$$

R^2 of Eq. (23) can be rewritten as

$$R^2 = \lim_{\lambda \to 0} \frac{\partial}{\partial \lambda} \ln Z(\lambda) \tag{26}$$

where

$$Z(\lambda) = \int D\mathbf{R} e^{\lambda \left[\int_0^L ds \left(\frac{\partial \mathbf{R}(s)}{\partial s} \right) \right]^2 - \beta H} \tag{27}$$

The $\ln Z$ term of Eq. (26) can be written as usual,

$$\ln Z = \lim_{n \to 0} \frac{\partial}{\partial n} Z^n \tag{28}$$

It follows from Eqs. (26) and (28) that

$$R^2 = \lim_{n, \lambda \to 0} \frac{\partial}{\partial n} \frac{\partial}{\partial \lambda} Z^n(\lambda) \tag{29}$$

Then Z^n is written as n replicas,

$$Z^n = \prod_{\alpha=1}^{n} \int D\mathbf{R}_\alpha \exp\left\{ \sum_{\alpha=1}^{n} \left[\lambda \left(\int_0^L ds_\alpha \frac{\partial \mathbf{R}_\alpha}{\partial s_\alpha} \right)^2 - \beta H_\alpha \right] \right\} \tag{30}$$

where

$$\beta H_\alpha = \frac{3}{2l} \int_0^L ds_\alpha \left(\frac{\partial \mathbf{R}_\alpha}{\partial s_\alpha} \right)^2 + \sum_{i=1}^{N} \int_0^L ds_\alpha X[\mathbf{R}_\alpha(s_\alpha) - \mathbf{r}_i] \tag{31}$$

When there are many obstacles randomly distributed, the obstacle density $\rho(\mathbf{r})$,

$$\rho(\mathbf{r}) = \sum_{i=1}^{N} \delta(\mathbf{r} - \mathbf{r}_i) \tag{32}$$

can be regarded as having a Gaussian distribution,

$$P[\rho] = \exp\left(-\frac{1}{2} \int d\mathbf{r} \frac{\rho^2(\mathbf{r})}{\rho_0} \right) \tag{33}$$

where

$$\rho_0 = \frac{1}{V} \int d\mathbf{r}\, \rho(\mathbf{r}) \tag{34}$$

is the mean obstacle density. Thus $R^2(\{\mathbf{r}_i\})$ is a functional of ρ and the averaging of Eq. (23) is over $P[\rho]$,

$$\langle R^2 \rangle = \frac{\int D[\rho] R^2[\rho] P[\rho]}{\int D[\rho] P[\rho]} \tag{35}$$

Substituting Eq. (29) for R^2 in Eq. (35), we get

$$\langle R^2 \rangle = \lim_{n,\lambda \to 0} \frac{\partial}{\partial n} \frac{\partial}{\partial \lambda} \frac{\int D[\rho] Z^n(\lambda) P[\rho]}{\int D[\rho] P[\rho]} \tag{36}$$

Specifically, consider the average of

$$\exp\left\{ -\sum_{\alpha=1}^{n} \sum_{i=1}^{N} \int_0^L ds_\alpha\, X[\mathbf{R}_\alpha(s_\alpha) - \mathbf{r}_i] \right\} \tag{37}$$

which appears in Eq. (36) through Eqs. (30) and (31). Since

$$\sum_{i=1}^{N} [\mathbf{R}_\alpha(s_\alpha) - \mathbf{r}_i] = \int d\mathbf{r}[\mathbf{R}_\alpha(s_\alpha) - \mathbf{r}]\rho(\mathbf{r}) \tag{38}$$

in view of Eq. (32), we obtain, from Eqs. (33) and (35),

$$\left\langle \exp\left\{ -\sum_{\alpha}^{n} \sum_{i}^{N} \int_0^L ds_\alpha [\mathbf{R}_\alpha(s_\alpha) - \mathbf{r}_i] \right\} \right\rangle$$
$$= \exp\left\{ \frac{\rho_0}{2} \sum_{\alpha}^{n} \sum_{\beta}^{n} \int_0^L ds_\alpha \int_0^L ds_\beta\, U[\mathbf{R}_\alpha(s_\alpha) - \mathbf{R}_\beta(s_\beta)] \right\} \tag{39}$$

where

$$U[\mathbf{R}_\alpha(s_\alpha) - \mathbf{R}_\beta(s_\beta)] = \int d\mathbf{r}\, X[\mathbf{R}_\alpha(s_\alpha) - \mathbf{r}] X[\mathbf{R}_\beta(s_\beta) - \mathbf{r}] \tag{40}$$

Since the interaction between the polymer segments and the obstacles are short ranged, we assume that $U(\mathbf{r})$ is also short ranged,

$$U(\mathbf{r}) = ul^4\delta(\mathbf{r}) \tag{41}$$

where u is a pseudopotential made dimensionless using the factor of l^4. Note that u is positive independent of the sign of the interaction between the obstacle and the polymer segment.

Combining Eqs. (29)–(31), (36), (39), and (41), we obtain

$$\langle \overline{R^2} \rangle = \lim_{n \to 0} \frac{\partial}{\partial n} \int \prod_{\alpha=1}^{n} D\mathbf{R}_\alpha \left[\sum_\alpha \left(\int_0^L ds_\alpha \frac{\partial \mathbf{R}_\alpha}{\partial s_\alpha} \right)^2 \right]$$

$$\times \exp\left\{ -\frac{3}{2l} \sum_\alpha \int_0^L ds_\alpha \left(\frac{\partial \mathbf{R}_\alpha}{\partial s_\alpha} \right)^2 \right.$$

$$- \sum_\alpha \int_0^L ds_\alpha \int_0^L ds_\alpha' W[\mathbf{R}_\alpha(s_\alpha) - \mathbf{R}_\alpha(s_\alpha')]$$

$$+ \frac{u\rho_0 l^4}{2} \sum_{\alpha,\beta} \int_0^L ds_\alpha \int_0^L ds_\beta \delta$$

$$\times [\mathbf{R}_\alpha(s_\alpha) - \mathbf{R}_\beta(s_\beta)] \right\} .$$

b. Variational Procedure. To evaluate the integrals of Eq. (41), we employ the Feynman variational procedure for the free energy of the chain. The free energy F_1 of the chain in the unreplicated system is given by

$$\exp(-\beta F_1) = \int D\mathbf{R} \exp(-\beta H) = Z(\lambda = 0) \tag{42}$$

The free energy $F(n)$ of the chain in n replicas is given by

$$\exp[-\beta F(n)] = Z^n(\lambda = 0) = \int \prod_{\alpha=1}^{n} D\mathbf{R}_\alpha e^{-\beta \sum_{\alpha=1}^{n} H_\alpha} \tag{43}$$

Therefore related by F_1 and $F(n)$ are related by

$$F_1 = \lim_{n \to 0} \frac{\partial}{\partial n} F(n) \tag{44}$$

We now proceed to establish the extremum condition for $F(n)$.

Let the trial Hamiltonian be

$$\beta H_0 = \frac{3}{2l} \sum_{\alpha=1}^{n} \int_0^L ds_\alpha \left(\frac{\partial \mathbf{R}_\alpha(s_\alpha)}{\partial s_\alpha}\right)^2 + Q \tag{45}$$

where

$$Q = \frac{q^2}{6l} \sum_{\alpha=1}^{n} \int_0^L ds_\alpha \, R_\alpha^2(s_\alpha) \tag{46}$$

with q the variational parameter. The free energy of the chain in n replicas given by Eq. (43) can be rewritten as (by adding and subtracting a term Q)

$$\exp[-\beta F(n)] = \int \prod_{\alpha=1}^{n} D\mathbf{R}_\alpha \exp(-\beta H_0 + \hat{X} + Q) \tag{47}$$

where

$$\hat{X} \equiv \frac{u\rho_0 l^4}{2} \sum_{\alpha,\beta=1}^{n} \int_0^L ds_\alpha \int_0^L ds_\beta \, \delta[\mathbf{R}_\alpha(s_\alpha) - \mathbf{R}_\beta(s_\beta)] \tag{48}$$

In general, the average of the function $\exp(-y)$ over some distribution function satisfies the inequality

$$\langle \exp(-y) \rangle_0 \geq \exp(-\langle y \rangle_0) \tag{49}$$

where the angular brackets with subscript zero indicate the averaging. Taking this averaging to be with respect to the trial Hamiltonian of Eq. (45), we obtain, from Eq. (49),

$$\langle \exp(\hat{X} + Q) \rangle_0 \geq (\langle \hat{X} \rangle_0 + \langle Q \rangle_0) \tag{50}$$

Substituting Eq. (50) into Eq. (47), we get

$$exp[-\beta F(n)] \geq$$

$$\exp(\langle \hat{X} \rangle_0 + \langle Q \rangle_0) \int \prod_{\alpha=1}^{n} D\mathbf{R}_\alpha e^{-\beta H_0} \equiv \exp[-\beta \tilde{F}(n, q)] \tag{51}$$

At the extremum of the free energy, $F_1 = \tilde{F}$,

$$\tilde{F} = \lim_{n \to 0} \frac{\partial}{\partial n} \tilde{F}(n, q) \tag{52}$$

the optimum value of the parameter q_0 is obtained from the condition

$$\left(\frac{\partial \tilde{F}}{\partial q}\right)_{q=q_0} = 0 \tag{53}$$

c. *Calculational Details.* The distribution function corresponding to the trial Hamiltonian is that of a harmonically localized random walk for each replica and is well known in the literature [63]. For each α,

$$G^{(\alpha)} \equiv \int_{\mathbf{R}_\alpha(s')}^{\mathbf{R}_\alpha(s)} \mathscr{D}[\mathbf{R}_\alpha(s)]$$

$$\times \exp\left[-\frac{3}{2l}\int_0^L ds \left(\frac{\partial \mathbf{R}_\alpha}{\partial s}\right) - \frac{q^2}{6l}\int_0^L ds\, \mathbf{R}_\alpha^2(s)\right]$$

$$= \left[\frac{q}{2\pi l \sinh(q|s-s'|/3)}\right]^{3/2}$$

$$\times \exp\left\{-\frac{q}{2l \sinh(q|s-s'|/3)}\right.$$

$$\times [(\mathbf{R}_\alpha^2(s) + \mathbf{R}_\alpha^2(s'))\cosh(q|s-s'|/3)$$

$$\left. - 2\mathbf{R}_\alpha(s)\cdot\mathbf{R}_\alpha(s')]\right\} \tag{54}$$

As shown by Feynman and Hibbs, $G(\alpha)$ can be expanded in exponential functions of $|s-s'|$ multiplied by products of eigenfunctions,

$$G^{(\alpha)} = \prod_{i=1}^{3} \sum_{m=0}^{\infty} \exp\left[-\frac{q}{3}\left(m+\frac{1}{2}\right)|s-s'|\right]\phi_m[\mathbf{R}_{\alpha i}(s)]\phi_m[\mathbf{R}_{\alpha i}(s')]$$

$$\approx \prod_{i=1}^{3} e^{-q|s-s'|/6}\phi_0[\mathbf{R}_{\alpha i}(s)]\phi_0[\mathbf{R}_{\alpha i}(s')] \tag{55}$$

where we have assumed that the ground state dominates the sum. Since

$$\phi_0[\mathbf{R}_{\alpha i}(s)] = \left(\frac{q}{\pi l}\right)^{1/2}\exp\left[-\frac{q}{2l}R_{\alpha i}^2(s)\right] \tag{56}$$

we obtain

$$G^{(\alpha)} \approx \left(\frac{q}{\pi l}\right)^{3/2}\exp\left\{-\frac{q}{2l}[R_\alpha^2(s) + R_\alpha^2(s')] - \frac{1}{2}q|s-s'|\right\} \tag{57}$$

Choosing $R_\alpha(0)$ to be the origin and integrating over $R_\alpha(L)$, we get

$$\int \prod_{\alpha=1}^{n} DR_\alpha e^{-\beta H_0} = e^{-nqL/2} \tag{58}$$

Furthermore, $\langle Q_0 \rangle_0$ can be readily obtained from the dominant ground-state eigenfunction as

$$\langle Q \rangle_0 = \frac{q^2}{6l} \sum_{\alpha=1}^{n} \int_0^L ds \frac{\int DR_\alpha(s) R_\alpha(s) R_\alpha^2(s) \phi_0^2[R_\alpha(s)]}{\int DR_\alpha(s) \phi_0^2(R_\alpha(s))}$$

$$= \tfrac{1}{4} nqL \tag{59}$$

Now we calculate the remaining term of Eq. (51). Parameterizing the delta function in Eq. (48), we obtain

$$\langle \hat{X} \rangle_0 = \frac{u\rho_0 l^4}{2} \sum_{\alpha,\beta=1}^{n} \int_0^L ds_\alpha \int_0^L ds_\beta \int \frac{d^3k}{(2\pi)^3} \langle \exp\{ik[R_\alpha(s_\alpha) - R_\beta(s_\beta)]\} \rangle_0 \tag{60}$$

Since

$$\langle \exp\{ik[R_\alpha(s_\alpha) - R_\beta(s_\beta)]\} \rangle_0$$

$$= \begin{cases} \exp\left[-\dfrac{k^2 l}{2q}(1 - e^{-q|s-s'|/3})\right] & \text{if } \alpha = \beta \\[2mm] \exp\left(-\dfrac{k^2 l}{2q}\right) & \text{if } \alpha \neq \beta \end{cases} \tag{61}$$

the coefficient of n in $\langle X \rangle_0$ is given by

$$\lim_{n \to 0} \frac{\partial \langle x \rangle_0}{\partial n} = \frac{u\rho_0 l^4}{2} \int_0^L ds \int_0^L ds'$$

$$\times \int \frac{d^3k}{(2\pi)^3} e^{-k^2 l/2q} \left[\exp\left(\frac{k^2 l}{2q} e^{-q|s-s'|/3}\right) - 1\right]$$

$$\equiv \Phi \tag{62}$$

Note that the right-hand side of Eq. (62) contains a renormalizable divergence, which is best treated by evaluating the integrals after the variation. Combining the results of Eqs. (51), (58), and (62), the replica free energy F becomes, from Eq. (52),

$$\beta \tilde{F} = \tfrac{1}{4} qL - \Phi \tag{63}$$

so that q_0 is given by [see Eq. (53)]

$$\frac{L}{4} = \frac{u\rho_0 l^4}{2} \int_0^L ds \int_0^L ds' \int \frac{d^3k}{(2\pi)^3} \left(\frac{k^2 l}{2q^2} \right)$$

$$\times \left\{ \exp\left[-\frac{k^2 l}{2q} (1 - e^{-q|s-s'|/3}) \right] \right.$$

$$\times \left[(1 - e^{-q|s-s'|/3}) - \frac{q|s-s'|}{3} e^{-q|s-s'|/3} \right] - e^{-k^2 l/2q} \right\} \tag{64}$$

$$= \frac{1}{4} \left(\frac{3\varepsilon l}{q} \right)^{1/2} u\rho_0 l^2 L$$

where

$$\varepsilon^{1/2} \equiv \left(\frac{3}{2\pi} \right)^{3/2} \int_0^\infty dx \left[\frac{1}{(1 - e^{-x})^{3/2}} - \frac{xe^{-x}}{(1 - e^{-x})^{5/2}} - 1 \right] \tag{65}$$

In going to Eq. (65), the integrations over s and s' are performed in the limit of $L \to \infty$. The value of $q(= q_0)$ obtained from the variational calculation follows from Eq. (65) as

$$q_0 = 3\varepsilon u^2 \rho_0^2 l^5 \tag{66}$$

It follows from Eq. (54) that the average mean-square end-to-end distance of the chain is

$$\langle R^2 \rangle = \frac{3l}{q_0} (1 - e^{-q_0 L/3}) = \frac{Ll}{z} (1 - e^{-z}) \qquad z \equiv \varepsilon u^2 \rho_0^2 L l^5$$

$$= \begin{cases} Ll & z \to 0 \\ \dfrac{1}{\varepsilon u^2 \rho_0^2 l^4} & z \to \infty \end{cases} \tag{67}$$

Therefore we find that there is a crossover for $\langle R^2 \rangle$ from the Gaussian result at $\rho_0 \to 0$ to a situation where $\langle R^2 \rangle$ is proportional to ρ_0^{-2} independent of the chain length as ρ_0 is increased. The dependence of q_0 on the space dimension d can be obtained by performing the k integral of Eq. (64) in d dimensions. It follows from Eqs. (64)–(67) that q_0 is proportional to $(u\rho_0)^{-2/(4-d)}$. Therefore $\langle R^2 \rangle$ is proportional to $(u\rho_0)^{-2/(4-d)}$ for $z \to \infty$ and $d < 4$.

A comparison between the predictions of Eqs. (76)–(77) and Monte Carlo simulation results is provided in Section II.B.3.

2. Mean-Field Theory: Flory Approach

The localization transition as developed and discussed in the previous section can also be qualitatively understood using simple Flory-type mean-field arguments. One starts from a generalized Flory–Imry–Ma free energy,

$$\beta F \simeq \frac{R^2}{l^2 N} + \frac{l^2 N}{R^3} + \frac{wl^3 N^2}{R^3} + \left\{ \begin{array}{c} -\dfrac{ulN}{R} \\[2mm] +u^2 N \log(R) \end{array} \right\} \tag{68}$$

The first and second term on the right-hand side of Eq. (68) describe the entropic contributions due to stretching and compression of a Gaussian random walk [see, e.g., 4). The third term accounts for self-intersections. The fourth term represents the disorder contribution from the porous media. The upper and the lower last terms are the proposed mean random fields by Cates and Ball [47, 53] and Edwards and Chen [49, 54], respectively. The basic idea is to estimate the saddle point of the partition function $\exp(-\beta F)$ by using scaling properties. Minimizing F with respect to R, w obtain

$$\frac{R^2}{l^2 N} - \frac{l^2 N}{R^2} - \frac{wl^3 N^2}{R^3} + \left\{ \begin{array}{c} \dfrac{uln}{R} \\[2mm] u^2 N \end{array} \right\} = 0 \tag{69}$$

For a random walk without self-avoidance ($w = 0$) the first term in Eq. (69) becomes irrelevant for $N \to \infty$ at $u \neq 0$, and we obtain the localization result $\nu = 0$, that is, independent of N and with amplitude

$$R \simeq \frac{l}{u} \qquad u \gg w\sqrt{N} \tag{70}$$

Without disorder ($u = 0$) the first and third terms balance for $N \to \infty$ and the usual Flory result for SAWs,

$$R \simeq l(wN^3)^{1/5} \qquad u < w \tag{71}$$

is recovered. The Flory solution [Eq. (71)] is spoiled by either of the disorder terms as soon as $u \neq 0$ if $N \to \infty$. Furthermore, the solution (70) only survives for $w < u$. For $w = u$ (in the case of the Cates–Ball potential) there is a balance between effects due to the random potential and the repulsive energy of the coil, and hence

$$R \simeq lN^{1/2} \qquad u = w \tag{72}$$

whereas in the case of the Edwards–Chen potential

$$R \simeq lN^{1/2} \qquad u\sqrt{N} < w \tag{73}$$

The physical origin of the above results is due to the fact that in the random potential the polymer "finds" a cavity and the probability that by Brownian motion it moves out of the cavity is less (by a factor proportional to N) than the probability of return. Effectively this leads to an attraction and hence to collapse.

It should be noted that the importance of the entropic localization effect for polymers in random media has recently been attempted to be generalized to random polymer networks by Edwards [55].

3. Simulations and Finite-Size Scaling

The analytical considerations as described in the last section have been initiated and stimulated by results from Monte Carlo simulations [46]. In the following section these and extended simulation results are summarized.

Consider a large cube of linear dimension L and a simple cubic lattice of lattice spacing a embedded in the cube such that $L = na$ and there are n cells in the cube. Each of these cells is then occupied with the probability $1 - p$ ($0 < p < 1$) according to the random-site percolation model [33]. The occupied cells constitute the solid particle and the unoccupied cells represent the fluid phase. Thus $1 - p$ is the volume fraction of the solid phase and p is its porosity (Fig. 1). For porosities $p \leq 0.6883$ ($= 1 - p_c$, where p_c is the site percolation threshold) the solid part of the medium is percolated. For $p \geq p_c$, the fluid phase is continuous. A polymer is introduced in the fluid phase. The model polymer may be represented by the "freely joined chain" consisting of $N - 1$ rigid links, each of length, say $l = 0.6a$, that are freely joined together (Fig. 1). Self-excluded volume effects are not included in the model. Ensembles of chain formation may be generated by a conventional kink-jump technique [46].

The unperturbed value of the mean-square radius of gyration $\langle S^2 \rangle_0$ in

Figure 1. Sketch of porous and flexible polymer chain as used in simulations.

the bulk $(p = 1)$ is given by

$$\langle S^2 \rangle_0 = \tfrac{1}{6} l^2 N \left(1 - \frac{1}{N^2} \right) \tag{74}$$

In Fig. 2, the ratio between the radius of gyration $\langle S^2 \rangle$ at various porosities $p < 1$ and the unperturbed value is presented. The angular bracket denotes the averaging over all possible polymer conformations as

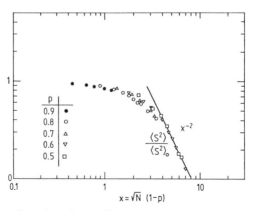

Figure 2. Log-scaling plot of normalized mean-square radius of gyration $\langle S^2 \rangle$ vs. scaling variable $\sqrt{N}(1 - p)$, where N is chain length and p is porosity.

well as different realizations of the porous medium at a given value of p (quenched averages). This value is plotted versus the scaling variable $\sqrt{N}(1-p)$ [46] as derived analytically in the previous section. This scaling variable can be qualitatively understood by the following argument using considerations of a dimensionless variable. As in a solution of many polymer chains, the only characteristic concentration relevant to the polymer chain is the overlap concentration C^* of the chain unperturbed by the porous medium. For the Gaussian chain, it is $C^* \sim 1/\sqrt{N}$. Writing the concentration C of the solid particles in units of C^*, the dimensionless variable therefore becomes $C/C^* \sim (1-p)\sqrt{N}$. Therefore the radius of gyration has a finite-size scaling form

$$\langle S^2 \rangle = \tfrac{1}{6} N l^2 f[(1-p)\sqrt{N}] \tag{75}$$

This scaling form provides a universal plot for all chain lengths and porosities; that is, all data fall on a universal curve as demonstrated in Fig. 2. For large values of the argument x, the scaling function $f(x)$ approaches the form $f(x) \sim x^{-2}$. Hence, a sufficiently long Gaussian chain in a porous medium is strongly collapsed such that $\langle S^2 \rangle$ is independent of N.

We now compare the predictions of Section II.B.1 [Eq. (67)] with the Monte Carlo simulation results. Since ρ_0 is proportional to $1-p$, where p is the porosity of the medium dealt with in the computer simulation and L is proportional to the number of segments N per chain, we observe that

$$\frac{\langle R^2 \rangle}{Ll} \approx \begin{cases} 1 & (1-p)N^{1/2} \to 0 \\ [(1-p)N^{1/2}]^{-2} & (1-p)N^{1/2} \to \infty \end{cases} \tag{76}$$

These asymptotic laws are precisely the results observed in Fig. 2. Furthermore, a comparison between the predictions of Eq. (67) and the simulation results in the crossover region can be attempted by identifying the pseudopotential u for the situation of the simulation, choosing u such that

$$z \equiv \frac{N(1-p)^2}{6.5} \tag{77}$$

From Eq. (67) we get

$$\frac{\langle R^2 \rangle}{Ll} = \frac{6.5}{N(1-p)^2} (1 - e^{-N(1-p)^2/6.5}) \tag{78}$$

which is plotted as the solid curve in Fig. 3. The data points are the same

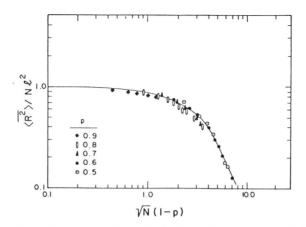

Figure 3. Comparison between Monte Carlo data and crossover formula (78). Log-log plot of mean-square end-to-end distance $\langle R^2 \rangle$ normalized by unperturbed value Nl^2 vs. $\sqrt{N}(1-p)$. Solid curve corresponds to Eq. (78). Data points are from Fig. 2.

as those in Fig. 2 corresponding to $6\langle S^2 \rangle / nl^2$ for different values of N and p. Since we expect $6\langle S^2 \rangle$ to be very close to $\langle R^2 \rangle$ numerically, we conclude that Eq. (67) agrees very well with the simulation results.

Of course, it is of interest to find whether this phenomenon still exists in the presence of a self-excluded volume of the polymer and what are the characteristics of the crossover between localized and strict self-avoiding conformations of the chain. A useful approach to investigate the crossover effects using simulations is to consider the Domb–Joyce polymer model [2, 64] on the cubic lattice. There each chain has a statistical weight

$$P_k = \exp\left[-w \sum_j n_k^2(j) \right] \qquad (79)$$

where $n_k(j)$ is the number of times the chain of the kth conformation visits the lattice site j. The dimensionless parameter w is the strength of the excluded volume interaction. The chain is Gaussian for $w = 0$, self-avoiding for $w > 0$, and strict self-avoiding for $w = \infty$. The results at intermediate values of w have been performed for porosities $p_c < p \leq 0.7$. As a particular example the results at $p = 0.4$ for the mean-square end-to-end distance $\langle R^2 \rangle$ are shown in Fig. 4. One observes a crossover from self-avoiding behavior at high w to collapsed ("localized") chain configurations at low w. At the critical point, where steric intramolecular repulsion and the media-induced compression of the chain compensate, the chain assumes unperturbed chain dimensions, $R^2 \sim N$. This can be

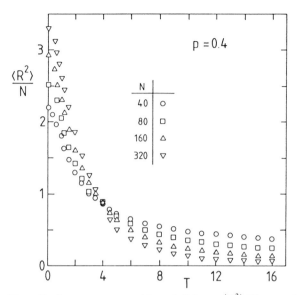

Figure 4. Normalized mean-square end-to-end distance $\langle R^2 \rangle / N$ vs. excluded volume parameter $T \equiv w$ for various chain lengths N at porosity $p = 0.4$.

used to estimate the "collapse" point w_c for various porosities and is shown for $0.4 \leq p \leq 0.7$ in Fig. 5. The collapse point, where $R^2 \sim N$, is roughly given by $1/w_c \approx 3.8/(-\ln p)$. To demonstrate the three distinct regimes (self-avoiding, Gaussian, and collapsed) and lead to characterize the crossover behavior [Fig. 6], it is useful to analyze the data using the following crossover scaling ansatz for $\tau \equiv |w - w_c|/w_c$:

$$\frac{R^2}{N} = g(\tau^{1/\phi} N) \quad \text{for } w < w_c \tag{80}$$

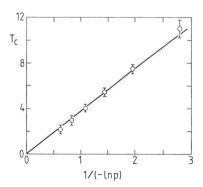

Figure 5. Estimated critical temperatures $T_c \equiv 1/w_c$ vs. porosity p.

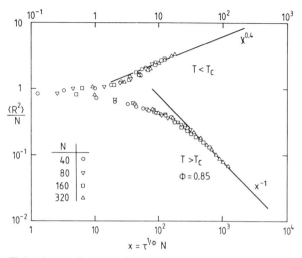

Figure 6. Finite-size scaling plot for normalized mean-square end-to-end distance $\langle R^2 \rangle / N$ vs. $x = \tau^{1/\phi} N$, where $\tau = (T - T_c)/T_c$ ($T \equiv 1/w$) at porosity $p = 0.4$.

with $g(x) \sim x^{-1}$ and hence $R^2 \sim \tau^{-1/\phi}$ for $x \gg 1$, but τ small, and

$$\frac{R^2}{N} = f(\tau^{1/\phi} N) \quad \text{for } w > w_c \tag{81}$$

with $f(x) \sim x^{(2\nu-1)}$ and hence $R^2 \sim N^{2\nu} \tau^{(2\nu-1)/\phi}$ for $x \gg 1$. The crossover exponent ϕ has been estimated by obtaining the best overlap between the various data points. However, it cannot be excluded that the crossover exponent is identical to the tricritical crossover exponent of the classical collapse transition $\phi = 0.5$ in three dimensions [65].

C. Disorder Effects on Phase Transitions

Since for sufficiently weak self-avoidance and strong disorder the polymer exhibits a collapse to a localized form, it is conceivable that the classical Θ collapse of a single polymer chain poor solvent may be altered by the influence of a quenched disordered environment. It is expected that the effect of disorder should shift the Θ point toward higher temperatures. It would be of interest to study this effect quantitatively by means of Monte Carlo simulations.

1. Adsorption on Heterogeneous Surfaces

The adsorption of a polymer on an "ideal" (i.e., planar and smooth) surface is now a classical subject and has been reviewed thoroughly, at least from a theoretical point of view, by Eisenriegler [66]. However, in reality most surfaces are not ideal but rather are heterogeneous and rough. The roughness can arise from physical or chemical origins. In the case of physical roughness the local curvature of the surface at different spatial locations can affect strongly the adsorption characteristics [67, 68]. In particular surfaces with a fractal physical roughness can strongly affect adsorption, the depletion layers between polymer solution and fractal surface, the grafting of polymer chains on such surfaces, and the segregation of binary polymer mixtures close to rough surfaces [69]. The case of chemical roughness, where certain regions of the surface have different affinity for the polymers as compared to the rest of the surface, is of particular interest as this problem is often encountered at some biological situations where adhesion of biopolymers on heterogeneous membranes plays a regulatory role in transmembrane signaling and shape transformation of membranes based on receptor clustering by multivalent ligands [70]. Several Monte Carlo simulations [71–73], mean-field theories [71, 74–77], and scaling arguments [75, 78] have contributed in embellishment and understanding of the polymer adsorption on heterogeneous surfaces.

However, as a first step in understanding the difference between the adsorption on a chemically rough surface and an ideal surface, it is useful to consider first the adsorption of a Gaussian polymer chain, that is, neglecting to a first approximation excluded-volume effects [71]. In the case of an ideal surface the basic physical origin of the criterion of polymer adsorption is the competition between the gain in potential energy obtained by the monomers by binding to the attractive surface and the loss in chain entropy associated with the reduction in the number of possible chain configurations of the adsorbed chain in comparison with that of a free chain. When the area of adsorption on the surface is randomly diluted, the chain has a tendency to be localized at regions of high concentrations of adsorptive sites. This may lead to a shrinkage or to an extension of the chain parallel to the surface depending on the ratio between chain length and a typical distance between adsorptive sites.

Following a simple Flory–Imry–Ma argument, the scaling form of the free energy of a chain confined in a two-dimensional plane with concentration v of nonadsorptive sites in the limit of $vN \to \infty$ is

$$F \sim \frac{N}{R_\parallel^2} + Nv \ln R_\parallel \qquad (82)$$

Combining this free energy with the adsorption free energy for a layer of thickness δ, we get, for the total free energy,

$$F \sim \frac{N}{\delta R_{\parallel}^2} + \frac{N}{\delta} v \ln R_{\parallel} + \frac{N}{\delta^2} - \frac{\epsilon N}{\delta} \tag{83}$$

where ϵ is the adsorption energy and $R_{\parallel}$ is the chain extension parallel to the surface. Minimizing F with respect to both δ and $R_{\parallel}$, the optimum values of the layer thickness and the chain extension parallel to the surface can be obtained. If the impurities on the surface actually stick out as poles and have spatial extensions beyond the typical size of the polymer, then the minimization procedure at each layer parallel to the surface gives

$$\delta \sim \frac{1}{\epsilon - v} \tag{84}$$

and

$$R_{\parallel} \sim v^{-1/2} \tag{85}$$

This situation is depicted in Fig. 7a. On the other hand, if the height of the impurities is less than the layer thickness δ, then a "mushroom-like" configuration as portrayed in Fig. 7b is expected. Here the layer thickness δ is the height of the mushroom and obeys Eq. (84). The core of the mushroom obeys the same law as Eq. (85), while the width of the cap of the mushroom must be the standard result of the extension of a Gaussian chain $R_{\parallel} \sim N^{1/2}$.

It must be noted that the result of Eq. (84) is independent of the height of the impurities. Since the adsorption energy per segment is $\epsilon \approx T_c - T$ and vanishes at the adsorption critical point in the absence of impurities, we infer from Eq. (84) that the critical temperature for adsorption is lowered by the presence of the impurities,

$$T_c(v) = T_c(v = 0) - v \tag{86}$$

where the numerical prefactor of the term linear in v is unknown.

Figure 7. Chemical roughness: polymer chain on adsorbing surface with repulsive barriers.

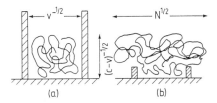

Therefore, it is harder to adsorb a Gaussian chain to a chemically rough surface than to an ideal one. Based on the above arguments, the lowering of the critical temperature is expected to be linear in the impurity density. The latter prediction is corroborated by results from Monte Carlo simulations [71] (Fig. 8).

However, these results are restricted to impurity concentrations well below the percolation threshold, $v < v_c$. Above v_c the adsorptive area consists of islands separated by nonadsorptive regions. In this case localization effects are expected to result similar to those depicted in Fig. 7a.

2. Disorder-Induced Adsorption

The phenomena of entropic localization have not yet been detected experimentally. The main difficulty is based on the dominance of the excluded-volume effect, which may mask the localization. However, proposals have been made for polymer networks [55] and for the coil–globule transition in porous solids [57, 58]. It has been argued [50, 57] that for sufficiently weak self-avoidance (SA) the disorder-induced localization should dominate the excluded volume. A regime where such weak SA is usually realized is near a Θ point. There the influence of disorder results in a shifting of the Θ point towards higher temperatures.

An even more promising situation where the localization effect could be observed is discussed below. In fact, it is possible to construct a polymer system [79] where the entropic localization is present at the *arbitrary* strength of self-avoidance. An interesting aspect of this model system is that it exhibits a critical adsorption transition that is caused solely by the quenched disordered structure of the polymer's environment. The main idea is to study an anisotropic system, where localization

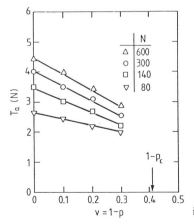

Figure 8. Shift of adsorption temperature with increasing concentration of nonadsorptive sites v.

of the chain takes place in $d' < d$ dimensions and simultaneous unfolding due to the SA constraint is expected in the remaining d–d' dimensions.

An appropriate choice of such a model environment is a disordered array of long parallel adsorbing rods ($d = 3$, $d' = 2$) (Fig. 9). As a consequence, one has anisotropic chain configurations with

$$\langle R_\perp^2 \rangle^{1/2} = A_\perp N^{\nu_\perp} \quad \text{in } 2 \perp -\text{Dim} \tag{87}$$

$$\langle R_\parallel^2 \rangle^{1/2} = A_\parallel N^{\nu_\parallel} \quad \text{in } 1 \parallel -\text{Dim} \tag{88}$$

Simultaneous presence of disorder and adsorption provides the following scenario:

At low temperatures the chain likes to adsorb on the rods and thus favors regions of *high* density of obstacles where they become localized. At high temperatures the chain is expected to move freely among the rods, thereby optimizing its entropy by preferring regions of *low density* of obstacles. What is the situation at intermediate temperatures? One important observation is that at a certain critical temperature both effects annihilate to lowest order and the chain is expected to behave approximately like a conventional SAW. It can be argued and is supported by Flory–Imry–Ma arguments that the two phenomena at low and high temperatures are based on two different mechanisms: At low temperatures the localization of the polymer by adsorption on the rods is dominating, whereas at high temperatures the localization by entropic repulsion, in formal analogy to the Anderson localization of quantum systems, is prevailing. The present model is currently the only one where the entropic localization is present at the arbitrary strength of self-

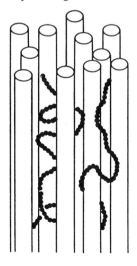

Figure 9. Polymer trapped among randomly distributed parallel rods.

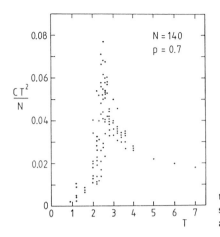

Figure 10. Specific heat C_v vs. temperature T for $N = 140$. Each data point corresponds to particular realization of disordered arrangements of rods.

avoidance. The existence of a critical temperature where the two competing mechanisms compensate each other is supported by analytical arguments and Monte Carlo results.

Monte Carlo estimates of the specific heat per monomer CT^2/N and the size $R_\perp$ of the polymer (Figs. 10 and 11, respectively) exhibit maxima at the temperature $T_{c1} \approx 2.6$, which indicates a phase transition.

For temperatures $T < T_{c1}$, the chain is strongly adsorbed such that $R_\perp$ is independent of N and $R_\parallel \sim N$. This is also demonstrated for $T = 2.0$, as an example, for various chain lengths N in Figs. 12 and 13.

The chain configurations are strongly unfolded and laterally collapsed with exponents

$$\nu_\perp = 0 \qquad \nu_\parallel = 1 \qquad T < T_{c1} \tag{89}$$

for the corresponding components of the radius of gyration. The adsorption preferentially takes place in regions of high obstacle density.

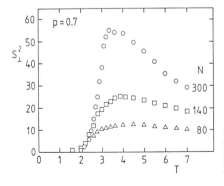

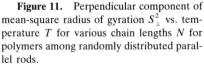

Figure 11. Perpendicular component of mean-square radius of gyration $S_\perp^2$ vs. temperature T for various chain lengths N for polymers among randomly distributed parallel rods.

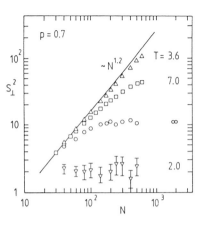

Figure 12. Perpendicular component of mean-square radius of gyration $S_\perp^2$ vs. temperature T for various chain lengths N for polymers among randomly distributed parallel rods.

At high temperatures, again the polymer chain is shrunk in the xy plane such that R is independent of N and $R_\parallel \sim N$,

$$\nu_\perp = 0 \qquad \nu_\parallel = 1 \qquad T > T_{c2} \qquad (90)$$

which is depicted in Figs. 12 and 13 for $T = 7.0$ and $T = \infty$, respectively. This effect is observed for all temperatures $T > T_{c2} \approx 3.6$, where T_{c2} is estimated from the maximum of $R_\perp$ (cf. Fig. 11). In this case, the adsorption energy is very small ($E/N = 0.36$ for $T = \infty$), which indicates a different mechanism for the shrinkage than that at low temperatures. In fact, at high temperatures the entropy loss due to the presence of obstacles causes this lateral collapse, which is a two-dimensional analogy of the localization of random walks in disordered quenched environments observed in Monte Carlo simulations of random-walk chains trapped in

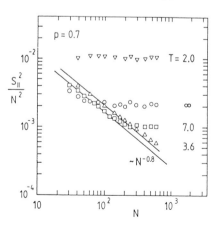

Figure 13. Parallel component of the radius of gyration, $S_\parallel^2$ vs. chain length N at various temperatures T.

three-dimensional percolated media [46, 51]. In the present case, this localization effect is imposed to the chain only perpendicular to the rods due to their disordered arrangements in the xy plane. Since the high-temperature compression is a purely *entropic* effect, it is still possible that $T_{c2} \neq T_{c1}$ even if only one peak in the fluctuations of the *energy* is seen.

Since our interpretation of the lateral compression $\nu_\perp = 0$ is independent of the SA-induced chain unfolding in the z direction, leading to $\nu_\parallel = 1$, the same phenomena should be observed in a two-dimensional model without SA. The results for $R_\perp^2$, depicted in Fig. 14, are very similar to our three-dimensional model: $R_\perp^2$ is almost independent of N at temperatures $T < T_{c1}$ and $T > T_{c2}$ and exhibits a maximum near $T = 2.9$.

The most prominent feature of Figs. 11 and 14, the appearance of an N-dependent maximum of $R_\perp^2$ at intermediate temperatures close to the adsorption transition T_{c1}, is more difficult to explain. Since this effect is observed in two and three dimensions, we must conclude that SA is not of primary importance for these phenomena. Rather, a subtle balance between attraction, leading at low temperatures to the adsorption of the chain onto the rods, and the repulsion between chain and rods, leading to localization at high temperatures, seem to be responsible for the emergence of conventional SA behavior at intermediate temperatures between two compressed states at low and high temperatures.

The existence of a single critical temperature

$$T_c = T_{c1} = T_{c2} \tag{91}$$

where repulsion and attraction are balanced such that

$$\nu_\perp = \nu_\parallel \sim \nu_F = \tfrac{3}{5} \qquad T = T_c \tag{92}$$

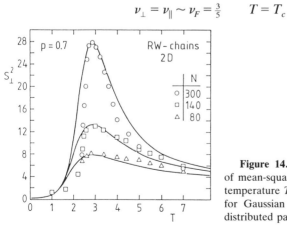

Figure 14. Perpendicular component of mean-square radius of gyration $S_\perp^2$ vs. temperature T for various chain lengths N for Gaussian polymers among randomly distributed parallel rods.

and $\nu = \frac{1}{2}$ in $d = 2$ is supported to lowest order approximation by mean-field arguments. One starts from a generalized Flory–Imry–Ma free energy for the present anisotropic model:

$$\beta F \simeq \frac{R_\parallel^2 + R_\perp^2}{l^2 N} + \frac{l^2 N}{R_\perp^2} + \frac{wN^2}{R_\parallel R_\perp^2} \left\{ \begin{array}{l} \dfrac{-ulN}{R_\perp} \\ +u^2 N \log(R_\perp) \end{array} \right\} \qquad (93)$$

The basic idea is to estimate the saddle point of the partition function $\exp(-\beta F)$ by using scaling properties. The first and second terms on the rhs of Eq. (93) describe the stretching and compression of a Gaussian random walk where we have generalized the common expression [see, e.g., 4] to our anisotropic situation. (It is easy to show that the longitudinal compression can be neglected.) The third term accounts for self-intersections wheres for the fourth term there are the two versions of the disorder contribution (refs. 50 and 48 and 54). Extremizing F with respect to $R_\parallel$, we obtain

$$R_\parallel^3 \simeq \frac{wl^2 N^3}{R_\perp^2} \quad \text{i.e.} \quad \nu_\parallel = 1 - \tfrac{2}{3}\nu_\perp \qquad (94)$$

Extremizing F with respect to $R_\perp$ and inserting Eq. (94), we obtain

$$R_\perp \frac{\partial}{\partial R_\perp}(\beta F) \simeq \frac{R_\perp^2}{l^2 N} - \frac{l^2 N}{R_\perp^2} - \left(\frac{w}{lR_\perp^2}\right)^{2/3} N + \left\{ \begin{array}{l} ulN/R_\perp \\ u^2 N \end{array} \right\} = 0 \qquad (95)$$

For a two-dimensional random walk without SA ($w = 0$) the first term in Eq. (95) becomes irrelevant for $N \to \infty$ at $u \neq 0$, and we obtain the localization result $\nu_\perp = 0$ with amplitude

$$R_\perp = \max(a, R_c) \quad \text{with} \quad R_c \simeq \frac{l}{u} \qquad (96)$$

Without disorder ($u = 0$) the first and third terms balance for $N \to \infty$ and the usual Flory result for SAWs,

$$R_\parallel = R_\perp \simeq (wl^2 N^3)^{1/5} \qquad u = 0 \qquad (97)$$

is recovered, in agreement with Eq. (95) when anticipating $u(T_c) = 0$, as shown below. The Flory solution [Eq. (97)] is spoiled by either of the disorder terms as soon as $u \neq 0$ if $N \to \infty$. Furthermore, the solution $R_\perp = R_c$ [Eq. (96)] only survives for $w < ul^3$. In our model, this condition is probably only fulfilled at very low temperatures where the adsorption becomes so strong that $R_c = a$, a regime that is hardly accessible to the Monte Carlo method. For strong SA, which we are interested in, SA is relevant on all length scales starting from the lattice constant, whereas disorder is irrelevant for $R < R_c$, and therefore $w > ul^3$. Consequently the only solution in the presence of disorder comes from balancing the third and fourth terms in Eq. (95), which yields, together with Eq. (94),

$$\nu_\perp = 0 \qquad \nu_\parallel = 1 \qquad u \neq 0 \qquad (98)$$

The temperature dependence of u^2 can be estimated as follows. The potential $U(\mathbf{r}_\perp)$ acting on a polymer segment at position $\mathbf{r}_\perp$ is represented by a sum over the positions $\mathbf{r}'_\perp$ of scattering sources (obstacles) times the interaction potential $\Phi(\mathbf{r}_\perp - \mathbf{r}'_\perp)$. Here, $\beta\Phi$ contains a temperature-independent hard-core repulsive and a temperature-dependent short-range attractive part with amplitudes v_0 and $v_1 = \Phi_0/k_BT$, The simplest way to proceed at this point is to assume a Gaussian probably distribution for the scattering strengths of the sources, but alternatively one obtains the same result as a lowest order approximation for our randomly distribution obstacles [10]. Then at the correlation reads

$$\langle U(\mathbf{r}_{\perp,1})U(\mathbf{r}_{\perp,2})\rangle = \rho_0 \int_{\mathbf{r}'_\perp} \int_{\mathbf{r}''_\perp} \Phi(\mathbf{r}_{\perp,1} - \mathbf{r}'_\perp)\Phi(\mathbf{r}_{\perp,2} - \mathbf{r}''_\perp)\delta^{d'}(\mathbf{r}'_\perp - \mathbf{r}''_\perp)$$

$$(99)$$

After performing the integration [e.g., for $\Phi(\Delta\mathbf{r}_\perp) = v_0\delta^{d'}(\Delta\mathbf{r}_\perp) - v_1\delta^1(|\Delta\mathbf{r}_\perp - a|)$] and taking the limit $a \to 0$, we obtain

$$u^2(T) = \rho_0(v_0 - v_1)^2 = \left(u_0 - \frac{\hat{u}_1}{T}\right)^2 \qquad (100)$$

where we have neglected higher order corrections. Independent of the specific potential used for the derivation, Eq. (100) represents a feature that is generally expected: For simple potentials with one form parameter, as in our case, there is exactly one zero $u(T_c) = 0$. This is basically our argument for Eq. (91) and for the appearance of a maximum in Figs. 11 And 14. From Eqs. (96) and (100) we have

$$R_c(T) = A_0(T)\left\|\frac{1}{T - T_c}\right\| \qquad (101)$$

with $A_0 = Tl/u_0$ and $T_c = \hat{u}_1/u_0$. Qualitatively, the mean-field result [Eq. (101)] describes very well the observed behavior. To demonstrate this, use the two-dimensional data with $u(T)$ given in Eq. (100). The curves shown in Fig. 14 for finite chain length N are obtained on the basis of the scaling function [47],

$$\langle R^2 \rangle = R_c^2 \left[1 - \exp\left(-\frac{l^2 N}{R_c^2} \right) \right] \tag{102}$$

where $R_c(T)$ was given in Eq. (101). All three curves are produced using a single set of parameters u_0, $\hat{u}_1$, l. Of course near the critical point T_c fluctuations will modify the critical behavior, which is certainly an interesting subject for future work.

Finally a remark concerning the thermodynamic limit of the present model system is necessary. Effects that might occur in exponentially large systems, such as Lifshitz tails or slow crossover behavior at marginality, have not been considered above. In fact, they are very difficult to find in Monte Carlo simulations as well as in real physical systems [63]. In addition, the critical properties of the adsorption transition have not been examined. Also, the characteristics of the chain configurations at T_c are not known exactly.

One might speculate about a suitable experimental situation where the disorder-induced adsorption could be observed. But one could also think of other geometries that have the same features as the above model system: disorder, adsorption, and anisotropy. Such a possible candidate would be a layered structure, which has a lateral dimension $d' = 1$, for example, a disordered stack of porous sheets.

3. Liquid Crystalline Polymers in Porous Media

The ordering of polymers with liquid crystalline groups either on the chain backbone or as side groups has been extensively studied [80–82]. There have been continued efforts [83–88] to propose minimal Hamiltonians for such stiff polymers to adequately describe their thermodynamic behavior. The leading theoretical works originated from a consideration of ordering by a collection of rodlike molecules.

The pioneering theoretical works [89–91] dealt with the extreme situation of rodlike polymers. Onsager [89] showed that a solution of hard rods undergoes a transition from an isotropic solution to an anisotropic solution at a certain critical concentration depending on the dimensions of the rod. Based on a mean-field lattice model with only repulsive excluded-volume interactions, Flory showed [90] that bulk systems of semiflexible chains undergo a first-order phase transition at a critical backbone stiffness and consequently a critical aspect ratio of the polymer. Alternatively, Maier and Saupe [91] showed that liquid crystals undergo first-order phase transition from a disordered state to an orientationally ordered state driven by the attractive orientational interactions between the molecules. These theoretical approaches are of textbook materials now [92, 93]. While the mean-field lattice theory of ordering of semiflexible chains has been criticized [94–98], improvements [99, 100] of the original theory have been made to account the statistics better and to include attractive orientational interactions between liquid crystalline segments. Flory's theory has also been extended to consider liquid crystalline blends [101].

There have been a number of discussions in the literature [92, 93, 102] to determine the molecular origin of the ordering phenomenon in bulk liquid crystalline polymers. While the Flory theory proposes that the phase transition is solely due to intramolecular forces, the Maier–Saupe theory proposes the origin to be solely intermolecular forces. In an effort to clarify this issue, several Monte Carlo studies [103–106] were reported. The current understanding of these simulation results is that it is necessary to keep both the intramolecular and intermolecular interactions between the various segments of the polymer in order to observe a phase transition between an isotropic and a long-range orientationally ordered state. This disorder–order phase transition has been shown to be first order in three dimensions. Although most theoretical and analytical calculations have so far centered around ideal homogeneous systems, it is of increasing interest to study the modification of this disorder–order transition by various disorders in the system.

In general, the experimental systems contain defects arising from surfaces, impurities, and other foreign species. Liquid crystalline polymer systems are some of the most impure systems due to disclinations, dust, and extraneous initiator and monomer from the polymerization. As liquid crystals are "soft," that is, the energy responsible for long-range order is small, disorder will have a significant effect on their thermodynamic properties. Although there are three common methods to introduce disorder into a model system, viz namely, dilution, random field, and random bond (as discussed in Section II.C), we now consider the effects of dilution on the liquid crystal transition. Before we address fully the Monte Carlo simulations [107] of this disordered problem, we first review the results of a model simulation pertinent to the pure case.

The model system studied without any quenched impurities consists of a set of chains of length N beads ($N - 1$ bonds) confined to a cubic lattice with lattice spacing equal to the bond length between any two consecutive beads in a chain. In particular, the case of trimers with $N = 3$ has been studied [107] carefully. To simulate an infinite system, the system is approximated as a set of infinitely many identical cells of length L with periodic boundary conditions. The density ρ of the system is defined as the fraction of trimers that occupy the lattice, $\rho = N_p N / L^3$, where N_p is the number of trimers and L^3 is the volume of the lattice. As pointed out earlier, it is necessary [106] to include both inter- and intramolecular interactions to simulate the isotropic–nematic transition. The intramolecular energy of two adjacent bonded segments is $\epsilon_b < 0$ if they are collinear and zero otherwise. The intermolecular energy between two neighboring nonbonded segments is $\epsilon_s < 0$ if they are parallel and zero otherwise. Although different values of ϵ_b and ϵ_s can readily be studied, consider now the case [107] where ϵ_b and ϵ_s are equal to ϵ. The steric interactions are considered as excluded volume; double occupancy of a given lattice site is forbidden. The Monte Carlo simulations were carried out using the Metropolis sampling technique. The system is characterized at each temperature by the heat capacity C_v and the orientational order parameter S.

S is defined as $S = \frac{1}{2}\langle 3f - 1 \rangle$, where f is the number of bonds that are in the preferred direction and $\langle \cdot \rangle$ denotes the ensemble average. The heat capacity C_v is defined as

$$C_v = \langle E^2 \rangle - \langle E \rangle^2 \qquad (103)$$

where E is the total normalized energy of the system given by

$$E = \left(1 - \frac{N_b}{N_p(N-2)}\right) + \frac{N_s}{4N_p} \qquad (104)$$

Here N_s is the number of parallel pairs of segments and N_b is the number of bends (gauche states) in the system.

The plot of the square of the order parameter S^2 for the pure system of size $L = 10$ against the reduced temperature $\tau = k_B T / \varepsilon$ is given in Fig. 15. The dependence of C_v on τ is given in Fig. 16a for the case of $L = 13$. It is usually a difficult task to determine the order of a transition of the type given in Figs. 15 and 16a.

One of the ways to determine the order is to study finite-size effects on the transition. For a first-order transition, the transition temperature T_c should depend on the lattice size L as

$$\frac{T_c - T_c^*}{T_c^*} \sim L^{-d} \qquad (105)$$

where T_c^* is the transition temperature for the infinite system, while the heat capacity maximum $C_v^{\max} \sim L^d$. On the other hand, for a second-order transition, the shift in the critical temperature scales as $L^{1/\nu}$ and

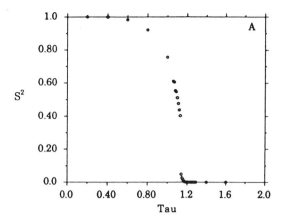

Figure 15. Dependence of square of order parameter S^2 on reduced temperature τ for $L = 10$.

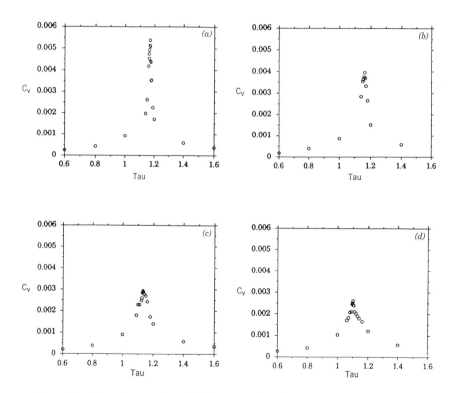

Figure 16. Dependence of specific heat on reduced temperature for $L = 13$. $(a-d)$ correspond to 0.0, 2.5, 5, and 7.5%, respectively.

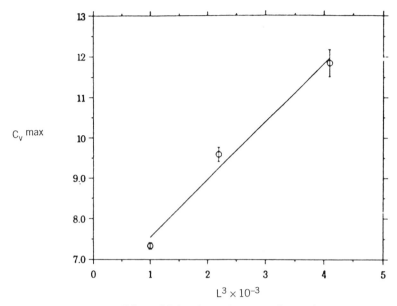

C_V max

L³ × 10⁻³

Figure 17. Effect of finite size on heat capacity maximum.

$C_v^{max} \sim L^{\alpha/d}$, where α and ν are the critical exponents. As shown in Figs. 17 and 18, the linear dependence of C_v^{max} on L^3 and that of τ_c on L^{-3} signify that the transition is first order. We now proceed to consider the effect of impurities on this transition.

Since the order–disorder transition temperature in simple magnetic systems is proportional to the field intermolecular interaction, the introduction of isolated random impurities will decrease the total inter-molecular interaction and therefore the transition temperature. In a similar way, we expect τ_c to be depressed by the addition of quenched impurities in the system. However, long-range order can only be attained if there exist sufficient pathways through which interactions can correlate. Above a certain impurity concentration, $1 - p_c$, where p_c is the porosity at the percolation threshold, these pathways cease to exist, and no transition can occur. Therefore, the transition temperature should de-crease with impurity concentration and eventually reach zero at the percolation threshold as sketched in Fig. 19.

The Monte Carlo simulations of the ordering of a collection of trimers in a quenched impure random medium were carried out by Dadmun and Muthukumar [107]. In these simulations, the initial configuration of the system is created as follows. A prescribed number of impurities are placed randomly in the cubic lattice via the percolation method. The

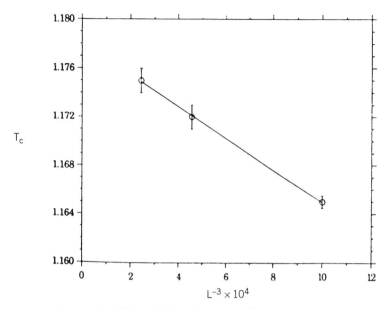

Figure 18. Effect of finite size on transition temperature.

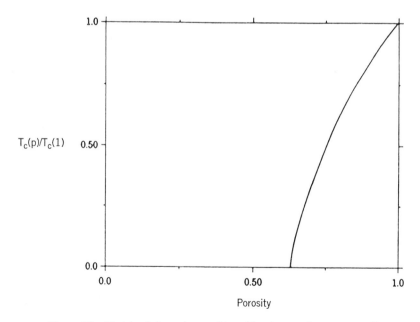

Figure 19. Sketch of dependence of transition temperature on porosity.

trimers are then added to the lattice by keeping each impurity site as forbidden for the occupancy by the segments of the trimers. Naturally there are some voids in the lattice after placing the impurities and trimers. The segments of the various trimers interact with intra- and intersegment interactions as mentioned above. There is no additional interaction between the polymer segments and the impurities beyond the self-avoidance condition. Different chain configurations are created by a modified reptation algorithm. A void is selected at random. If one of the randomly selected nearest-neighbor lattice points is the end of a trimer, then the trimer is moved into the void and the other end is vacated, thereby displacing the void and creating the new configuration for the chain. The new configuration is accepted according to the Metropolis sampling technique. The orientation order parameter S and the heat capacity C_v are monitored at different reduced temperatures τ.

The dependence of C_v on τ for the system with $L = 13$, the trimer segment density $\rho = 81.25\%$, and the impurity concentrations being 0, 2.5, 5.01, and 7.51% is presented in Figs. 16a–d. Inspection of these plots shows that as the impurity concentration increases, the transition temperature τ_c decreases. There is also a decrease in the specific heat maximum as well as a rounding of the specific heat peak as the impurity concentration increases. The dependence of τ_c on the porosity of the medium is given in Fig. 20. It is found that the transition temperature is depressed by the impurities and the transition temperature decrease linearly with the impurity concentration only in the impure region. This suggests that the presence of impurities changes the nature of the phase transition. To further investigate this, finite-size analysis has been performed on a system with impurity concentration of 5.01% with $L = 10, 13, 16$. The finite-size effects on C_v^{max} and τ_c are given in Figs. 21 and 22.

The observed nonlinear dependence in these figures and a comparison with Figs. 17 and 18 show that the order–disorder liquid crystalline transition in the random medium is not a first-order transition.

Thus the quenched impurities have a strong effect on the nematic-to-isotropic transition of a model liquid crystalline polymer system. Specifically, the transition temperature is lowered, the heat capacity peak is rounded, and the order of the transition is changed. These simulation results are in agreement with the theoretical work [108–111].

Further work is needed to study longer chains and much larger systems. Furthermore, in typical experimental conditions the surfaces of the random medium can interact with the liquid crystalline groups in a specific way [112]. For example, the liquid crystalline segments can align with the surface at a preferred angle. In practice, there can be a

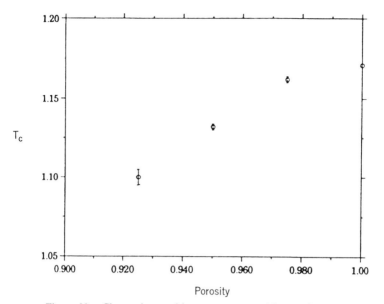

Figure 20. Change in transition temperature with porosity.

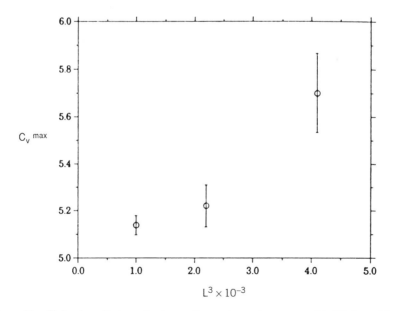

Figure 21. Finite-size effects on heat capacity maximum for system with 5% impurities.

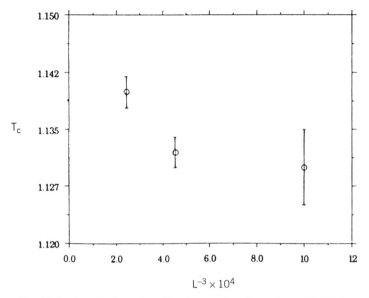

Figure 22.   Finite-size effects on transition temperature for system with 5% impurities.

competition between the surface effects and the finite volume of the cavities in the random medium. If the surface is an orienting surface for the liquid crystals, then for very low concentrations of the impurities, the transition temperature will be increased. But it will be eventually decreased at higher concentrations of the impurities, because the volume effects will dominate over the surface effects in the latter conditions [113]. These issues are becoming of current interest in the context of polymer-dispersed liquid crystals as viable display devices [114, 115].

III. DIFFUSION AND TRAPPING

Many experimental conditions, such as gel permeation chromatography, gel electrophoresis, enhanced oil recovery, and ultrafiltration, are controlled by transport properties of polymers in porous materials [116].

The diffusion of a single particle in porous media has been studied extensively using the "ant in a labyrinth" model [117–124]. On length scales over which the pore space is fractal one expects to observe anomalous diffusion. Related crossover behavior and corresponding exponents have been explained in terms of the fractal and spectral dimensions of the medium [122–124].

The analogous problem of a polymer in a labyrinth is more complex, since the effects due to entanglements [125, 126] of long chains with the constituents of the media may be in competition with pure disorder effects. In addition another particular polymer-specific aspect makes this problem even more complex. In the porous environment the chain is forced to squeeze through narrow channels in order to move to regions that are less occupied by obstacles and hence entropically more favorable. The motion of the chain between these different "entropic traps" is slowed down significantly by the bottlenecks connecting these regions. To distinguish between effects related to antlike anomalous diffusion and effects due to entropic barriers, this section is divided into two corresponding parts.

In contrast to small molecules, the diffusion of a macromolecule is more complex due to various intriguing effects. One can distinguish between three possible competitions: the anomalous diffusion due to the structure of the material, the activated diffusion due to the presence of entropic barriers, and in the case of very long chains, entanglements with the environment.

Various analytical approaches concerning anomalous diffusion have been published [127–130]. Most notably, in the first paper, by Martinez-Mekler and Moore [127], the diffusion constant is estimated using a one-loop renormalization group method. Although the authors claimed that their approach was quite inappropriate to calculate the diffusion constant, their ansatz has been recently reconsidered more thoroughly and successfully by Ebert et al. [130] and compared to Monte Carlo results [131].

One of the most interesting aspects of polymer diffusion in disordered media is its potential to elucidate effects complementary to the reptation idea, which is essentially based on geometric (i.e., topologic) considerations and ignores the entropic (and eventually energetic) constraints based essentially on short-ranged correlated density fluctuations in polymer liquids [132]. This will be discussed in the second part of this section.

A. Anomalous Diffusion

Diffusion of atoms or small molecules in percolated media has been intensively studied numerically for more than 10 years [118–120]. One of the main observations is the appearance of a regime of anomalous diffusion, which is a consequence of the self-similar structure of the media. The crossover between the intermediate anomalous and the long-time classical diffusion is now well understood [122].

Recently these investigations have been extended to the diffusion of a single polymer chain in disordered media [46, 133–136]. Since the interpretation of the data from computer simulations are quite intriguing, in Section III.A.1 the essential results are summarized for Gaussian polymers and for polymers with excluded volume, both of them in two and in three dimensions. The interpretation of the results and their comparison with the theory of anomalous diffusion on disordered lattices are given in Section III.A.2.

1. Monte Carlo Simulations

The model for a porous solid in three dimensions consists of small, hard cubes distributed at random with probability $1 - p$ on the cubic lattice with mesh size a (Fig. 23). For porosities $p \leq 0.6883$ ($= 1 - p_c$, p_c is the site percolation threshold [33]) the solid is percolated. The polymer is modeled by the freely joined chain [137] consisting of $N - 1$ rigid links of length $l/a = 0.6$ freely joined together. The dynamics is associated to the model by the conventional kink-jump technique [137, 138]: A chain configuration is changed locally by trying rotations of two successive links around the axis joining their end points by an angle ϕ chosen randomly from the interval $(-\Delta\phi, \Delta\phi)$. The parameter $\Delta\phi$ is chosen arbitrarily. If an end point of the chain is chosen, the terminal link is rotated to a near position by specifying two randomly chosen angles (ϕ, θ) in three dimensions, with $\cos\theta$ being equally distributed in the interval $-1 <$

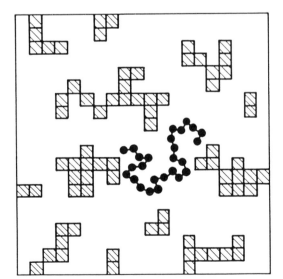

Figure 23. Two-dimensional representation of model porous media with a pearl-necklace polymer chain.

$\cos \theta < 1$. We rejected all rotations that lead to an overlap with the solid cubes constituting the disordered medium. When rejection takes place, we retain the old configuration and count it as the new one.

The diffusion constant can be estimated as usual from the slope of the mean-square displacement of the center of mass $R^2(t)$ of the chain,

$$R^2(t) = \langle [\mathbf{R}_{cm}(0) - \mathbf{R}_{cm}(t)]^2 \rangle \tag{106}$$

according to

$$D = \lim_{t \to \infty} \frac{R^2(t)}{6tl^2} \tag{107}$$

Estimates of the mean-square displacement of a bead relative to the center of mass of the chain can be obtained according to

$$r^2(t) = \langle [\mathbf{r}_k(0) - \mathbf{R}_{cm}(0) - \mathbf{r}_k(t) + \mathbf{R}_{cm}(t)]^2 \rangle \tag{108}$$

where $\mathbf{r}_k(t)$ is the position vector of the kth bead at time t.

The configurational correlation time of the polymer τ_p can be estimated in two ways. First, τ can be identified with the time for the onset of time-independent behavior of $r^2(t)$, that is, $r^2(t) = \text{const}$ for $t \geq \tau_p$, and second,

$$\tau_p = \int_0^\infty dt \, \phi(t) \tag{109}$$

where

$$\phi(t) = \frac{\langle S^2(0)S^2(t) \rangle - \langle S^2 \rangle^2}{\langle S^4 \rangle - \langle S^2 \rangle^2} \tag{110}$$

is the configurational correlation function of the square radius of gyration at time t, $S^2(t)$. The ensemble average has to be taken over chain configurations as well as over different realizations of the porous media.

a. Gaussian Polymers in Three-Dimensional Porous Media. Typical results of $R^2(t)$ and $r^2(t)$ for a Gaussian polymer chain in three dimensions (cube lattice) are presented in Fig. 24 for various chain lengths N and at a porosity $p = 0.6$ [46]. The displacements for a particular bead of

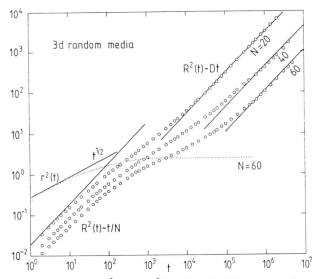

Figure 24. Log-log plot of $R^2(t)$ and $r^2(t)$ of localized Gaussian polymers in three dimensions for various chain lengths N at porosity $p = 0.6$.

the chain are very close to the Rouse dynamics [137, 139],

$$r^2(t) \sim t^{1/2} \tag{111}$$

and the configurational relaxation time is in agreement with a free-draining polymer, that is, $\tau_p \sim N^2$ (Table III). The latter finding is quite important since the longest relaxation time τ defined as the time characterizing the crossover from anomalous to classical diffusion

$$R^2(t) \sim Dt \quad \text{for } t > \tau \tag{112}$$

TABLE III
Normalized Configurational Correlation Time τ_p/N^2 of Localized Gaussian Polymers in Three Dimensions at Various p and for Chain Lengths N

p	$N = 20$	$N = 40$	$N = 60$	$N = 80$	$N = 100$
0.9	0.20	0.22	0.22	0.21	0.20
0.8	0.25	0.28	0.23	0.34	0.30
0.7	0.38	0.32	0.42	0.38	0.36
0.6	0.40	0.36	0.40	0.40	0.35
0.5	0.38	0.32	0.33		

Note: Error $\leq 10\%$.

TABLE IV
Normalized Terminal Relaxation Time τ/N^2 of Localized Gaussian Polymers in Three
Dimensions at Various p and for Chain Lengths N

p	$N = 20$	$N = 40$	$N = 60$	$N = 80$	$N = 100$
0.9	0.4	0.9	1.4	—	2.0
0.8	4.0	10.0	—	20.0	24.0
0.7	12.5	37.5	39.8	—	170.0
0.6	25.0	130.0	180.0		
0.5	100.0	312.0	440.0		

Note: Error $\leq$ 10–20%.

is orders of magnitude larger than τ_p. For comparison, Table IV contains the Monte Carlo data of τ/N^2 for various porosities p and chain lengths N.

The interpretation of the data and their extrapolation to large N is somewhat ambiguous. It can be argued that τ could follow either a power law [45],

$$\frac{\tau}{N^2} \sim (1-p)^4 N^2 \quad \text{for } (1-p)\sqrt{N} \gg 1 \tag{113}$$

or an exponential law of Arrhenius type due to entropic barriers [136, 135] (which is discussed in detail in the following section),

$$\frac{\tau}{N^2} \sim \exp(AN) \tag{114}$$

where A is a constant. Corresponding interpretations of Monte Carlo data for the diffusion coefficient D have been given as well, following either a power law,

$$DN \sim (1-p)^{-4} N^{-2} \quad \text{for } (1-p)\sqrt{N} \gg 1 \tag{115}$$

or an exponential law,

$$D \sim \frac{1}{N} \exp(-BN) \tag{116}$$

where B is another constant.

The dynamics of Gaussian polymer chains in disordered media has an additional complication: The equilibrium configuration of such a chain for $(1-p)\sqrt{N} \gg 1$ is, as discussed in Section II.C, strongly collapsed ("localized random walk") such that the size of the polymer is independent of chain length,

$$\langle S^2 \rangle \sim (1-p)^{-2} \quad \text{for } (1-p)\sqrt{N} \gg 1 \tag{117}$$

It has been argued [47] that the chain might not only exhibit a localization in configurational space but also be localized in real space and diffusion should be expected to be suppressed. However, this limiting case might be correct in a strict sense, but its documentation by experiments or computer simulations is hardly accessible.

b. *Gaussian Polymers in Two-Dimensional Porous Media.* Similar findings have been found during corresponding simulations of a Gaussian polymer in two dimensions moving among hard squares of width a [133]. The squares are distributed at random with probability $1 - p$ ($0 \le p \le 1$) on the square lattice with mesh size a. Here the polymer chain is modeled by a slightly different freely joined chain consisting of $N - 1$ rigid links freely joined together. Dynamics is associated to the model by a "bead–spring" technique especially suitable for polymer motions in two dimensions. It consists in randomly selecting a new position r_k of the th bead under the constraint $|r_k - r_{k+1}| = |r_k - r_{k+1}| + |r_k - r_{k-1}|$; that is, all possible new positions r_k are located on an ellipse defined by r_{k-1}, r_{k+1}, and $|r_k - r_{k+1}| + |r_k - r_{k-1}| = \text{const}$. If an end point of the chain is selected, the terminal link is rotated to a new position by specifying a randomly chosen angle and keeping the length of the link fixed. This type of freely joined polymer model has a fixed contour length $\langle l \rangle (N - 1)$ with an average length $\langle l \rangle$ of the link that has been chosen as $\langle l \rangle / a = 0.6$. Of course, it has been verified by Monte Carlo methods that, in the case of absence of obstacles, the present model exhibits the expected properties of Gaussian statistics and Rouse dynamics. During the simulations all attempts were rejected that led to an intersection of the polymer and the obstacles. Simulations have been performed for $20 \le N \le 100$ and $p = 0.75$. The results are presented in Fig. 25.

The first crossover of $R^2(t)$ from Rouse to anomalous diffusion appears at $t \simeq \tau_0$, which, according to our data, is independent of N and characterizes the local rearrangements of monomers at scales of the order of the bond length.

In the intermediate time regime, the motion of the chain is restricted by the presence of the obstacles and undergoes anomalous diffusion $R^2(t) = D^* t^{2k}$, where $k = 0.33 \pm 0.03$ and the anomalous diffusion coefficient $D^* \sim N^{-1.0 \pm 0.1}$ (Fig. 26). In the third regime the polymer undergoes ordinary diffusion and exhibits $R^2(t) \sim Dt$ with $D \sim N^{-2.0 \pm 0.2}$ (Fig. 26). The third regime at $t > \tau$ is characterized by the terminal relaxation time τ

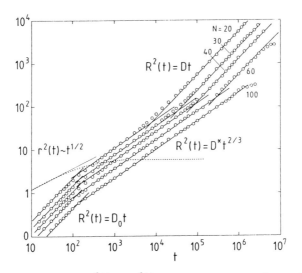

Figure 25. Log-log plot of $R^2(t)$ and $r^2(t)$ for Gaussian polymers in two dimensions at porosity $p = 0.75$.

that terminates the regime of anomalous diffusion. This time is estimated from the Monte Carlo data to $\tau \sim N^{3.0 \pm 0.4}$, which is shown in Fig. 27. Similar findings have been reported in the previous section for Gaussian polymers in three dimensions that imply some common theme for Gaussian polymers in disordered media. This will be discussed in Section III.A.2.

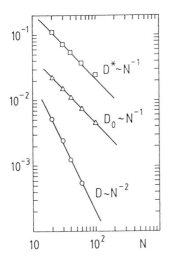

Figure 26. Log-log plot of anomalous coefficient D^* and classical diffusion coefficients D and D_0 of Gaussian polymers in two dimensions at $p = 0.75$.

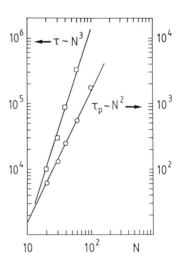

Figure 27. Log-log plot of polymer configurational relaxation time τ_p and longest relaxation time τ for Gaussian polymers in two dimensions at $p = 0.75$.

It is important to note that the configurational correlation time of the polymer itself is still that of a Rouse chain, $\tau_p \sim N^2$, characterizing the onset of the time independence of $r^2(t) \sim N$ for $t > \tau_p$ (Fig. 27). Some consequences of the difference between τ and τ_p are discussed in Section III.A.2.

According to $r^2(t)$ (Fig. 25), there are no indications for a "defect"-like motion leading to $r^2(t) \sim t^{1/4}$, as expected according to the reptation model [125]. Rather the dynamics is closer to the Rouse model $r^2(t) \sim t^{1/2}$. Of course, since one expects entanglements between the chain and the obstacles, the appearance of reptation for much longer chains as an additional dynamic mechanism, in addition to effects due solely to the randomness of the medium, cannot be strictly excluded.

c. Self-avoiding Polymers in Three-Dimensional Media. The self-avoiding polymer is modeled by the "pearl-necklace" chain consisting of N hare spheres that are freely joined together by rigid links [135]. The dynamics is achieved by random kink-jumps along the chain. All rotations are rejected that lead to an overlap with either a sphere of the chain (self-excluded volume) or the solid cubes forming the porous medium.

It is interesting to compare the effects coming from a disordered medium with the one in the case of a periodic medium. In the latter case the medium is organized in such a way that the small cubes form a regular periodic network consisting of infinitely long rods pointing in all three directions according to the geometry of the simple cubic lattice. Using, for example, a porosity of $p = \frac{1}{2}$, as expected, the dynamics of the chain is

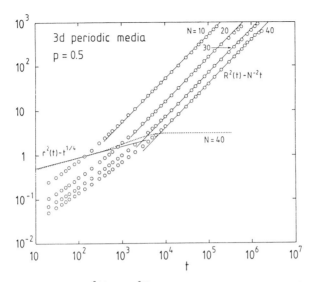

Figure 28. Log-plot of $R^2(t)$ and $r^2(t)$ for self-avoiding polymers in three dimensions among periodic obstacles.

that of the *reptation* model (Fig. 28),

$$r^2(t) \sim t^{1/4} \quad \text{for } t < \tau \tag{118}$$

and $D \sim N^{-2}$, $\tau \sim \tau_p \sim N^3$ [125, 126]. On the other hand, if the medium consists of randomly distributed small cubes, the dynamics (Figs. 29–31) are quite different from reptation. Rather they are similar to the results presented above for Gaussian chains with two important exceptions: (1) The chains are not collapsed but still self-avoiding, that is, $\langle S^2 \rangle \sim N^{6/5}$, and (2) the configurational relaxation time τ_p has a similar N dependence as the terminal relaxation time τ (Tables V and VI), although they differ considerably with respect to their amplitudes, $\tau \gg \tau_p$.

According to Figs. 29–31, the displacements $r^2(t)$ of a particular bead of the chain is very close to the Rouse dynamics [139] $r^2(t) \sim t^{1/2}$; there are no indications for a "breathing" or "defect"-like motion of $r^2(t) \sim t^{1/4}$ according to the reptation model [125, 126]. This has also been observed in all other cases studied by Monte Carlo [45, 133–135].

There are three regimes of center-of-mass behavior. At short times, one observes the Rouse behavior $R^2(t) \sim tN^{-1}$ due to the unhindered displacement over distances smaller than the average separation of the obstacles. At intermediate times, the medium restricts the motion of the chain, which undergoes anomalous diffusion $R^2(t) \sim t^{2k}D^*$, where the

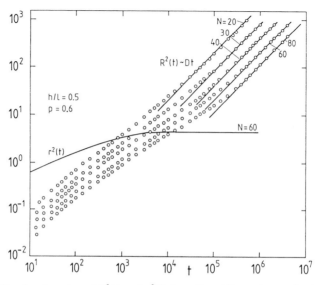

Figure 29. Log-log plot of $R^2(t)$ and $r^2(t)$ for self-avoiding polymers in three dimensions at porosity $p = 0.6$.

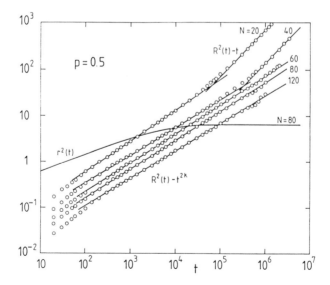

Figure 30. Log-log plot of $R^2(t)$ and $r^2(t)$ for self-avoiding polymers in three dimensions at porosity $p = 0.5$.

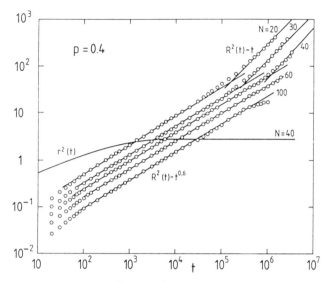

Figure 31. Log-log plot of $R^2(t)$ and $r^2(t)$ for self-avoiding polymers in three dimensions at porosity $p = 0.4$.

TABLE V

Normalized Correlation Time of Polymer Configuration $10^3 \times \tau_p/N^3$ for various p and Chain Lengths N

p	$N = 20$	$N = 30$	$N = 40$	$N = 60$	$N = 80$	$N = 120$
0.8	7.5	—	5.3	5.1	5.3	
0.7	8.1	7.8	6.9	6.5	5.9	
0.6	9.4	9.3	10.0	10.6	9.4	13.0
0.5	12.5	—	14.8	13.0	14.5	19.1
0.4	13.7	—	20.0	50.0		

Note: Error ≤ 10%.

TABLE VI

Normalized Correlation Time of Polymer Configuration $10^3 \times \tau_p/N^3$ for Various p and Chain Lengths N

p	$N = 20$	$N = 30$	$N = 40$	$N = 60$	$N = 80$
0.8	0.5	0.7	0.8	0.7	0.8
0.7	1.2	1.5	1.5	1.4	1.6
0.6	2.5	2.6	2.8	2.8	3.0
0.5	12.5	—	13.0		

Note: Error ≤ 20%.

anomalous diffusion exponent is estimated to be $2k = 0.60 \pm 0.05$ and the anomalous diffusion coefficient is $D^* \sim N^{-1.0 \pm 0.2}$. Finally, the chain resumes ordinary diffusion $R^2(t) \sim tD$ at long times and $D \sim N^{-2.0 \pm 0.3}$ (Table VII).

Interpreting the data of Tables V and VI as power laws, one gets approximately $\tau_p \sim N^{3.0 \pm 0.4}$ and $\tau \sim N^{3.0 \pm 0.7}$. However, it is important to note that the termination time τ, which is defined as the time characterizing the crossover of the center-of-mass motion $R^2(t)$ from anomalous to classical, is much larger than the configurational relaxation time τ_p given by the crossover of the bead displacement $r^2(t) = \text{const}$ at $t \geq \tau_p$.

d.   Self-avoiding Polymers in Two-Dimensional Media. We consider the situation of a self-avoiding chain that moves among infinitely long parallel repulsive rods of diameter a [134]. Such an anisotropic arrangement of rods may be realized in nematic fluids or in uniaxial stretched networks. The rods pointing in the z direction are distributed in the xy plane at random with probability $1 - p$ on a square lattice with mesh size a. For $p \leq p_c$ ($p_c = 0.59$) the medium is percolated [36]. Similar as described in the previous section, the polymer is modeled by the pearl-necklace chain. The configurational properties of the polymer are similar to the unperturbed isotropic case ($p = 1$) and the mean-square radius of gyration exhibits, as usual, $\langle S^2 \rangle \sim N^{1.2}$.

As expected, the chain dynamics parallel to the rods do not differ significantly from ordinary Rouse dynamics. However, the dynamics in the xy plane are quite different. Then perpendicular component of the mean-square displacement of the center of mass and one monomer, $R^2(t)$ and $r^2(t)$, respectively, are depicted in Fig. 32. There are three regimes of center-of-mass motion. At short times, free-draining motion according to $R^2(t) \sim tN^{-1}$ is observed, which results from unrestricted displacements over distances smaller than the average separation of the rods, $\sim p^{-1/2} \simeq 1.1a$. At intermediate times, the rods restrict the motion of the chain,

TABLE VII
Normalized Diffusion Coefficient DN^2 for Various Porosities p and Chain Lengths N

p	$N = 20$	$N = 30$	$N = 40$	$N = 60$	$N = 80$
0.8	1.11	1.20	1.33	1.51	1.60
0.7	0.67	0.74	0.74	0.72	0.76
0.6	0.35	0.37	0.31	0.33	0.29
0.5	0.13	—	0.11		

Note: Error $\leq 20\%$.

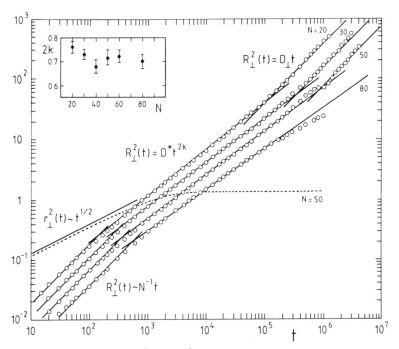

Figure 32. Log-log plot of $R^2(t)$ and $r^2(t)$ for self-avoiding polymers in three dimensions at porosity $p = 0.75$.

which undergoes anomalous diffusion $R^2(t) \sim t^{2k}D^*$, where $2k = 0.70 \pm 0.05$ (see insert in Fig. 32), and the anomalous diffusion coefficient $D^* \sim N^{-0.85\pm0.2}$ (Fig. 33). Finally, the chain resumes classical diffusion $R^2(t) \sim tD$ at long times and $D \sim N^{-2.0\pm0.2}$ (Fig. 33).

The model motions of the chain are described approximately by the Rouse model; that is, the mean-square displacement of one particular sphere of the chain is close to $r^2(t) \sim t^{1/2}$ (Fig. 32). There is no evidence for defect like motions $r^2(t) \sim t^{1/4}$ according to the reptation model.

On the other hand, estimates of the configurational relaxation time of the polymer and its termination time τ for anomalous diffusion indicate power laws on the order of $\sim N^3$ (Fig. 34), which seems to be in agreement with reptation [125, 126]. However, it should be emphasized that the termination time is about three orders of magnitude larger than the configurational relaxation time, which means that anomalous diffusion is essentially *decoupled* from configurational relaxation. This cannot be understood based on the reptation model, but rather seems to require

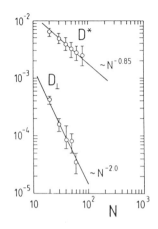

Figure 33. Log-log plot of anomalous diffusion co-efficient D^* and classical diffusion coefficients D and D_0 for self-avoiding polymers in two dimensions at porosity $p = 0.75$.

an explanation based on the structure of the porous medium, which is discussed below. In fact we expect that the range of anomalous diffusion becomes larger as p approaches p_c.

Simulations at a much lower porosity $p = 0.65$, closer to the percolation threshold $p_c = 0.59$, are presented in Fig. 35. The results for $R^2(t)$ confirm the appearance of anomalous diffusion found also for $p = 0.75$. The corresponding anomalous exponent, which is estimated as $2k = 0.67 \pm 0.06$, is in variance within the statistical error with the exponent found for $p = 0.75$. The estimate of the anomalous diffusion exponent gives $D^* \sim N^{0.9 \pm 0.2}$.

2. Fractality and Crossover Scaling

The reasons for the appearance of the anomalous diffusion of a polymer are not well understood. Therefore it is important to note that anomalous

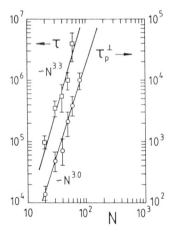

Figure 34. Log-log plot of polymer configuration-al relaxation time τ_p and longest relaxation time τ for self-avoiding polymers in two dimensions at porosity $p = 0.75$.

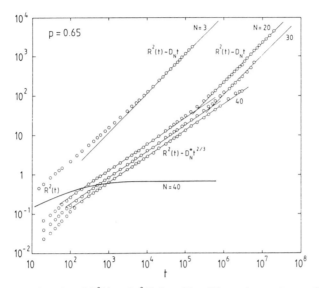

Figure 35. Log-log plot of $R^2(t)$ and $r^2(t)$ for self-avoiding polymers in two dimensions at porosity $p = 0.65$.

diffusion has *not* been observed for *periodically* distributed obstacles [46], but rather reptation. This obviously supports the idea that the disorder of the medium plays a dominant role in the appearance of anomalous diffusion by a polymer Chain. With respect to this fact and remembering the well-known case of anomalous diffusion of a single particle in percolating clusters, one is tempted to attribute the anomalous polymer diffusion to the disordered structure of the media.

In fact, estimates given above for the anomalous exponent $2k$ for polymers are in reasonable agreement with estimates for the "ant" [38, 118]: 0.69 in $d = 2$ and 0.57 in $d = 3$. The particular difference between an ant in a labyrinth and a "polymer in a labyrinth is due to the chain length dependence. The time interval during which anomalous diffusion is observed is larger the longer the molecule. In fact, diffusion of an ant in those media, which have been considered above for polymeric dynamics, that is, $0.4 \le p \le 1$ for three-dimensional media and $0.65 \le p \le 1$ for two-dimensional media, exhibit only weak diffusional anomalies, because the percolation correlation length is still quite small.

So the question remains why polymers exhibit anomalous diffusion over a much larger regime than single particles do. A possible simple explanation can be given in analogy to the problem. In particular, the crossover scaling from anomalous to ordinary diffusion of a polymer can be constructed as a generalization of the well-known crossover scaling for a single particle ($N = 1$) on an infinite cluster. The mean-square displacement of an ant is given by [122] $R^2(t) = t^{2k}g(t/\xi^{1/k})$, where $\xi \sim |p - p_c|^\nu$ is the percolation connectedness (or correlation) length and with the two limiting cases $g(s) = $ const for $t \ll \xi^{1/k}$ and $g(s) \sim s^{1-2k}$, and hence $R^2(t) \sim t\xi^{2-1/k}$ for $t \gg \xi^{1/k}$. The anomalous exponent is related to the fractal and the spectral dimensions d_f and d_s, respectively, by $2k = d_s/d_f$ [122–124].

Assuming that the displacement for the ant is $R^2 \approx \sigma^2 t^{2k}$, where $\sigma < a$ is the local jump distance, and assuming that the center of mass of the polymer chain follows essentially the same trajectory as the ant but needs, according to the Rouse model, a time of the order of $\tau_0 \approx \sigma^2 N$ in order to move the distance σ, leads to the displacement for the center of mass $R^2(t) \sim N^{-1}t^{2k}$. In general, we assume

$$R^2(t) = D^* t^{2k} f\left(\frac{t}{\tau}\right) \tag{119}$$

where $D^* \sim N^{-x}$ is the anomalous diffusion coefficient and $\tau \sim N^z \xi^{1/k}$ is the termination time. Since $f(s) \sim s^{1-2k}$ for $s \gg 1$ and $f(s) = $ const for $s \ll 1$, we obtain the scaling relation

$$w = x + z(-2k) \tag{120}$$

and a diffusion coefficient $D \sim N^{-w}\xi^{2-1/k}$ for classical diffusion $R^2(t) \sim Dt$ at $t \gg \tau$. In fact, estimates from Monte Carlo simulations seem to be in agreement with Eq. (119):

(a) $x = 0.9 \pm 0.2$, $2k = 0.68 \pm 0.05$, $z = 3.3 \pm 0.6$, and $w = 2.0 \pm 0.2$ for self-avoiding chains in two dimensional media [134];

(b) $x = 1.0 \pm 0.2$, $2k = 0.6 \pm 0.05$, $z = 3.0 \pm 0.7$, and $w = 2.0 \pm 0.3$ for self-avoiding chains in three-dimensional media [45]; and

(c) $x = 1.0 \pm 0.1$, $2k = 0.66 \pm 0.06$, $z = 3.0 \pm 0.4$, and $w = 2.0 \pm 0.2$ for Gaussian chains in two-dimensional media [133].

Using the relations $r^2(t) \sim \langle S^2 \rangle \sim N^{2\nu}$ for $t \geq \tau_p$ and $r^2(\tau_p) \approx R^2(\tau_p) \sim \tau_p^{2k} N^{-x}$, one obtains the configurational correlation time $\tau_p \sim N^{(2\nu+x)/2k}$. For *self-avoiding* polymers with $2\nu = 1.2$, $x \approx 0.9$, and $2k \approx 0.68$, $\tau_p \sim N^{3.1}$, which is in reasonable agreement with direct estimates. For *Gaussian* polymers in two dimensions [133] one has $2\nu \approx 0$, $x \approx 1.0$, and $2k \approx 0.66 \pm 0.6$, which yields $\tau_p \sim N^{1.50 \pm 0.15}$, whereas from direct estimates from Monte Carlo data according to $r^2(t \geq \tau_p) \approx $ const, one gets $\tau_p \sim N^{2.0 \pm 0.4}$.

Since there are two different time scales, τ_p and τ, there are in addition two corresponding different length scales, $\langle S^2 \rangle \sim N^{2\nu}$ and $\Lambda^2 \equiv R^2(\tau) \sim N^{2kz-x}\xi^2$, respectively. If $z > (2\nu + x)/2k$, anomalous diffusion is prevailing for all N, but if $z < (2\nu + x)/2k$, anomalous diffusion will vanish for large N. Since $\langle S^2 \rangle$ = const for *Gaussian* polymers, anomalous diffusion should always be present in this case, whereas the situation for *self-avoiding* polymers are less clear, at least from the point of view of available Monte Carlo data.

It should also be noted that the anomalous exponent k of a polymer might be different from the ant's exponent, because only the "backbone" of the infinite cluster is practically accessible to the chain, at least in the case of a self-avoiding polymer, since branches of the cluster would act as "dead ends," causing long-time retractive motions of the polymer. In this case, in calculating the anomalous exponent $2k = d_s/d_f$, where d_s and d_f are the spectral and the fractal dimensions, respectively, one probably has to replace the fractal dimension by the dimension d_b of the backbone.

It is certainly a formidable task to calculate the exponents x, z, and k, and an interesting question is whether they are related, similar as in the case of an ant, to percolation exponents.

However, effects solely due to the disorder of the medium on the polymer dynamics can be strongly masked if the length of the chains becomes much larger than those investigated so far. One effect might be due to the presence of randomly distributed bottlenecks ("entropic barriers") in the medium; this is discussed below in the next section. The other effect might be due to entanglements, leading eventually to reptation, especially if the medium consists of long rods [134] or hard squares in two dimensions [133].

In fact, the coincidence of diffusivity and relaxation time exponents, $w \approx 2$ and $z \approx 3$, with the corresponding exponents from reptation theory is surprising, despite the fact that local reptational correlations $r^2(t) \sim t^{1/4}$ are not observed, in contrast to the case of a periodic network [46], where reptation is clearly observed. But this coincidence can only be fortuitous, because in contrast to reptation the dynamics of a polymer in a random media has two different time scales, the configurational correlation time τ_p and the termination time τ with $\tau \gg \tau_p$ characterizing the onset of isotropic motion of the chain over large distances.

However, in the special case of a polymer moving among randomly distributed rods or between hard squares in two dimensions, it is conceivable that for very long chains entangled with rods or squares, the reptation effect is in addition to effects solely due to the randomness of the medium. Eventually reptation might dominate the dynamics in these cases.

It has been argued [48] that Gaussian chains might also be localized in space and diffusion is essentially suppressed. But this limiting case is hardly accessible by simulations, because very large N and media of large sizes are an unavoidable prerequisite to prove this conjecture. However, for small N one should recover the ant limit where diffusion is always present.

Entropic barriers probably have a very important influence on the polymer dynamics in disordered solids and is discussed in the next section. There the chain is forced to squeeze through narrow channels in order to move to regions that are less occupied by obstacles and hence entropically more favorable. The motion of the chain between these different "entropic traps" is slowed down significantly by the bottlenecks connecting these regions. Of course, the concept of entropic barriers is not useful in explaining the anomalous diffusion reported above, rather it is an *additional* effect of random media on the dynamics of polymers. Finally it should be noted that interaction between polymer and random media, other than pure repulsive interactions (e.g., adsorption), may influence the dynamics of polymers as well [140].

B.   Entropic Barrier Model for Polymers

1.   *Diffusion in Random Media and Polymer Liquids*

The dynamics of a collection of very long polymer chains in a highly entangled state has attracted tremendous effort in terms of both experimental and theoretical research [126]. This area still continues to be an active research topic [132]. Although a rigorous mathematical description of this problem can be posed [141], rigorous analytical solutions cannot be achieved due to the complexity of the problem. In view of this difficulty in analytical calculations, let us consider an extreme scenario, as originally envisaged by de Gennes [4, 125], where the model problem is analytically solvable. Let us assume the following:

(a)   The configurations of all chains except any one labeled chain are frozen permanently so that the labeled chain moves around in a rigid background. This assumption is a more stringent constraint than the condition that should be present in a polymer solution or a melt where all chains move equally. In view of the severity of the assumption, this model should lead to upper bounds for the various laws of chain dynamics in comparison with the actual results for polymer solution and melts.

(b)   The immobile environment confining the labeled chain has such a strong topological correlation that only the ends of the chain control the chain dynamics and the dynamic degrees of freedom of the monomers other than the end monomers are substantially quenched in order to just follow the motion of the ends.

These two assumptions allow us to imagine the melt or a concentrated solution to be a system whose extreme limiting behavior is given by a chain trapped in a random system of obstacles with full topological correlation. For each obstacle is supposed to mimic the uncrossability constraint of an entanglement. By full topological correlation, we imply that a sufficient number of obstacles are placed in the system in such a fashion that the labeled chain is confined in a tubelike region and the chain can move only through its ends. This mode of chain motion is referred to as the reptation. Thus a dense polymer system of very long chains is approximated by a scenario given in Fig. 36.

For such a simple model, where an impossible three-dimensional problem has been converted into a one-dimensional random-walk problem, the scaling behavior of the molecular weight (M) dependencies of the diffusion coefficient D of the center of mass and the relaxation time τ of the chain are found to be M^{-2} and M^3, respectively. Since the zero-shear-rate viscosity η of an entangled polymer system is proportional to τ, it follows then that $\eta \sim M^3$ if reptation is the dominant mechanism. By writing explicit stochastic equations for a chain undergoing reptation, Doi and Edwards [126] have derived formulas for the various viscoelastic properties of polymer solutions and melts and diffusion coefficient and relaxation time of a chain.

The experimental results on the M dependence of η is extremely clear. For long chains above the entanglement molecular weight M_e and at high enough concentrations, $\eta \sim M^{3.4}$, in disagreement with the reptation results. For $M < M_e$, the Rouse law $\eta \sim M$ is observed. The experimental situation for D is not so clear. While the earlier research provided substantial support for the reptation prediction $D \sim M^{-2}$, the most recent investigations [142–155] show that $D \sim M^{-\alpha}$, where α can be anywhere between 2 and 3. These results are summarized in Table VIII.

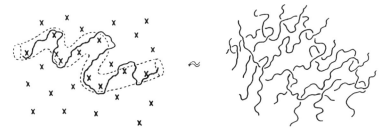

Figure 36. Approximation of melt by model where chain is constrained by fixed environment with full topological correlation.

TABLE VIII
Comparison between Experimental Results and Predictions of Models for Diffusion
Coefficient of Labeled Chain D and Viscosity η

	Unentangled Melt or Polymer Solution with Hydrodynamic Screening		Entangled Melt or Polymer Solution or Chains Trapped in Gels	
Quantity	Experiment	Rouse model	Experiment	Reptation model
D	M^{-1}	M^{-1}	$M^{-\alpha}$	M^{-2}
η	M	M	$M^{3.4}$	M^3

Note: Exponent is $2 \leq \alpha \leq 3$.

For polymer melts of high molecular weight, $\alpha = 2$. For polymer solutions and for polymers trapped in gels α is found to be significantly higher than 2, in the range of 2.5–2.9. The experimental results on the concentration dependence of D in polymer solutions are also not in accord with the prediction based on the reptation model.

It is to be noted that the reptation model builds in a stronger constraint than in the melt or concentrated polymer solutions. The topological constraints are taken to be completely frozen in the model. In spite of the severe constraint, the predicted exponents are weaker than the experimentally observed results for both D and η. This clearly indicates that there must be another dominant mechanism determining the chain dynamics.

The failure of reptation results for concentrated polymer solutions and chains in swollen gels simply implies that the basic assumptions of the model (viz., frozen environment and sufficient topological correlations) are not valid for these conditions. It therefore becomes necessary to consider a model system where topological correlations of the background for the chain diffusion are not complete. In other words, let us look for the systematic deviations from the Rouse behavior by introducing spatial constraints to the motion of a chain. Motivated by these arguments, Muthukumar and Baumgärtner [135, 136] studied a model described below.

a. Model Simulation. Consider a cubic lattice of lattice spacing a. The linear size of the lattice is taken to be $300a$ so that there are 300^3 cells in the lattice. Each of these cells is then occupied with probability $1 - p(0 < p < 1)$ according to the random site percolation algorithm, as already described. The occupied cells constitute the solid particles and the unoccupied cells represent the fluid phase. Thus $1 - p$ is the volume fraction of the solid phase. Only the values of $p > p_c = 0.3117$ need to be considered here so that the fluid phase is continuous. It must be emphasized that the solid phase is not made up of points. It is constituted by cubes each of volume a^3. These cubes are connected to each other depending on the value of p. For $p > 0.6883$, these cubes are present as lattice animals. For $p \leq 0.6883$, the solid phase is percolated.

The discretized version of the system is used only to prepare the solid phase in a random way. After obtaining the solid phase for a chosen value of p the fluid phase is taken to be a continuum. A polymer chain is introduced in the continuous space representing the fluid phase. The model for the polymer chain is the pearl-necklace model. This consists of N hare spheres of diameter h that are freely joined together by $N - 1$ rigid lengths of length l. The chain dynamics is followed using the dynamical Monte Carlo algorithm described earlier.

b. Results for Model Random Medium. The mean-square radius of gyration $\langle S^2 \rangle$ is found to be proportional to $N^{1.2 \pm 0.06}$ for the above values of parameters of the model for all values of p in the range $0.5 \leq p \leq 1$. Thus the self-avoiding walk statistics are not altered by the presence of impurities. This is in contrast to the case of a random-walk chain or a chain with attractive interbead potentials. In the latter case, as discussed in Section II.B, the chain is found to shrink due to an effective attractive interaction induced by the repulsive interactions between the beads and the impurities. Furthermore, this shrinkage alters the chain dynamics significantly. Therefore, enormous care must be exercised in obtaining dynamic information from simulations of a random-walk chain in random media.

When there are no impurities $(p = 1)$, a log-log plot of the time- (t) dependent mean-square displacement of the center of mass of the chain, $R^2(t)$, versus t is a straight line with a slope of 1 for each value of N. The double log plot of the intercepts of these lines versus N gives a straight line with a slope of -1, thus observing the Rouse behavior. The time-dependent mean-square displacement of the center of mass of the chain $R^2(t)$ is plotted against the Monte Carlo time t in Fig. 37 for $p = 0.6$. The plot of the mean-square displacement of the $(N/2)$th bead with respect to the center of mass of the chain, $r^2(t)$, is also included in the figure for $N = 60$.

There are three distinct regimes for the time evolution of $R^2(t)$. In the early and later time regimes, there is classical diffusion, $R^2(t) \sim t$. The duration of the intermediate crossover regime is longer as the chain length is increased and p is decreased. The mean-square displacement of

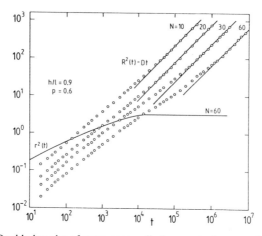

Figure 37. Double-log plot of mean-square displacement of center of mass $R^2(t)$ and of one bead relative to center of mass $r^2(t)$ vs. time t for different chain lengths and $p = 0.6$.

a monomer relative to the center of mass $r^2(t)$ shows an approximate Rouse behavior of $t^{1/2}$ for short times and smoothly crosses over to time independence. The diffusion coefficients D_0 and D in the early and late regimes of classical diffusion for $R(t)$ are obtained as the intercepts of the lines shown in Fig. 37. The diffusion coefficients D_0 in the short-time regime for all values of p are the same for a given N and the N dependence is found to be

$$D_0 \sim N^{-1 \pm 0.1} \tag{121}$$

in accordance with the Rouse law. The values of D in the late regime are plotted against N for different values of p in Fig. 38. It is clear from Fig. 38 that the reptation law $D \sim N^{-2}$ is not supported by the simulation data. Instead, D appears to obey an apparent law

$$D \sim N^{-3} \tag{122}$$

for sufficiently large values of N. It must be emphasized that this power law is only an apparent behavior. A different functional form such as an exponential may equally fit the data. Nevertheless we must conclude that the applicability of the reptation law to the simulation data is undoubtedly ruled out and that the apparent value of the exponent α is larger than 2. At the risk of overemphasis, it must be pointed out that this

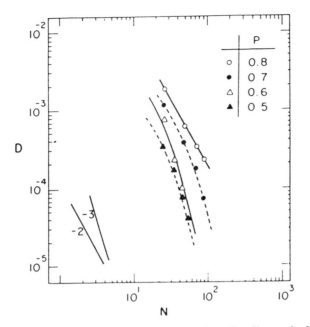

Figure 38. Log-log plot of D vs. N for different values of p. Slopes of -2 and -3 are shown for guidance.

result does not prove that reptation is wrong but only that the reptation mechanism is not valid for the model studied here.

c. Results for Regular Tight Networks. Let us consider the same simulation as above, but now the impurities are placed in a regular network fashion as in Fig. 39 instead of randomly. For $p = \frac{1}{2}$, the solid phase is organized in the form of a regular network of infinitely long rods on the cubic lattice with spacing $2a$. The obtained results in these

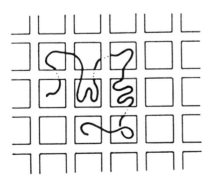

Figure 39. Simulation of chain in regular network with full topological correlation.

simulations are in agreement with the predictions of the reptation model,

$$R^2(t) \sim N^{-2}t \qquad t \to \infty \tag{123}$$

$$r^2(t) \sim t^{1/2} \qquad t < N^3 \tag{124}$$

The different dynamic laws observed in the case where the impurities are distributed randomly are due to the lack of the full topological correlation of the impurities.

The slow dynamics of the chain in a random array of impurities, as expressed by a stronger value of α than 2, is due to the presence of a distribution of "cavities" connected through "bottlenecks," as sketched in Fig. 40. When the chain is in one of these cavities, it has gained in entropy due to its freedom to assume very many configurations. In its trajectory from one cavity to another, the chain encounters a bottleneck that squeezes the chain, thus reducing its entropy. Therefore, the bottlenecks set up entropic barriers in random locations and the cavities act like entropic traps.

d. Entropic Barriers. In an effort to understand the role of entropic barriers on the chain diffusion, a toy Monte Carlo simulation has been studied [135] to follow the escape of a self-avoiding chain of N beads from a well-characterized cubic cavity through the gates at the centers of the walls of the cavity, as shown in Fig. 41. For this well-defined model problem, the original scaling arguments of Daoud and de Gennes, Brochard and de Gennes, and Guillot et al. have been generalized

Figure 40.   Creation of entropic barriers by bottlenecks.

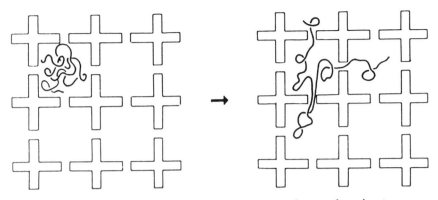

Figure 41. Escape of self-avoiding chain from entropic traps through gates.

[156–159]. First,

$$D = D_0 \exp\left(-\frac{\Delta F}{k_B T}\right) \tag{125}$$

where D_0 in the diffusion coefficient of a single chain without any geometric constraints. The change in the confinement free energy, ΔF, arising from the entropic barriers needs to be evaluated for a typical process depicted in Fig. 41. If f is the fraction of monomers in one of the bottlenecks and z is the average number of cavities that contain the $(1 - f)N$ unconfined monomers per bottleneck,

$$\Delta F = fF_2 + \left(\frac{1-f}{z} - 1\right)F_1 \tag{126}$$

where F_1 and F_2 are the confinement free energies for fully containing the chain in the cavity and the bottleneck, respectively. The above result is only an average result. The precise value of the prefactor for the second term should account for the simultaneous residence of the chain in several cavities and bottlenecks, and the number of cavities need not be equal to the number of bottlenecks.

According to the scaling argument, F_2 and F_1 are proportional to $Nc^{-1/\nu}$ and $NL^{-1/\nu}$, respectively, where c and L are the linear sizes of the bottleneck and the cavity, as shown in Fig. 41. The fraction of monomers in a bottleneck is a crossover function and depends on the size of the bottleneck. If the bottleneck is large enough to confine the chain, then $f = 1$. On the other hand, if the volume of the bottleneck is small in comparison with the size of the chain, the equilibrium value of f

approaches $N^{-1}c^{(1/\nu)-1}$. Thus

$$f = 1 \qquad\qquad \zeta \leq 1 \qquad\qquad (127)$$

$$f = N^{-1}c^{(1/\nu)-1} \qquad \zeta \gg 1 \qquad\qquad (128)$$

where $\zeta \sim R_\perp^2 R_\parallel / c^2 \lambda$ with $R_\perp$ and $R_\parallel$, respectively, being the components of the radius of gyration of the chain in a plane perpendicular and parallel to the axis of the bottleneck. Here, λ is the extension of the bottleneck as defined in Fig. 41. Substituting the results of the above argument, we get

$$\frac{D}{D_0} = \exp\left\{ -N\left[fc^{-1/\nu} + \left(\frac{1-f}{z} - 1\right)L^{-1/\nu} \right] \right\} \qquad (129)$$

Thus D is decreasing exponentially with N for the limiting case $f \to 1$. For a realistic situation where ζ is greater than unity but not infinitely larger, there are N-dependent contributions from the square bracket of the above equation, and consequently D would appear as an exponential of an apparent power of N different from unity. The scaling form of the above equation can be tested by rewriting it as

$$N^{-1}\ln\left(\frac{D}{D_0}\right) = A - sN^{-1} \qquad\qquad (130)$$

where s is proportional to $c^{-1}[1 - z^{-1}(c/L)^{1/\nu}]$ and A is negative and proportional to $(1/L)^{1/\nu}$. The Monte Carlo data for the above described simulation are in agreement with the above equation, as shown in Fig. 42. The slope increases as c is decreased, in agreement with the scaling analysis.

It should be noted that similar conclusions have been obtained from the transport of a hard sphere of radius a, representing the effective size of a macromolecule, through a porous network with Gaussian-distributed pore sizes r, $f(r) = 2\alpha^2 r \exp(-\alpha^2 r^2)$, and mean effective size $r_m = \pi^{1/2}/2\alpha$ of the pores, which leads to $D \sim \exp(-a/r_m)$ [160].

The problem of transport of flexible polymers in solution through conical pores driven by hydrodynamic flows has been considered by Daoudi and Brochard [161]. They found that above a critical current of the solvent the polymers are sufficiently stretched to enter the pore. In semidilute solutions the critical current is predicted to vary with concentration c as $c^{-15/4}$.

We shall compare the predictions of the entropic barrier theory of D with the experimental results on the chain diffusion in controlled pore glasses and polymeric networks in the following section.

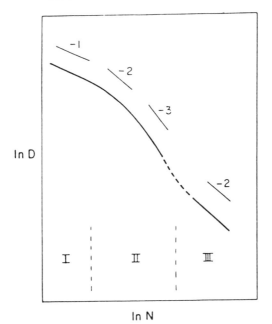

Figure 42. Plot of $N^{-1}\ln D/D_0$ vs. N^{-1} for different values of c/h.

e.   Comparison with Experimental Results. The simulations and scaling analyses described above show that there exists a new dynamic regime intervening the Rouse and reptation regimes. In this intermediate regime, the chain dynamics arises from the entropic traps created by spatial inhomogeneities and lack of topological correlations for full entanglement. When the entropic barriers contribute significantly to the chain dynamics, D has an apparent power law $N^{-\alpha}$ with α around 3. Instead of such an apparent power law, an exponential law $D \sim \exp(-N)$ can also be approximately fitted to the simulation results. Irrespective of the nature of the fitting function, the molecular weight dependence of D in this regime is more pronounced than the reptation law. This behavior is illustrated in Fig. 43. In regime I, Rouse law $D \sim N^{-1}$ is observed. Regime II corresponds to a situation where the chain dynamics is dominated by the entropic barriers present in the system. For infinitely long chains at any given concentration of the obstacles we would expect the reptation to be present, and this corresponds to regime III. The duration of regime II depends on the chain length, the concentration of the obstacles, and the topological correlation of the obstacles. For example, while we observe an apparent law of $D \sim N^{-3}$ for a random distribution of obstacles, the reptation result $D \sim N^{-2}$ is observed for a

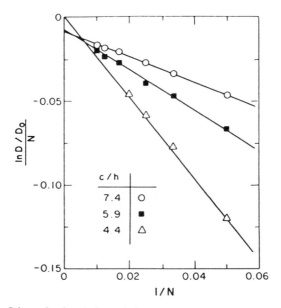

Figure 43. Schematic description of three dynamic regimes for molecular weight dependence of coefficient. Different slopes have been identified for guidance. I and III are respectively Rouse and reptation regimes. II is entropic barrier regime. Apparent slope less than -2 can easily be observed in regime II depending on nonuniversal features of experimental system.

topologically correlated network of obstacles, although the concentration of the obstacles if fixed at the same value. It is obvious from Fig. 43 that any slope smaller than -1 (even less than -2) can be observed in regime II depending on the range of N for a given particular system. The apparent exponent that is extracted in this regime is nonuniversal and depends on the specific problem. In fact the different values of the exponent α observed experimentally by different investigators are a direct result of their experimental situation belonging to this crossover regime. The fortuitous agreement between the reptation prediction and the initial experiments in the semidilute regime is indeed the early part of the entropically activated dynamic regime and not actually due to reptation. The same comment is pertinent to the most recent extensive simulations of Kolinski et al. [162] and Kremer and Grest [163]. It would be interesting to explore any possible connection between the entropic barrier model and the coupling scheme of Ngai and co-workers [164].

The conclusion that there is a pronounced entropically activated intermediate regime between the Rouse and reptation regimes is consistent with all known experimental data on D for widely different

systems. The tracer diffusion coefficient of linear polystyrenes in entangled polymer solutions of polystyrene at different total polymer
concentrations (from 13 to 40.6 wt %) was found [152] to scale as $N^{-2.5}$,
while in the melt the reptation law $D \sim N^{-2}$ has been observed. Similarly
the dynamic light-scattering study by Lodge and Rotstein [154] on the
diffusion of linear polystyrene in poly(vinylmethylate) gels at a concentration of $0.235 \, g/cm^3$ showed $D \sim N^{-2.9}$. On the other hand, Sillescu
[155] and co-workers found the reptation law for chain diffusion in
extremely tight networks. In polymer solution or swollen gels there are
significant spatial inhomogeneities allowing for lack of complete topological correlation. Such a scenario allows for entropically activated
dynamics so that an apparent exponent of $\alpha = 2.5, 2.9$ is possible. On the
other hand, when the concentration of the polymer solution approaches
100 wt %, the topological correlations are strong for long chains and the
reptation mechanism could become dominant if the entanglement constraints are long-lived. Also, when the network is tight, the spatial
inhomogeneities of a swollen gel are squeezed out and strong topological
correlation exists in order to facilitate the reptation mode to become
dominant. It is to be noted that the actual value of D in regime II is
higher than the value if reptation were present, although the N dependence in regime II is stronger than in the reptation regime. Furthermore,
if we assume that the characteristic lengths for the cavities (which act as
the entropic traps) and the bottlenecks (which create the entropic
barriers) are proportional to the correlation length ξ for the density–
density correlations in the semidilute solutions and swollen gels, then
Eqs. (125)–(128) yield

$$\frac{D}{D_0} \sim \exp\left\{-N\left[fc^{1/(3\nu-1)} + \left(\frac{1-f}{z}-1\right)c^{1/(3\nu-1)}\right]\right\} \tag{131}$$

$$f \sim 1 \qquad\qquad \zeta \le 1 \quad \text{(weak confinement)} \tag{132}$$

$$f \sim N^{-1}c^{(1-\nu)/(3\nu-1)} \qquad \zeta \gg 1 \quad \text{(strong confinement)} \tag{133}$$

where the scaling result $\xi \sim c^{-\nu/(3\nu-1)}$ is used with c being the polymer
concentration and ν being the radius of gyration exponent. For a given
value of ζ, the result of Eqs. (131)–(133) can be forced into an apparent
stretched exponential that has been empirically found [165]. The entropic
barrier theory is supported further by dynamic light-scattering studies of
Easwar [166] and Guo et al. [167] on the diffusion of linear polystyrene
molecules in controlled pore glasses. Generalizing Eqs. (123) and (124)

for this problem,

$$\frac{D}{D_0} = A \exp\left\{-N\left[fc^{-1/\nu} + \left(\frac{1-f}{z} - 1\right)L^{-1/\nu}\right]\right\} \tag{134}$$

where f is given in Eqs. (127) and (128). The prefactor $A < 1$ is present to normalize the diffusivity of a chain within a pore space to the macroscopic diffusion coefficient. Rewriting this equation, we obtain

$$\frac{1}{N}\ln\frac{D}{D_0} = \frac{1}{N}\ln A - (c^{-1/\nu} - L^{-1/\nu}) \tag{135}$$

for the weak confinement limit ($\zeta \leq 1$) and

$$\frac{1}{N}\ln\frac{D}{D_0} = \frac{1}{N}\left(\ln A - \frac{b}{c}\right) + (L^{-1/\nu}) \tag{136}$$

for the strong confinement limit ($\zeta \gg 1$), where b is some unknown numerical prefactor.

Therefore a plot of $1/N \ln D/D_0$ versus $1/N$ is expected to show two straight lines with the following features. For small N, entropic barriers are not present so that Eq. (135) is valid with the slope $\ln A$ being negative. As the molecular weight is increased, we get into the strong confinement limit where Eq. (136) is valid with the slope $\ln A - b/c$ being more negative. Also the crossover point between these two straight lines is the molecular weight at which the entropic barriers become operative. In view of Fig. 43, this molecular weight should be that molecular weight at which deviations from the Rouse law for D should occur. Finally, the difference in slope between the weakly confined and strongly confined regimes should be inversely proportional to the pore size of the porous medium. All of these predictions have been observed [166, 167] experimentally providing additional support for the existence of a distinct regime where the chain dynamics is dominated by entropic barriers.

2. Electrophoresis

Gel electrophoresis [168] consists of a sample of charged polymers with different molecular weights being introduced into a gel with an external electric field applied across the gel. The separation of different polymers is achieved due to the differing rates with which charged polymers with different molecular weights migrate through the gel. The problem has recently attracted considerable interest in theoretical understanding as well as computer simulation, in view of the practical importance of gel electrophoresis.

In the past, the reptation model of chain dynamics has been used to explain the data for high-molecular-weight polymer in gel electrophoresis. The reptation theories [169–173] of gel electrophoresis assume that a polymer of uniform charge Q is confined to a tortuous tube of length L formed by the topological constraints of the gel. The motion of the chain, constrained to move only along the contour of the tube, is taken to be "biased" in the field direction, leading to the average electrophoretic mobility μ, given by

$$\mu = \frac{Q\langle R_x^2\rangle}{fL^2} \sim N^{2\nu-2} \tag{137}$$

where $N = L/a$ is proportional to the molecular weight of the polymer, while a is the average distance between the constraints created by the gel; a is proportional to the familiar primitive path step length or the tube diameter; f is the translational friction coefficient $(f \sim N)$ of the chain; Q is also proportional to N; $\langle R_x^2\rangle$ is the mean-square end-to-end distance of the chain in the direction of the applied electric field over the duration of the chain migration; and ν is the scaling exponent for the molecular weight dependence of $\langle R_x^2\rangle^{1/2}$. For a Gaussian chain $\nu = \frac{1}{2}$, and therefore, according to reptation theory, the mobility will scale as N^{-1}. In some experiments with long DNA fragments, the $\mu \sim 1/N$ behavior has been observed. Assuming that the chain statistics of a very long DNA molecule are Gaussian, this experimental result for the mobility is taken as the verification of reptation. However, other workers [174–177] were not able to fit their results with a simple $1/N$ dependence for the mobility. In general, we expect the size of a charged polymer to be significantly modified by an external electric field, even if the field is weak. Furthermore, if the electric field is strong, nonlinear effects set in and the mobility increase with the electric field and becomes insensitive to the molecular weight. It is of interest to monitor the individual components of the reptation theory of gel electrophoresis, namely $\langle R_x^2\rangle$ and μ. In other words, it is necessary to measure simultaneously the mobility, size, and shape of the polyelectrolyte chain as it moves across the gel in the presence of the applied electric field. Several computer simulations [178–180] of this problem have already been reported in the literature. However, all of these simulations suffer from serious difficulties. These investigations either build in reptation to start with or consider self-intersecting chains without any intersegment potential interactions or ignore three-dimensional effects or lack of gel heterogeneity or a combination of all of these. The reader should refer to ref. 181 for a comparison of the different simulation models used for this problem.

The results of a Monte Carlo simulation [181] on a model that is a generalization of the model described in the preceding section are given below accounting for the gel heterogeneity, space dimension of 3, and potential interaction between segments and between segments and gel. Furthermore, the experimental regime where the size of the chain is comparable to the average pore size is particularly probed. It is this regime, in which polymers are separated by electrophoresis by a constant electric field, is the most efficient.

The random medium is taken to be the same model as in Section II.B, consisting of cubic obstacles distributed at random in space. The density of these obstacles is such that they make a collection of polydisperse lattice animals or a percolating cluster. The polymer modeled as a pearl-necklace chain of N beads of diameter h and charge q separated by a distance l. The beads i and j separated by a distance r_{ij} interact via a Debye–Hückel potential

$$V(r) = \frac{q^2 e^2}{4\pi\epsilon r}\exp(-kr) \qquad r > h \tag{138}$$

$$\kappa^2 = \frac{e^2 N_A \, \Sigma_i \, q_i^2 c_i}{\epsilon k_B T} \tag{139}$$

$$V(r) = \infty \qquad r \leq h \tag{140}$$

where e is the electron charge, ϵ is the dielectric constant of the medium, N_A is the Avogadro number, c_i is the concentration of the ith ion of charge q_i, and κ^{-1} is the screening length that includes the contribution of counterions of the polyelectrolyte that are not part of the condensed fraction as well as added low-molecular-weight salts. The interaction between the chains and the random medium is taken to be purely excluded-volume effect. The chain moves in the empty space and performs off-lattice motion in the presence of an applied electric field. The chain configurations are generated using the kink-jump technique. A newly generated configuration is checked for violation of the excluded-volume condition with the random medium. If this condition is violated, the new configuration is rejected and the earlier configuration is counted as the new configuration. If the excluded-volume conditions with the random medium is not violated, the Metropolis test is used to accept the new configuration. For the movement of the mth bead to a new position, with the electric field along the x direction, it is necessary to calculate

$$\Delta E_m = \sum_{i<j} V_{\text{new}}(r_{ij}) - \sum_{i<j} V_{\text{old}}(r_{ij}) + qeE(X_{m,\text{new}} - X_{m,\text{old}}) \tag{141}$$

where X_m is the position of the mth bead along the x direction. If $\Delta E_m < 0$ the move is accepted; otherwise the move is accepted if

$$\exp\left(-\frac{\Delta E_m}{k_B Y}\right) \geq w \tag{142}$$

where w is a random number in the internal $(0, 1)$. If the trial is not accepted, the bead is returned to its original position and the configuration is counted as the new one. Chains with $10 \leq N \leq 70$ have been studied for the following values of the parameters:

$$\frac{qEl}{k_B T} = 0.12 \tag{143}$$

$$\kappa l = \tfrac{2}{3} \tag{144}$$

$$\frac{q^2 e^2}{4\pi\epsilon k_B Tl} = 8.33 \times 10^{-3} \tag{145}$$

$$p = 0.7 \tag{146}$$

With the electric field along the x direction, the mobility is given by $\mu = \dot{X}_{cm}/E$, where $\dot{X}_{cm}$ is the steady-state velocity of the x component of the center of mass of the polymer. The steady-state velocity was obtained from the slope of the plot of the root-mean-square displacement of the x component of the center of mass against Monte Carlo time. To approach a steady state, the simulations were allowed to "warm up" for 100,000 Monte Carlo steps for the shorter chains and up to 600,000 Monte Carlo steps for the longer chains. One Monte Carlo step is defined as N attempted moves for a chain of N beads. The resulting mobility was averaged over five realizations of the porous media. The average quantities, such as the mean-square radius of gyration, the x component of the mean-square end-to-end distance, and the time-dependent mean-square displacement of the center of mass of the chain, were obtained by averaging over the various polymer configurations and different realizations of the porous media.

Thus, in these simulations, the chain dynamics are determined by the chain connectivity, interbead screened Coulomb interaction, the potential interaction between the beads and the external constant electric field, the excluded-volume interaction between the beads and the random medium, and the fluctuations in the chain entropy introduced by the spatial heterogeneities of the random medium. The key results are summarized here and the reader should refer to ref. 181 for more details:

(a) For a given chain length, as the strength of the constant electric field is increased, both the chain mobility and the component of the mean-square radius of gyration along the field direction reach a plateau. See Fig. 44. This demonstrates the coupling between the average size of the chain and its mobility.

(b) For weak electric fields (e.g., $qEl/k_BT = 0.12$) the N dependence of the mobility is found to be in qualitative agreement with the actual experimental data. See Fig. 45. Three regimes may be recognized.

(c) In regime 1, chains are smaller than the average pore size and the mobility of the chain is that of structureless particles, influenced primarily by collisions with obstacles.

(d) In regime 2, the chain size is comparable to the average pore size so that the chain mobility is determined by entropic barriers created by the spatial constrictions for the whole chain.

(e) In regime 3, the chain size is much larger than the average pore size and the entropic barriers are created for even higher Rouse modes of a single chain. Now the chain can be considered to be a set of blobs tethered together. The blobs are captured by the entropic traps and the tethering is done via the bottlenecks, as in Fig. 41. In fact, the chain is more readily stretched in this regime by the electric field. For example,

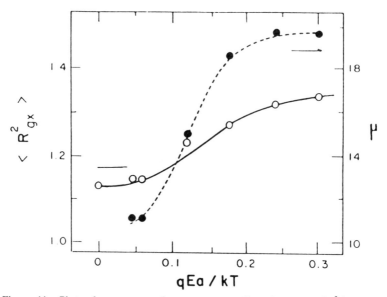

Figure 44. Plots of component of mean-square radius of gyration $\langle R_{gx}^2 \rangle$ along field direction and electrophoretic mobility μ against dimensionless electric field qEa/k_BT for $N = 10$ and other values of parameters given in Eqs. (143)–(146).

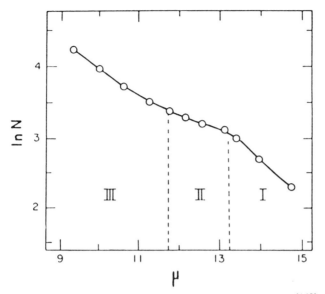

Figure 45. Plot of log N vs. for values of parameters given in Eqs. (143)–(146).

the mean-square radius of gyration along the x direction is found to be proportional to $N^{1.9}$ for $qEl/k_BT = 0.12$. Using Eq. (137) with this result of $\langle R_x^2 \rangle$, the reptation prediction is $\mu \sim N^{-0.1}$. Instead it was observed that $\mu \sim N^{-0.27 \pm 0.04}$ from these simulations, thus showing that reptation is not the dominant mechanism. This conclusion is further supported by actually following the position of the beads of chain as a function of time. The two-dimensional (x, y) projections of a chain of 54 beads are given in Fig. 46 at two different Monte Carlo times t. Only those obstacles that are within a distance $2l$ from a bead on the chain over the time period in the figure are shown. Further, to enhance the greater visibility of the chain, the size of the obstacles has been reduced in this figure by a factor of 2. In 60 Monte Carlo time units, the chain represented in Fig. 46a has moved to a configuration represented in Fig. 46b. These representative trajectories of the chain show that all beads at all length scales of the chain move and contribute to the chain mobility instead of just the ends of the chain. A close scrutiny of these trajectories shows that the chain is actually carried by the entropic traps corresponding to the cavities of the medium. The conclusion from these simulations is that the dominant dynamics in electrophoresis experiments is not reptation and that the entropic traps and barriers arising from the spatial inhomogeneities of the random medium or the gel play a crucial role in determining the chain mobility.

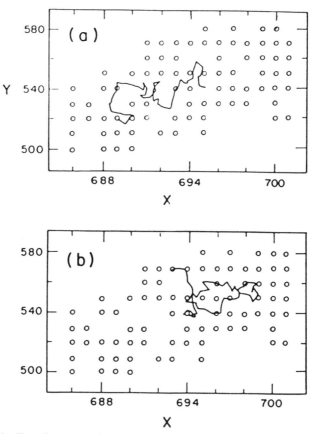

Figure 46. Two-dimensional (x, y) projections of chain of 54 beads at two times (a) and (b) separated by 70 Monte Carlo time units. Electric field is in x direction.

As in the case of the molecular weight dependence of D, three regimes can be identified for the molecular weight dependence of the electrophoretic mobility, analogous to Fig. 43. Taking $\mu \sim N^{(1-\alpha)}$, where α is the effective exponent discussed in the preceding sections, the N dependence of μ in the intermediate regime dominated by entropic barriers is expected to be steeper than the Rouse or reptation regime results. There is clear experimental evidence to this argument. Calladine and co-workers [175] found that $\mu \sim N^{-2}$ for DNA in various gels and Arvanitidou and Hoagland [176] found that $\mu \sim N^{-1.4}$ for poly(styrene sulfonate) in polyacrylamide gels, both in the intermediate regime. Furthermore, if the free-energy barriers dominate the polymer dynamics, we expect the topology of the polymer to play an insignificant role in the molecular weight dependence of μ. Smisek and Hoagland [177] have found that the mobility of flexible polyelectrolytes depends strongly on the molecular weight and not on the topology of the polymer. This experimental observation has recently been discussed [182] in terms of the entropic barrier mechanism.

There have been extensive reports by Zim and co-workers [183–185] on careful simulations of the electrophoretic mobility and elegant analysis based on the same theme of dominance of entropic barriers described above. The reader is urged to study these references as well.

REFERENCES

1. P. J. Flory, *Statistical Mechanics of Chain Molecules*, Interscience, New York, 1969.

2. C. Domb, *Adv. Chem. Phys.* **15**, 229 (1969).

3. C. Domb, *J. Stat. Phys.* **30**, 425 (1983).

4. P. G. De Gennes, *Scaling Concepts in Polymer Physics*, Cornell University Press, Ithaca, 1979.

5. J. des Cloizeaux and G. Jannink, *Polymers in Solution*, Clarendon Press, Oxford, 1990.

6. M. Mezard, G. Parisi, and M. Virasoro, *Spin Glass Theory and Beyond*, World Scientific, Singapore, 1987.

7. K. H. Fischer and J. A. Hertz, *Spin Glasses*, Cambridge University Press, 1991.

8. L. M. Gierasch and J. King, eds., *Protein Folding*, Amer. Assoc. Adv. Sci., Washingtron, 1990.

9. M. Kardar, *Nucl. Phys. B* **290**, 582 (1987).

10. D. S. Fisher and D. A. Huse, *Phys. Rev. B* **38**, 386 (1988).

11. B. Derrida, *Physica A* **163**, 71 (1990).

12. T. Hwa and D. S. Fisher, *Phys. Rev. B* **49**, 3136 (1994).

13. B. Duplantier, *Phys. Rev. A* **38**, 3647 (1988).

14. D. Thirumalai, *Phys. Rev. A* **37**, 269 (1988).

15. A. Baumgärtner and B. K. Chakrabarti, *J. Phys.* **51**, 1679 (1990).

16. S. M. Bhattacharjee and B. K. Chakrabarti, *Europhys. Lett.* **15**, 259 (1991).

17. F. Seno and A. Stella, *Phys. Rev. Lett.* **65**, 2897 (1990).

18. P. G. de Gennes, *Phys. Lett. A* **38**, 339 (1972).

19. D. S. McKenzie, *Phys. Rep.* **27C**, 25 (1976).

20. M. Muthukumar and B. G. Nickel, *J. Chem. Phys.* **80**, 5839 (1984); **86**, 460 (1987).

21. J. C. Guillou and J. Zinn-Justin, *Phys. Rev. Lett.* **39**, 95 (1977).

22. A. J. Guttmann, *J. Phys. A* **22**, 2807 (1989).

23. M. E. Fisher, *Phys. Rev.* **176**, 257 (1968).

24. A. B. Harris, *J. Phys. C* **7**, 1671 (1974).

25. B. K. Chakrabarti and J. Kertesz, *Z. Phys. B* **44**, 221 (1981).

26. A. B. Harris, *Z. Phys. B* **49**, 347 (1983).

27. Y. Kim, *J. Phys. C* **16**, 1345 (1983).
28. A. K. Roy and B. K. Chakrabarti, *Phys. Lett. A* **91**, 393 (1982).
29. P. M. Lam and Z. Q. Zhang, *Z. Phys. B* **56**, 155 (1984).
30. M. Sahimi, *J. Phys. A* **17**, L379 (1984).
31. Y. Meir and A. B. Harris, *Phys. Rev. Lett.* **63**, 2819 (1989).
32. A. K. Roy and B. K. Chakrabarti, *J. Phys. A* **20**, 215 (1987).
33 K. Kremer, *Z. Phys. B* **45**, 149 (1981).
34. S. B. Lee and H. Nakanishi, *Phys. Rev. Lett.* **61**, 2022 (1988).
35. S. B. Lee, H. Nakanishi, and Y. Kim, *Phys. Rev. A* **39**, 9561 (1989).
36. D. Stauffer, *Introduction to Percolation Theory*, Taylor and Francis, London, 1985.
37. R. Rammal, G. Toulouse, and J. Vannimenus, *J. Phys.* **45**, 389 (1984).
38. D. Lhuiller, *J. Phys.* **49**, 705 (1988).
39. A. Aharony and A. B. Harris, *J. Stat. Phys.* **59**, 1091 (1989).
40. A. K. Roy and A. Blumen, *J. Stat. Phys.* **59**, 1581 (1990).
41. S. Havlin and D. Ben-Avraham, *Adv. Phys.* **36**, 695 (1987).
42. P. M. Lam, *J. Phys. A* **23**, L831 (1990).
43. I. Chang and A. Aharony, *J. Phys. I* **1**, 313 (1991).
44. S. de Queiroz, F. Seno and A. L. Stella, *J. Phys. I* **1**, 339 (1991).
45. A. Baumgärtner and M. Muthukumar, *J. Chem. Phys.* **87**, 3082 (1987).
46. S. F. Edwards and M. Muthukumar, *J. Chem. Phys.* **89**, 2435 (1988).
47. M. E. Cates and R. C. Ball, *J. Phys.* **49**, 2009 (1988).
48. J. F. Douglas, *Macromolecules* **21**, 3515 (1988).
49. S. F. Edwards and Y. Chen, *J. Phys. A* **89**, 2963 (1988).
50. J. D. Honeycutt and D. Thirumalai, *J. Chem. Phys.* **90**, 4542 (1989).
51. J. Machta and R. A. Guyer, *J. Phys. A* **22**, 2539 (1989).
52. J. Machta, *Phys. Rev. A* **40**, 1720 (1989).
53. T. Nattermann and W. Renz, *Phys. Rev.* **40**, 4675 (1989).
54. S. F. Edwards, *Springer Proc. Phys.* **42**, 11 (1989).
55. S. F. Edwards, in *Polymer Physics*, S. M. Bhattacharjee, ed., World Scientific, Singapore, 1992, p. 201.
56. M. Muthukumar, *J. Chem. Phys.* **90**, 4594 (1989).
57. A. Baumgärtner, *Springer Ser. Chem. Phys.* **51**, 141 (1989).
58. T. Vilgis, *J. Phys.* **50**, 3243 (1989).
59. E. A. DiMarzio, *Phys. Rev. Lett.* **64**, 2791 (1990).
60. S. P. Obukhov, *Phys. Rev. A* **42**, 2015 (1990).
61. R. A. Guyer and J. Machta, *Phys. Rev. Lett.* **64**, 494 (1990).
62. T. M. Nieuwenhuizen, *Phys. Rev. Lett.* **62**, 357 (1989).
63. R. P. Feynmann and A. R. Hibbs, *Quantum Mechanics and Path Integrals*, McGraw-Hill, New York, 1965.
64. C. Domb and G. S. Joyce, *J. Phys. C* **5**, 956 (1972).
65. M. E. Fisher, *Rev. Mod. Phys.* **46**, 597 (1974).

66. E. Eisenriegler, *Polymers near Surfaces*, World Scientific, Singapore, 1993.

67. P. G. de Gennes, *C. R. Acad. Sc. Ser. II* **299**, 913 (1984).

68. D. Avnir, D. Farin, and P. Pfeifer, *Nature* **308**, 261 (1984).

69. F. Brochard, *J. Phys.* **46**, 2117 (1985).

70. A. S. Perelson, in "Cell Surface Dynamics", ed. by A. S. Perelson, C. Delisi, F. W. Wiegel, Marcel Dekker, New York, 1984, page 223.

71. A. Baumgärtner and M. Muthukumar, *J. Chem. Phys.* **94**, 4062 (1991).

72. A. Balazs, K. Huang, and P. McElwain, *Macromolecules* **24**, 714 (1991).

73. A. Balazs, M. C. Gempe, and Z. Zhou, *Macromolecules* **24**, 4918 (1991).

74. T. Odijk, *Macromolecules* **23**, 1875 (1990).

75. D. Andelman and J. F. Joany, *Macromolecules* **24**, 6040 (1991).

76. K. L. Sebastian and K. Sumithra, *Phys. Rev. E* **47**, R32 (1993).

77. C. van der Linden, B. van Lent, F. Leermakers, and G. J. Fleer, *Macromolecules* **27**, 1915 (1994).

78. T. M. Niewenhuizen and G. Forgacs, *Europhys. Lett.* **15**, 837 (1991).

79. A. Baumgärtner and W. Renz, *J. Phys.* **51**, 2641 (1990).

80. A. Blumshtein, ed., *Liquid-Crystalline Order in Polymers*, Academic Press, New York, 1978.

81. A. Cifferi, W. R. Krigbaum, and R. B. Meyer, eds., *Polymer Liquid Crystals*, Academic Press, New York, 1982.

82. A. N. Semenov and A. R. Khokhlov, *Sov. Phys. Usp* **31**, 988 (1988).

83. K. F. Freed, *Adv. Chem. Phys.* **22**, 1 (1972).

84. M. G. Bawendi and K. F. Freed, *J. Chem. Phys.* **83**, 2491 (1985).

85. A. ten Bosch and P. Sixou, *J. Chem. Phys.* **83**, 899 (1985).

86. S. M. Bhattacharjee and M. Muthukumar, *J. Chem. Phys.* **86**, 411 (1987).

87. J. B. Lagowski, J. Noolandi, and B. Nickel, *J. Chem. Phys.* **95**, 1266 (1991); **96**, 3362 (1992).

88. A. M. Gupta and S. F. Edwards, *J. Chem. Phys.* **98**, 1588 (1993).

89. L. Onsager, *Ann. N. Y. Acad. Sci.* **51**, 637 (1949).

90. P. J. Flory, *Proc. R. Soc. London Ser. A* **234**, 60 (1956); *Proc. Natl. Acad. Sci.* **79**, 4510 (1982).

91. W. Maier and A. Saupe, *Z. Naturforsch. Teil A* **14**, 882 (1959); **15**, 287 (1960).

92. P. G. de Gennes and J. Prost, *The Physics of Liquid Crystals* Oxford University Press, Oxford, 1994.

93. S. Chandrasekhar, *Liquid Crystals*, Cambridge University Press, Cambridge, 1992.

94. J. F. Nagle, *Proc. R. Soc. London Ser. A* **337**, 569 (1974).

95. M. Gordon, P. Kapadia, and A. Malakis, *J. Phys. A* **9**, 751 (1976).

96. A. Malakis, *J. Phys. A* **13**, 651 (1980).

97. P. D. Gujrati, *J. Phys. A* **13**, L437 (1980); *J. Stat. Phys.* **28**, 441 (1982).

98. P. D. Gujrati and M. Goldstein, *J. Chem. Phys.* **74**, 2596 (1981).

99. E. A. DiMarzio, *J. Chem. Phys.* **35**, 658 (1961).

100. P. J. Flory and R. Ronca, *Mol. Cryst. Liq. Cryst.* **54**, 311 (1979).

101. M. Ballauff, *Makromol. Chem. Rapid Comm.* **7**, 407 (1986).

102. W. M. Gelbart, *J. Phys. Chem.* **86**, 4298 (1982).

103. A. Baumgartner and D. Y. Yoon, *J. Chem. Phys.* **79**, 521 (1983).

104. A. Baumgartner, *J. Chem. Phys.* **81**, 484 (1984).

105. A. Baumgartner, *J. Phys. A* **17**, L971 (1984).

106. A. Baumgartner, *J. Chem. Phys.* **84**, 1905 (1986).

107. M. Dadmun and M. Muthukumar, *J. Chem. Phys.* **97**, 578 (1992).

108. Y. Imry and S. K. Ma, *Phys. Rev. Lett.* **35**, 1399 (1975).

109. Y. Imry and M. Wortis, *Phys. Rev. B* **19**, 3580 (1979).

110. K. Hui and A. N. Berker, *Phys. Rev. Lett.* **62**, 2506 (1989).

111. M. Aizenmann and J. Wehr, *Phys. Rev. Lett.* **62**, 2503 (1989).

112. P. Sheng, *Phys. Rev. Lett.* **37**, 1059 (1976); *Phys. Rev. A* **26**, 1610 (1982).

113. M. D. Dadmun and M. Muthukumar, *J. Chem. Phys.* **98**, 4850 (1993).

114. J. W. Doane and Z. Yaniv, Liquid Crystal Chemistry, Physics, and Applications, Proceedings, SPIE, **1080**, 1989.

115. T. Kajiyama, H. Kikuchi, and A. Takahara, Liquid Crystal Materials, Devices, and Applications, Proceedings, SPIE, **1665**, 1992.

116. F. A. Dullien, *Porous Media, Fluid Transport and Pore Structure*, Academic Press, New York, 1979.

117. P. G. de Gennes, *La Recherche* **7**, 919 (1976).

118. R. B. Pandey, D. Stauffer, A. Margolina, and J. G. Zabolitsky, *J. Stat. Phys.* **34**, 427 (1984).

119. J. W. Haus and K. W. Kehr, *Phys. Rep.* **150**, 263 (1987).

120. J. P. Bouchaud and A. Georges, *Phys. Rep.* **195**, 127 (1990).

121. A. Aharony, in *Scaling Phenomena in Disordered Systems*, R. Pynn and A. Skjeltrop, eds., Plenum Press, New York, 1985.

122. Y. Gefen, A. Aharony, and S. Alexander, *Phys. Rev. Lett.* **50**, 77 (1983).

123. S. Alexander and R. Orbach, *J. Phys. Lett.* **43**, L625 (1982).

124. R. Rammal and G. Toulouse, *J. Phys. Lett.* **44**, 13 (1983).

125. P. G. de Gennes, *J. Chem. Phys.* **55**, 572 (1971).

126. M. Doi and S. F. Edwards, *The Theory of Polymer Dynamics*, Clarendon Press, Oxford, 1986.

127. G. C. Martinez-Mekler and M. A. Moore, *J. Phys.* **42**, L413 (1981).

128. R. Loring, *J. Chem. Phys.* **88**, 6631 (1988).

129. S. Stepanow, *J. Phys. I* **2**, 273 (1992).

130. U. Ebert and L. Schäfer, *Europhys. Lett.* **21**, 741 (1993).

131. U. Ebert, L. Schäfer, and A. Baumgärtner, *Phys. Rev. E* (in press).

132. T. P. Lodge, N. A. Rotstein, and S. Prager, *Adv. Chem. Phys.* **79**, 1 (1990).

133. A. Baumgärtner, *Europhys. Lett.* **4**, 1221 (1988).

134. A. Baumgärtner and M. Moon, *Europhys. Lett.* **9**, 203 (1989).

135. M. Muthukumar and A. Baumgärtner, *Macromolecules* **22**, 1941 (1989).

136. M. Muthukumar and A. Baumgärtner, *Macromolecules* **22**, 1937 (1989).

137. A. Baumgärtner and K. Binder, *J. Chem. Phys.* **71**, 2541 (1979).
138. A. Baumgärtner, *Ann. Rev. Phys. Chem.* **35**, 419 (1984).
139. P. E. Rouse, *J. Chem. Phys.* **21**, 1273 (1953).
140. O. Lumpkin, *Phys. Rev. E* **48**, 1910 (1993).
141. M. Muthukumar, *J. Non-Crys. Solids* **131–133**, 654 (1991).
142. J. Klein, *Nature (London)*, **271**, 143 (1978); J. Klein and B. J. Briscoe, *Proc. R. Soc. London, Ser. A* **365**, 53 (1979).
143. L. Leger, H. Hervet, and F. Rondelez, *Macromolecules* **14**, 1732 (1981).
144. C. R. Bartles, B. Crist, and W. W. Graessley, *Macromolecules* **17**, 2702 (1984).
145. M. Antonietti, J. Coutandin, R. Grutter, and H. Sillescu, *Macromolecules* **17**, 798 (1984); M. Antonietti, J. Coutandin, and H. Sillescu, *Macromolecules* **19**, 793 (1986).
146. B. Smith, E. T. Samuski, L-P. Yu, and M. A. Winnik, *Macromolecules* **18**, 1901 (1985).
147. P. F. Green, P. J. Mills, C. J. Palmstrom, J. W. Mayer, and E. J. Kramer, *Phys. Rev. Lett.* **57**, 2145 (1984).
148. E. D. von Meerwall, E. J. Amis, J. D. Ferry, *Macromolecules* **18**, 260 (1985).
149. H. Kim, T. Chang, J. M. Yohanan, L. Wang, and H. Yu, *Macromolecules* **19**, 2737 (1986).
150. H. Watanabe and T. Kotaka, *Macromolecules* **20**, 530 (1987).
151. N. Nemoto, T. Kojima, T. Inoue, M. Kishine, T. Kirayama, and M. Kurata, *Macromolecules* **22**, 3793 (1989).
152. L. M. Wheeler and T. P. Lodge, *Macromolecules* **22**, 3399 (1989).
153. N. Nemoto, M. Kishine, T. Inoue, and K. Oseki, *Macromolecules* **23**, 659 (1990).
154. T. P. Lodge and N. A. Rotstein, *J. Non-Cryst. Solids* **131–133**, 671 (1991).
155. H. Sillescu, *J. Non-Crys. Solids* **131–133**, 593 (1991).
156. M. Daoud and P. G. Gennes, *J. Phys.* **38**, 85 (1977).
157. F. Brochard and P. G. de Gennes, *J. Chem. Phys.* **67**, 52 (1977).
158. G. Guillot, L. Leger, and F. Rondelez, *Macromolecules* **18**, 2531 (1985).
159. F. Brochard and E. Raphael, *Macromolecules* **23**, 2276 (1990).
160. M. Sahimni and V. L. Jue, *Phys. Rev. Lett.* **62**, 629 (1989).
161. S. Daoudi and F. Brochard, *Macromolecules* **11**, 751 (1978).
162. A. Kolinski, J. S. Skolnick, and R. Yaris, *J. Chem. Phys.* **86**, 1567, 7164, 7174 (1987).
163. K. Kremer and G. S. Grest, *J. Chem. Phys.* **92**, 5057 (1990).
164. K. L. Ngai, A. K. Rajagopal, and T. P. Lodge, *J. Polym. Sci. Part B: Polym. Phys.* **28**, 1367 (1990); K. L. Ngai and J. Skolnick, *Macromolecules* **24**, 1561 (1991).
165. G. D. J. Phillies, and P. Paczak, *Macromolecules* **21**, 214 (1988).
166. N. Easwar, *Macromolecules* **22**, 3492 (1989).
167. Y. Guo, K. H. Langley, and F. E. Karasz, *Macromolecules* **23**, 2022 (1990).
168. A. T. Andrews, *Electrophoresis: Clinical Applications*, Oxford, University Press, New York, 1986.
169. L. S. Lerman and H. L. Frisch, *Biopolymers* **21**, 995 (1982).
170. O. J. Lumpkin and B. H. Zimm, *Biopolymers* **21**, 2315 (1982).
171. O. J. Lumpkin, P. Dejardin, and B. H. Zimm, *Biopolymers* **24**, 1573 (1985).
172. G. W. Slater, J. Noolandi, *Phys. Rev. Lett.* **55**, 1579 (1985); *Biopolymers* **25**, 431 (1986); G. W. Slater, J. Rousseau, and J. Noolandi, *Biopolymers* **26**, 863 (1987).

173. J. L. Viouy, *Electrophoresis* **10**, 429 (1989).

174. H. Hervet and C. P. Bean, *Biopolymers* **26**, 727 (1987).

175. C. R. Calladine, C. M. Collins, H. R. Drew, and M. R. Mott, *J. Mol. Biol.* **221**, 981 (1991).

176. E. Arvanitidou and D. Hoagland, *Phys. Rev. Lett.* **67**, 1464 (1991).

177. D. L. Smisek and D. Hoagland, *Science* **248**, 1221 (1990).

178. M. Olvera de la Cruz, J. M. Deutsch, and S. F. Edwards, *Phys. Rev. A* **33**, 2047 (1986).

179. J. M. Deutsch and T. L. Madden, *J. Chem. Phys.* **90**, 2476 (1989).

180. E. O. Shaffer and M. Olvera de la Cruz, *Mcromolecules* **22**, 1351 (1989).

181. J. Melenkevitz and M. Muthukumar, *Chemtracts* **1**, 171 (1990).

182. D. Hoagland and M. Muthukumar, *Macromolecules* **25**, 6696 (1992).

183. B. H. Zimm and S. D. Levene, *Quart. Rev. Biophys.* **25**, 171 (1992).

184. B. H. Zimm and O. Lumpkin, *Macromolecules* **26**, 226 (1993).

185. D. L. Gosnell and B. H. Zimm, *Macromolecules* **26**, 1304 (1993).

AUTHOR INDEX

Numbers in parentheses are reference numbers and indicate that the author's work is referred to although his name is not mentioned in the text. Numbers in *italic* show the pages on which the complete references are listed.

276–278(60), 292(58, 85), *295–296*
Daube, K. A., 526(303), *622*
Daul, C., 531(323–324), *623*
Davidson, N. S., 69(3), *157*
Davis, E. A., 348(105b), *615*
Davis, R. M., 52(103), *65*
Dawson, K., 25(46), *63*
de Gennes, P., 183(146), *258*
de Gennes, P. G., 2(4), 5–7(14), 9(4, 20–21), 28(58), 30(4), 34(69), 36(14), 37(77), 52(14), *62–64*, 84–86(55), 98(55), 116(55), 120(121), 133(134), 135(136), 144(142), 152(184), *159, 161–163*, 166(14–15), 168(14–15, 42, 51–53, 56–57, 59–61), 186(52), *254–256*, 268(36), 272(49), 282(36), 287(36), 288(36, 81), 290(36), *295–296*, 370(133), *616*, 625–626(4), 627(4, 18), 649(67), 656(4), 659(92), 666(117), 667(125), 674–675(125), 684(4, 125), 691(156–157), *703, 705–707*
de la Torre, J. G., 88–89(76), 94(76), *160*
De Laretta, E., 510(271a), *621*
De Levie, R., 448(207), 497(264), *618, 621*
de Queiroz, S., 630(44), *704*
De Rooji, N. F., 585(351), *624*
de Vallra, A. M. B. G., 106(117), *161*
De Vitis, M., 83(54), *159*
De'Bell, K., 83(48), *159*
Degiorgio, V., 240(202), *259*
Deiss, E., 531(324), *623*
Dejardin, P., 697(171), *707*
Delouis, C., 510(272c), *621*
Delsanti, M., 19(35), *63*
Demeter, K., 310(36a), *612*
Denny, M. S., 88(74), 96(74), 98(74), *160*
Deplessix, R., 288(81), *296*
Derrida, B., 626(11), *703*
des Cloizeaux, J., 2(5), 5(5), 7–8(5), 25(48), *62, 64*, 272(50), 290(83), *295–296*, 626(5), *703*
Deslouis, C., 532(334), *623*
Deutsch, J. M., 697(178–179), *708*
DeVault, D., 323(63c), *613*
Dewald, J., 71(19), 152(19), *158*
Dexpert, H., 526(287), *622*
Dhoot, S., 250(213), *260*
di Meglio, J.-M., 168(58), *256*
Dian, G., 470(230h), *619*
Diaz, A. F., 442(194, 196, 199), 444(194, 199), *618*
Dickman, R., 169(113–115), *257*

Dignam, M. J., 422(179–180), *617*
DiMarzio, E., 193(150), *258*
DiMarzio, E. A., 633(59), 659(99), *704–705*
Disko, M. M., 127(132), 152(132), *162*
Doane, J. W., 666(114), *706*
Dobbs, H. T., 20(38), *63*
Doblhofer, K., 341–342(99a), 486(262a–b), *615, 620*
Dobrynin, A., 32(63), *64*
Dobrynin, A. V., 35(75), 52(105), 53(75), 55(75), *64–65*
Doi, M., 2(6), 44(89), 47–49(6), *62, 65*, 143(141), 144(143), 148(143), *162*, 667(126), 675(126), 684(126), *706*
Dolan, A. K., 169(92–93), 180(92–93), *256–257*
Domb, C., 625(2–3), 627(2), 646(2, 64), *703–704*
Dominey, R. N., 526(301), *622*
Dong, S., 532(328), *623*
Donnan, F., 27(50), *64*
Dormidontova, E. E., 44(87), *65*
Douglas, J., 127(128), *161*
Douglas, J. F., 89(82), 94(82), *160*, 221(176), *259*, 633(48), 656(48), 683(48), *704*
Downard, A. J., 470(225), *619*
Dozier, W., 99(105), 108–109(105), 119–120(105), *161*
Drew, H. R., 697(175), 702(175), *708*
Dreyfuss, M. P., 263(21), 273(21), 276(61), *294–295*
Dreyfuss, P., 150(170), *163*, 263(21), 276(21, 61), *294–295*
Drifford, M., 19(35), *63*
Druger, S. D., 313(48–49, 52–53), *613*
Dubois-Violette, E., 133(134), *162*
Duffit, G. L., 498(268b), *621*
Dullien, F. A., 666(116), *706*
Dumas-Bouchiat, J. M., 531(313–314, 316), *623*
Dünweg, B., 99(98), *160*
Duplantier, B., 83(52), *159*, 626(13), 632(13), *703*
Duplessix, R., 120(121), *161*
Durand, D., 102(114), *161*
Durand, R., 526(296), *622*
Düweg, B., 4(12), 57(12), *62*

Eales, R. M., 421(168), 427–429(168), *617*
Eastwick-Field, V., 526(295b), *622*
Eastwood, J. W., 88(73), *160*

SUBJECT INDEX